Cognitive Science
Volume III

The International Library of Critical Writings in Psychology

Series Editors: A.J. Chapman
Professor of Psychology
University of Leeds
N.P. Sheehy
Professor of Psychology
School of Psychology
The Queen's University of Belfast

1. Industrial and Organizational Psychology (Volumes I and II)
   Cary L. Cooper
2. The Psychology of Gender (Volumes I, II, III and IV)
   Carol Nagy Jacklin
3. Social Psychology (Volumes I, II and III)
   Elliot Aronson and Anthony R. Pratkanis
4. The Psychology of Memory (Volumes I, II and III)
   Peter E. Morris and Martin A. Conway
5. The Psychology of Attention (Volumes I and II)
   Geoffrey Underwood
6. Cognitive Science (Volumes I, II and III)
   Noel P. Sheehy and Antony J. Chapman
7. Ergonomics and Human Factors (Volumes I and II)
   David Oborne

Cognitive Science Volume III

Edited by

Noel Sheehy

Professor of Psychology
The Queen's University Belfast, UK

and

Antony J. Chapman

Professor of Psychology
University of Leeds, UK

THE INTERNATIONAL LIBRARY OF CRITICAL WRITINGS IN PSYCHOLOGY

An Elgar Reference Collection
Aldershot, UK

Published by
Edward Elgar Publishing Limited
Gower House
Croft Road
Aldershot
Hants GU11 3HR
UK

British Library Cataloguing in Publication Data
Cognitive Science. – (International
Library of Critical Writings in
Psychology; Vol. 6)
I. Sheehy, Noel II. Chapman, Antony J.
III. Series
153.4

ISBN 1 85278 698 1 (3 volume set)

Printed in Great Britain by Galliard (Printers) Ltd, Great Yarmouth

Contents

Acknowledgements

The editors and publishers wish to thank the following who have kindly given permission for the use of copyright material.

Ablex Publishing Corporation for articles: Philip R. Cohen and C. Raymond Perrault (1979), 'Elements of a Plan-Based Theory of Speech Acts', *Cognitive Science*, **3**, 177–212; Terry Winograd (1980), 'What Does It Mean to Understand Language?', *Cognitive Science*, **4**, 209–41; David E. Rumelhart and Donald A. Norman (1982), 'Simulating a Skilled Typist: A Study of Skilled Cognitive-Motor Performance', *Cognitive Science*, **6**, 1–36.

Academic Press, Inc. for article and excerpt: Roger C. Schank (1972), 'Conceptual Dependency: A Theory of Natural Language Understanding', *Cognitive Psychology*, **3**, 552–631; M.F. Garrett (1975), 'The Analysis of Sentence Production', in Gordon H. Bower (ed.), *The Psychology of Learning and Motivation: Advances in Research and Theory*, Chapter 9, 133–77.

Association for Computational Linguistics for article: Barbara J. Grosz and Candace L. Sidner (1986), 'Attention, Intentions, and the Structure of Discourse', *Computational Linguistics*, **12** (3), July–September, 175–204.

Association for Computing Machinery, Inc. for article: Lee D. Erman, Frederick Hayes-Roth, Victor R. Lesser and D. Raj Reddy (1980), 'The Hearsay-II Speech-Understanding System: Integrating Knowledge to Resolve Uncertainty', *Computing Surveys*, **12** (2), June, 213–53.

Cambridge University Press for excerpt: Donald A. Norman (1991), 'Cognitive Artifacts', in John M. Carroll (ed.), *Designing Interaction: Psychology at the Human–Computer Interface*, Chapter 2, 17–38.

Elsevier Science Publishers B.V. for articles: John Kimball (1973), 'Seven Principles of Surface Structure Parsing in Natural Language', *Cognition*, **2** (1), 15–46; Steven Pinker (1979), 'Formal Models of Language Learning', *Cognition*, **7**, 217–83.

Kathleen Grice for excerpt: H.P. Grice (1975), 'Logic and Conversation' in Peter Cole and Jerry L. Morgan (eds), *Syntax and Semantics*, *Volume 3: Speech Acts*, 41–58.

Kingston Press Services Ltd for article: Michael Studdert-Kennedy (1980), 'Speech Perception', *Language and Speech*, **23** Part 1, 45–65.

MIT Press for excerpt: Mitchell P. Marcus (1979), 'A Theory of Syntactic Recognition for Natural Language', in Patrick Henry Winston and Richard Henry Brown (eds), *Artificial Intelligence, An MIT Perspective, Volume 1: Expert Problem Solving, Natural Language Understanding, Intelligence Computer Coaches, Representation and Learning*, 195–229.

Every effort has been made to trace all the copyright holders but if any have been inadvertently overlooked the publishers will be pleased to make the necessary arrangement at the first opportunity.

In addition the publishers wish to thank the Library of the London School of Economics and Political Science and the Photographic Unit of the University of London Library for their assistance in obtaining these articles.

Introduction

A language is a mapping from some physical form, such as a sound, to a representation of meaning, and *vice versa*. The role of linguistics within cognitive science is to provide a formal account of linguistic knowledge. Cognitive scientists are equally interested in understanding how such grammars are related to those aspects of knowledge that are recruited in the *comprehension and production* of language itself.

Language Comprehension

A first approach to the mapping of physical form to meaning is to assume that the mapping is direct: that linguistic forms and semantic forms are related to a more or less straightforward fashion. One school of thought – the grammar-driven view – holds that grammars are essential for understanding how language is understood and produced. This view is consonant with Fodor's arguments that human information-processing systems are componential or modular; (1985) reprinted in Volume I as Chapter 4: the grammar or representational system used in language processing has little or no access to representational forms in other modules. The papers in Volume III by Kimball (1973, reprinted as Chapter 2), Marcus (1979, reprinted as Chapter 4) and Pinker (1979, reprinted as Chapter 5) are classic expositions and refinements of this view. Efforts to develop computer programs that could understand natural languages, as opposed to formal languages, have relied heavily on this view of language processing. However the pioneering efforts of Winograd (1980, reprinted in Volume III as Chapter 6) provided clear indications that computational language processors would have to make use of 'real-world' knowledge if they were to approximate human capabilities.

An alternative view – the 'meaning-driven' view – contends that the mapping between linguistic form and meaning can be accomplished with limited recourse to grammatical representations. According to this view general knowledge and contextual information are essential for understanding language processing. Language is for communication. Children learn to communicate before they learn to speak, and by the time children develop an ability to utter simple sentences they are skilled communicators. They can indicate their wishes, they can refer to objects by pointing, and they have a good understanding of order and turn-taking in communicative exchanges. The use of language for communication is characterized by sets of pragmatic rules and conventions that regulate a large part of language understanding. Grice (1975, reprinted in Volume III as Chapter 3) has argued that dialogues are guided by conversational maxims, such as 'be informative', 'be relevant' and 'do not lie'. These maxims are assumptions about the intentions and beliefs of speakers and are essential to normal language understanding at many levels. They have found a place in the development of computer-based language understanding systems where it became clear that effective human–computer collaboration placed requirements on the computer to infer something of the intentional and belief states of the user. In fact the majority of computer programs do this ineffectively. For example, the majority require a user to signal explicitly to the computer, usually by means of a key stroke, that they have stopped 'talking' and that it is the computer's 'turn'.

Computational models of language processing have focused on written forms of language. This has reflected technological limitations – it is only in relatively recent times that the technology has become available to recognize spoken words. However, this has led to a restricted view of language processing. Most language is spoken, not written, and it is obvious that written language is different from the spoken word. Studdert-Kennedy's (1980, reprinted in Volume III as Chapter 7) review and analysis of speech perception considers a broad set of theoretical issues for natural-language processing that were largely neglected in cognitive science. She considers the fundamental evidence bearing on the relationship between acoustic cues and phonetic categories and identifies a number of perceptual phenomena (such as categorical perception) which appear to be confined to speech perception. Although most cognitive scientists would agree that speech perception must require processing mechanisms different from those implicated in general auditory perception, the need for a special 'speech perception module', and the particular perceptual phenomena which demonstrate the activity of special processes, are contentious.

Speech-based natural language processing inspired the development of a new kind of computer program. In the HEARSAY II speech recognition programme (Erman, Hayes-Roth, Lesser and Reddy, 1980, reprinted in Volume III as Chapter 8) a continuous stream of speech data is fed to a computer program, comprising numerous 'experts' of speech information. Some of the experts can process low-level speech segments, such as phonemes, others process syllables and words while others process high-level structures such as grammatically-valid phrases. The HEARSAY II architecture creates a control style involving both top-down and bottom-up processing. The control architecture is formulated around a *blackboard*, a central repository of information that is monitored by all expert processes. This kind of control scheme contrasts with means-end analysis, developed by Newell and Simon (1963, reprinted in Volume II as Chapter 7) and was the antithesis of the standard programming style of using explicit procedural calls. The parallel nature of HEARSAY II inspired the elaboration of control architectures using the blackboard concept.

Cognitive scientists explain language understanding in terms of its dependence on: (i) grammatical representation; (ii) pragmatic functions of language communication; and (iii) general or 'common sense' knowledge about the world. Their explanations are concerned with the comprehension of language and say little about the production of language. Because the initiation of a linguistic utterance is thought to begin with 'thought' itself, including intention and motivation, a proper understanding of language production places demands on cognitive science to address psychological processes which Norman (1980, reprinted in Volume I as Chapter 1) pointed out were largely neglected. For cognition to be communicated it must be physically enacted. A complete account of a cognitive system must explain how information is transmitted into the surrounding environment, as well as how information is taken in.

Language Production

Action is characterized at all levels by a surplus of options and choices. This is sometimes referred to as the 'degrees of freedom' problem. The degrees of freedom of a system can be thought of as the least number of independent coordinates required to specify the position of system elements without violating any geometrical constraints. Indeterminacies arise when the number of degrees of freedom of the system carrying out a task exceeds the number of degrees of freedom required to specify the execution of a particular task. In such circumstances there are multiple solutions to the control problem. For cognitive science the problem is to articulate how an intelligent system discovers successful patterns of action in the thickest of possibilities.

Most skilled action involves sequences of movements. The classical argument held that skilled action comprised serially-structured sequences of simple action units. However, the classical explanation fails to explain why temporal context can influence separate action units. In speech, for example, the articulatory configuration associated with the production of a given phoneme change depends on the identity of phonemes nearby in the sequence. In order to understand better the role of context sensitivity in the control of action, cognitive scientists have used production errors to make inferences about the underlying architecture of production. Garrett's (1975, reprinted in Volume III as Chapter 9) exemplary investigations of speech errors – he found that words that interact are commonly from the same syntactic category – have led cognitive scientists to speculate about the parallel structure of action production systems. Rumelhart and Norman's (1982, reprinted in Volume III as Chapter 11) model of typing specifies a degree of parallelism in a sequential process – skilled typing – that accounts for a variety of phenomena in typing, including the context-sensitive nature of movement towards the keys.

Theory and enquiry into comprehension and production have often proceeded as if the two were independent, modularized sub-systems, with an interface that has yet to be precisely defined. This has led to a schism in theory with the result that our understanding of comprehension processes has been little informed by understanding of production, and *vice versa*. The need to achieve a closer integration is pursued in Cohen and Perrault's (1979, reprinted in Volume III as Chapter 10) arguments for adopting a theory of speech acts based on a view of communication as purposeful action. Understanding communication, including linguistic forms of communication, can be developed from the observation of others' actions and the adoption of beliefs about their plans and intentions. These ideas are extended in Grosz and Sidner's (1986, reprinted in Volume III as Chapter 12) analysis of the influence of discourse context on the interpretation of phrases and clauses within an utterance. They argue that discourse structure is a composite of three interrelated structures: linguistic structure, the structure of intentions and the structure of attentional states. On the one hand this is an issue for speech comprehension. On the other it has to do with the actions speakers perform in order to influence the attentional and intentional stance of both themselves and their listeners. Participation in a dialogue requires interactants to represent the joint plans of multiple agents. Grosz and Sidner show how such plans cannot be defined solely in terms of the private plans of individual information-processing agents.

Cognitive science makes use of the technology of the day to build intelligent information processors and to implement and evaluate their theories of cognition. It is also concerned to exploit the insights of technology in a very practical way by endeavouring to make the technology easier to use. The study of human computer communication, considered as a branch of cognitive science, is concerned to make new intelligent technologies more accessible to both specialist and non-specialist users. Much of the early work on human computer communication (or HCI – human computer interaction) was concerned with improving the ergonomics of the interface to the technology as encountered by users, and with the development of design guidelines and principles. Theoretical statements about the relationship between computers and their users were largely confined to hypotheses concerning users' plans and beliefs about the performance of particular tasks, such as record retrieval. Norman's (1991, reprinted in Volume III as Chapter 13) review and critique signals the significant advances that have been made while also drawing attention to the fact that we still lack a general theory to explain how and why some human computer collaborations are found to be fruitful and others useless. Much greater importance should be attached to the intentional stance adopted by users of computing technology, and that brings us back to some fundamental issues posed by Dennett and Kinsbourne (1992, reprinted in Volume I as Chapter 6).

Conclusion

Cognitive scientists must pursue the implications of their view of intelligent systems as essentially adaptive mechanisms. A science of intelligent systems is confronted with the challenge that the characteristics of the systems they seek to explain are not immutable. In the long run almost any part of an intelligent information-processing system can change adaptively. In this respect cognitive science is likely to attach greater importance to the cultural, historical and social contexts in which human cognition is constructed and through which it evolves.

Noel Sheehy and Antony J. Chapman

Part I
Comprehension

[1]

COGNITIVE PSYCHOLOGY 3, 552–631 (1972)

Conceptual Dependency: A Theory of Natural Language Understanding[1]

ROGER C. SCHANK

Computer Science Department and Committee on Linguistics, Stanford University, Stanford, California 94305

1. INTRODUCTION

The goal of the research described here is to create a theory of human natural language understanding. While this goal may seem to be virtually unattainable, it is possible to divide the process of creating such a theory into subgoals which seem somewhat more attainable. In creating such subgoals it is reasonable to inquire how one would know when these subgoals had been achieved. In other words, it may be fine enough to create a theory of natural language understanding in a vacuum, but how does one select any given theory over any other? It seems to me that there are two possible general criteria by which such a theory can stand or fall. One is the ability to stand up under the tests that psychologists can devise. This is dependent on the claims that are made by such a theory being made explicit enough to be proved wrong. The second testing possibility comes from the field of computer science. It is highly desirable for computers to understand natural language. An explicit theory of human language understanding should, in principle, be able to be put on a machine to enable computers to communicate with people in natural language.

The question of natural language understanding by computers is so enormous that those who have made attempts to solve the problem have had to severely limit the domain of the particular problem that they were trying to solve, and sacrifice theoretical considerations for programming considerations. Previous attempts at natural language understanding have neglected the theory of an integrated understanding system for the practicality of having a system that operates. Winograd

[1] This research is supported by Grant PHS MH 06645-10 from the National Institute of Mental Health, and by (in part) the Advanved Research Projects Agency of the Office of the Secretary of Defense (SD-183). The views and conclusions contained in this document are those of the author and should not be interpreted as necessarily representing the official policies, either expressed or implied, of the Advanced Research Project Agency or the U. S. Government.

(1970) and Woods (1967), for example, have created programs that could actually do something, i.e., affect manipulations of blocks with a robot or answer questions about a highly restricted data base, and thus can legitimately claim to have achieved some measure of understanding. But a general theory of language understanding is not attempted nor is there any reason to believe that less restricted data bases could not drastically alter their approaches.

Whereas it would seem reasonable to create understanding programs that can interact with humans, psychiatric interviewing programs that have been written, Colby and Enea (1967), and Weizenbaum (1968), have not attempted to do much more than demonstrate that it is possible for men to have conversations with machines.

Quillian's (1966) semantic memory, while attempting to model a human memory, has presented no theory of the processing of linguistic information within his model and has basically restricted himself to modeling a small portion of the understanding process, namely, the relating of tokens in a memory for the purpose of disambiguation of word senses.

What is needed, and what has been lacking, is a cohesive theory of how humans understand natural language without regard to particular subparts of that problem, but with regard to that problem as a whole. The theory that is described here is also intended to be a basis for computer programs that understand natural language. But what is described here is a theory not a program. A program does, in fact, exist that does many of the things that are described here. But the task as I see it is primarily to understand human understanding and secondarily to create computer programs to actually understand. Since programs are always far behind the theory, except when the program is the theory; and since programming practicalities often cause the alteration of theories, extant and future programs will only be briefly referred to here. What will be discussed is the theory of such a program or explicit procedure. Such theorizing is neither traditionally part of computer science or psychology but is rather in between somewhere. (If that is the case then so be it.) As computers become more and more powerful it will be necessary to create theories of this type in order to handle the enormous problems that we should like to solve.

2. THE THEORETICAL FRAMEWORK

2.1 *Conceptual-Base Theory*

The theory presented here has as its initial premise that the basis of natural language is conceptual. That is, I claim that there exists a conceptual base that is interlingual, onto which linguistic structures in a given language map during the understanding process and out of which such structures are created during generation.

Thus, the first subgoal that the theory presented here attempts to achieve is the representation of the conceptual base that underlies all natural languages. The simple fact that it is possible for humans to understand any given natural language if they are immersed in it for a sufficient amount of time and to be able to translate from that language to whatever other natural language with which they are well acquainted, would indicate that such a conceptual base has psychological reality. People fluent in many languages can pass freely from one to another, sometimes without even being overtly aware of what language they are speaking at a given instant. What they are doing is invoking a package of mapping rules for a given language from the conceptual base. The conceptual base has in it the content of the thought that is being expressed. This conceptual content is then mapped into linguistic units via realization rules. We will not discuss realization rules in any detail in this paper. The primary purpose here will be to explain what such a conceptual base looks like and how it functions during the understanding process.

There is evidence that such an interlingual conceptual base exists in people's heads. Both Lenneberg (1967) and Furth (1966) found that thinking is not all impaired by a lack of language. Inhelder and Piaget (1958) noted that logical thinking does not find its base in the verbal symbol. Anderson (1971) noted that subjects tend to remember the conceptual content of an utterance rather than a visual image or a more linguistic representation. Bower (1970) noted the need for hypothesizing such a conceptual base in order to handle similarities in visual and verbal processing.

What I am suggesting then is that such a conceptual base exists; that its elements are concepts and not words; that the natural language system is stratified with the actual language output being merely an indicator of what conceptual content lies beneath it; and that the conceptual apparatus that we tend to call thinking functions in terms of this conceptual base, with concepts and the relations between these concepts as the operands.

I shall refer in this paper to the conceptual content underlying an utterance, that is, what would be represented in the conceptual base, as the meaning of an utterance. The theory that is proposed here is doing its job if two linguistic structures, whether in the same or different languages, have the same conceptual representation if they are translations or paraphrases of each other. Furthermore, another subgoal for this research and test of the theory, is the question of the existence of explicit procedures for formally realizing two such linguistic structures as the one conceptual structure.

Most linguistic theories have not had these goals and hence need not be discussed at any length here. Since the transformational-generative grammar proposed by Chomsky (1957, 1965) is syntax-based it cannot seriously be proposed as a theory of human understanding nor is it intended as such. Linguists who have tried to modify traditional generative grammars have either remained within the problem of trying to account for syntax [e.g., Fillmore (1968)] or else have tried to reorient generative grammar to be semantics-based [e.g., Lakoff (1969) and Chafe (1970)]. These approaches tend to suffer because they are primarily interested in generation of random sentences; are word-based; and in the case of Chafe, fail to recognize the distinction between the identity of conceptual content and the intention of that content within a given situation. There is then, a need for a theory of natural language understanding that is conceptually based. Furthermore, there is a need for an understanding model that recognizes that syntax is not a relevant object of study in its own right, but should be studied only as a tool for understanding of meaning. Early attempts at computer programs that used natural language were primarily concerned with syntactic analysis [e.g., Kuno & Oettinger (1962)]. Whereas no one would claim today that syntactic analysis of a sentence is sufficient for programs which use natural language, it may not even be necessary. This is not to say that syntax is not useful: it most certainly is. But its function is as a pointer to semantic information rather than as a first step to semantic analysis as had been traditionally assumed. This will be discussed at length below.

It is also important to realize in attempting to understand human understanding, that humans certainly do not regularly engage in the process of understanding isolated sentences. They understand sentences with respect to the linguistic and situational contexts in which they were uttered and any theory of natural language understanding must account for this too.

Lastly, it is necessary to recognize that an important part of the understanding process is in the realm of prediction. Kuno and Oettinger (1962) realized this in the predictive analysis program for syntactic analysis. However, humans engaged in the understanding process make predictions about a great deal more than the syntactic structure of a sentence, and any adequate theory must predict much of what is received as input in order to know how to handle it.

The point of this paper then, is to explicate a theory of natural language understanding that:

(1) is conceptually based;
(2) has a conceptual base that consists of a formal structure;
(3) can make predictions on the basis of this conceptual structure;
(4) is not limited to the understanding of isolated sentences;
(5) has formal rules for analyzing natural language utterances into the conceptual base.

In addition to the explication of this theory, this paper will also briefly touch upon the problems of:

(1) analysis of the intention of an utterance;
(2) prediction of the elements of communication that lie outside the conceptual content of an utterance;
(3) the connection of conceptual structures and the understanding process with an underlying memory structure.

2.2 *Conceptual Dependency*

The conceptual base is responsible for formally representing the concepts underlying an utterance without respect to the language in which that utterance was encoded. A given word in a language may or may not have one or more concepts underlying it. We seek to extract the concepts that the words denote and relate them in some manner to those concepts denoted by other words in a given utterance.

We are dealing here with two distinct levels of analysis that are part of a stratified system [cf. Lamb (1966)]. On the sentential level, the utterances of a given language are encoded within a syntactic structure of that language. The basic construction of the sentential level is the sentence. The words of the sentence are related to each other in certain specified ways that we shall briefly discuss in section 3.

The next highest level in the system that we are presenting is the conceptual level. We call the basic construction of this level the conceptualization. A conceptualization consists of concepts and certain formal relations that exist between these concepts. We can consider that both of these levels exist at the same point in time and that for any unit on one level, some corresponding realizate exists on the other level. This realizate may be null or extremely complex, (see Lamb (1964) for discussion of this general idea). The important point is that underlying every sentence in a language there exists at least one conceptualization. Conceptualizations may relate to other conceptualizations by nesting or other specified relationships, so it is possible for a sentence in a language to be the realization of many conceptualizations at one time. This is like saying that one sentence can express many complete ideas and the relation of those ideas.

The basic unit of the conceptualization is the concept. There are three elemental kinds of concepts. A concept can either be a nominal, an action, or a modifier. Nominals are considered to be those things that can be thought of by themselves without the need for relating them to some other concept. That is, a word that is a realization of a nominal concept tends to produce a picture of that real world item in the mind of the hearer. We thus refer to nominal concepts as PP's (for picture producer). A PP then, is the concept of a general thing for example, a man, a duck, a book, or a pen; or of a specific thing, for example, John, New York, or the Grand Canyon.

An action is that which a nominal can be said to be doing. There are certain basic actions (henceforth referred to as ACT's) that are the core of most verbs in a language, but this will be explained in section 4.

A modifier is a concept that makes no sense without the nominal or action to which it relates. It is a descriptor of the nominal or action to which it relates and serves to specify an attribute of that nominal or action. We refer to modifiers of nominals as PA's (picture aiders) and modifiers of actions as AA's (action aiders).

It should be emphasized here that what we have said so far about concepts refers to their conceptual properties and not their sentential ones. While it is possible to have a sentence without a verb, or without a subject, for example, or to have an adjective without a noun, conceptually, the corresponding things cannot exist.

Each of these conceptual categories (PP, ACT, PA, and AA) can relate in specified ways to each other. These relations are called dependencies. They are the conceptual analog of syntactic dependencies used by Hays (1964), Klein (1965), and others. A dependency relation between two conceptual items indicates that the dependent item predicts the existence of the governing item. A governor need not have a dependent, but a dependent must have a governor. The rule of thumb in establishing dependency relations between two concepts is whether one item can be understood without the other. A governor can be understood by itself. However, in order for a conceptualization to exist, even a governor must be dependent on some other concept in that conceptualization.

PP's and ACT's are inherently governing categories, while PA's and AA's are inherently dependents. However, any governor can also be dependent, and in order for a conceptualization to exist this must be the case for at least two governors. Conceptualizations can themselves be dependent on certain concepts and/or other conceptualizations in certain ways. This will be explained more fully in section 4.

We represent the conceptual base by a linked network of concepts and dependencies between concepts that is called a conceptual dependency network (which we abbreviate 'C-diagram'). Let us look at what such a network is like. Consider sentence (1):

(1) John hit his little dog.

'John' is the name of an object, so it represents a concept which can be understood by itself, and it is thus a PP. 'Hit' represents a concept of action. Each of these concepts is necessary to the conceptualization in the sense that if either were not present, there would be no conceptualization. Thus, we say that a two-way dependency exists between them. That is, they each act as governors which can be understood by themselves but which must both be present in order to form a conceptualization. We denote the two-way dependency by $\leftrightarrow$. The words 'his' and 'little' both represent dependent concepts in that in order to understand them it is necessary to hold them in waiting until what they modify appears. That is, they cannot be understood alone. 'Dog' is the name of a concept which is a PP and is, therefore, a governor. The PP 'dog' is conceptually related to the ACT 'hit' as object. That is, it is dependent on 'hit' in that it cannot be understood with respect to the conceptualization except in terms of 'hit.' We denote objective dependency $\overset{o}{\leftarrow}$. We thus have the following network so far:

$$\text{John} \leftrightarrow \text{hit} \overset{o}{\leftarrow} \text{dog}$$

We can now add the dependents that were waiting for 'dog' as governor. 'Little' represents a PA that is dependent on 'dog.' We call this attributive dependency and denote it by $\uparrow$. The concept given by 'his' would appear to be dependent on 'dog' as well, and it is, but it is not a simple concept. 'His' is really another syntactic representation of the PP 'John,' that is being used in the syntactic form that indicates possession. What we have is one PP acting as dependent modifier to another PP. We denote prepositional dependency by $\urcorner$ between the two PP's involved, and a label indicating the type of prepositional dependency. (Here POSS-BY indicates that the dependent possesses the governor.) The final network is then:

$$\begin{array}{l} \text{John} \leftrightarrow \text{hit} \overset{o}{\leftarrow} \text{dog} \\ \qquad\qquad\qquad\quad \uparrow\urcorner \ \text{POSS-BY} \\ \qquad\qquad\quad\ \ \text{little John} \end{array}$$

A conceptual dependency network may be treated as a whole unit by reference to the two-way dependency link. Thus, the time of the events of this conceptualization may have been 'yesterday.' This would be indicated by use of an attributive dependency between the time-PP (PP_T) 'yesterday,' and the ↔ as follows:

```
                      o
John ↔ hit ← dog
         ↑           ↑↑  POSS-BY
yesterday     little John
```

Similarly conceptualizations can relate to other conceptualizations and to ACT's through the two-way dependency link. This will be described in section 4.

It might be pointed out here, that while it appears as if each conceptual category is a variant of one other syntactic category, this is not at all the case. It is quite usual for nouns to be realized as ACT's for instance. This will be seen shortly.

2.3 *Components of the System*

Basically we can view the analysis of a sentence as proceeding through the following processors:

(A) *Syntactic Processor*

The function of the syntactic processor is to bracket the sentence into syntactically dependent units. This is done in conjunction with the following components:

(1) a dictionary of words and their syntactic categories; (2) a dictionary of verbs that chooses verb senses based on the surrounding syntactic and semantic environments; (3) a syntactic expectation maker that predicts the syntactic category most likely to follow the syntactic category of the present word; (4) a list of the kinds of syntactic dependencies that may exist; (5) a routine that finds the main noun and verb in a given sentence.

Syntactic processing is by no means done in isolation from the rest of the processes. It would be nonsense to consider that syntactic processing is completed before conceptual processing begins. The two processors 'talk together' while they work.

(B) *Conceptual Processor*

(1) a dictionary of the possible conceptual realizates of a word; (2) a dictionary of ACT's that is used to select the correct ACT realizate of a given verb given the semantic environment on the conceptual level; (3) a conceptual expectation maker that predicts conceptual information most likely to follow given the present incomplete conceptual structure; (4) a list of the different conceptual dependency rules that may exist;

(5) heuristics for searching through syntactic structures to find the main PP and ACT candidates for a conceptualization.

Conceptual processing is also not done in isolation but is reliant on the memory structure operating above it in order to make some of its decisions.

(C) *Conceptualization-Memory Interface*

(1) a dictionary that relates two concepts or conceptual structures based on the individual experience of the hearer; (2) a dictionary that finds the function of a given PP in the real world; (3) a dictionary that recognizes the implications or connotations of a given word or concept and relates them to the intermediate memory; (4) a conceptualization expectation maker that predicts the next likely conceptualization; (5) a list of possible relations among conceptualization types; (6) a routine for searching through conceptualizations for information to be stored in long-term memory.

The Interface operates on the conceptualizations in order to know how to deal with them in intermediate memory so as to generate a response.

(D) *Intermediate Memory*

(1) core beliefs; (2) recent input conceptualization; (3) information about topic under discussion from long-term memory; (4) emotional variables dependent on input conceptualizations; (5) routine to modify long-term memory information; (6) response-initiator.

The Intermediate Memory is the operating memory whose function it is to begin responses to input conceptualizations and to affect the long-term memory. The long-term memory is the depository for information about everything.

(E) *Long-Term Memory*

(1) information about particular subject and experiences.

Considerable work has been done on parts A and B and a fair amount has been done on C. This will be the main subject of this paper. We are beginning to work on D and E but are not prepared to do more than barely touch on them here.

3. THE SENTENTIAL LEVEL

3.1 *The Syntactic Processor*

While it is unnecessary to do a complete syntactic analysis of a sentence in order to process it conceptually, it is necessary to be aware of the syntactic relationships that exist between words. The analysis method described here is conceptually based. Since that is the case, the syntactic bracketing that we do here is often not necessary. That is, syntactic routines could fail for one reason or another, and the correct analysis could still be obtained by the conceptual processor. It is sometimes possible to understand sentences (in a foreign language, for instance) where the words are known but the syntax is not. This is because the organization of the conceptual base is interlingual and thus the conceptual processing is identical to that done when the input is in the speaker's native language. But certainly it is the case that not all sentences can be understood without regard to syntax and so it is worthwhile to have a syntactic processor.

The primary function of the syntactic processor is to find the main noun and main verb in a sentence. Of secondary importance is the establishment of syntactic dependencies between the main noun and verb and the other parts of the sentence.

In order to find the main noun and main verb in a sentence a set of heuristics are employed that make hypotheses about where these items are to be found and proceed to check these hypotheses. When a decision is made as to the most likely candidates for main noun and main verb, the verb dictionary is called. The verb dictionary is entered with a triple that consists of the following: (Subject, Verb + Verb type, Object). The object can be

(a) Noun
or (b) Noun, Noun
or (c) nil

The correct verb sense is chosen by finding the most explicit semantic category of the nouns in subject and object slots that applies for the correct verb type.

At this point, the main responsibility for processing is shifted over to the verb–ACT dictionary. This dictionary is the interface of the syntactic and conceptual systems, and initiates the creation of the conceptual networks. The verb–ACT dictionary will be described at length.

3.2 *The Verb–ACT Dictionary*

The verb-ACT dictionary is intended to sort out the different senses of a verb, so the appropriate one can be selected, and then to establish candidates for conceptual dependency. These candidates are accepted, rejected, or rearranged in a new conceptual order by the ACT part of the dictionary which will be discussed in section 4.

In this discussion, a verb-sense is defined as a different conceptual realizate for a given verb. So, if we say that a verb has four senses, what we mean is that the main noun (Subject), main verb, and object noun are arranged conceptually in four different conceptual configurations. These different configurations may in fact include entirely different ACT's. That is, we are saying that the given verb has four possible meanings. We select the thing that was meant in this instance by reference to the other information in the entering triple, namely, the verb-type, and the semantic information that is part of the given nouns.

Consider sentence (2):

(2) The big man took the book.

The syntactic processor will analyze this into a triple that is (man, take-vt, book). Furthermore, the syntactic dependency 'man' has been established.

↑
big

The verb-type vt means that a transitive verb has been encountered, that is, a verb that has one noun as object.

The verb–ACT dictionary entry for 'take' is as follows: (For purposes of clarity, the conceptual networks that are actually given in the verb–ACT dictionary as the realizates of the triple will be written in English format in this section. The actual conceptual networks will be presented in section 4.)

take

verb-type	conceptual meaning	semantic category of X	of Y	of Z
vt	X take Y to the possession of X	human	physobj	
vx	X take Z to the possession of Y	human	human	physobj
vt	X ingest Y to the innards of X	human	medication	

Since we were given the verb-type of vt by the syntactic processor, we consider only those entries that begin with vt. (Here X indicates the subject noun of the triple and Y indicates the object noun. Z indicates the second object noun if that exists.) The problem of conceptually analyzing this triple is then one of selecting between the alternative vt meanings. This is done by reference to the semantic categories of X and Y. These semantic categories are found as part of the entry of the dictionary for each noun. Often there are different senses of a noun and different categories to go with those senses. In that case all the senses are considered as possible candidates to be chosen by the verb dictionary. These semantic categories are similar in function to the selectional restrictions discussed by Katz and Fodor (1963) and Chomsky (1965). That is, these semantic categories exist at the sentential level. They can be considered the sentential semantics, or rules for combination of syntactic entities. What we are doing here is trying to accept the most likely interpretation of a given sentence by reference to the meaning of the elements of that sentence. Thus, the purpose here is different than that of the generative linguist whose idea it is to reject nonsensical utterances. We are interested here in rejecting only relatively nonsensical utterances. That is, we are looking for that combination which makes the most sense with respect to the alternatives so that we can select it.

The semantic category of man is 'human' and of 'book' is 'physobj.' The semantics accepts a 'human' as the potential actor (X) for both senses of the vt, but the object (Y) must be a 'medication' in the second vt. In the first, a physobj is required, which is what we have. Thus, we are now finished with the syntactic processing of this sentence, unless the conceptual processor rejects what has been output as conceptually unacceptable.

The other syntactic verb types that are used are as follows:

vx–*double-object verb.* A vx verb is present when two nouns follow that verb without the mediation of a preposition or the appearance of a new verb. (For example 'I gave the girl candy.')

vi–*intransitive verb.* A vi verb is present when no noun follows the verb and the only preposition-noun complexes that follow utilize prepositions such as, in, on, and other locations.

vio–*indirect object verbs.* A vio verb takes 'to' or 'from' or some other preposition that indicates direction before the noun that follows it.

vs–*state verb.* A vs verb is one that requires a verb or an entire sentence as object, such as 'like' in 'I like going' or 'wish' in 'I wish that he was here.'

4. THE CONCEPTUAL LEVEL

4.1 *Conceptual Rules*

The Conceptual Level is intended to represent the concepts and relations between concepts that underlie natural language utterances. The Conceptual Processor relies on the predictions that it can make from what it knows about the input it has already received in order to know how to deal with new input. The key to making these predictions is in the conceptual rules. The conceptual rules are the list of permissable conceptual dependencies. That is, there exist formally defined dependency relations between given categories of concepts. These conceptual dependencies, and only these, make up the formal organization of the conceptual networks of the conceptual level. These conceptual rules are, in essence, the rules for relating different categories of items in the universe based on the original assumptions in conceptual dependency theory. Thus, conceptual categories can relate in the following ways:

1. PP ↔ ACT:
Ex: John ↔ sleep
Sent: John is sleeping

The heart of any conceptualization is the two-way dependency between PP's and ACT's. This rule states that PP's can ACT and when they do there is a mutual dependence. This rule does not say that all PP's can ACT or that any PP can do any ACT. We define the particular possible relations between two categories as the semantics of that conceptual relation. That is, Rule 1 is a rule about the syntax of the conceptual level. It describes the possible combination of conceptual categories. The semantics of the conceptual level states which PP's can do which ACT.

Thus, the sentential level has syntax and semantics for the categories and particular words of that level. The conceptual level has its own syntax and semantics for the categories and concepts of that level.

2. PP ⇄ PA
Ex: John ⇄ tall
Sent: John is tall

A conceptualization that utilizes a ⇄ is called an attributive conceptualization. Such a conceptualization exists when an attribute is being predicated about a given PP. It is the predication of that attribute that makes the dependency two-way. That is, in order for these items to exist as a conceptualization, each is equally necessary.

3. PP ⇄ PP
Ex: John ⇄ doctor
Sent: John is a doctor

This rule is similar to Rule 2 and is a set inclusion-type predication conceptualization.

4. PP
↑
PA
Ex: man
↑
tall
Sent: The tall man

This is a dependency between a concept and an attribute of that concept that has already been predicated. In discourse, conceptual attributes are predicated, either explicitly or implicitly, before they are used to differentiate concepts of the same linguistic name. In other words, 'the tall man' would only be used to differentiate two men whose relative height is either visually apparent or has been previously remarked upon in a predication.

5. PP
   ↑
   PP

Ex: man ↑ LOC New York; dog ↑ POSS John

Sent: The man in New York . . .
John's dog . . .

Two conceptual objects in the world can be related to each other in various fashions. The three principal ones are containment, location, and possession and these are marked on the ↑ arrows. The direction of dependency indicates the concept which is being attributively differentiated, as in Rule 4. That is, 'in New York' is information about 'man' but 'man in' is not information of a useful kind about 'New York.'

6. $\text{ACT} \overset{o}{\leftarrow} \text{PP}$

Ex: $\text{hit} \overset{o}{\leftarrow} \text{boy}$

Sent: (He) hit the boy

This is objective dependency. The PP is related as object to the ACT which governs it. Government here is quite literal in that the semantics of the conceptual level works from ACT to the objective PP. That is, a given ACT may require a PP of a given type as its objective dependent. Certain ACT's do not require an objective PP, but those that do must have an object conceptually even if none was actually mentioned in the input discourse. That is, if an ACT like 'hit' is present, an object for 'hit' exists conceptually even if it is not explicitly in the sentence. This will be further elaborated upon in the discussion of 'case' dependencies (Section 4.3).

The six rules given thus far are enough to express the conceptual representation of the conceptualization underlying sentence (1). We will present other conceptual rules as they are needed for the examples.

4.2 *Underlying ACT's*

Until this point, it may have seemed that what we are passing off as conceptual ACT's are really no more than verbs in a different guise. This is only partially true. Actually, ACT's are whatever is verbal in nature on the sentential level, rewritten into a primitive form. 'Love,' for example, is an ACT, even if it is possible to nominalize it in English. In other words, it is the responsibility of the conceptual level to explicate underlying relationships that speakers know to exist. Although 'love' might be a noun in a given sentence, all speakers know that somewhere a subject and an object (for the right sense of 'love') must exist. It is the purpose of the conceptual rules to mark an ACT such as 'love' as requiring conceptual rules (1) and (6). Thus, when 'love' is encountered in a sentence, it is discovered to be the realizate of the ACT 'love' and immediately the question of the PP's that are its actor and object is raised. The conceptual processor can then search through the sentence to find the candidates for these positions. It knows where to look for them by the sentential rules given earlier and what it is looking for by the syntax and semantics on the conceptual level. Thus, it is the predictive ability of the formulation of the conceptual rules that makes them powerful tools.

It is interesting at this point to look at ACT's that do not have direct English realizates in order to more clearly pinpoint the problem of conceptual representation. Consider sentence (3):

(3) The man took a book.

Since 'man' is the actor here and 'book' is the object of the action 'took,' it might seem to be appropriate here to conceptually analyze this sentence as:

$$\text{man} \overset{p}{\leftrightarrow} \text{take} \overset{o}{\leftarrow} \text{book}$$

(We write a 'p' over the ↔ to denote that the event being referred to occurred in the 'past.')

However, in attempting to uncover the actual conceptualization underlying a sentence, we must recognize that a sentence is often more than its component parts. In fact, a dialogue is usually based on the information that is left out of a sentence but is predicted by the conceptual rules. For example, in this sentence, we know that there was a time and location of this conceptualization and furthermore that the book was taken from 'someone' or 'someplace' and is, as far as we know, now in the possession of the actor. We thus posit a two-pronged recipient case, dependent on the ACT through the object. The recipient case is used to denote the transition in possession of the object from the originator to the recipient. Thus, we have the following network:

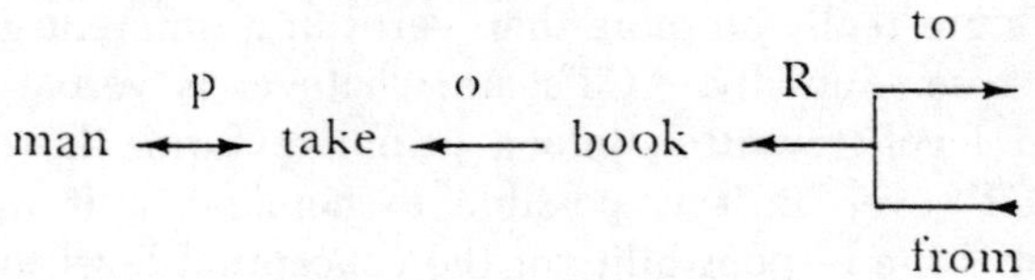

Now, suppose we were given sentence (4):

(4) I gave the man a book.

It is interesting to compare the underlying representations of these two sentences. 'Give' is like 'take' in that it also requires a recipient and an original possessor of the object. Then we have:

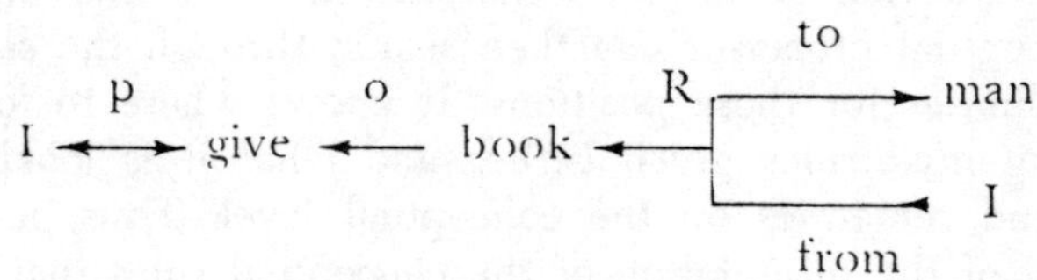

Note that these two conceptualizations look very much alike. They differ in the identity of actor and recipient in (3) and actor and originator in (4), and in the verb. But, is there any actual reason that the verbs should be different? Are they actually different? It would appear that conceptually the same underlying action has occurred. What is actually different between (3) and (4) (assuming that in (3) the originator was also 'I') is that the initiator of the action, the actor, is different in each instance. The action that was performed, namely, transition of possession of an object, is the same for both. We thus, conceptually realize both 'give' and 'take' by the ACT 'trans.' Thus, the conceptualizations underlying (3) and (4) are:

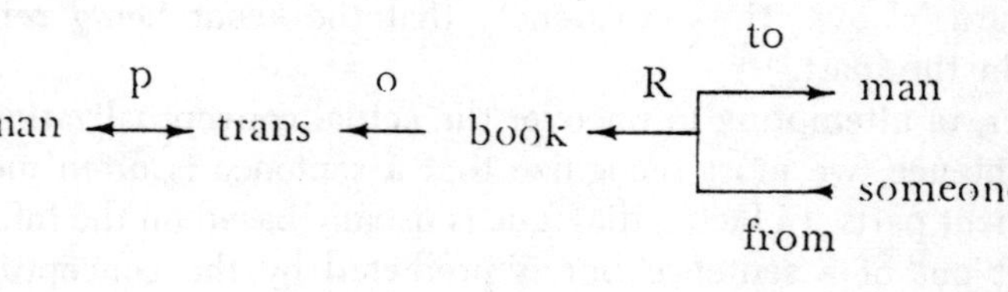

(3)

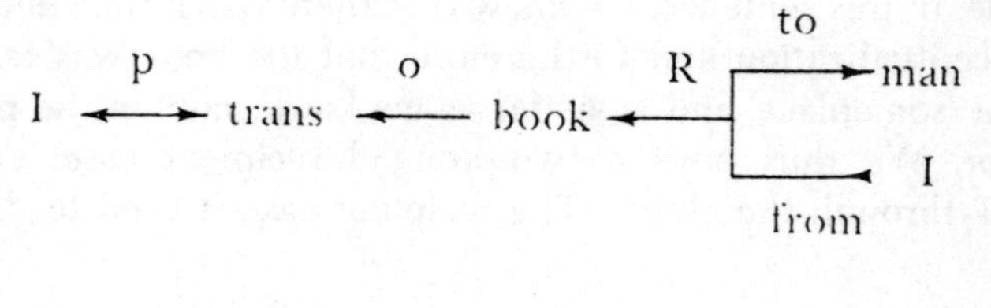

(4)

'Give' is then defined as 'trans' where actor and originator are identical, while 'take' is 'trans' where actor and recipient are identical.

Important here, is that a great many other verbs besides 'give' and 'take' are realized as 'trans' plus other requirements. For example, 'steal,' 'sell,' 'own,' 'bring,' 'catch,' and 'want,' all have senses whose complex realizates include as their ACT, the ACT 'trans.' It is this conceptual rewriting of sentences into conceptualizations with common elements that allows for recognition of similarity or paraphrase between utterances.

The recognition of paraphrase is central to the problem of conceptual representation. If two sentences are agreed to have the same meaning, one conceptual dependency diagram must suffice to represent them. We shall see in section 5 how the three sentences

(5) I want a book.
(6) I want to get a book.
(7) I want to have a book.

can be analyzed into one C-diagram representative of the meaning that is shared by one interpretation of each of them.

4.3 *Conceptual Cases*

We have seen in the last section a new conceptual rule which we label (7):

7. 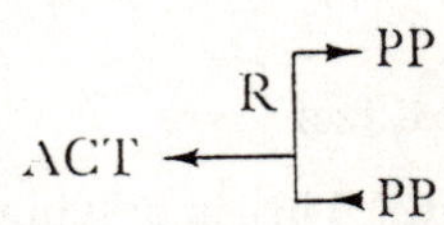

This conceptual rule states that certain ACT's require a two-part recipient in a dependency similar to that of objective dependency. The similarity lies in the fact that this type of dependency is demanded by certain members of the category ACT. If it is present at all, it is because it was required. There is no option and no ACT's can have a recipient dependency without having required it. Thus, this dependency can in a sense be considered to be a part of the ACT itself.

We call those dependents that are required by the ACT, conceptual cases. There are four conceptual cases in conceptual dependency, namely, OBJECTIVE, RECIPIENT, DIRECTIVE, and INSTRUMENTAL. We use conceptual case as the basic predictive mechanism available to the conceptual processor. That is, if I say 'I am going,' a very reasonable inquiry would be 'Where?' Dialogues are often partly concerned with the filling in of the case slots in a conceptualization. People do not usually state all the parts of a given thought that they are trying to communicate. Usually this is because the speaker was trying to be brief and leave out assumed or unessential information or simply information that he did not want to communicate. A conceptual case often may not be realized in a given sentence. The sentence will nevertheless appear well-formed syntactically. But, a C-diagram that contains only the sententially realized information will not be well-formed conceptually. That is, a conceptualization is not complete until all the conceptual cases required by the ACT have been explicated. The conceptual processor makes use of the unfilled case slots to search for a given type of information in a sentence or larger unit of discourse that will fit the needed slot. In an interactive situation, some measure of the desire to fill such slots is necessary in order to direct the response initiator to form an inquiry as to the nature of the missing case concepts when such information has not been previously stated.

Linguists who are interested in syntactic well-formedness have begun to look into the possibility of using cases in their representations recently (e.g., Fillmore (1968)). However, the cases used by these linguists are syntactic cases and should not be confused with conceptual cases. Syntactic cases have to do with the well-formedness of a sentence given a particular verb. Conceptual cases have to do with the well-formedness of a conceptualization given a particular ACT. The two are not always the same. Consider for example, the problem of instrument in sentence (8):

(8) John grew the plants with fertilizer.

Here, 'fertilizer' is the instrument of the verb 'grow.' This is a syntactic case relation and is treated as such by certain linguists including Fillmore. But, conceptually the job is not so simple. Why? Simply because 'grow' is not something that 'John' can do to something else. The conceptualization

$$\text{John} \overset{p}{\leftrightarrow} \text{grow}$$

is perfectly, all right, but it means that 'John grew.' If we mean 'the plants grew' then the representation of that is:

$$\text{plants} \overset{p}{\leftrightarrow} \text{grow}$$

But where is 'John'? In the conceptualization underlying this sentence, 'John' was doing something that caused these plants to grow. What was he 'doing'? We don't know exactly, but we do know that he did it with the fertilizer. We might be tempted to posit a conceptual instrument here and have a conceptualization with no particular ACT in it (represented by a dummy ACT—'do'). Thus, we could have (allowing a new conceptual rule $(\text{ACT} \overset{\text{I}}{\leftarrow} \text{PP})$):

$$\text{John} \leftrightarrow \text{do} \overset{\text{I}}{\leftarrow} \text{fertilizer}$$

We then could relate this conceptualization to the previous conceptualization. They relate to each other casually since John's action caused the plants to grow. We denote causality by ↑↑ between the two-way dependency links. This indicates that it was not the actor or the action by itself that caused the new conceptualization, but rather it was the combination of the two that caused a new actor–action combination. The causal arrow is a dependency, and consequently the direction of the arrow is from dependent to governor or in this instance from caused to causer. That is, the caused conceptualization could not have occurred without the causer occurring so it is dependent on it. Thus, we have (placing an 'i' over the ↑↑ to denote intentional causation):

$$\begin{array}{c} \text{John} \overset{\text{p}}{\leftrightarrow} \text{do} \overset{\text{I}}{\leftarrow} \text{fertilizer} \\ \uparrow\uparrow\text{i} \\ \text{plants} \leftrightarrow \text{grow} \end{array}$$

Now, while this is roughly a characterization of what is going on here, it is not correct. In fact, 'fertilizer' was not the instrument of the action that took place. It was the object. Consider what probably happened. John took his fertilizer bag over to the plants and added the fertilizer to the ground where the plants were. This enabled the plants to grow. This is conceptually another instance of 'trans.' What we have is this:

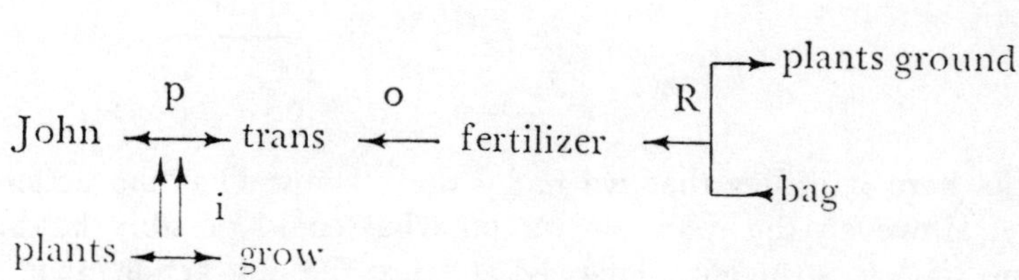

Thus, what appeared to be an instrument syntactically and then conceptually turned out to be an object of an action after all. This, as it turns out, is what always happens to a conceptual instrument. That is not to say that there is no conceptual instrumental case, but simply that a single PP cannot be a conceptual instrument.

Consider sentence (9):

(9) Fred hit the boy with a stick.

This sentence means that the boy was hit by a stick. We assume that the stick didn't do the action of its own accord, so it is reasonable to have the actor be 'Fred.' But we have just stated PP's cannot be conceptual instruments. Thus,

$$\text{Fred} \overset{p}{\leftrightarrow} \text{hit} \overset{o}{\leftarrow} \text{boy} \overset{I}{\leftarrow} \text{stick}$$

is no good. What actually happened here is that Fred threw the stick or swung the stick or performed some other action with the stick as object. In other words, what we call 'Fred hitting the boy' is really another action entirely which we interpret as Fred hitting the boy. We thus allow a new conceptual rule (8).

8. ACT $\overset{I}{\leftarrow}$ ↗ This rule states that an entire conceptualization can be the instrument of a given ACT. Since this is a case dependency, the new conceptualization can be considered to be part of the original ACT and thus the original conceptualization in some sense subsumes the instrumental conceptualization.

Thus, the underlying conceptualization for sentence (9) is:

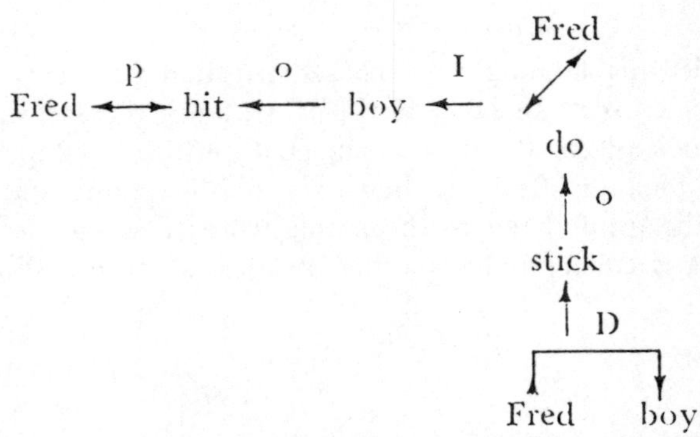

The 'do' here indicates that we really don't know what the action was exactly. However, the requirements on what could possibly fit the 'do' slot are rather stringent. That is, it must be an action that takes OBJECTIVE and DIRECTIVE cases (DIRECTIVE case is denoted by

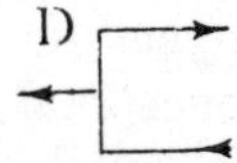

and will be discussed below) and furthermore the particular semantic categories of 'boy' and 'stick' must be allowed. It turns out that 'throw' and 'swing' are about the only ACT's that will fit those requirements.

The instrumental conceptualization is written vertically simply because we have no Z—coordinate to write in. The instrumental conceptualization should be considered to be like the other cases, a mainline dependent of the ACT.

Syntactic instruments are nearly always realized conceptually as the object in the instrumental conceptualization. Usually, these syntactic instruments are so stated because of brevity and because of the possible endless nesting of instruments. All conceptualizations that take RECIPIENT or DIRECTIVE case and most that take OBJECTIVE case, take INSTRUMENTAL case as well. Thus, every time a conceptualization is used instrumentally, it is likely that there is an instrument of that instrumental conceptualization. For example, in the instrumental conceptualization for (14) the instrument there could be something like 'Fred grabbed the stick.' In turn, the conceptualization's instrument could be something like 'Fred moved his hand to the stick.' In other words, there is a great deal of information underlying the information actually stated in a given sentence. Which of this information must be retrieved is of concern to the individual in the dialogue situation. Our concern here is to make explicit the relationships that exist between concepts that have been referenced and to be able to discover those we deem it necessary to know.

One more thing must be made clear about sentence (9) however. It would not be incorrect to say that 'The stick hit the boy.' Since conceptual dependency is supposed to handle paraphrase it is necessary that this information be derivable from the above representation. It turns out that it is.

One of the important principles in conceptual dependency representation is that actors can only be animate or some natural phenomenon. If we had the sentence (10):

(10) The stick hit the boy.

the conceptual representation would be:

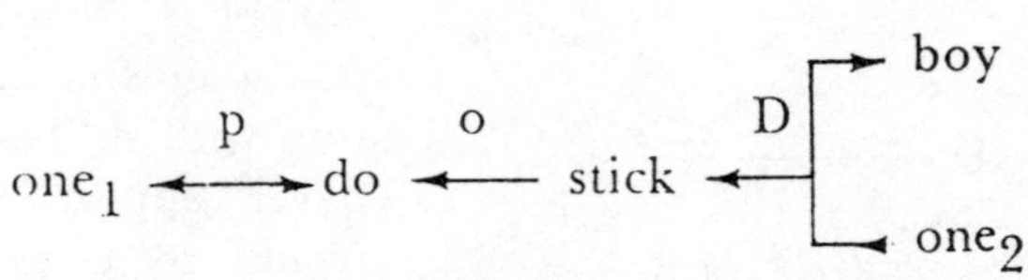

Here, 'one₁' and 'one₂' are dummy PP's which could be 'Fred' and 'Fred' or could be 'gravity' and 'tree.' No matter which it was, the actual hitting is represented by the contact of $\overset{o}{\leftarrow}$ PP, and

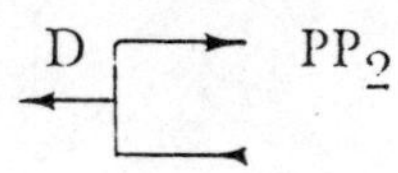

That is, when an object and a directive object exist in the same conceptualization they can be said to have made physical contact. By the same token, when $\overset{o}{\leftarrow} PP_1$, and

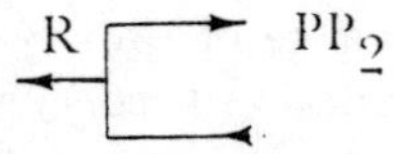

exist in the same conceptualization they can be said to have made social and possibly physical contact. That is, object and recipient, for example, 'book' and 'Fred' in (4) represent the possessive relationship when they exist in the same conceptualization. All this is of interest for operation on these conceptual structures and will not be further touched upon in this section. In other words, there are implications that can be drawn from these conceptualizations. We shall see an extension of this in section 8.

It may not have been obvious why, in conceptual rule (7) for example, the recipient case is written as dependent on the ACT and not on the object as it is written in the C-diagram. These cases are equally dependent on the ACT. They are written in no particular order except that it is necessary to write the object near the recipient or directive because of the implications that can be made from the existence of these two cases simultaneously. Thus, we should consider that the conceptual cases are retrievable through the ACT itself rather than through each other. But it should not be surprising if it turns out that the recipient calls to mind the object or vice versa.

9. 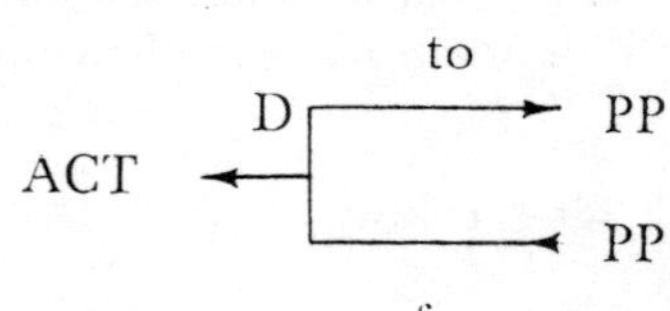

Ex: 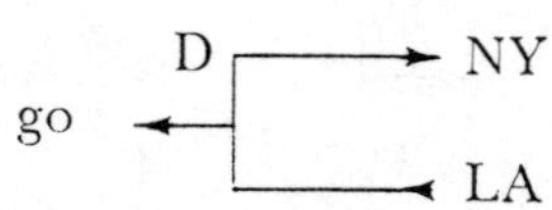

Sent: (He) went to N.Y. from L.A.

This rule indicates that two PP's can be dependent on the ACT as DIRECTIVE case. The directive case is extremely similar to the recipient in form and is almost in complementary distribution with it. That is, the two never appear together and would seem to be different forms of the same phenomenon. The most common ACT that takes directive case is 'go.' Like 'trans,' 'go' is a primitive that is actually the underlying ACT for most verbs of motion.

Rules 6, 7, 8, and 9 constitute the conceptual cases of conceptual dependency theory. Thus, there are only four cases, of which there can be as few as none or as many as three for a given ACT. Each ACT category (explained in section 4.5) requires a certain set number of cases. Thus, any given ACT requires a given number of cases and, no matter what the English realization, will have exactly that number in the C-diagram.

The notion of syntactic case is not used at all in conceptual dependency. Traditional linguists' ideas about case are only about so-called case languages. English is not traditionally a case language so the syntactic processor for English has no need of case. Although Fillmore's (1968) remarks on the underlying syntactic cases in English are well taken, his ideas are within the framework of generative grammar and, therefore, do not pertain here. That is, we are not interested in what is, and what is not, an acceptable utterance in English. Syntactic rules that explain what syntactic cases can be omitted in a syntactic construction shed no light on how the understanding of natural language can best be effected.

We are concerned, however, with recognizing when a syntactic construction is reflective of an underlying conceptual case. The four cases are usually realized in English by the following constructions following the verb (where N is the PP for that case, or in the instance of instrument, the N is the ACT of the instrumental conceptualization if it can be so realized or else it is the objective PP).

OBJECTIVE—(nil) N
RECIPIENT—(to, from) N
INSTRUMENTAL—with N, by N(-ing), using N
DIRECTIVE—(to, from, toward) N

In addition, there are quite a few other constructions which realize conceptual cases. For example, 'by the N' as object of a motion verb (such as 'go') is a realizate of directive case, as is 'at the N.' But it is important to understand that since case requirements are established by the ACT, prepositional phrases are mapped into those cases that are needed by the ACT if the semantic category of the PP is all right.

As examples of the conceptual cases we have the following sentences and their conceptual realizates:

OBJECTIVE

(11) I ate an apple.

$$\text{I} \overset{p}{\leftrightarrow} \text{eat} \overset{o}{\leftarrow} \text{apple}$$

(12) The man was eaten by a tiger.

$$\text{tiger} \overset{p}{\leftrightarrow} \text{eat} \overset{o}{\leftarrow} \text{man}$$

INSTRUMENTAL

(13) I ate the pudding with a spoon.

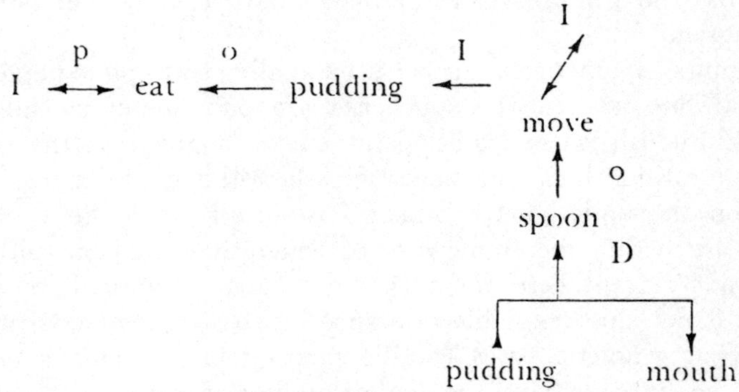

(14) I used a knife to stab him.

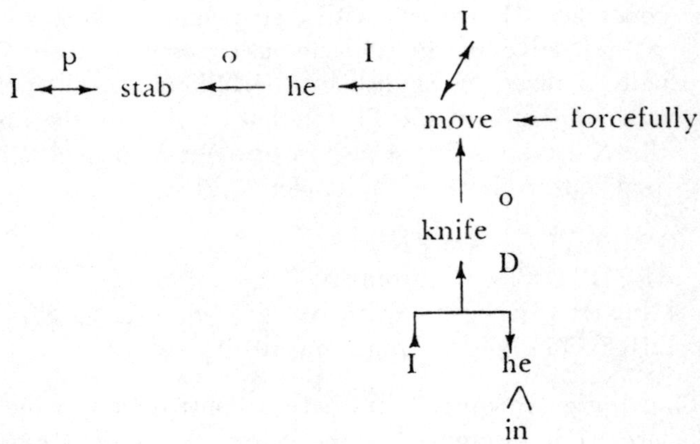

RECIPIENT

(15) I sent mother money.

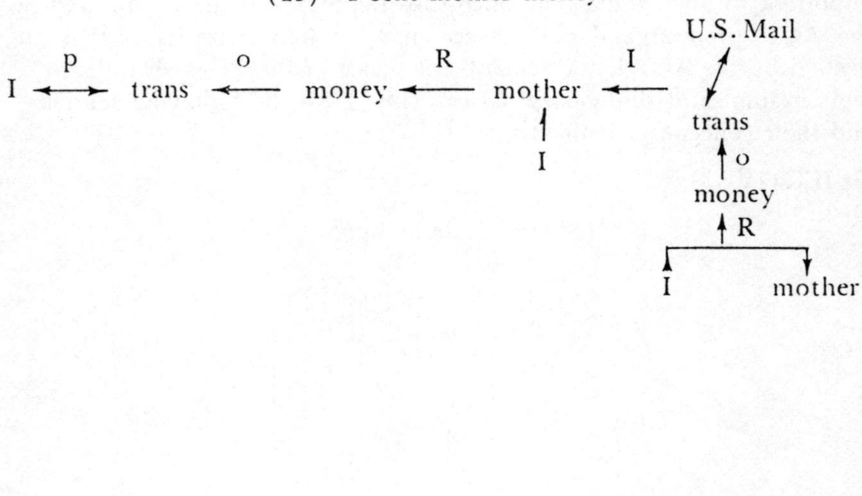

(16) It was the book that he took.

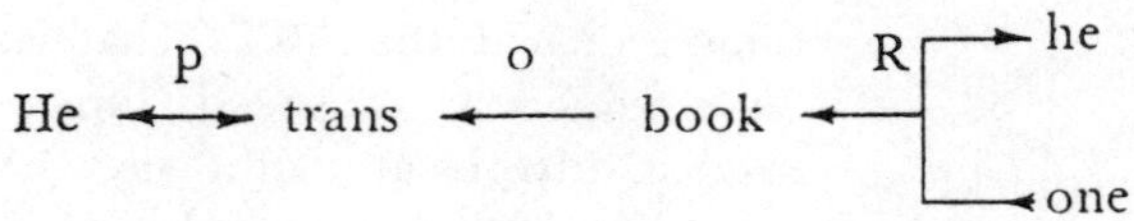

DIRECTIVE

(17) He went home.

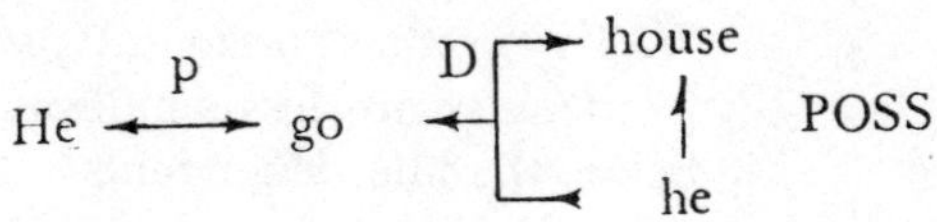

(18) John took a plane from New York to Los Angeles.

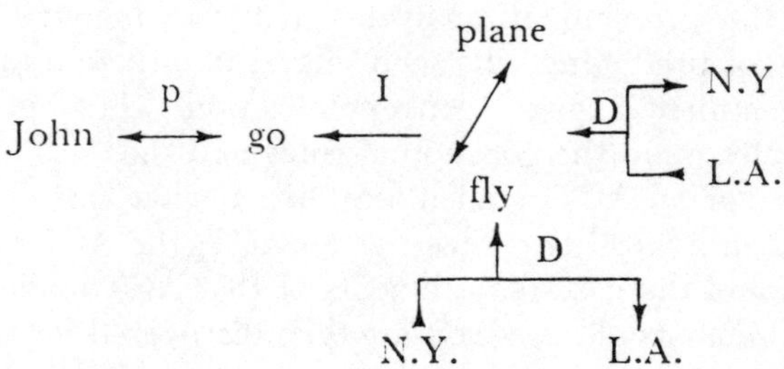

The C-diagrams given here often make assumptions that may seem to be unwarranted. Rather than put no conceptual realizate in certain slots (e.g., 'U.S. Mail' in (15)) I put in what I thought might have been right. Sentences are not usually spoken in isolation so it is often easier to know what fits in these slots. But sometimes guesses like this are warranted, although I do not mean to imply that the given C-diagrams are the only possible ones for a given sentence. Rather, the cases and semantic requirements on those cases are the only ones possible for a given sentence, but the actual concept in those slots, with the exception of the ACT, if not explicitly stated is just assumed by me. Similarly in (18), it may be that the intended meaning is that 'John stole the plane.' Certainly the C-diagram presented does not represent that interpretation. If a sentence is ambiguous, the C-diagram of the sentence that is given here is only what I consider to be the most likely. A computer program must have a mechanism for making such educated guesses also, and we will discuss a limited version of such a thing in section 7.

In (14), we have used a new conceptual rule for modifying ACT's.

10. ACT
↑
AA

Ex: move ; run
↑ ↑
forcefully quickly

Sent: thrust; run quickly

This rule provides for the modification of particular ACT's according to the degree or the dimensions of the ACT. That is, ACT's can be modified as to speed, force, and others. English adverbs are not always AA's. 'Happily' describes a PP for example not an ACT. The upward direction of the arrow in this rule indicates that the AA is a minor aspect of the conceptualization. That is, the main line of the conceptualization consists of the PP's and ACT's on the horizontal plane. Vertical dependencies are less significant and are called below-the-line dependents.

4.4 *ACT Categories*

Central to the problem of analyzing natural language conceptually is the problem of predicting what conceptual information must be found in order to complete a given conceptualization. These predictions are based principally upon the conceptual rules and the ACT categories.

The category of an ACT is an indication of what cases an ACT must take. Thus, when a word is realized as an ACT, the ACT's category can be looked up, and the case requirements of that ACT made known. The semantic requirements of each case are then discovered for the individual ACT. This allows for powerful predictions to be made and changes the basic analysis process from bottom-up to top-down and bottom up.

There are the following ACT-types:

PACT—Physical ACT–PACT's require an objective case and an instrumental case. They are representative of the traditional actor-action–object-type constructions. Example: hit, eat, touch.

EACT—Emotional ACT–EACT's require an objective case only, and are abstract in nature. Example: love.

TACT—Transfer ACT–TACT's require objective, recipient, and instrumental cases and express alienable possession. Example: trans.

CACT—Communication ACT–CACT's also require objective, recipient, and instrumental cases. The object, however, is never concrete and, in fact, is never a PP but always a conceptualization. Example: communicate, say, read. In order to do this, we need a new conceptual rule:

11. CACT ← $\overset{o}{\nearrow}$

This rule states that a particular type of ACT, a CACT takes a conceptualization in the objective slot rather than a PP. Whereas most other conceptual rules about

ACT's hold for all ACT's, this one does not, but rather supercedes the corresponding general rule when the ACT is a CACT.

An example of this type of ACT can be seen in the conceptual construction underlying sentence (19):

(19) John told Mary that he loves her.

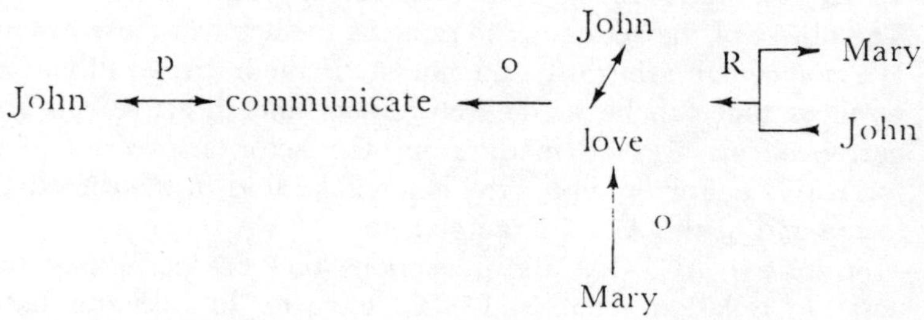

This conceptual construction says that 'John communicated to Mary that he loves her.' Actually, John could have done that in any of a number of ways not involving speech. We say that 'tell' means conceptually 'communicate by saying,' i.e., you can communicate ideas by saying words. Thus, there is an instrument missing here whose object must be the actual words. Since the object consists of words and not concepts as is usual, we write the ideas in quotes to indicate that they actually are not part of the C-diagram:

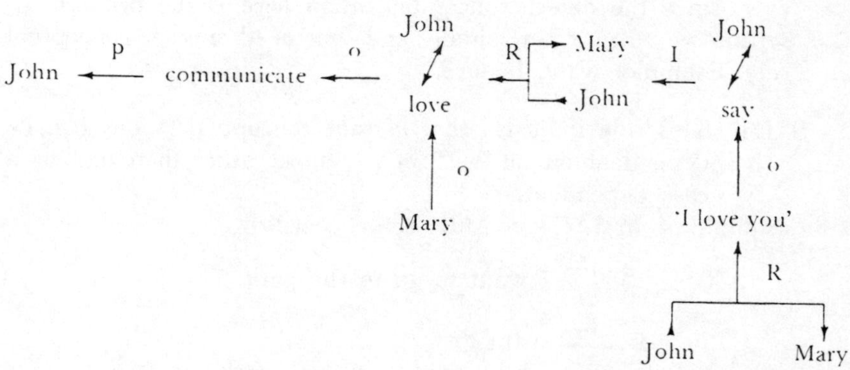

Here again, this might not be exactly what was said in (19) since the instrument could have been left out of the sentence. The sentence could have ended with 'by kissing her' and then that would belong in the instrument conceptualization. We use the C-diagram shown above until given reason to do otherwise. That is, we make assumptions about the information that was not explicitly stated. This is something that humans do, and therefore something that a theory of understanding and an interactive computer program must do. We now return to the ACT categories.

DACT—Direction ACT–DACT's take directive case, objective case, and instrumental case and express motion of objects that are inanimate. Example: move.

RACT—Reflexive ACT–RACT's take directive and instrumental case. The object of the action is the same as the actor in these instances. We somewhat arbitrarily do not state them twice although inferences that can be made from object and directive case's concurrence can also be made from the actor and directive case for RACT's. For example, 'the actor is located in the directive PP at the end of the ACT.' Example: go.

IACT—Intransitive ACT–IACT's are actions that are performed by an actor in isolation. That is, IACT's have to do with the state of existence of the actor and nothing else. Thus, IACT's take no case at all. Example: sleep, be.

SACT—State ACT–SACT's are a special kind of ACT in that they serve to introduce another conceptualization that they are not a part of. That is, whereas instrumental conceptualizations are actually part of the complete process of which the main ACT is the name, SACT's describe the actor's relationship to the nested conceptualization that the SACT requires. We can consider the nested conceptualization of the SACT to function as objective case. However, since the object conceptualization here is the primary assertion we write it horizontally and thus need another conceptual rule. Example: want, believe.

12. SACT ↑ ↔ This rule indicates that certain conceptualizations can be dependent on an ACT in a manner other than that of a case dependent.

As an example of an SACT consider sentence (20):

(20) I want to go to the park.

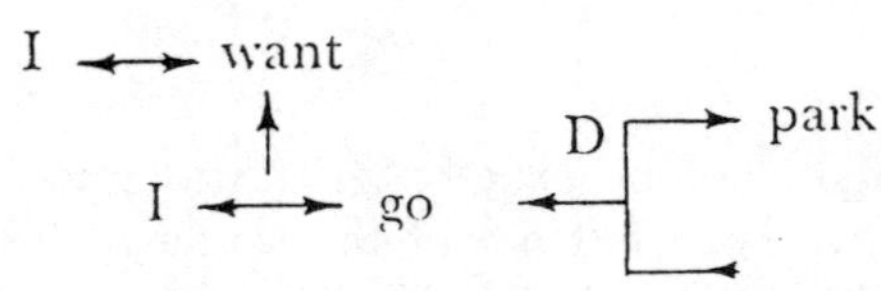

The usefulness of the prediction of a conceptualization made by an SACT is seen in analyzing a sentence such as (21):

(21) The baby wants his mother.

The syntax of English leads one to believe that this is a simple actor–action–object construct. But 'want' is the realization of the SACT 'want.' Thus, there must be an object conceptualization. So, we have:

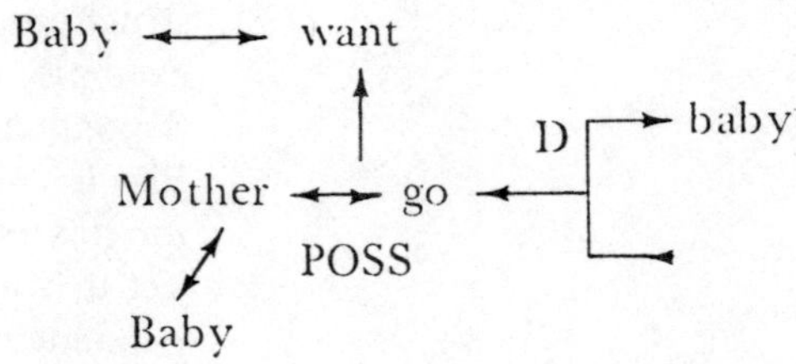

The 'go' conceptualization is gotten from the verb–ACT, dictionary for 'want' as will be seen in section 5.

Just as two sentences which look the same can be conceptually quite different, so those that look different can be conceptually quite similar. Consider sentences (22) and (23):

(22) Sam believes that John is a fool.
(23) Sam believes Fred.

(22) is:

Sam ↔ believe
↑
John ⇄ fool

(23) is again an instance of an SACT that is realized sententially as a transitive verb. However, what we mean by (23) is that 'Sam believes that what Fred said is true.' Thus, we have:

Sam ↔ believe
↑
Fred $\overset{p}{\leftrightarrow}$ communicate $\overset{o}{\leftarrow}$ ⤢ ⇄ true

This brings up three conceptual rules that need to be introduced:

13. ⤢ ⇄
Ex: John
⤢ ⇄ nice
hit
↑o
Fred
Sent: It is nice that John hit Fred.

This rule says that a conceptualization can be the subject of a prediction about that conceptualization. It is this rule that is used to handle 'John's love of Mary was beautiful' as will be seen in section 5.

14.

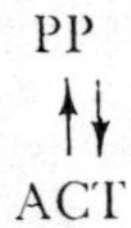

Ex:

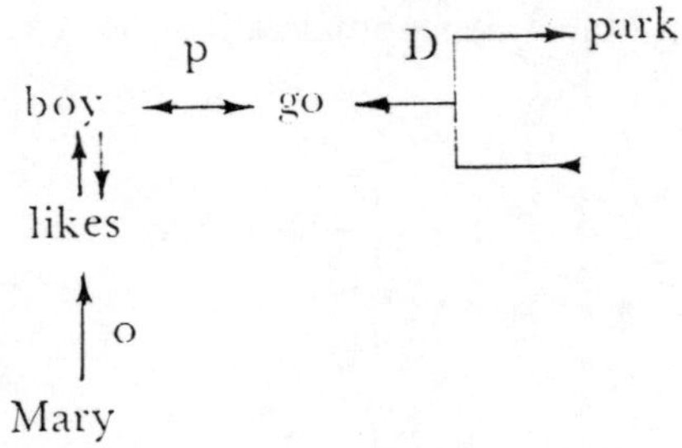

Sent: The boy who likes Mary (went to the park).

Julian

Rule 14 indicates that a conceptualization can be used to modify a PP rather than to predicate something about it. This takes care of 'who' constructions in English for example. It is a two-way dependency in that the PP is needed as actor for this conceptualization. Yet it is an attributive dependency in that it further specifies which PP is being talked about.

15.

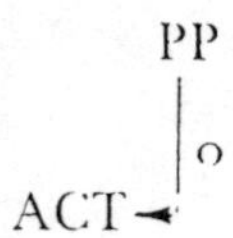

Ex:

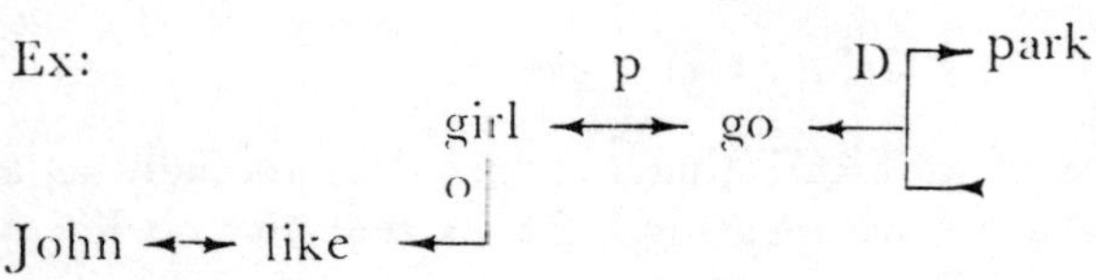

Sent: The girl John likes (went to the park).

It is possible for PP's to have dependent attributive conceptualizations in which they are the object rather than the actor. That is, the fact that they were the object of some conceptualization serves to identify them. Actually this is not a new rule, but is rather a new way of writing an old rule to indicate that a conceptualization relates attributively to a part of the conceptualization being predicated. When a ↔ appears in a case slot it can also be treated in this way. This is what was done for (23). A paraphrase for (23) would thus be (24):

(24) The statement that Fred made is believed to be true by Sam.

4.5 *Conceptual Relations*

So far we have been discussing the conceptualization. The conceptualization is the basic construction of the conceptual level. However, it is the fact that conceptualizations can relate to other conceptualizations that makes natural language useful and makes thought interesting. The dependencies that denote the way conceptualizations relate to other conceptualizations are called conceptual relations.

The most important conceptual relation is that of causality.

16. $\overset{\leftrightarrow}{\underset{\leftrightarrow}{\uparrow\uparrow}}$ Causality is denoted by ↑↑, and indicates that the governor caused the dependent to exist. This is not an implication arrow and should not be read as that logical symbol. It is a dependency arrow and signifies that the governor's existence preceded that of the dependent, and, in this instance, caused the dependent.

The causal arrow is realized in English in a great number of ways. The simplest form uses some variant of the word 'cause' as in sentence (25):

(25) John cried because Mary left him.

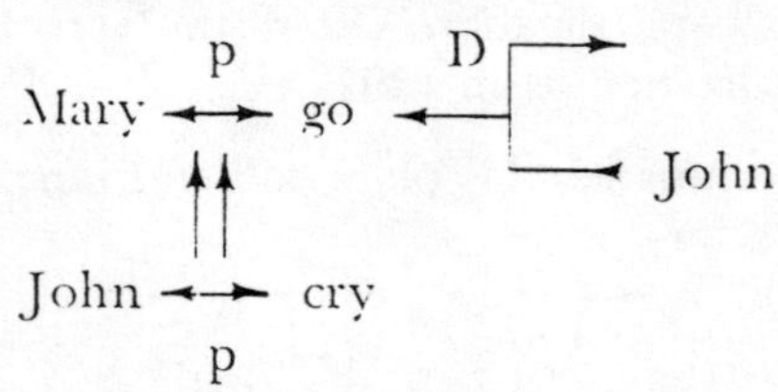

Another common causal takes sentential form using 'when.'

(26) When Fred gave Mary his notebook she tore it up.

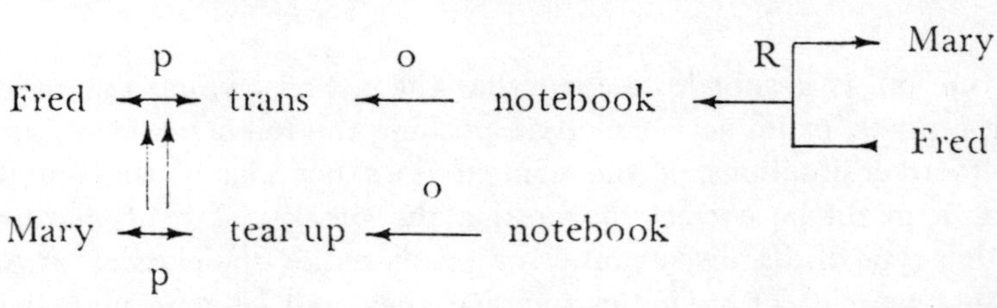

Causal is often expressed with only a single verb expressing both the causal and the end result.

(27) John killed his teacher.

$$\begin{array}{c} \text{John} \overset{p}{\leftrightarrow} \text{do} \\ \uparrow\uparrow \\ \text{teacher} \leftrightarrow \text{die} \\ \uparrow \overset{p}{\text{POSS-BY}} \\ \text{John} \end{array}$$

The 'do' used here is a dummy standing for an ACT which was left unstated. If the ACT were stated it would be realized sententially by use of a 'by' construction as in (28):

(28) John killed his teacher by shooting him in the head.

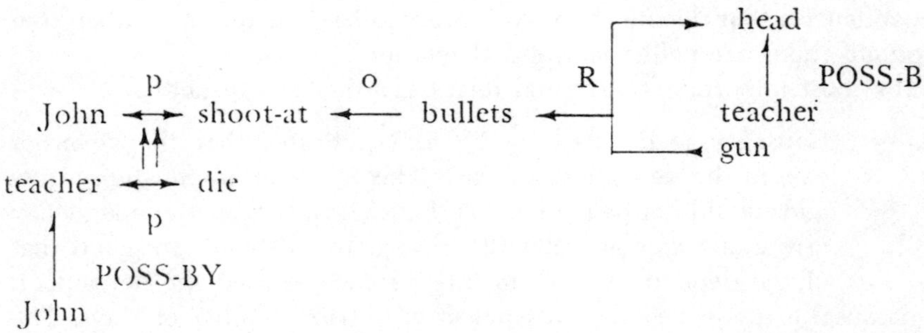

'Kill' is a member of a class of transitive verbs that we call pseudo-state verbs. They all have the property that the object of the verb is the actor of the dependent conceptualization. Often, the verb is the ACT of that dependent conceptualization as in (29):

(29) Sam flew his plane to San Francisco.

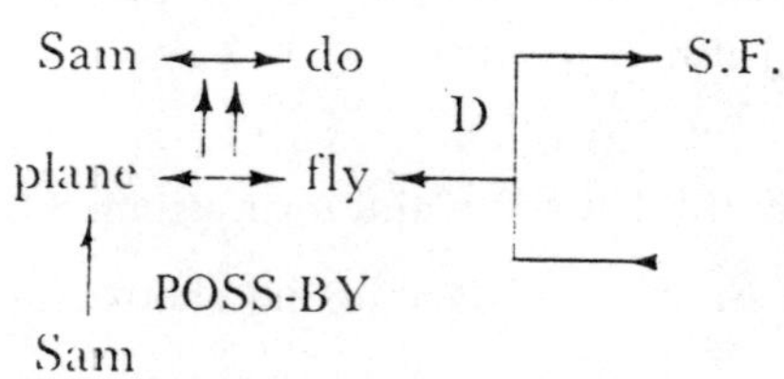

Here, the 'do' is a substitute for either the actions which are required to pilot a plane, or an action such as 'pushing the remote control button.' In interactive situations, if the context does not clarify the nature of the 'do,' it might be correct to question the speaker as to its nature.

Another type of disguised causal is represented by the class of transitive verbs that we call ZPA's indicating that they will become mental state PA's. ZPA's are quite similar to pseudo-state verbs in that they realize causal conceptualizations where the object of the sentence is the actor of the dependent conceptualization.

(30) John comforted Mary.

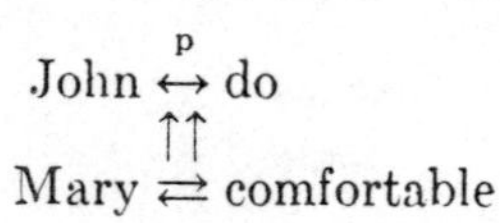

'Comfortable' is a PA. John's specific actions are unknown but their result was a mental state for Mary.

The nature of the causal relation represents a step away from the notion that semantic representation must operate with the same elements as syntactic representations do.

There is no particular reason why what is realized in English as a verb must by necessity be represented semantically by something that is verb-like (e.g., the PRO-verbs of Lakoff (1970)). Often, a linguistic entity can best be realized by a relation rather than a concept. A representation must consist of concepts and relations. There is no intrinsic reason why these relations should not be realized in a language by a word. The justification for doing so has to do with the basic syntax of such a representation schema. That is, there are rules for making a certain relationship between concepts in a representation. If these rules are cohesive, then certain predictions can be made on the basis of the absence of some item from the representation. If relations predict different things than do concepts, a decision on what can be an adequate representation of a word or group of words is made on the basis of the predictive power of the choice of the different representation. Clearly, if the predictions of a relation are more useful in a given situation than those given by a concept, a relation is chosen.

This brings up the question of what makes a relation different from a case. The answer is that a case is part of an underlying ACT and is predicted by that ACT. A relation is a rule for connecting different conceptualizations. Thus, relations serve as connectors within a memory whereas other types of dependency connect things within a conceptualization. That, by the way, is the reason that SACT's are not considered to take objective case. SACT's are on the border between cases and relations. While they require a conceptualization as part of the ACT, they are at the same time relating two conceptualizations.

Conceptual relations predict other conceptualizations, so, to the extent that a conceptual relation is expressed by a word of a language, predictions can be made about future conceptualizations on the basis of that word.

An important reason for treating relations differently than ACT's (as in 'cause' for example) has to do with the inherent properties of a relation as opposed to an ACT. A relation is used to relate dependencies not concepts. Thus, it is the relational quality of 'cause' that allows sentences like (31):

(31) Since smoking can kill you, I stopped.

This sentence contains two conceptualizations related by a causal and a causal relating that causal to a third conceptualization. Such a thing is nearly impossible to handle in more traditional linguistic representations.

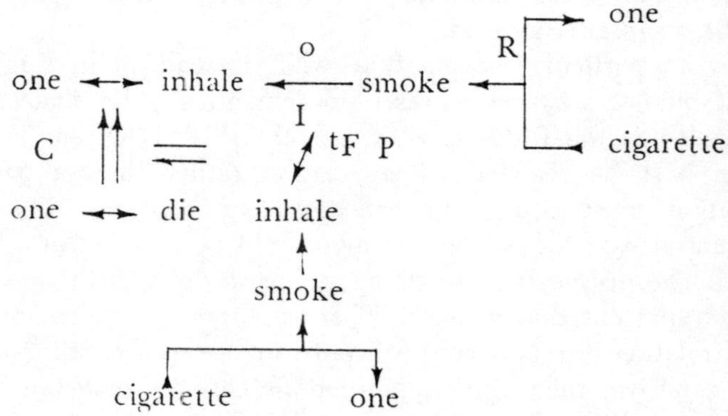

In this example, I am just assuming cigarettes to be the donor whereas the computer system would put in 'smokeable object' of which cigarette is a member. Two new symbols have been introduced in this C-diagram and they bear an interesting relationship. Any conceptualization can be modified by certain conceptual tenses of which 'p' for past is one. The others are:

F—future
C—conditional
t—transition
t_S—beginning transition
t_F—finished transition
k—continuing
?—interrogative
/—negative
nil—present
△—timeless

These tenses modify a conceptualization as a whole. The English word 'stop' for example is actually an instance of the conceptual tense 't_F' and thus predicts an ACT. That ACT was unstated in (31) but was actually 'inhale.'

The interesting point here is that causal links behave as conceptualizations rather than as ACT's in that they also can be modified by conceptual tense. Thus 'can' in (31) is realized as a 'c' over the causal link generated by 'kill.' What is being expressed there is that the causal connection is potential rather than actual. Thus, the entire conceptual construction states that the actor did something because of a potential causality.

The other conceptual relations are time and location. The time of a given conceptualization can be another conceptualization. We have two conceptual rules expressing time.

17. T
↓
↔

Ex: yesterday
↓
↔

Sent: (It happened) yesterday.

This rule relates another conceptual category ('T' for concepts like 'yesterday' or '12 o'clock') and a conceptualization. Most conceptualizations require a time. The time of something modifies the entire conceptualization and not any particular item in it. Thus, it is the time of the joining together of the actor and the ACT.

18. ↔
↓
↔

This rule states that one conceptualization can be the time of another. The dependency is usually realized in English by 'while.'

Thus, we have sentence (32):

(32) While going home I saw a frog.

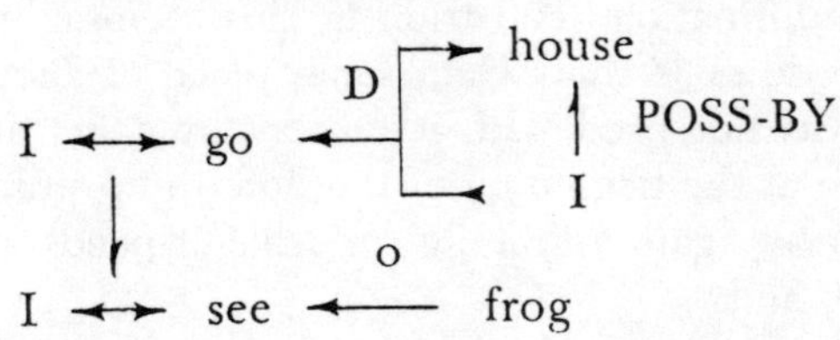

The other conceptual relation is location. Location is not considered to be a case simply because it refers, as do all relations, to the conceptualization as a whole rather than the ACT in particular. The location of a conceptualization is the location of each dependent together with the actor and ACT. All conceptualizations that have time have location.

19. PP
↓
↔

Ex: park
↓
↔

Sent: In the park (I saw an elephant).

The symbol for the location relation is ↿ between the location and the conceptualization. The location modifies the entire conceptualization, although there does exist a rule (Rule 5) which will allow individual PP's to have their own attributive locations.

A sentence that utilizes different types of location is (33):

(33) Yesterday, the boy in that chair hit the boy on the piano in the mouth in the park.

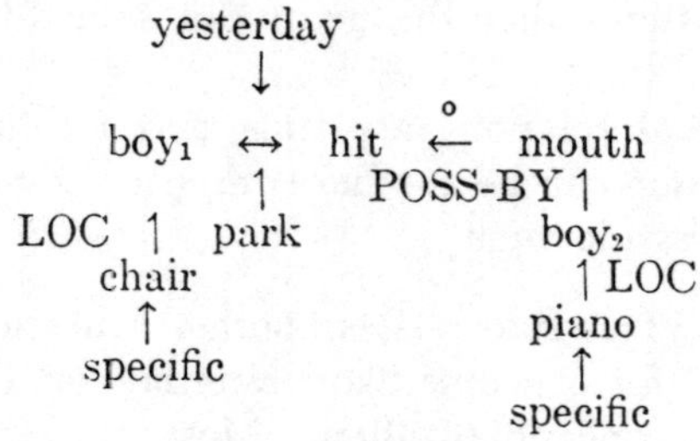

5. THE CONCEPTUAL PROCESSOR

5.1 *Processing*

The conceptual processor receives as input the triple consisting of the main noun, main verb, and the verb category, and the direct object(s) of the verb. From this information it must decide upon a conceptual configuration that will be valid by using the verb–ACT dictionary, the conceptual rules, and the conceptual semantics. Furthermore, it must make predictions as to what information is required by the conceptual construction and proceed to search the unprocessed part of the sentence for this conceptual information. The trick to this seemingly top-down part of the analysis process is that when some piece of the sentence is encountered that was not predicted, it must shift gears and process that piece as if it were at the beginning of the bottom-up process. It can then finish the process by again searching for what it needs after having disposed of what it finds.

Let us consider sentence (34):

(34) I want to go to the park with the girl.

Upon entering the verb–ACT dictionary, we find for 'want':

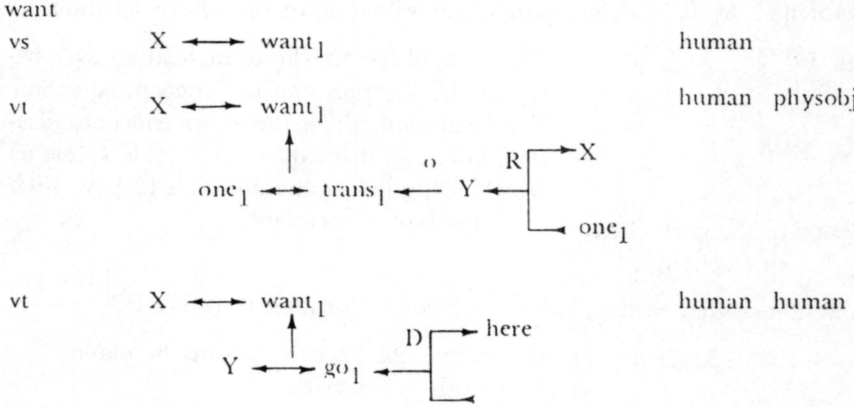

The verb–ACT dictionary is actually a list of potential conceptual realizates for given verbs in given sentential environments. Thus, entire conceptual structures can be formed when a triple containing a verb and its sentential environment is discovered to have a unique conceptual realizate. One more part of the process remains after the correct conceptual realizate is discovered and that is the looking up of the newly given ACT (in the verb–ACT dictionary, concepts are subscripted and words are not) in a dictionary of ACT's. This dictionary sets up the case requirements on the ACT and the semantic requirements on the cases. In addition, the actor is connected to the ACT given that conceptual semantic requirements on conceptual actor are the same as those on X (the main noun) sententially.

The entry in the ACT dictionary for 'want$_1$' is simply:

	ACT-category	actor
want$_1$	SACT	animal

Since we had a vs, the first realizate for 'want' is the only possible one. Thus, 'want$_1$' is looked up in the ACT dictionary and the conceptual construction indicated there is set up:

$$\begin{array}{c} I \leftrightarrow \text{want} \\ \uparrow \\ \leftrightarrow \end{array}$$

Since SACT's demand a conceptualization, we prepare for it by placing the ↔ in the right position. The main conceptualization is established as soon as the ACT dictionary entry is found to be compatible with the information with which it was presented.

The triple for the next verb in the sentence (I, go–vio, nil) is encountered and the verb–ACT dictionary is entered, but this time with the information that a conceptualization is expected and when found that it will be placed in the awaiting dependent spot. The entry for 'go' is:

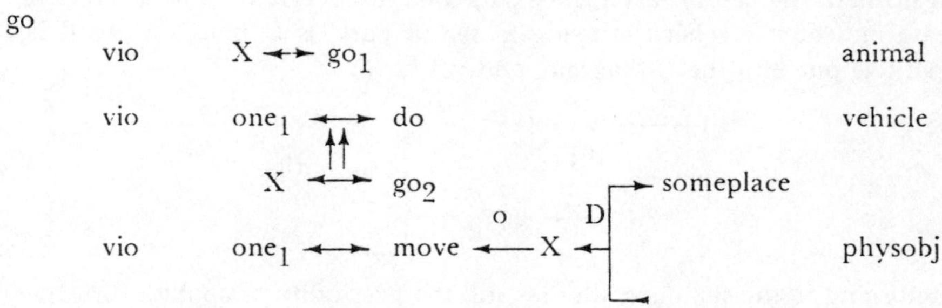

Since X here is 'I,' the first entry is correct, and the ACT dictionary for 'go$_1$' is entered:

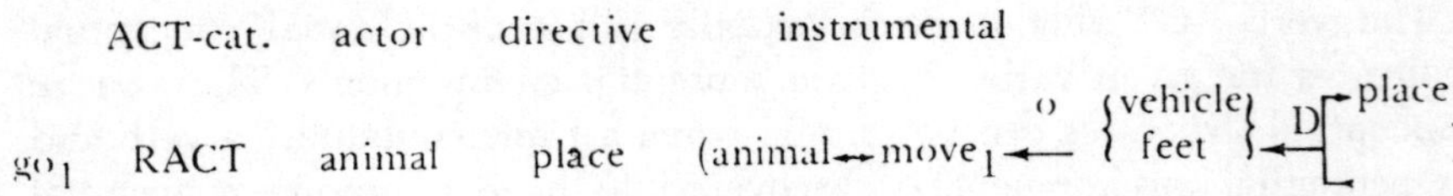

A conceptualization is set up including the conceptual semantic requirements of the cases whose specific concepts are unknown (written in quotes):

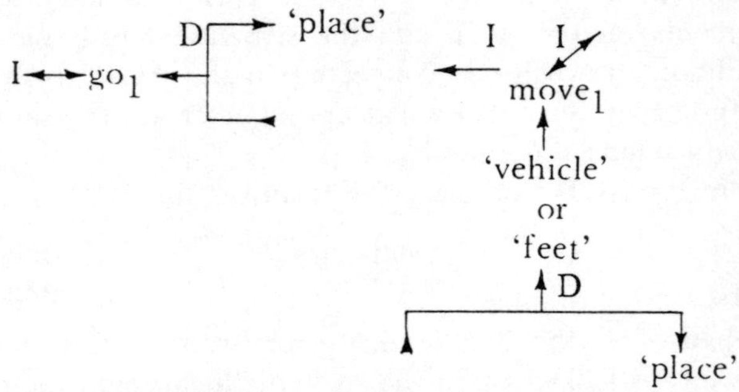

This construction is then placed in the spot that required it, so we have:

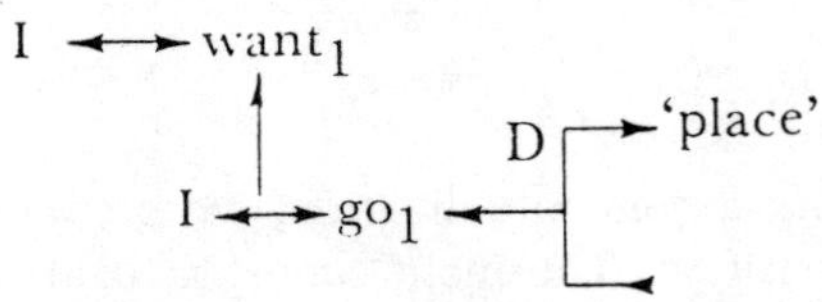

(For simplicity, we will leave the instrument out of the C-diagram until we need it.)

Now the processor goes back to the sentence to pick up what is left. It is looking for a directive case realizate primarily and an instrumental case realizate secondarily (this is because instruments are the least often realized of the cases). It finds '$\overset{\text{to}}{\leftarrow}$ park' and discovers 'to' to be a directive case indicator. A check is made to see if 'park' is a 'place.' Since it is, 'park' is put into the C-diagram, and we have:

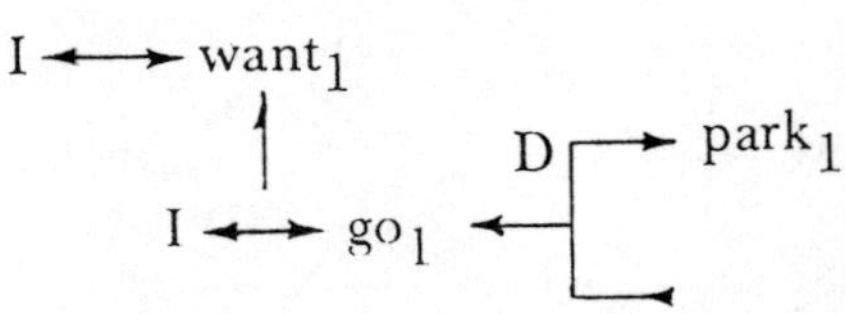

Returning to the sentence, there is still the possibility of another directive case realizate or an instrumental realizate. The syntactic dependency '$\overset{\text{with}}{\leftarrow}$ girl' is found. 'With' is known to be an object of instrumental case realizate. So, the attempt is made to place 'girl' as object of the instrumental case. Since 'girl' is neither 'vehicle' or 'feet,' this fails. Now the expectations are exhausted and the question is where else will a 'with' construction fit? It is easy to see from examination of the following sentences that 'with' is a many ways ambiguous word:

(35) I went with a book to the park.
(36) I went to the park with the playground.
(37) I went with the girl to the park.
(38) I hit the boy with the bat.
(39) I hit the boy with the girl.
(40) I hit the boy with vengence.

Basically there exist four conceptual realizations for 'with PP.'

(1) PP is object of instrumental case.
(2) PP is an additional actor of the conceptualization.
(3) PP is an attribute of PP immediately preceding.
(4) PP is an attribute of the actor of the conceptualization.

These potential realizates are checked for conceptual validity in roughly the order given. The 'roughly' refers to the interface with memory that the conceptual processor must deal with. That is, the conceptual processor only processes in isolation for certain things. It is often the case that the memory must be consulted in order to go on (this will be further discussed in section 8).

The first possibility has been eliminated by the conceptual semantics. But the question of whether 'park with a girl' is a recognizable unit has been begged. That is, is there, in the intermediate memory, some park that has been previously described as containing a girl? If such a unit exists it must be recognized before processing on it takes place. That is, even before 'with' is checked for possible indication of instrument. This is precisely the same process as that which must be followed for idioms. If the unit as a whole is recognized by memory as having recently occurred before, then the previously assigned conceptual realizate takes precedence over any processed ones. (We have not yet investigated the possibility of the learning of idiomatic expressions in sufficient detail to eliminate processing of anything but wholly new phrases, but it presents an interesting possibility.) Thus, a previous reference to 'the park with a girl' can interrupt the processing.

Assuming no such reference is found, then the next possibility is the main line place for 'girl' as additional actor. Except for the above-mentioned circumstance, main line concepts are searched for before the attributes of those concepts are. Thus, since 'with' indicates a possible additional actor, the conceptual semantics are checked to see if 'girl' can 'go,' i.e., if 'girl' is an 'animal.' Since it is, we are done, and we have (using the logical 'and' to indicate two actors):

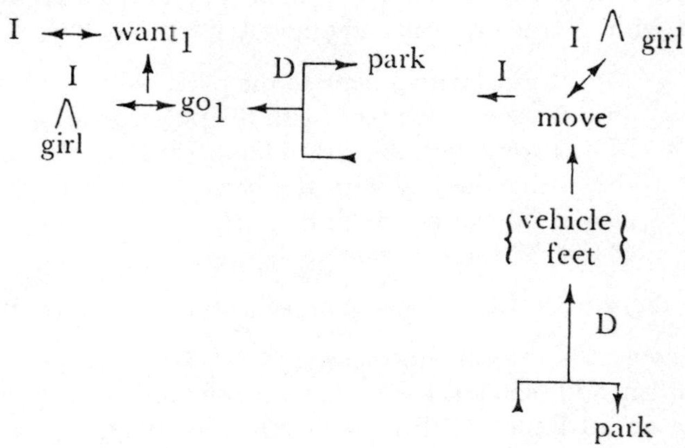

If 'girl' had not fit as actor it would have been checked as a possible attribute of 'park' as would be done for (36).

The order of checking things presented here is not ad hoc. It is necessary in conceptual processing to go from generality to specificity. If 'girl' had first been checked as an attribute of 'park,' it would have been virtually impossible to eliminate it. We cannot really hope to put in a memory all the things that a 'park' can contain, simply because that list is infinite. Thus, we check conceptual dependencies according to the predictions made by the conceptual rules (as for the cases) first, by the likelihood of such a syntactic construction existing second, and lastly by the compatibility with the world model for a given concept.

5.2 *Conceptual Semantics*

So far we have made use of the idea of conceptual semantics without actually explaining it. The conceptual semantics are the delimitations on possible conceptual dependencies. That is, whereas the conceptual categories of two concepts may allow their combination in some specified way, the actual concepts may not make sense when combined. It is the job of the conceptual semantics to keep out the nonsense.

Every PP has at least one semantic category of which it is a member. These semantic categories are the same for the sentential semantics and the conceptual semantics. Their job is to reduce the amount of information necessary to produce a simplistic world model. It is important to recognize that the conceptual semantics can be wrong, that is, what is ruled out may be just what was meant. Sometimes this can be corrected by lack of choice and sometimes by the interactive situation. Consider sentence (41):

(41) The boy ate a book.

The verb–ACT dictionary entry for 'eat' is:
eat

vt	$X \leftrightarrow eat_1 \leftarrow Y$	animal	food
vi	$X \leftrightarrow eat_1 \leftarrow$	animal	

'Book' is listed as having the semantic category 'readable object,' and thus the only vt sense that exists is not satisfied. The only alternative here is go back to the syntactic processor and see if there were any ambiguous paths. Clearly there were not, so the same triple is again presented to the verb–ACT dictionary. It must be passed on to the ACT dictionary when a triple is presented for the second time and this is done. The ACT entry for 'eat,' is:

	ACT-cat	actor	object	instrument
eat_1	PACT	animal	food	(animal $\leftrightarrow trans_1 \leftarrow$ food $\overset{R}{\leftarrow}$ [→ mouth, →])

Here again the semantics fail to accept the input. Ordinarily the syntactic processor and verb–ACT dictionary would be asked for another interpretation. But it is already known that none is available. So, the input is accepted and this C-diagram is produced:

$$boy \overset{p}{\leftrightarrow} eat_1 \overset{o}{\leftarrow} book^*$$

The asterisk is an indicator to memory to do one of two things. Either the conceptual semantics must be altered to include 'book' as 'food' or the system must understand that something very odd has occurred and react accordingly. The decision as to which to do is part of another problem and will not be discussed in this paper.

The other situation occurs in sentences of the type of (42) and (43):

(42) I hit the boy with a stick.
(43) I hit the boy with long hair.

The problem here occurs in the semantics of the ACT 'hit':

	ACT-cat	actor	object	instrumental
hit_1	PACT	human	physobj	(human $\leftrightarrow$ do $\overset{o}{\leftarrow}$ weapon $\overset{D}{\leftarrow}$ [→ physobj, →]

The semantics for instrument require a 'weapon' as object, and since 'stick' is a weapon, there is no problem in the analysis of (42). However, in (43) 'hair' is not a weapon so the possibility of dependence on 'boy' is tried. It is found that 'boy' is 'animal' and 'hair' is a 'body part.' The information that animals possess body parts is part of the conceptual semantics for animals and thus this construction is all right. The two analyses are:

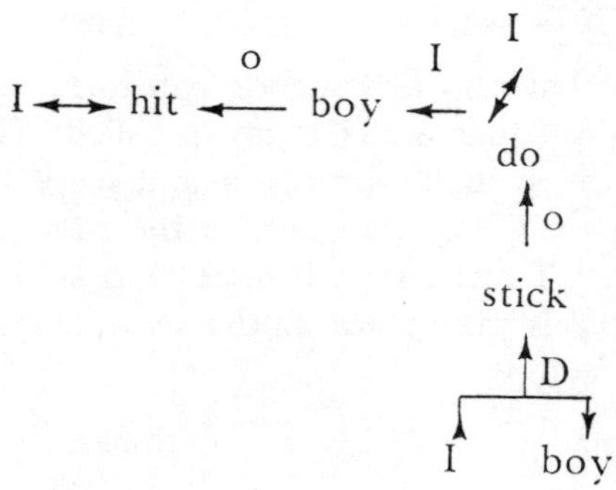

(43)

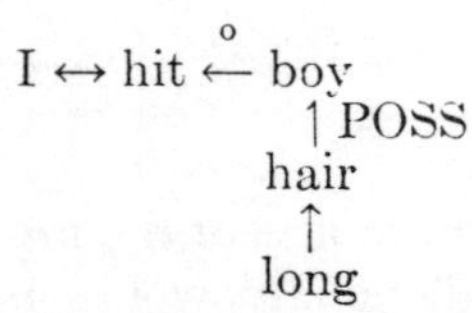

The 'do' in (42) can then be found to be either 'throw,' or 'swing,' by searching for ACT's that require directive case and weapons as object. That problem will not be discussed here. What is of interest is that either of these analyses could be wrong. That is, the boy could have been carrying a stick in (42) and in (43) he could have actually been hit with long hair.

There is no way of always being correct in natural language understanding simply because natural language is full of ambiguity. What we strive for is the right guesses for the right reasons. If we then fail, we can correct ourselves. Were either of the alternate interpretations the intended ones in (42) or (43) we would not have wanted to change our analysis procedure. The only change that we would want to make, upon finding a mistake to have been made would be to add to the meaning of 'hair' that it is a potential weapon. Such category learning would be however a thing that required reinforcement. That is, if such a use of hair did not turn up for a long while, the new category would be forgotten.

It can be seen then that the conceptual semantics are weighted with respect to the listing of the most usual occurrences first. In addition, the conceptual semantics are hierarchic. That is, food, readable object and weapon are all physical objects as are animals. The hierachy is not the usual type however. For example, 'human institution' is not in the hierachy, yet it is a semantic category. We are in the process of replacing the hierarchic system with a feature system, so we shall not dwell on the semantic categories being used here. A feature system offers the latitude of referring to the actual features that are necessary for a category (i.e., a weapon is above all 'rigid' and 'holdable') so that one can gauge the effectiveness of things performing in certain slots (i.e., 'hair' is not 'rigid' and thus will not serve well as an instrument of 'hit' even if it is so used). This system has been worked out by Sylvia Weber Russell but is not documented yet (except see Weber (1971)).

The remaining question is how do the conceptual semantics differ from the sentential semantics? Although they seem quite similar that is only because often the subject and object of a sentence perform the same function in a conceptualization. But, the job of the sentential semantics is basically to enable the conceptual processor to determine which syntactic construction was found. The sentential semantics serve to more clearly spell out the sentential environment and the selection is made by choosing the most specific environment (or lowest semantic category in the hierarchy) in the verb–ACT dictionary and selecting the conceptual realizate for that environment.

The conceptual semantics have the job of arranging the given concepts so that they make sense with respect to the world model. Thus, the conceptual semantics are interlingual, i.e., the same no matter what language you started with. They represent the rules for organizing the world conceptually. The sentential semantics have to do with the particular language only. Thus, sententially it is all right for 'man' to be the actor of 'fly' because you can say this in English. Conceptually, 'men' have yet to 'fly' as such, so the conceptual semantics would find this invalid. But actually, the finding of an invalidity rarely occurs. The verb–ACT dictionary would have found 'go,' for the verb 'fly' and the conceptual semantics for 'go,' would be those checked.

Thus, the conceptual analysis process is concerned with finding what is valid as a conceptual realizate for a sentential part. The conceptual level is interlingual, the sentential level and the list of realizations of sentential pieces are part of a given language.

5.3 *Syntactic Ambiguity*

One of the biggest problems that has plagued computational linguists who build parsing systems is that of syntactic ambiguity. The traditional approach has been to find all the possible syntactic analyses for a given sentence. Kuno and Oettinger's (1963) 'Time flies like an arrow' is famous in this regard.

Such an approach disregards the problem of simulating human understanding. Certainly humans do not find all the syntactic interpretations of a sentence before they consider the meaning of the sentence. Human frequently fail to see alternative interpretations of a sentence even if prompted that a second interpretation exists. On the other hand, it can be shown that humans try wrong paths in understanding and are forced to back up and try another analysis.

Whereas it is possible to have a conceptual analyzer use rules of syntactic structure to produce conceptual dependency diagrams, it is not really advisable. Initial versions of the computer implementation of conceptual dependency theory, (Schank & Tesler (1969)) simply reversed the realization rules that mapped conceptual structures into linguistic ones. The result of parsing by such syntactic rules was that syntactic ambiguity was resolved in an ad hoc manner. That is, if a sentence is two ways ambiguous there exist two different mapping rules that map the two conceptual structures into the one sentential structure. In analysis, the problem of choosing which realization rule to try first could only be decided randomly. Decisions of which rule to apply in a given situation can be shown to be not random. Since our computer system is intended to mimic a human engaged in the same task, it is important for us to find an analysis procedure that at least does what we know a human does even if we cannot verify all of the things it does in this way.

Consider sentence (44):

(44) I saw the Grand Canyon flying to New York.

This sentence is interesting because of what would be considered its poor syntactic structure. Because the instantaneous image of the 'Grand Canyon flying' is called into the mind of the hearer, often this sentence evokes a chuckle when told to an audience. The fact is that most people attempt to attach 'fly' to 'Grand Canyon' fail, and then find the correct analysis of this sentence. We require that our processing system will do the same.

The syntactic processor passes the triple (I, see-vt, Grand Canyon) to the conceptual processor when it encounters this sentence. The verb-ACT dictionary entry for 'see' is:

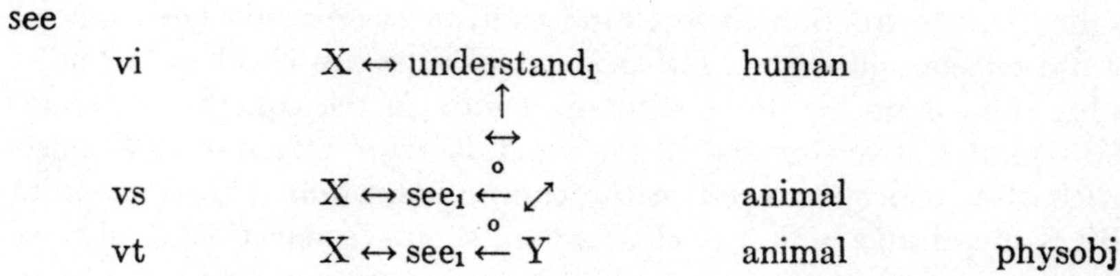

The vt sense outputs the given structure for the ACT see_1. The ACT 'see_1' is given as follows:

	ACT-cat	actor	object	recipient
see_1	CACT	animal	{physobj / conceptualization}	animal
			instrument	
			(animal ↔ $look\text{-}at_1$ $\overset{o}{\leftarrow}$ {physobj / conceptualization})	

Thus, the following conceptual structure is found:

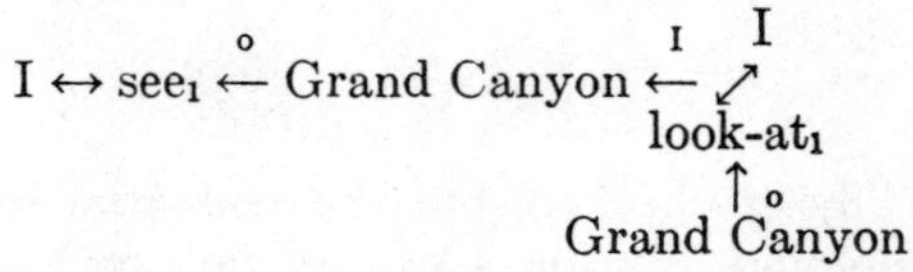

The next word in the sentence is 'flying.' Since 'flying' is realized by an ACT only, it is the candidate for the verb part of a new triple to be sent up to the conceptual processor. Since a prepositional phrase beginning with 'to' follows, it is a vio verb. However the problem exists of what the subject of the triple for 'flying' is. This problem is not divorced from the problem of where to attach the new conceptualization that will be formed from this triple. Previously, when we entered the conceptual processor with a triple, there was a place waiting for it in the old conceptualization or else there was no old conceptualization, in which case there was no problem. But here we are within a sentence and we already have all the main line elements of a conceptualization filled in.

These two problems have the same basis for solution. There is no place in the right side of the conceptualization to place a new conceptualization. But, there is also no immediate candidate for the position of subject of the new conceptualization. The rule in situations such as this is always the same. If there is no PP available as subject, the last PP which was placed in the conceptualization from the sentence is the prime candidate. This takes care of the problem of the entering triple for the conceptual processor, but still does not explicate where the new conceptualization is to be placed. The actual problem is where to place the 'fly' construction since Grand Canyon has already been placed in the old conceptualization. The conceptual rules are checked to find a place for connecting 'fly' to something already in the conceptualization. Again, 'Grand Canyon' is the prime candidate for attaching 'fly' since it was the last concept placed in the conceptualization. The conceptual rule PP is found and now it is clear where such a connection would go.

$$\begin{array}{c} PP \\ \uparrow\downarrow \\ ACT \end{array}$$

So the verb–ACT dictionary is entered with the triple (Grand Canyon, fly–vio, nil). The entry for fly is:

fly

vio	$X \leftrightarrow fly_1$	bird plane insect	
vt	$X \leftrightarrow do$ $\uparrow\uparrow$ $Y \leftrightarrow fly_1$	human	plane
vio	$X \leftrightarrow go_1 \overset{I}{\leftarrow} \nearrow$ plane, $\swarrow$ fly_1	human	

The vio entry for 'fly' is checked, but the sentential semantics does not allow places to function as subject of the vio- 'fly.' Thus, the triple is rejected and sent back to the syntactic processor. In other words, the attempt has been made to connect 'fly' to 'Grand Canyon' in a construction that could be roughly paraphrasable as (45):

(45) I saw the Grand Canyon which was flying to N. Y.

Something else must be tried. Again we are faced with the problem of finding a suitable PP for subject and a place to put the new conceptualization. It is important to bear in mind that, although conceptual information is being used, it is the syntactic processor that is operating here in an interface with the conceptual processor. I mention this because decisions that are made at this point are rules of English that are conceptually based. They are not interlingual rules.

The only other subject that is available for the triple with 'fly' is 'I' because it is the next to most recent PP that was processed (it is also the only other PP that was processed). Now we could apply the same reasoning that was used for Grand Canyon with respect to where to attach the conceptualization that will be found for this triple. However, there is a rule of English for operating on conceptual structures that states that it is not permissable to cross the two-way dependency link from right to left after the two-way link that has been placed in the

C-diagram. If a PP is needed from the left side of the conceptualization it must be the subject of a time conceptualization that modifies the original conceptualization. This is a rule of English and has no universal validity.

Thus, we can enter the verb–ACT dictionary with the new triple (I, fly–vio, nil) and we have a place to put it in the C-diagram, namely, as the time of the first conceptualization.

This time the verb–ACT dictionary accepts the triple, and consulting the ACT dictionary sets up the directive case, so we now have:

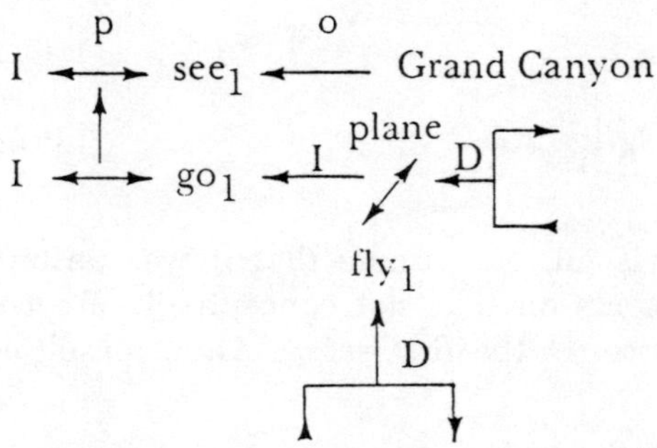

Again 'to' is found to indicate directive case and 'New York' is placed in the 'to' part of both directive cases in the C-diagram.

A more traditional example of syntactic ambiguity is given by the classic sentence (46):

(46) Visiting relatives can be a nuisance.

This sentence has three possible interpretations roughly paraphrased by (47), (48) and (49):

(47) Relatives who visit can be a nuisance.
(48) To visit relatives is a nuisance.
(49) When relatives visit, it can be a nuisance.

I have stated previously that whereas we want to be able to represent the three different meanings of sentences like this conceptually, it does not necessarily follow that a conceptual analyzer should find all three meanings at once. Rather, we want to find the most likely first and so on. Since all speakers of English don't find the same meanings first, we are dealing with an individual model. That is, we want to naturally find one meaning and then another, without claiming that this is the way all speakers operate, but stating that it might be the way many of them operate.

Let us see how this sentence is processed. The syntactic processor is immediately faced with the ambiguity decision because the triple created by 'visit' can have the noun following it as either subject or object when there is no other normal modifier. Often syntactic agreement cleans this up, but for 'can' this doesn't work. If we choose the object slot for the noun, 'relatives' we will get (48). Thus, we have the triple (nil, visit–vt, relatives). We find for 'visit':

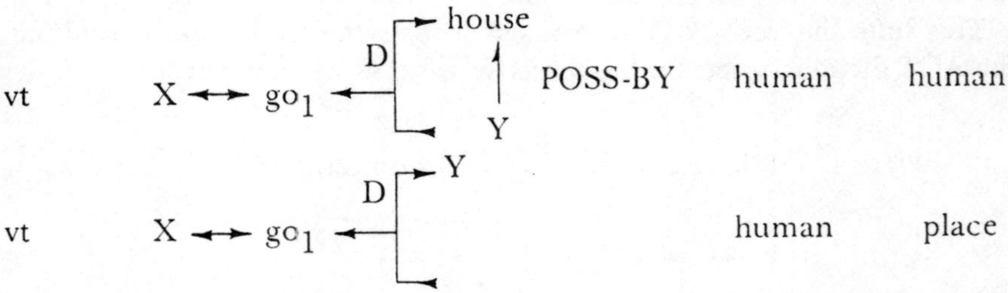

Since the subject is nil, we assume that it was unstated and that the semantic requirements on that slot conceptually are correct. 'Relatives' are human so we select the first sense. After consulting the ACT dictionary we have:

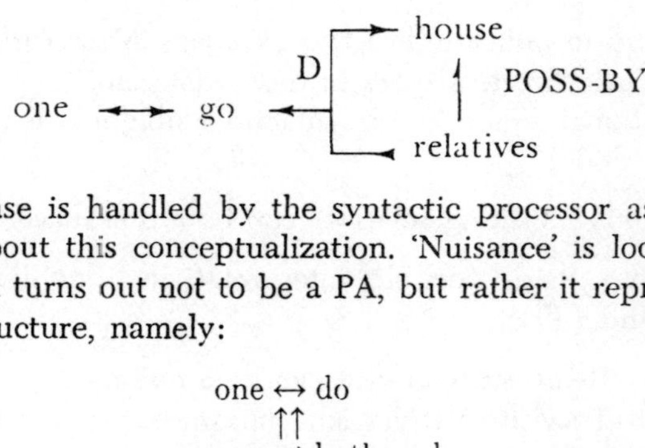

The next phrase is handled by the syntactic processor as a conditional predication about this conceptualization. 'Nuisance' is looked up in the dictionary and turns out not to be a PA, but rather it represents a larger conceptual structure, namely:

one ↔ do
↑↑
one ⇄ bothered

'Bothered' is a mental state PA. Thus, we have a command to the conceptual processor to combine these two structures. Since the previous conceptualization is an example of 'one ↔ do' and this is all one sentence, adding the conditional we have:

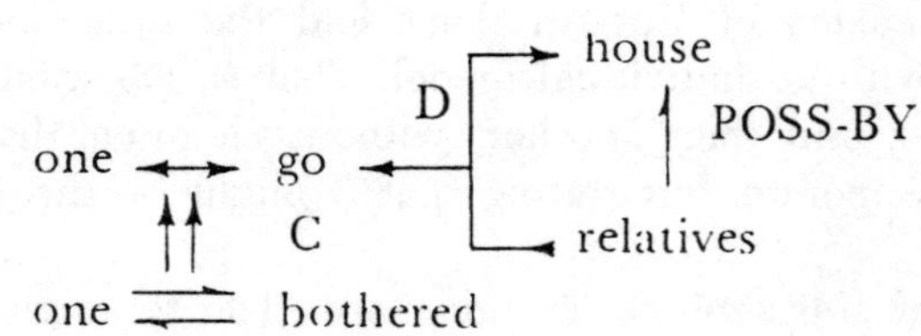

It is easy to see that (49) would be analyzed nearly the same way but the initial triple would have reversed subject and object. Thus, the result would be:

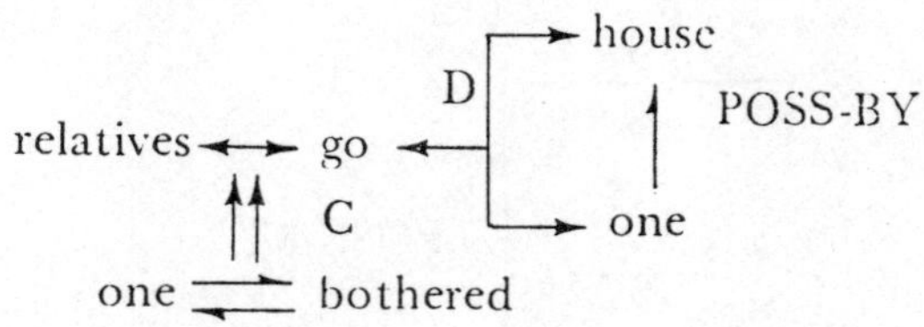

The third interpretation (47) comes from the nature of the relationship between PP's and ACT's. Basically there are three conceptual rules that deal directly with PP and ACT: PP $\leftrightarrow$ ACT, ACT $\overset{o}{\leftarrow}$ PP, and $\begin{matrix}\text{PP} \\ \uparrow\downarrow \\ \text{ACT}\end{matrix}$. The first two interpretations utilized the first two conceptual rules. The third conceptual rule can also be realized in English as 'V-ing N.' When this syntactic interpretation is tried, the same triple as occurred for (49), (relatives, visit–vt, nil) is obtained. The same processing occurs, only this time the two-way link is not $\leftrightarrow$ but $\uparrow\downarrow$, indicating that another predication will be made about the main PP. Thus, we have:

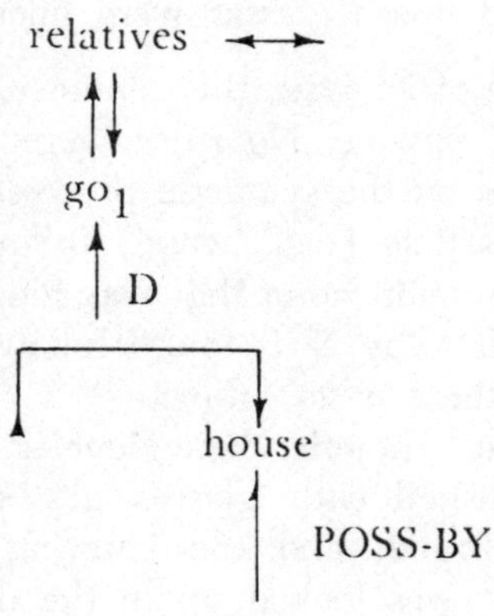

When 'can be a nuisance' is encountered it is treated the same as before, that is, causally dependent on the $\leftrightarrow$. We still need an ACT for relatives, but there is no more sentence, so we insert the dummy 'do':

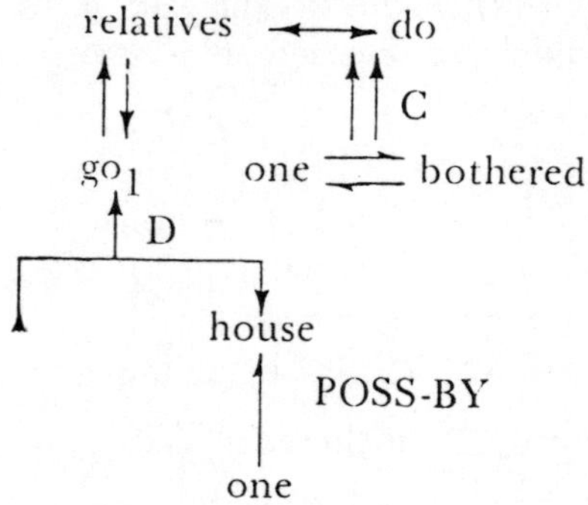

This conceptualization represents that the relatives who are visiting can bother you by doing something other than visiting. For example, they might be a nuisance because they eat a lot. This interpretation of this sentence is rarely found in linguist's discussion of this sentence, but, it would seem to me to be the most valid interpretation. That is, one is tempted to inquire after hearing this sentence as to what the relatives do that is so bothersome. That is expressed by the last C-diagram. We would thus, orient a computer analyzer to look for this interpretation first.

5.4 *Semantic Ambiguity*

A classic example of semantic ambiguity is expressed in sentence (50):

(50) The old man's glasses were filled with sherry.

From the point of view of analysis, this sentence is different than others presented since it is a passive. No rules were given for passive constructions in the section on the syntactic processor, but a simple rule is that when a passive particle (e.g., 'were') followed by a verb in past tense, is found then the main noun that was found is placed in the object slot of the triple. If a 'by N' is found following the verb then it is the subject, otherwise there is no subject.

In this sentence, the syntactic dependencies between 'old,' 'man's,' and 'glasses' are established with 'glasses' at the head. The passive is encountered, turned around, and the entering triple is (nil, fill–vt, glasses). When 'glasses' was looked up in the dictionary it was found to be a noun. However, conceptually, 'glasses' is semantically ambiguous. That is, the word 'glasses' represents two different concepts, 'drinking glasses,' and 'spectacles.' The ambiguity is carried along syntactically until it can be resolved conceptually.

The entering triple calls 'fill' from the verb-ACT dictionary:

fill

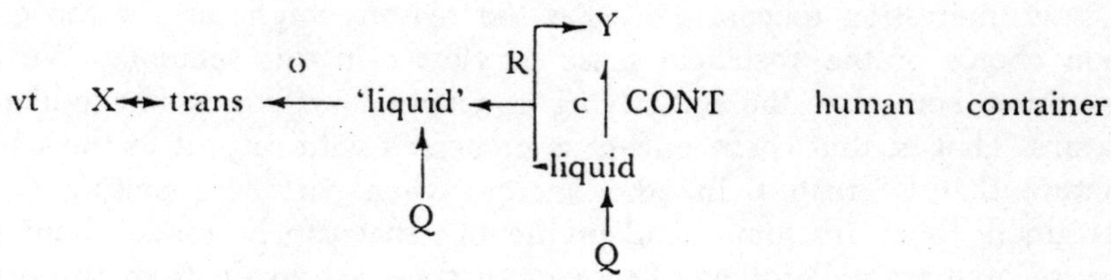

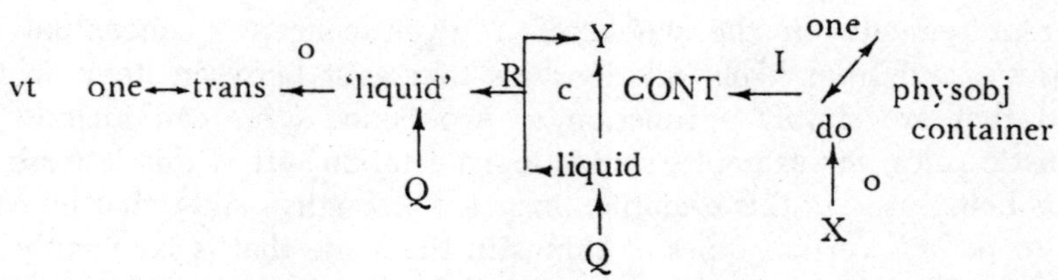

The C-diagram for the first sense of 'fill' indicates that this means that a person will put an amount of liquid, Q, into a container that can contain at most amount Q of liquid. Since the semantic requirements on the sentential object need a 'container,' here is where the choice between the different senses of 'glasses' is made. 'Glasses$_2$' whose semantic category is container is chosen. Here, the 'with sherry' construction looks like a candidate for instrument. But once the conceptual realization for 'fill' has been selected, the conceptual processor switches to a top-down mode allowing for it to look for what it needs. It most clearly needs a 'liquid' as object and this is what is searched for. Since 'sherry' is the next PP to be treated, and is a liquid, it is immediately placed in the C-diagram. Here again, we see that all conceptual cases are not equal in their demand to be filled. Object is always first if it is required and instrument is always last. Since the subject is nil, the final C-diagram is:

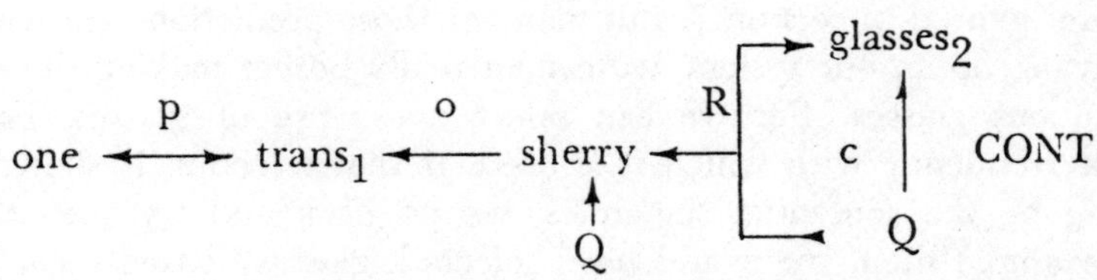

The most interesting aspect of this sentence, however, has not yet been discussed. Namely, it is nearly always the case that when English speakers hear this sentence they select 'glasses$_1$' first meaning spectacles, and then revise their choice after hearing 'fill.' This is certainly not what we did here, and this is a problem for a system that is supposed to be a simulation of human understanding.

It is interesting to consider what the reasons might be for the common choice of the spectacle sense of glasses in this sentence. We are assuming here that the underlying conceptual system is image-like in nature. That is, that each concept is processed with respect to the whole picture that is created. In other words, when part of a picture is ascertained, there are some valid predictions that can be made about the nature of what will follow. These predictions are made from the organization of the parts of the conceptual whole (conceptual rules); from the requirements on the whole of a given concept (conceptual semantics); and from likely relationships that exist between items in the world that are simply a function of experience (this can include all linguistic rules, for example, syntactic expectation). It is this latter type that is being used in this example. Speakers of English know that humans tend to possess certain types of things in the sense that is commonly referred to. That is, whereas, a man could have either type of glasses as his legal possession, we tend to speak about his possessions which are pseudoinalienable, i.e., his car, his house, and his clothing. So, when a possessive occurs in English, we can predict that the possessive refers to things that we would tend to associate with that human. If, for example, the phrase had been 'the bartender's glasses' we would likely have chosen the container, or in any event the choice would be somewhat more difficult. But no matter what, we would have chosen something. That is to say, understanders do not carry along alternative parses of a sentence or alternative senses of a word. Rather, they select one, and if they are wrong, they go back. Their selection is based on context and the context in this case is the prediction made by possession. We are not, in this system, at a point where we can be making all the predictions that a human makes in a given situation. A human is making a great number of predictions at every point in an analysis. Some of those predictions are crucial and we will speak more about those that we have heretofore ignored in section 7. But many of those predictions are wrong or irrelevant. So, in our system, we cannot really bother making the prediction about glasses. But we can select one sense of glasses, based upon the frequency with which it is used. If that selection is shown to be wrong by the sentential semantics, we go back and try the other. In this example then, we would have selected 'glasses,' based on a frequency count of our own experience, and then have gone back and selected the container sense of glasses. It is important to point out that this theoretical procedure is correct in that alternative senses of a word are not carried along. However, even though the correct result is obtained, the actual procedure employed is not based on human processes. We thus consider this procedure to be only partly theoretically valid. At such time that contextual prediction can be effected we shall implement that. Right now, most predictions of this kind would be useless, which seems to point out that such a procedure should be avoided until a tight theory regarding this problem can be worked out.

5.5 *Syntactic Similarity*

Consider sentences (51) and (52):

(51) John's love of Mary was beautiful.
(52) John's can of beans was edible.

While these two sentences conform to identical syntactic patterns, they are quite different conceptually. We would expect then, that they should be processed in syntactically identical fashion while being processed quite differently on the conceptual level.

The main verb in each of these sentences is 'be.' Conceptually, however, there really is no predicated action here and consequently the verb–ACT dictionary is never called. There is a syntactic rule that states that if a 'be' particle is discovered, then if an adjective or noun immediately follows, the verb–ACT dictionary is not consulted. Instead a special attributive conceptualization is set up (denoted by $\rightleftarrows$ in conceptual dependency). This attributive conceptualization requires a PP on the left side of the link and either a PA or PP on the right side. Consequently, what is passed to the conceptual processor is a double consisting of the main noun found by the same procedure as before and the predicated noun or adjective.

For (51) and (52) then all that is required by the conceptual processor is the marking of the syntactic dependencies, and the double (main noun, $\left\{\begin{matrix}\text{adjective}\\\text{noun}\end{matrix}\right\}$). The syntactic rules given above mark 'love' as the main noun in (51) and 'can' as the main noun in (52). 'Can,' in English, is syntactically ambiguous. That is, 'can' can be a verb, a noun, or an auxiliary. It is not an auxiliary because there is no verb following. It is necessary for the syntactic processor to predict a verb when a potential auxiliary has been spotted. This prediction fails to materialize. However, another interesting problem crops up in (52). It is possible to eliminate the verb possibility for 'can' because of syntactic placement. That is, after encountering 'John's' the prediction (as made implicitly in the syntactic rules) is that a noun will follow. Both 'can' and 'canning,' in this syntactic slot must be considered to be nouns since that is the only thing predicted by the syntax. So, no matter what, the main noun configuration is:

```
                can
possession ↑↑ of
           John  beans
```

But, conceptually 'can' is also ambiguous. Not surprisingly, the fact that 'can' can take these two syntactic forms is reflective of the fact that sentential 'can' is conceptually both the PP 'can' (the object) and the ACT complex which means 'can' (to place in a can). Because there is no '-ing' ending, the 'can' here is a PP.

This ambiguity does not exist conceptually in (6). That is, 'love' is an action no matter what syntactic form it comes in. But, here 'love' is processed as a noun in the same way as 'can' and for the same reasons. So we have:

```
                love
possession ↑↑ of
           John's  Mary
```

But, it is important to note, that before the syntactic processing is complete, all syntactic–conceptual interface problems must have been met. Since in (51) 'love' is realized as an ACT, a triple is formed for entering the verb–ACT dictionary that has 'John' as Subject and 'Mary' as object. The triple (John, love, Mary) is then passed on to the verb–ACT dictionary.

It is interesting to notice here, that it is only the very superficial aspects of these two sentences that are similar. It is for that reason we expect the processing of the two sentences to bear only little similarity. From this point in the analysis process, all similarity is lost since their superficial similarity has no meaning to the conceptual system. The point made earlier about this should be emphasized. While syntax is useful for finding the parts of the sentence that are to be conceptually processed, of itself it has no value. A conceptual analyzer uses the results of certain syntactic observations based on the success or failure of syntactic predictions. The syntactic rules are simply heuristics for finding things. We make the claim that this is all that need be done in an analysis.

The verb–ACT dictionary processes the triple for (51) into:

$$\text{John} \overset{p}{\leftrightarrow} \text{love} \overset{o}{\leftarrow} \text{Mary}$$

since love is an EACT (takes object only). Now, the predication marked by 'was PA' causes an attributable dependency to be established with the entire conceptualization as subject. Since 'beautiful' is a PA, and thus fits the conceptual rule requirements for attributive two-way dependency, the final analysis is:

```
      John  p
   p   ↗    ⇄ beautiful
      love
       ↑ o
      Mary
```

Sentence (52) is a bit more interesting. We have the syntactic dependency group:

```
      can
  Poss ↑↑ of
  John  beans
```

but, we have no entering triple. Thus, we have, a PP, 'can' and the predication 'was edible.' Looking up edible we find that it is not a PA or a PP. Therefore, the prediction that an attributive two-way dependency was present fails. 'Edible' is in the dictionary as follows:

```
              c        o
  'animal' ↔ eat ← 'food'
```

This definition means that 'edible' represents a conceptual construction about the possibility of eating something which is assumed to be a food. 'Not edible' would indicate that the object of the action is not a 'food.' In other words, this is at the same time a definition of the semantic category 'food.' (The quotation marks indicate realizable semantic categories.)

Thus, we now have a conceptualization, given by the definition of 'edible.' We are faced now with the problem of filling in the unrealized slots in this conceptualization. We begin a search for the cases and look for a 'food.' The only place that there is to look at all is the syntactic dependency complex given above. The head of that construction is 'can' and 'can' is not a 'food.' So we must search the dependents and here we find 'beans' is a food. We thus have:

```
              c        o
  'animal' ↔ eat ← beans
```

We must now conceptually realize all the elements that are part of the syntactic complex that 'beans' is in. 'Beans' is related to 'can' by 'of' and we find that this is a containment relation (this is handled by a program that is not described here, see Weber (1971)). But, if 'can' is used here, 'John's' must be used also since it is related to can as a modifier of possession. The final C-diagram is then:

```
        c       o
  one ↔ eat ← beans
               ↑ CONT-BY
              can
               ↑ POSS-BY
              John
```

Now it is possible that this was not what was meant here. It could be that, the 'can' was being referred to as 'edible' as was predicted by the syntax and rejected by the semantics. Here again, we are making no effort to guess what could possibly have been meant but rather what was most probably meant. We can always correct any mistake in an interactive situation, and we can alter our probabilities.

It now remains to discuss sentences (10), (11), and (12) mentioned earlier:

(10) I want a book.
(11) I want to get a book.
(12) I want to have a book.

The assumption here is that these sentences all mean the same thing. While it can be argued that they do not because of varying contexts or connotations, we assume the context where the conceptual content of all three sentences is the same and disregard for the moment implications.

Using the verb–ACT definition of 'want' given in section 5.1, we see that for the vt sense that exists in (10), an entire structure is the realizate of the idea of 'want a physobj,' namely, for (10):

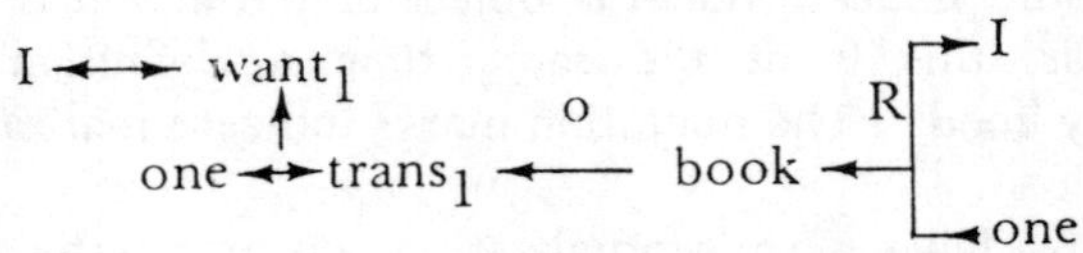

The idea here is that what is wanted is that somebody give this book to the subject in any unspecified manner.

The same structure can be realized for (11) given that this is the vs sense of want and that the verb–ACT entry for 'get' is:

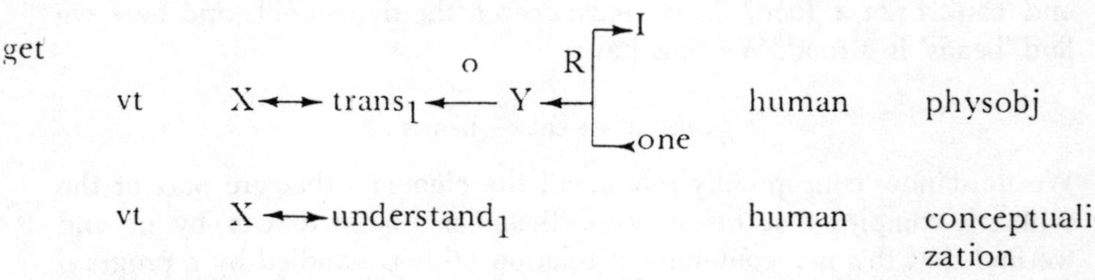

Thus, given that the SACT 'want' requires a conceptualization and that 'book' is a physobj, we have.

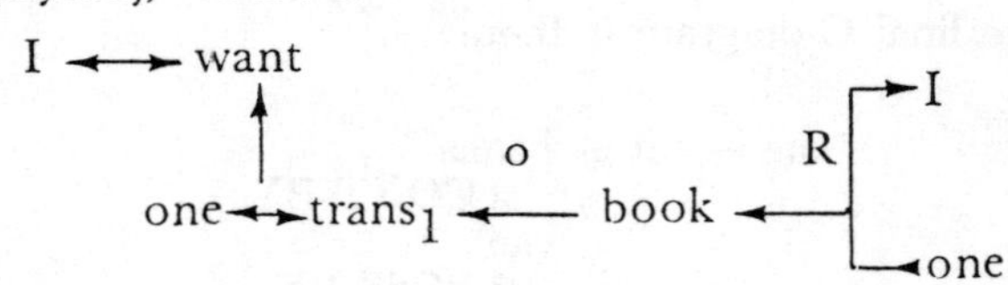

which is the same as the above conceptual construction.

We have not, until this point, discussed the verb 'to have.' Since 'have' in its alienable sense represents the result of a conceptualization that involves acquiring the object in the first place, we consider 'have' to be an instance of 'trans.' After the 'trans' has been completed, 'have' can be assumed to be an implication derived from the 'trans' about a relationship between two PP's. But it is important to understand that 'have' presupposes this 'trans' relationship for alienable possession. If such is the presupposition then it is part of the conceptual content and therefore must be represented in the conceptual diagram. The verb–ACT entry for 'have' is:

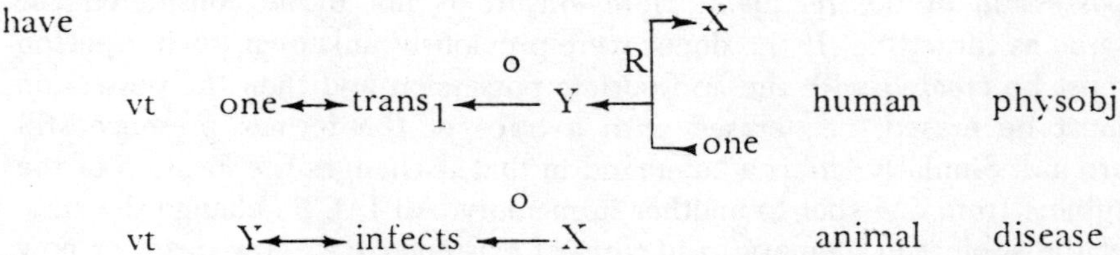

Thus, this is virtually the same as the previous example and the SACT, 'want,' requires the first conceptualization:

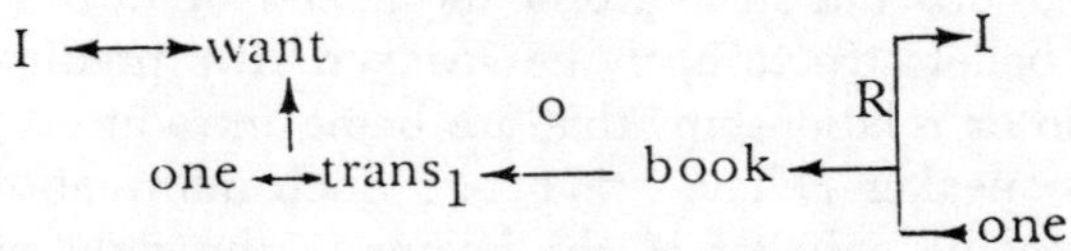

There exists, at the Stanford Artificial Intelligence Project, a computer program that will do all that has been described so far in this paper. Of course, all of the details of the conceptual processing that takes place in that program have not been described here, but the actual conceptual analyzer works roughly in the above manner.

6. CONCEPTUAL DEPENDENCY REVIEWED

It is perhaps reasonable to inquire at this point about the nature of conceptual dependency.

We can consider that a theory such as conceptual dependency is doing its job based upon the following things:

(a) *Predictive ability.* The theory must predict accurately what kinds of conceptual items have not been explicitly mentioned from the sentence or group of sentences being processed. This predictive ability as to type (e.g., conceptual or semantic category), allows for two things. First, it furthers the analysis process by directing the search through data for the information that is needed, or by formulating inquiries as to the needed information when such inquiries are called for. Second, such predictions form the basis of the system's interaction with memory. For example, certain kinds of information are left out of a sentence largely because the information is redundant within context. It is necessary to predict the existence of this redundant information in the first place and then to find its appropriate manifestation in context. (We will explore this in more detail in section 7.)

(b). *Responses.* The conceptual dependency networks must form the basis for the memory structure by treating certain constructions as commands to the memory. For example, 'trans' is a command to the memory to erase the object from the possession of the donor and place it in possession of the recipient. Here erasure is not to be considered the same as forgetting. If the donor were previously unknown, such a person must be created with the appropriate possession and then the possession must be erased, i.e., erased with a trace of the former presence still around. Similarly 'go' is a command in that it changes the location of the subject from one spot to another in memory. All IACT's change the state of the subject in memory as to form of existence, e.g., now dead, or now asleep. In a sense IACT's are like PA's in that they change an attribute of the subject in memory. Other conceptualizations are commands to add a belief to the belief structure of the hearer or to the hearer's conception of the belief structure of the speaker. Any predication, that is, new information or relationships that are being introduced go into a file that states 'this speaker believes that . . . .' Such information can be operated on to become a belief of the hearer in the right circumstances. (See Colby & Smith (1969) for a computer implementation of this.)

In turn, the internal responses generate external responses based upon the coherence of the new conceptualization with previous memory information. We have just begun to work on this and thus leave the response aspect of the problem for the time being.

(c) *Paraphrase.* If two sentences that express the same conceptual content are analyzed into the same conceptual structure we have done the job that we have been attempting. As long as the paraphrase relation always holds and no input utterance that is agreed to be meaningful cannot find a conceptual representation, the major aspect of the theory has been realized.

(d) *Computer programs.* If (c) holds true, and the procedure has been explicitly stated clearly enough, it should be in principle possible to create a computer program to do these analyses. This would, of course, be of immense usefulness. It may be necessary to modify traditional computer programming approaches in order to accomplish the task set out here. The bottom-up parsing technique must be switched to a top-down one and back again, in this system. This may well require a nondeterministic algorithm of the kind of which Floyd (1967) spoke. Such an attempt is currently being worked out at Stanford.

(e) *Psychological models.* Conceptual dependency makes some psychological claims that have yet to be tested. It states that the basis for language is an interlingual conceptual structure; that there exists a network of concepts and relations of which one relation is the central core (↔); that an action is made up of from zero to three principal dependents one of which is itself another action construction, and that it is, therefore, virtually impossible to succinctly describe an action in words; that these dependents are retrievable through the ACT and that it is probably not possible to think of an actor and a case dependent without somehow thinking of the action that links them. Furthermore, the claim is made, that while some dependents are retrievable through each other (object and recipient for example) this is because of an immediate implication that is generated by their simultaneous existence and is, thus, an item retrieved from memory that is derived from the conceptualization. It claims that, while the case dependents are retrievable through the ACT, many things are retrievable as relations connected to the core of the conceptualization (↔) or the simultaneous occurrence of an actor and action. Thus, time and location of a conceptualization should function more like causals than like cases. And it claims that conceptualizations relate to other conceptualizations as either subsuming them (as instruments) or causing them in a specified way.

These remarks are intended only to make clear some of what we are claiming to the psychologist. We anxiously await results of tests that could prove such claims to be in error.

7. ANALYSIS OF INTENTION

7.1 *Expectations*

We have been, so far in this paper, treating sentences as if they existed in isolation, a procedure not uncommon in most discussions by linguists. But, people do not understand sentences in isolation and it is necessary to recognize this in any theory of understanding. It is, however, possible to extend some of the techniques of prediction described so far in this paper to the problem of predicting more than just syntactic or conceptual types that are likely to occur.

There are also predictions that can be made that are based on criteria other than that directly derivable from the stratified linguistic system that comprises conceptual dependency theory. Consider the following situation and conversation:

John meets his friend Fred on the street. Fred is holding a knife. John is angry because his wife Mary has yelled at him.

Fred: Hi.
John: What are you doing with that knife?
Fred: Thought I'd teach the kids to play mumbletypeg.
John: I could use a knife right now. (agitated tone)
Fred: What's the matter?
John: Damn Mary, always on my back. She'll be sorry.
Fred: I don't think a knife will help you.
John: You're just on her side. I think I ought to . . . .

Now what can Fred expect that he will hear next? There are at least six distinct types of information with which we can answer this question. Sententially, Fred expects a verb. Conceptually, since 'ought' is an SACT, a conceptualization is expected. But we can also make predictions based on context. According to the context, there are only a certain set of concepts which will fit into the needed conceptualization such that the conceptualization makes sense in context. We most certainly would be surprised if the next piece of information would be 'I think I ought to have fish for dinner.' It is knowing what we do and do not expect at any point in any analysis which allows us to be surprised, shocked, or whatever other emotional attribute by a piece of information. You are not able to be surprised if you don't anticipate and it is, therefore, necessary for a system such as this to anticipate.

What we anticipate here are the following four types of statements based on their contextual likelihood: (1) hurt someone, (2) end relationship with somebody, (3) go to someplace, and (4) emote.

These are classes of actions. We don't know which sentential form 'hurt,' 'go,' or 'emote' will take but we can estimate the likelihood of the class on the basis of the conceptual category and the prevailing semantic categories that have been used in context. All of these above actions are predicted on the strength of their likely consequences. That is, a desired consequence is known (John feel better) and the above actions would each lead to John's feeling better, but each in a different way.

A fourth type of expectation or prediction is conversational. That is, people talk for a reason, usually to communicate something or to gain some desired effect in the hearer. Here, it is either to arouse sympathy or to inform about something he is about to do. But the use of ought implies he might not do this, so that his probable reason in making this statement has to do with the effect which it will create on the hearer. Thus, we can predict what kind of effect is intended to be made by the speaker and then expect certain types of utterances.

A fifth kind of expectation information has to do with a world view of the situation based on his own individual memory model. Thus, if he knows John to be a convicted murderer, his expectation of John's completion of this sentence ought to be different from his expectation if John were an avowed pacifist.

A sixth type of expectation is based on a memory structure that is common to the cultural norm rather than the particular language or particular individual. The results of this kind of expectation have to do with the options that Fred can take as a result of the expected input from John. That is, the conversation is heading toward death (this idea will be explained below) and Fred's expectation of this can avert the situation by appropriate action, either physical informative conversational or emotional conversational. It is his expectation that decides the appropriate action.

Basically then, we must recognize that any complete processing system for a natural language utterance takes place within a context that is extremely complex because there are humans in the conversation. Each has a complex memory to begin with and is now in a new complex situation. Part of this problem is being able to anticipate. The anticipation with respect to linguistic processing then is a function of a set of different types of expectations at any point in the analysis. These expectations are of various kinds and aid the basic analysis capability tremendously. Our predictions are of the following type:

(1) sentential—what syntactic category is likely to occur
(2) conceptual—what conceptual element is needed at this point in the parse to help complete the C-diagram
(3) contextual—what information type fits in the structure created for information (kind of C-diagram) at this point in the overall parse of the conversation
(4) conversational—similar to (3) but 'what answers a question' or responds to the input in any of the ways mentioned on the previous page would fit here
(5) world view—what we expect of the substance of the information rather than its type, dependent on the presuppositions and biases of the hearer
(6) memory structure—total correlation of this memory model. The process of 'living' is important here, i.e., is this statement tending to describe information or events that will 'satiate' or 'hunger.' That is, 'Am I pleased by this?'

The first two expectation types are an intrinsic part of the sentential and conceptual levels, respectively. Contextual expectations are part of the conceptualization–memory interface in that they aid the formation of complete conceptualizations by using information from memory. The other three are a part of the intermediate memory structure that deals with input and decides what to do with it with respect to generating internal and external responses.

7.2 *Functional Associations*

The simplest type of prediction that can be made as to the nature of a missing action or object in a conceptualization is from the function of that item. Consider sentence (53);

(53) I like books

$$\text{I} \leftrightarrow \text{like} \overset{o}{\leftarrow} \text{books}$$

This analysis is not an allowable construction conceptually because the ACT 'like' is of two possible conceptual types, each with its own semantic restriction. The EACT sense of 'like' (like_1) allows conceptual objects but requires that these objects be of the class 'animal.' The other sense of 'like' (like_2) is conceptually an SACT which requires an entire conceptualization as object. Since a conceptualization must have an actor and an ACT and we are thus faced with the problem of uncovering these in the analysis of the above sentence. We have then:

$$\begin{array}{ccc} \text{I} & \leftrightarrow & \text{like}_2 \\ & & \uparrow \\ & & \leftrightarrow \end{array}$$

We know that 'books' is part of this conceptualization and by the heuristics of the conceptual dependency system we know that 'I' is as well. The problem is what arrangement and what ACT is correct.

Now we, as speakers, know that the most reasonable answer to this problem is:

$$\begin{array}{ccccc} \text{I} \leftrightarrow \text{like}_2 & & & & \\ \uparrow & & & & \\ \text{I} & \leftrightarrow & \text{read}_1 & \overset{o}{\leftarrow} & \text{books} \end{array}$$

The question is how the system is to arrive at such a conceptualization.

Consider the dictionary entry in an ACT dictionary for 'read_1.' The ACT 'read_1' is listed in our system as requiring a 'human' subject and an object that is chosen from the set of objects that have been made by men for exactly the purpose of 'reading' them. That is, while we could list all possible such objects (books, newspapers, etc.) or categorize them in some artificial hierarchical structure, conceptually the object of 'reading' is 'that which is read.' Specifically, this class could include anything with printing on it or whatever. The point here is that we can call the potential object a member of the class 'read$_1$-PP$_o$.' Then, in any listing of the elements of the world, their semantic category would be the place that fit in our ACT-based model. 'Book' would be:

book: N; read$_1$-PP$_o$;

where 'read$_1$-PP$_o$' denotes that it is the conceptual object of the ACT 'read.' Then our diagram must become:

$$
\begin{array}{l}
I \leftrightarrow \text{like}_2 \\
\qquad \uparrow \\
\qquad \leftrightarrow \text{read}_1 \overset{o}{\leftarrow} \text{book}
\end{array}
$$

The only thing missing is the actor, which is 'I' due to the heuristic which governs these situations.

The categorization of 'book' as 'read$_1$-PP$_o$' is not ad hoc. 'Book' belongs to no other semantic category, and this particular semantic category is in the hierarchy as a physobj. That is, because there are other ACT's that can be done to 'books,' this does not include them as part of the category of 'book.' This category, then, is indicative of the function for which the object was created in the world. The fact that secondary functions may exist is of no concern since what we want to do here is predict redundant information. Usually, if formation was left out, it is because it can be predicted simply on the basis of such natural function information. If a mistake is made, then it can be discovered by inquiry in interactive mode. We wish to make the most likely guesses for the given context rather than the 'right' ones.

Thus, many other objects can be categorized in this manner. For example, consider 'knife.' 'Knife' is the object of the ACT in the instrument of cutting. A funny way to say 'I sliced the meat with a knife' is to say 'I cut the cuttee with a cutter.' Now, of course, the specificity of the particular concepts is lost with this paraphrase, but a 'knife' is a potential 'cutter' and that is what is important here. That is, when 'knife' or any other cutting instrument is mentioned the association with 'cut' or some specific 'cut' term must be made. A context aids this process considerably, but regardless of context some association will be made by the human and must, therefore, fit in with any theory of a system of expectation for conceptual predictive analysis.

Thus, we can say that 'knife' is an instance of 'cut-PP$_I$.' This means that it serves as the object in the conceptual instrumental case in conceptualizations involving 'cut.' More accurately, we can say that a 'knife' can be expected to be used in this way and also that conceptualizations involving 'cut' will have as instrument a member of the class 'cut-PP$_I$.'

The primary point is that there are associations between objects and the action for which they are made which are fairly straightforward that are an important part of the process of expectations, and more importantly, are part of what has been understood from an utterance.

7.3 *The Relationship Between SACT's and ZPA's*

SACT's are interesting in that they communicate a feeling about a conceptualization. When they express the notions of 'wanting' or 'fearing' they are at the same time expressing a mental state attribute which the actor believes will come to be true of himself if the conceptualization being referred to occurs. In other words for sentence (54)

(54) John wants Mary to come home.

a likely implication is that 'John' will be pleased by this action's occurrence. That is, we can generate a

$$\begin{array}{c} \uparrow\uparrow \text{cf} \\ \text{John} \leftrightarrow \text{pleased} \end{array}$$

off the two-way dependency in 'Mary ↔ go' that expresses that this could result in John being pleased. We claim that is part of the meaning of 'want.'

Now consider sentence (55):

(55) I fear bears.

As in (54), 'fear' is an SACT which requires a conceptualization as object. Since 'bears' are not made for a use they qualify as actors in the needed conceptualization. 'Bears' can do many things, so we can't be sure what it is they do that is frightful. Thus, we have:

$$\begin{array}{c} \text{I} \leftrightarrow \text{fear} \\ \uparrow \\ \text{bears} \leftrightarrow \text{do} \end{array}$$

This expresses that 'fear' of objects is not related to the object itself, but rather to the action that a 'bear' might do. We can make a similar implication from 'fear' as we did from 'want.' Namely, 'fear' implies possibly 'harm.' Thus, we have:

$$\begin{array}{c} \text{I} \leftrightarrow \text{fear} \\ \uparrow \\ \text{bears} \leftrightarrow \text{do} \\ \uparrow\uparrow \text{cf} \\ \text{I} \rightleftarrows \text{hurt} \end{array}$$

In other words, you 'fear' consequences, not particulars.

The main point here is that there is more to the understanding process than ascertaining the conceptualizations expressed directly by a sentence. It is necessary also for a theory of understanding to account for concepts and conceptualizations expressed indirectly.

The relationship that we have made between an SACT and a ZPA mediated by another conceptualization would appear to be valid whenever an SACT or ZPA are present. Furthermore, certain ACT's have consequents that are either good or bad for either the actor or object. For example, 'eat,' 'drink,' 'love,' 'fight,' and 'hit' are all positive with respect to the subject. That is, when these actions occur we can assume that the actor wanted them to happen and that the attribute 'became more satisfied' would apply to the actor. On the other hand 'eat,' 'hit,' 'cut,' and 'attack' are negative with respect to the object. That is, we can expect that the object was damaged in some way from them.

7.4 *Intentions*

Consider a sequence such as this:

(56) Q: Do you want a piece of chocolate?
(57) A: I just had an ice cream cone.

In a model of natural language understanding, it is unreasonable to claim that the system has understood the utterance unless it is capable of producing for (56) not only a conceptual diagram of the information just stated, but also something like the answer 'no I don't want a piece of chocolate.' That is what a human could understand in the above situation and it is incumbent upon any so-called understanding model to understand the same.

We can actually do this as follows. The conceptual dependency analysis of (56) is:

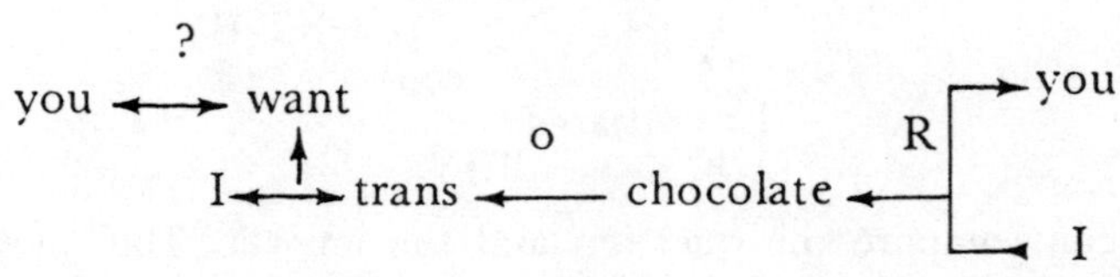

However, the model that we have been discussing would be charged with taking the conceptual representation of the input and drawing the necessary implications that can be said to be understood implicitly. In this case, chocolate is discovered in the dictionary to be an 'eat: PP_o.' The association between 'want' and 'eat' fits into the SACT–ACT–ZPA paradigm and yields the implications that the ZPA 'satiate' is caused by the connection with respect to the subject of 'want.' This gives us:

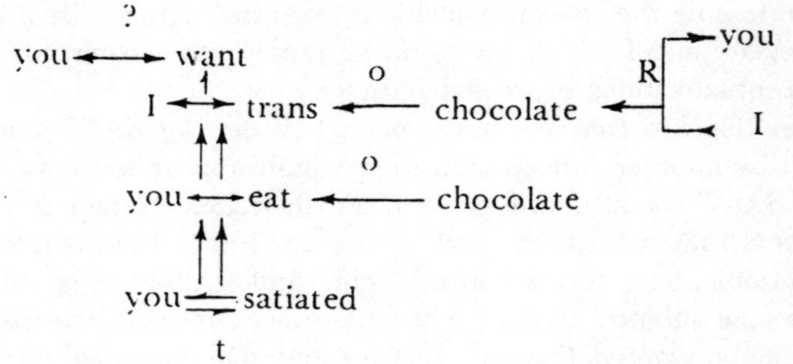

Now it is also true that people eat for reasons other than satiation, particularly, for pleasure. So the causal connection

$$\begin{array}{c}\Uparrow\\ \text{you} \rightleftarrows \text{pleased}\\ t_f\end{array}$$

is also a consequence of the 'eat' conceptualization. But this is not necessary here.

Now we are ready to analyze the answer (A). The conceptual diagram associated with the input is:

$$\begin{array}{l}\text{I} \leftrightarrow \text{eat} \overset{o}{\leftarrow} \text{ice cream}\\ \quad t_f \qquad\qquad\quad \uparrow \text{CONT-BY}\\ \qquad\qquad\qquad\quad \text{cone}\end{array}$$

This diagram is obtained by utilizing the dummy quality of the verb 'have' and finding the ACT associated with 'ice cream,' again 'eat.' Here again, 'eat' implies the causal for satiation and we have:

$$\begin{array}{l}\text{now}\\ \downarrow\\ \text{I} \leftrightarrow \text{eat} \overset{o}{\leftarrow} \text{ice cream}\\ \quad t_f \qquad\qquad\quad \uparrow \text{CONT-BY}\\ \Uparrow \qquad\qquad\qquad \text{cone}\\ \text{I} \rightleftarrows \text{satiated}\\ \quad t_f \qquad\qquad \text{PHYS}\end{array}$$

Now we can compare the question and the answer. The question can be matched with the answer by looking at:

$$\begin{array}{l}\qquad ?\\ \text{you} \leftrightarrow \text{want}\\ \qquad\quad \uparrow\\ \qquad\quad \vdots\\ \text{you} \rightleftarrows \text{satiated}\\ \quad t_f \qquad\qquad \text{PHYS}\end{array}$$

from the question, and:

$$I \rightleftarrows_{t_f} \text{satiated} \quad \text{PHYS}$$

from the answer. Since 'you' and 'I' represent the same token in memory, the answer to a question about desired transition (t) has been answered with a statement of completed transition (t_f). In other words, we can assume that we have, 'do you want to be satiated?,'–'I have just been satiated.' Thus, we have the simple implied negative.

The point here is that the implications that are to be found here are part and parcel of the understanding process and, in fact, make little sense without them. We can expect that a natural language-analysis system must be continually making these associative implications in order to be able to use them when they are needed.

Essentially we are setting up a peculiar kind of world model here. We are saying that people do things for reasons and that people say things for reasons and understanding these reasons is an important part of understanding natural language utterances. It is the analysis of the intention of an utterance or ACT that is the primary element necessary to correctly respond to that utterance or ACT.

8. CONNECTION WITH MEMORY

8.1 *Idiosyncratic Definitions*

If we begin to talk about the meaning of a sentence with reference to the intention of the utterance, we open a box of problems that have as a general solution the setting up of an entrance into the memory structure and the defining operations within that structure. Previously, we have said that the conceptual diagrams are intended to convey the content of what was said. But it has been seen that what was said is often not quite what was 'meant' in the intentional sense of that word. In order to have an effective program that can converse with a human, it is necessary for the program to know what the speaker 'means' at any given point. This notion could be carried to the logical absurdity of trying to figure out what the speaker 'really' meant. Was he lying for example. This, in fact, is what a sophisticated psychiatric interviewing program must do. But a program which is intended to simulate language understanding ability must start at the beginning, that is, it must first carry out the logical implication and inferences that a normal speaker performs. What are these then?

The answer to that question is manifold. The first part of the problem is what structure or type the solutions conform to. That is, are conceptual dependencies the representation of the entire situation? Here we must make the differentiation between the meaning of the sentence and the meaning of the speaker. That is, conceptual dependency representation is a characterization of the conceptual content of a sentence (i.e., 'what was actually said'). The meaning of the speaker is the intention of that content. This corresponds to the statement 'I don't understand what you mean.' We often say this, when the conceptualizations that we have derived from what a speaker has said do not fit in with our previous experience or do not have enough information to let us know how to interpret the utterance. We thus make the distinction between interpretation and understanding (as have others, e.g., Deese (1967)). Understanding is capable of being characterized by conceptual dependency networks. The interpretation process utilizes these networks in conjunction with the overall memory structure so as to produce the impetus for the generation routine. That is, the end result of the interpretation process is used in conjunction with the 'reasons for talking,' and the structure of the conversation in order to begin the generation of a response either verbal or physical. For example, the meaning of (58) is:

```
                           tk
(58) Fire.         something ↔ burn
                              ↑
                            here
```

but the interpretation process utilizes this network in addition to its knowledge of what happens when 'something ↔ burn' to produce a result. Here we have:

```
                   something ↔ burn
                             ↑↑
                   something ⇄ die

                             x
                             ↓ LOC
and                something ↔ burn
                             ↑↑
                   something ↔ burn
                             ↑ LOC
                 ↑      ↑    x
               other  combustible

and              human ⇄ combustible
```

For the hearer this means:

```
                            tk
                   something ↔ burn
                             ↑↑
                        hearer ↔ die
                               ct
This coupled with    hearer ↔ want
                              ↑
                          hearer ↔ live
                                 k
```

yields

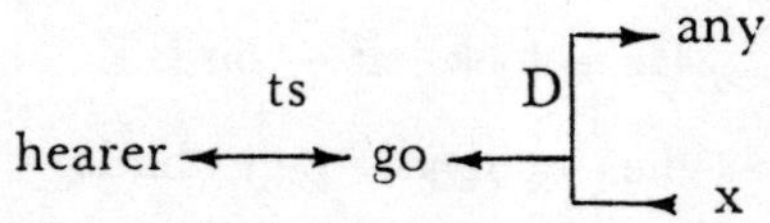

That is, 'Fire' initiates the response in the hearer of getting out of the vicinity of the fire.

Sometimes then, there is a clear distinction between the meaning of the sentence and the meaning of the speaker. This distinction becomes slightly fuzzy in a sentence such as (59):

(59) He acts like Harry.

We can say that a sentence such as this is meaningless in the case that the hearer has no idea who 'Harry' is or how 'Harry acts.' This conforms to the oft heard statement 'I hear what you're saying but I don't get what you mean.' That is, the hearer doesn't understand.

In terms of conceptual structures, the network representation is 2-fold. First, we have the notion that 'he' acts in some manner (x):

he ↔ do
↑
x

For 'Harry' we have the same conceptualization:

Harry ↔ do
↑
x

Now, this is virtually meaningless, as is this sentence unless we know something about Harry. If we do, then the second conceptualization acts as a pointer into memory to retrieve the set of ACT's known to be associated with Harry. For instance, we could have in memory:

Harry ⇄ cat
Harry ↔ pounce on ←(o) birds
↑↑
birds ↔ die
Harry ↔ eat ←(o) cat food
one ↔ pet ←(o) Harry
↑↑
Harry ↔ purr

Here we see that the statement 'He acts like Harry' is still meaningless unless the range of remembered ACT's about Harry is delimited. In actual conversation the delimitation is often made by the context. That is, if I note that 'he is acting like Harry' when I see him watching birds, I might be saying that:

$$\begin{array}{l} \text{He} \leftrightarrow \text{look—at} \overset{o}{\leftarrow} \text{birds} \\ \quad\ \uparrow\uparrow \\ \text{he} \leftrightarrow \text{pounce—on} \overset{o}{\leftarrow} \text{birds} \\ \quad\ \uparrow\uparrow \\ \text{birds} \leftrightarrow \text{die} \end{array}$$

and furthermore that

$$\begin{array}{c} \text{he} \\ = \\ \text{Harry} \end{array} \leftrightarrow \text{look—at} \overset{o}{\leftarrow} \text{birds}$$

That is, the first set of conceptualizations are obtained from a direct memory search, where information about 'Harry' is retrieved with respect to the particular context. This is the partial intention of the speaker. We can assume that the rest of the statement is intended to draw the parallel between 'he' and 'Harry.' The second conceptualization equates 'Harry' and 'he' with respect to a particular action. (This uses the representation of comparatives (where = can be realized sententially as 'like') described in Schank (1969).)

Often the kind of memory retrieval to which we are referring can be assumed to be directly derivable as the meaning of the sentence when the contextual delimitation needed for such a retrieval is provided by the sentence itself. Consider (62): 'He is doglike in his devotion.' This sentence is effectively a command to memory to seek out any knowledge of the devotion of dogs. Clearly, this statement is meaningless if such knowledge is lacking. But, assume an item in memory about the behavior of dogs, e.g.,

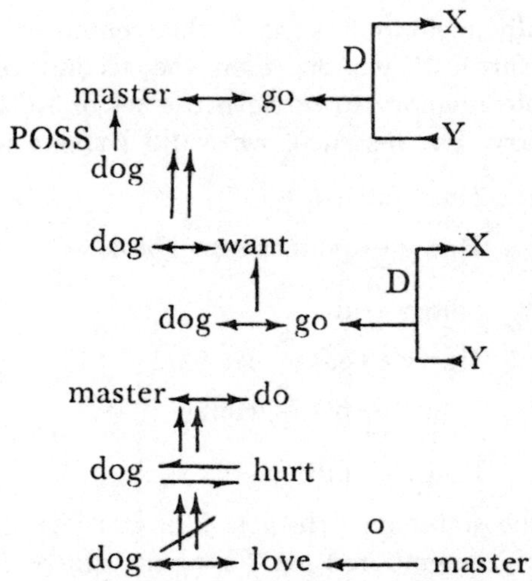

Now, 'devotion' is an English word which keys into this memory structure for 'dog.' That is, in some sense it can be said to be an English idiom with a conceptual heuristic for a realizate.

So the meaning of the above sentence is that 'he' should be substituted in the memory structure keyed by 'devotion–dog.' A plausible inference then is

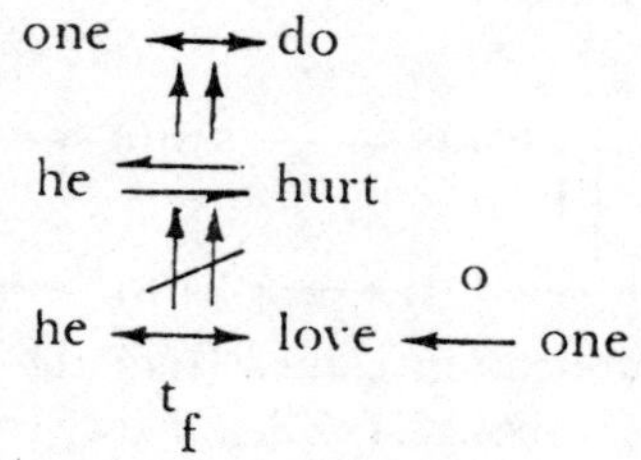

Thus, certain words and word-pairs can cause procedures in the memory to be called into operation. Usually, the operation is based on the immediately previous established context, but often this context is established by the common particular structure in the long-term memory of each conversant. That is, cultural definitions of how dogs behave with respect to devotion facilitate communication for this sentence. If the conversation were conducted in English by members of two radically different cultures, it might be possible that communication would be stymied. Consider, for example, a hypothetical culture where dogs are a constant threat to children and food supply and are thus hated and feared. Clearly, to such a person the above statement, when heard, would be misunderstood with respect to the intention of the speaker. Thus, a definition of 'devotion' would not match in structure any item in the memory of the hearer having to do with 'dogs.'

8.2 *Interaction With Belief Structures*

It is now reasonable to go back to the types of expectation with which we were concerned earlier. In the conversation between 'John' and 'Fred' we noted that the context predicts what kinds of conceptualizations are likely to be asserted. That is, what do we expect him to say that would fit in with the contextual situation?

We answered by claiming that what was likely was that John would say 'I think I ought to $\left\{\text{kill}\left\{\begin{matrix}\text{Mary}\\\text{you}\end{matrix}\right.\ (\text{Fred})\right\}\right\}$' or 'I think I ought to $\left\{\text{end my relationship with}\left\{\begin{matrix}\text{Mary}\\\text{you}\end{matrix}\right.\ (\text{Fred})\right\}\right\}$.' It should be clear that the particular words that would be used here are not at issue, but only their conceptual content. Now, the question is, how do we get a machine to make these predictions?

The problem is one of derivation. That is, where would this information come from? Consider the statement made by John previous to the one under discussion ('I could use a knife right now'). This is represented as:

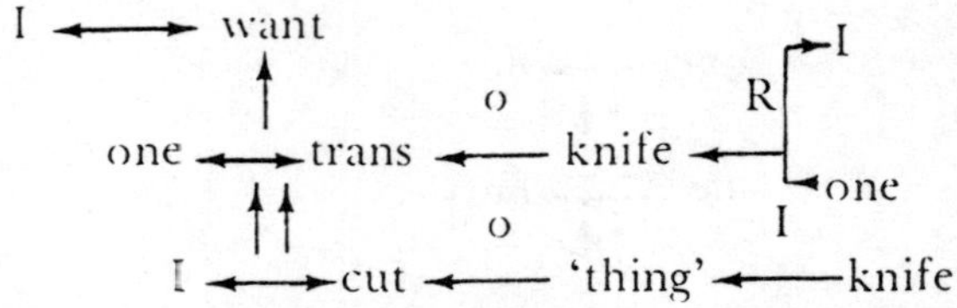

Here the first causal implication comes from the SACT–ACT–ZPA paradigm, or, in this case—'want–ACT–ZPA.' Now, we can say that we have a conceptualization in the intermediate memory that will affect the context. That is,

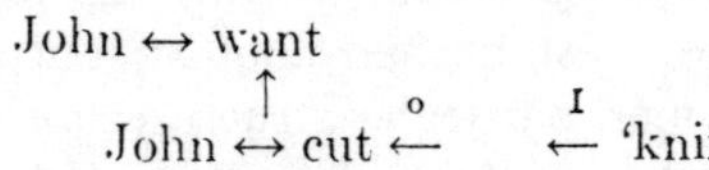

In order to make accurate use of this information, it is necessary to have at the system's disposal a belief that could be characterized as part of the world view expectation. This belief is of the general order:

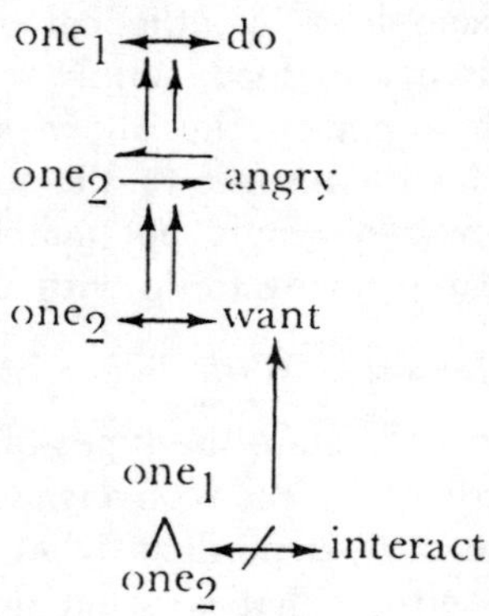

That is, this rule explains that if one is angry at someone that means that one doesn't want to interact with that person right now. Now there could also be a rule that says:

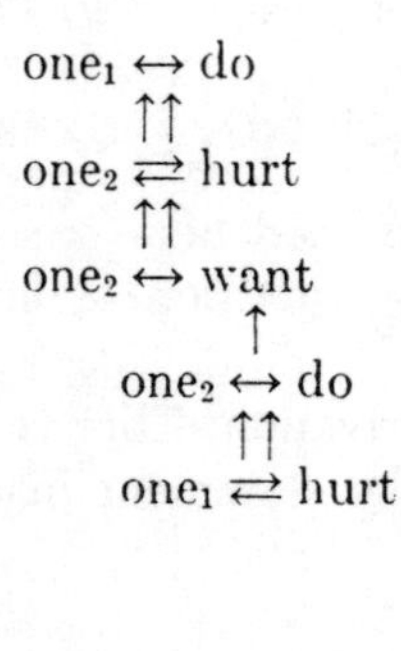

In other words, if one is hurt one wants to retaliate. Now, of course, this rule is not always true for every individual. We would like to note the conditionality of this rule by placing a 'c' over the causal link preceding the 'one$_2$ $\leftrightarrow$ want' and then using the rule if it is the case that in our memory of the individual to whom we are talking we have, for example:

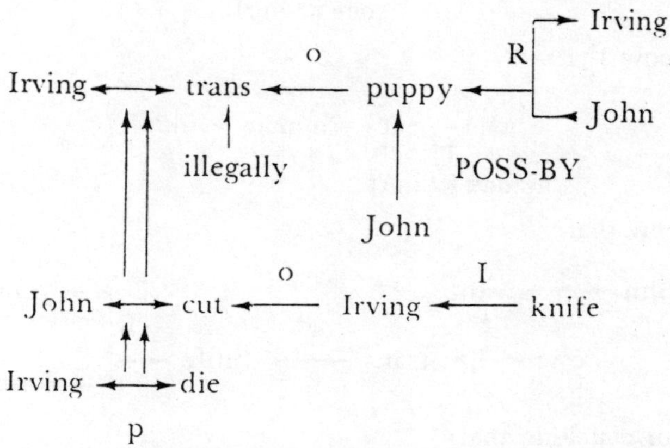

That is, if we know that John already killed for some reason like this, we might guess that John will retaliate again. On the other hand we might have the rule from memory:

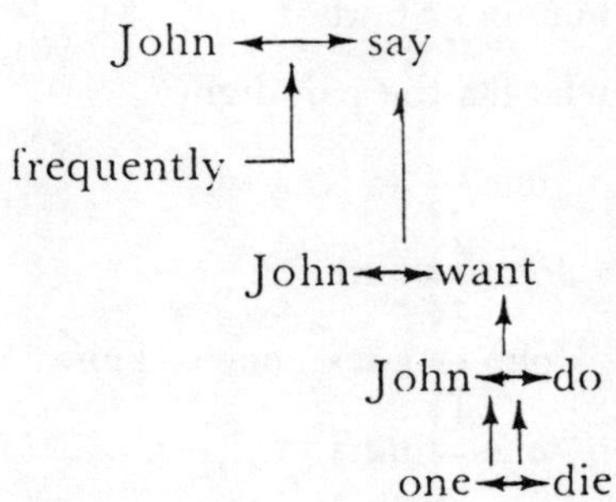

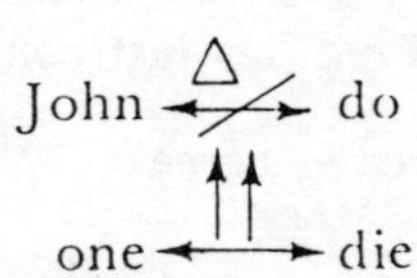

That is, John talks about killing people but never has done it.

The point is that if we can decide that it is the case that John will at least say that he believes that:

one $\leftrightarrow$ do
$\uparrow\uparrow$
John $\rightleftarrows$ hurt
$\uparrow\uparrow$
John $\leftrightarrow$ want
$\uparrow$
John $\leftrightarrow$ do
$\uparrow\uparrow$
one $\rightleftarrows$ hurt

and, we know that:

one $\leftrightarrow$ cut $\overset{o}{\leftarrow}$ 'human' $\overset{I}{\leftarrow}$ 'knife'
$\uparrow\uparrow$
human $\rightleftarrows$ hurt

and, we know that:

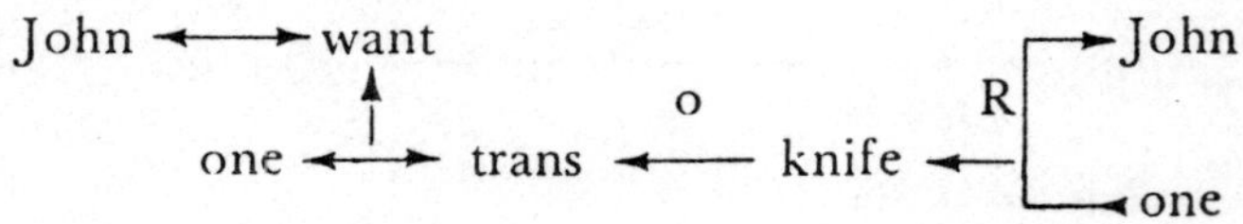

then we can conclude that:

John $\leftrightarrow$ want
$\uparrow$
John $\leftrightarrow$ cut $\overset{o}{\leftarrow}$ 'human' $\overset{I}{\leftarrow}$ knife
$\uparrow\uparrow$
human $\rightleftarrows$ hurt

Now the question is, who fits the paradigm:

one $\leftrightarrow$ do
$\uparrow\uparrow$
John $\rightleftarrows$ hurt
$\uparrow\uparrow$
John $\leftrightarrow$ cut $\overset{o}{\leftarrow}$ one $\overset{I}{\leftarrow}$ knife
$\uparrow\uparrow$
one $\rightleftarrows$ hurt

Since John has said that Mary angered him, she fits the paradigm by definition of 'angered.' Since Fred has just convinced John that:

Fred $\leftrightarrow$ agree
Mary $\uparrow$
$\swarrow\!\!\nearrow$ $\rightleftarrows$ good
do

we can say that 'Fred' and 'Mary' are in the same situation in the paradigm. This is done by yet another belief that says:

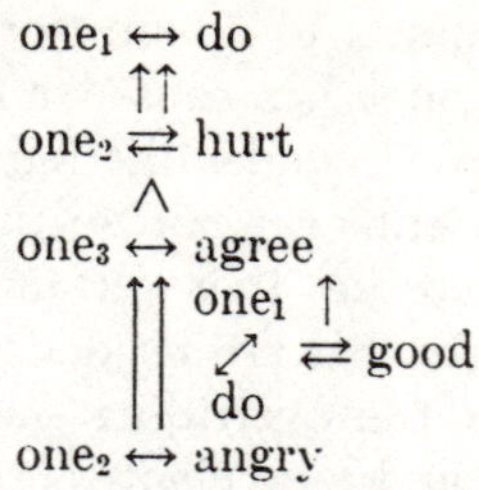

That is, if one sides with one's enemy then one is angry at the enemy's compatriot also. Thus, we can say that John is likely to say that he will kill either Fred or Mary. Also, we can say that the context of the knife aside, he is likely to say that he doesn't want to interact with either Fred or Mary.

The important point here is that it is possible to make contextual predictions as to the content of expected conceptualizations, but that this process of prediction is based on a belief system that includes generalized rules for operating in the world, and idiosyncratic beliefs about the behavior of an individual in the world based upon one's view of people and the particular person under discussion.

9. CONCLUSION

The goal of the research described here is to develop a model which accurately simulates a human being in a natural language understanding situation. Whereas, I am primarily interested in how humans understand, the computer serves as a useful tool for forcing one to make explicit the processes that one describes as being part of understanding. Computer simulation then, turns out to be a good impetus to the generation of theories about human behavior which are testable. The model described here can be tested only in conjunction with some other computer model of a particular kind of interaction. A cohesive system is being developed that will test this model as part of two different psychological models, one of a psychiatric interview where the computer functions as psychiatrist and the other of a psychiatric interview where the computer models a paranoid patient (see Colby *et al.* (1971)).

Since our desire is to enable computers to use natural language in any manner that one might want them to, it is necessary to understand how it is that people do these things that we would like our machines to do. In order to achieve this goal, I claim that it is not possible to separate language from the rest of the intelligence mechanisms of the human mind. Language simply does not work in isolation. It is a nice idea that one should in principle be able to fully describe and characterize language by itself as most linguists are trying to do, but in fact it is just not possible. The ability of linguists to ignore this while trying to separate language into neat formal rules has caused an unbelievable number of unrealistic studies to take place under the banner of linguistics. People neither randomly generate sentences nor do they attempt to assign syntactic markers to input discourse. It is certainly true that humans may perform some of the subtasks that are needed in order to have a formal model do these things, but the overriding question is one of purpose. What are we trying to do, and might not there be a better way?

The answer to these and other like questions is that language exists as a medium for expressing thoughts. In order to deal with thoughts on a machine it is necessary to characterize them in some way. That is, we must extract the inherent ideas from the linguistic input and characterize these ideas in some fashion so to be able to use them. It is the use of these ideas that has been sorely neglected by linguists, yet it is precisely the use of ideas that is the communication process. In order to claim that we have understood what somebody has been telling us, we must process the received input in a certain way. Now this does not mean that we must react to the input in the correct way in order to claim to understand it. If I say 'go get me a pickle,' the hearer's lack of motion does not indicate that he doesn't understand me. It may simply indicate his recalcitrance at being ordered around. But, if the above statement is the punchline to a joke, and the hearer does not laugh, it might well mean that he has not made the correct inferences necessary to 'understand' my joke. In some sense, even a different sense of humor is an inability to understand my speech. This is, in fact, what communication is all about. Certain pairs of people find it harder to communicate than other pairs. This is indicative of a lack of certain common memory structures and inference relations. We cannot 'understand' somebody whose initial assumptions and cultural background are radically different from our own, even if we share a common language. That is, understanding language is a misnomer or at least is only a small part of the problem. Understanding what one has heard is a complex process that necessitates connecting words with certain conceptual constructions that exist in one's memory. The entire linguistic process uses the output of such understanding and interpreting mechanisms in order to produce reasonable replies (verbal or not). What constitutes a reasonable reply is an intrinsic part of the linguistic process, but yet is still a conceptual process and is, therefore, I suppose, out of the domain of traditional linguistics. Yet it is unreasonable for it to remain in that scientific no-man's-land. A computer model must respond as well as understand. Of course, its response must be connected to a powerful responding mechanism that is, in fact, the point of the entire computer program, that is, why the program was written in the first place. These then are the problems of computer understanding of natural language.

In a complete automatic linguistic system the responses that are generated will be dependent on the corroboration of the predicted input as compared to the actual input and the memory structure. That is, we respond differently to different people saying the same things, and differently to the same people saying the same things in different contexts. These contexts include, physical, conversational, and time contexts. In other words, no person is really the same at any given point in time as he was at some other time with respect to the viewer's own memory model of that person. So, in some sense, the context is always different and the responses should always be potentially different according to the time of the conversation. It is precisely the predictive ability that permits this difference in response. And, the difference in response is caused by the difference in analysis. That is, in order to effectively analyze a given linguistic input, it is necessary to make predictions as to what that input might look like, compare the actual input to the expected input, and coordinate both with the memory model. Understanding is, therefore a complicated process which cannot be reasonably isolated into linguistic and memory components but must be a combined effort of both.

I have avoided discussing conversational predictions in this paper because the work that we have begun to do on them is still even more sketchy than that presented here and, more importantly, is of a very different character.

To a large extent the conversational expectations are dependent on generative mechanisms, that is, a sort of 'why are we talking' apparatus must be used. As an example of the kind of mechanism that we are talking about, an illustration from some recent research that we have been doing will help.

We spent some time with a child of age 2.2 years, talking to her and endeavoring to understand her understanding mechanism. During discussions with her we obtained some interesting examples of unusual answers to questions that suggest a model for generation of responses in the child.

Mother: Did you go to the toilet?
Child: I go home.
Mother: Did Peter go to the bathroom?
Child: Peter cry.
Interviewer: Did you ever go in a plane?
Child: I go in a bus.

Responses of this kind indicate an answering mechanism in this child that has as its primary purpose the making of true statements. A procedure that would generate this behavior would simply check the proposed conceptualization with memory, and if it is not found, check to see if a new case dependent will fit in the conceptualization and make it true. Two factors are important in choosing this new object: first, the immediate context or intermediate memory is checked for similar conceptualizations; if that fails, a true conceptualization with the same semantic category is checked for. If no possible case dependent is found, the case is eliminated altogether and a new ACT is looked for, again by the same standards.

Another interesting insight into the understanding mechanisms of this child was provided by the following two exchanges:

Interviewer: What's that? (pointing to a picture of butter)
Child: Butters.
Interviewer: And what do you do with butter?
Child: Eat it.
Interviewer: How do you eat it?
Child: On a spoon.
Interviewer: On bread?
Child: Yes.

The association between 'bread' and 'butter' was then used by the child in the following sequence that occurred half an hour later.

Mother: (to another child) What did you eat for lunch?
Other child: Sandwich
Interviewer: What do you want to eat?
Child: Sandwich too.
Mother: What kind of sandwich?
Child: Butter on it.
Mother: She has never asked for butter before.

Here we see the child making statements that may, in fact, not even be true for her. She might very well object to receiving a butter sandwich if one was made for her. She seems here to be testing the new association (butter–bread) that she heard, and is, in fact, expanding it (butter–bread–sandwich). We hope to be able to build a program that can learn, as a child does, how to do what we have described in this paper instead of being spoon-fed the tremendous information necessary.

In order to do this it might be necessary to await an effective automatic hand–eye system and an image processor. Until then we shall attempt to put in the huge amounts of information that is necessary in order to more effectively test our model.

REFERENCES

ANDERSON, J. *Recognition confusions in sentence memory*, unpublished paper, Stanford University, Stanford, California, 1971.

BOWER, G. Imagery as a relational organizer in associative learning. *Journal of Verbal Learning and Verbal Behavior*, 1970, **9**, 529–533.

CHAFE, W. *Meaning and the structure of language*. University of Chicago, Illinois: Chicago Press, 1970.

CHOMSKY, N. *Syntactic structures*. Mouton and Co., 1957.

CHOMSKY, N. *Aspects of the theory of syntax*. M.I.T. Press, 1965.

COLBY, K., & ENEA, H. Heuristic methods for computer understanding of natural language in context-restricted on-line dialogues. *Mathematical Biosciences*, 1967, **1**, 1–25.

COLBY, K., & SMITH, D. Dialogues between humans and an artificial belief system. *In* Walker & Norton (Eds.), Proceedings of the International Joint Conference on Artificial Intelligence, 1969.

COLBY, K., WEBER, S. & HILF, F. Artificial paranoia. *Artificial Intelligence*, 1971, **2**, 1–25.

DEESE, J. *Some contributions of psycholinguistic studies to content analysis*, presented to National Conference on Content Analysis, Philadelphia, 1967.

FILLMORE, C. The Case for Case. *In* Bach & Harms (Eds.), *Universals in linguistic theory*, Holt, Rinehart and Winston, 1968.

FLOYD, R. Nondeterministic algorithms. Journal of the ACM, **14**, No. 4. Pp. 636–644.

FURTH, H. *Thinking without language*. Free Press, 1966.

HAYS, D. Dependency theory: A formalism and some observations. *Language*, 1964, **40**, 511–525.

INHELDER, B., & PIAGET, J. *The growth of logical thinking from childhood to adolescence*. Basic Books, 1958.

KATZ, J., & FODOR, J. The structure of a semantic theory. *Language*, 1963, **39**.

KLEIN, S. Automatic paraphrasing in essay format. *Mechanical Translation*, 1965. Pp. 68–83.

KUNO, C., & OETTINGER, A. "Multiple path syntactic analyzer" in *Information Processing* 1962, North Holland.

LAKOFF, G. On generative semantics. *In* Steinberg and Jakobovits (Eds.), *Semantics: An interdisciplinary reader in philosophy, linguistics, anthropology and psychology*. Cambridge University Press, 1971.

LAKOFF, G. *Linguistics and natural logic*. Department of Linguistics, University of Michigan, 1970.

LAMB, S. The sememic approach to structural semantics. *American Anthropologist* (special issue), 1964.

LAMB, S. *Outline of stratificational grammer*. Georgetown University Press: Washington, D. C., 1966, pp. 57–73.

LENNEBERG, E. *Biological foundations of language*. John Wiley, 1967.

QUILLIAN, R. *Semantic Memory*. Cambridge, Massachusetts: Bolt Beranek and Newman, 1966.

SCHANK, R. *A conceptual dependency representation for a computer-oriented semantics*. AI Memo-83, Computer Science Department, Stanford University, Stanford, California, 1969.

SCHANK, R., & TESLER, L. A conceptual dependency parser for natural language. *Statistical Methods in Linguistics*, 1970, **6**, 33–51.

WEBER, SYLVIA *The processing of conceptual nominals in a conceptual dependency system*. Proceedings of the 1971 International Conference on Computational Linguistics, Debrecen, Hungary (forthcoming).

WEIZENBAUM, J. ELIZA—a computer program for the study of natural language communication between man and machine. *Communication of the ACM*, 1966, 9, 36–45.

WINOGRAD, T. *Procedures as a representation for data in a computer program for understanding natural language*. Technical Report MAC TR-84, M.I.T., 1971.

WOODS, W. *Semantics for a question-answering system*. Report No. NSF-19, Harvard University Computation Center, 1967.

[2]

Seven principles of surface structure parsing in natural language*

JOHN KIMBALL

Indiana University

Abstract

In generative grammar there is a traditional distinction between sentence acceptability, having to do with performance, and sentence grammaticality, having to do with competence. The attempt of this paper is to provide a characterization of the notion 'acceptable sentence' in English, with some suggestions as to how this characterization might be made universal. The procedure is to outline a set of procedures which are conjectured to be operative in the assignment of a surface structure tree to an input sentence. To some extent, these principles of parsing are modeled on certain parsing techniques formulated by computer scientists for computer languages. These principles account for the high acceptability of right branching structures, outline the role of grammatical function words in sentence perception, describe what seems to be a fixed limit on short-term memory in linguistic processing, and hypothesize the structure of the internal syntactic processing devices. The operation of various classes of transformations with regard to preparing deep structures for input to parsing procedures such as those outlined in the paper is discussed.

1. Introduction

In grammar there is a distinction between those sentences which are rejected by speakers on grounds of grammaticality versus those rejected for performance reasons. Thus, there is a quadripartite division of the surface structures of any language: (a) Those sentences which are both grammatical and acceptable, e.g. 'It is raining';

* I am indebted to Frank DeRemer for many discussions on the material in this paper, particularly for providing information on current computer work in parsing, and for suggestions concerning the principles discussed in section 4; also to Cathie Ringen, Jorge Hankammer, Paul Postal and the reviewers for reading and commenting on an earlier version.

(b) those sentences which are grammatical but unacceptable, e.g. 'Tom figured that that Susan wanted to take the cat out bothered Betsy out'; (c) those which are ungrammatical but acceptable, e.g. 'They am running'; and (d) those which are both ungrammatical and unacceptable, e.g. 'Tom and slept the dog'.

Part of the problem facing linguists, in particular psycholinguists, is to find a characterization of the processing of linguistic experience, i.e. performance, adequate to distinguish unacceptability from ungrammaticality. In the following paper I will present an interconnected set of principles of surface structure parsing in natural language which is intended to provide a characterization of the notion 'acceptable sentence'. In particular, these hypotheses will explain the difficulty experienced by native speakers of English when sentences of the classical problem types such as in (1) are encountered.

(1) a. That that that two plus two equals four surprised Jack astonished Ingrid bothered Frank.
 b. Joe believes that Susan left to be interesting.
 c. The boat floated on the water sank.
 d. The girl the man the boy saw kissed left.

Further, these hypotheses will explain why a sentence like (2b) is not normally interpreted as (2a) in the same way that (3b) can be interpreted as (3a).

(2) a. The woman that was attractive took the job.
 b. The woman took the job that was attractive.

(3) a. The woman that was attractive fell down.
 b. The woman fell down that was attractive.

The first section of the paper will be concerned with the traditional hypotheses that have been presented to account for the unacceptability of sentences like (1). In Section 3 I will look at some of the parsing techniques designed by computer scientists for processing the sentences of programming languages. Of particular interest will be those techniques which allow a string to be parsed top-down left-to-right building a phrase structure tree over the string as it is read, since this is the process employed by speakers of natural languages. We will find that there are restrictions on the class of languages which allow this kind of parsing, and it will be possible to examine the claim that part of the function of transformations is to put deep structures in a form to allow limited memory left-to-right parsing. It will also be possible to examine the claim (*cf.* Chomsky, 1965, chapter 1) that transformations function to construct a surface tree which is optimally parsed by those techniques available to native speakers. In Section 5 we will see that the known transformations of English divide themselves into three distinct classes with respect to the kinds of output structures that are produced.

2. Some previous accounts of surface structure complexity

It is interesting to consider first attempts to explain the low acceptability of a sentence like (1d) 'The girl the man the boy saw kissed left'. The surface parse tree for this sentence is shown in (4).

(4)

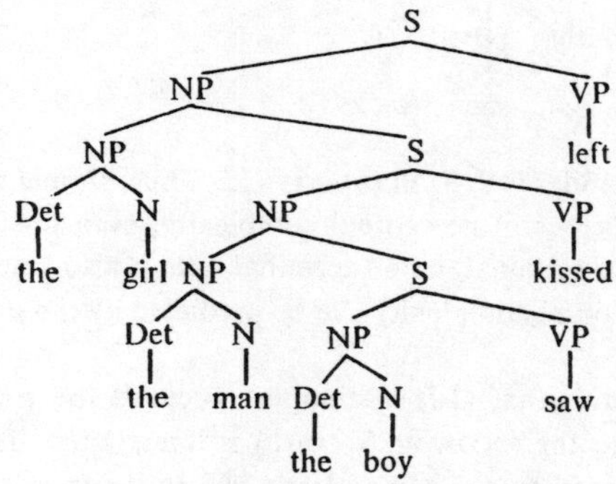

One hypothesis (Fodor and Garrett, 1967) is that such sentences are perceptually complex due to the low proportion of terminal symbols (words) to sentences. In general, one of the effects of transformations seems to be to reduce the non-terminal to terminal ratio in surface structures (Chomsky, 1965) as compared to deep structures; i.e., to map relatively tall structures such as (5a) into relatively flat structures such as (5b).

(5) a. b.

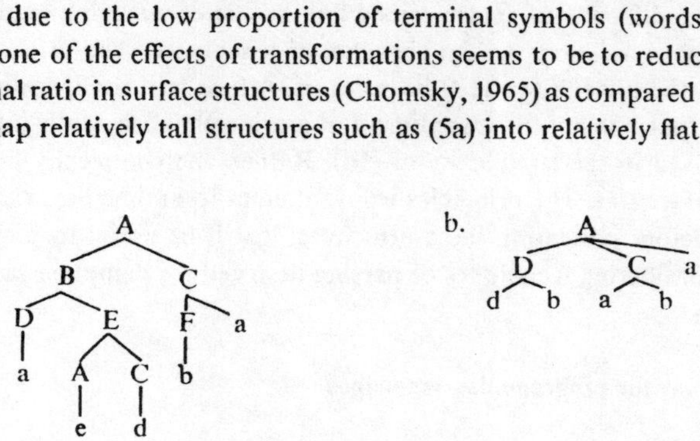

It is argued that flat structures are preceptually less complex, although no explanation of this fact has been offered. Why it should be that flat structures are easier to parse than deeply embedded structures is accounted for by Principle Two of Section 4 below.

The Fodor and Garrett hypothesis seems to be a specific form of this more general hypothesis. This specific hypothesis is, however, easily falsified by considering sentences which have fewer terminals per S than (1d), but which are perceptually much less complex. (6) is such a sentence.

(6)

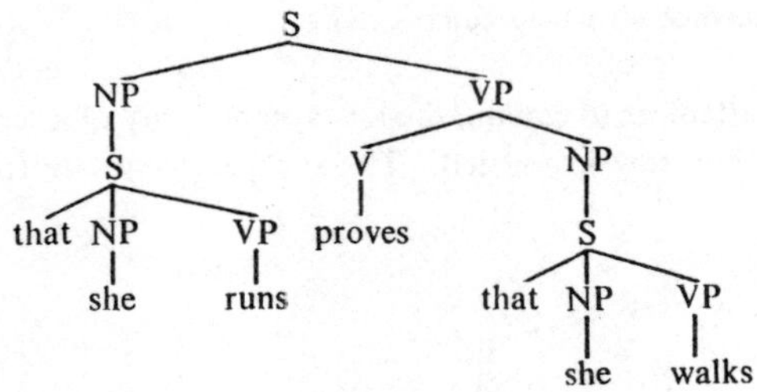

In (1d) the ratio of terminals to S-nodes is 9:3; in (6) it is 7:3. Thus, simple terminal to S ratio will not account for all cases of perceptual complexity, even those it was designed to cover. The more general terminal to non-terminal ratio is also inadequate. That this ratio is relevant to perceptual complexity can be predicted by the principles presented below in Section 4.

Chomsky and Miller (1963) claim that (1d) is complex because the perceptual strategy of assigning a N (noun) to the following V (verb) is interrupted. Thus, the subject-verb relation must be assigned to *the boy* and *saw* before it can be assigned to *the man* and *kissed*, and this before it can be assigned to *the girl* and *left*. This hypothesis treats the difficulty of center embedding as one of assigning grammatical and, perhaps, semantic relations to the elements of a sentence. The principles discussed below treat the difficulty in (1d) in terms of surface tree configuration only. I conjecture, then, that it is not the impediment of a principle of semantic interpretation which is involved in the complexity of (1d). Rather, the complexity lies in the structure of the surface tree. The principles below attempt to outline the exact nature of this difficulty. Before presenting these principles, it will be useful to prepare the way for them by considering techniques of parsing designed for computer languages.

3. Parsing algorthms for programming languages

Programming languages are often based on context free languages; the problem of designing parsers for such languages is solved by constructing a parser from the CF grammar for the language. Since the parser 'knows' the productions of the language, its job is then defined as that of selecting which productions had to have applied to generate a particular input string. (A usual requirement for programming languages is that their grammar be unambiugous, so the solution to the parsing problem for any string is always unique.) The parser must reconstruct the derivational history of a string from context. For example, if a grammar contains two productions with identical right hand sides, say $A \rightarrow X$, and $B \rightarrow X$, where X is an arbitrary string of terminals and non-terminals, then when the parser reads X, it must decide from context which production in fact applied. For example, *A* might be introduced by a rule C → D A, D → d, and *B* by the rule C → E B, E → e. Thus, when X is to be parsed as an *A*, it will always be preceded by a *d* and when it is to be parsed as a *B*, it will be preceded by an *e*.

Given that this is the general type of problem that must be solved by a parser, then two types of questions may be asked concerning a particular language. First, is it possible to parse the string deterministically from left-to-right as it is read into the computer. And second, what is the size of the largest context that must be examined at any time by the parser to decide how to build the parse tree? For reasons of efficiency, an optimal programming language can be considered to be one in which it is possible to parse left-to-right, and for which there is a fixed finite bound on the size of the forward context which must be examined at any point.

There are two general strategies used in parsing algorithms (*cf.* McKeeman *et al.*, 1970). In the first, a tree is built for an input string by starting with the initial symbol of the grammar (that which is topmost in all trees generated by the grammar) and building a tree downwards to the terminal symbols. Such procedures are called *top-down*. The operation of such a procedure on an input string $a_{1_1} \ldots a_{1_n}$ for a grammar with an initial symbol *S* is illustrated below:

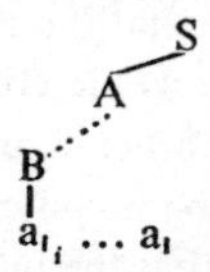

The first step of the algorithm is to build a tree down to the first symbol of the input string, a_{1_1}. The language must be such that it is always possible to do so uniquely by examining no more than *k* symbols ahead of a_{1_1} (where *k* is fixed for the language) if the parser is to be deterministic. At this point, *B* acts as the new 'initial' symbol for some substring, and the parser operated to complete *B*, again working top-down. When B is filled out, the next higher node in the pathway down from *S* is the new initial symbol, for the purposes of the parser. The class of grammars which permit parsing by an algorithm such as that outlined above is called the class of LL(k) grammars. For such languages trees can be built top-down, and it is never necessary to look more than *k* symbols ahead of the given input symbol to determine what action is to be taken by the parser.

The second type of parsing procedure involves building a tree from the bottom-up. The first action of such a parser is to assign the first *m* input symbols to some node, which is then placed at the top of a stack. Thus, in parsing $a_{1_1} \ldots a_{1_n}$, if the first *m* symbols are dominated by *B*, the parser will operate as illustrated below:

	Stack	*Input string*
Stage 1:		$a_{i_1} \ldots a_{i}a_{i_{m+1}} \ldots a_{i_n}$
Stage 2:	B	$a_{i_{m+1}} \ldots a_{i_n}$

The parse is completed when the initial symbol is the only symbol in the stack, and the input string has been completely read in. Those languages which can be parsed bottom-up by looking ahead in the input string no more than *k* symbols are called LR(k) languages; the grammars which generate such languages are called LR(k) grammers (*cf.* Knuth, 1965).

In general, parsing of computer languages differs from that in natural languages in two significant ways. The first involves the fact that programming language grammars are unambiguous; a parser for a programming language yields a unique tree for each string. Not only this, but also the behavior of the computer parser must be deterministic. That is, its action at any given string of input terminals and stack configuration must be uniquely determined. A model of parsing in natural language must allow for more than one parse and should predict on the basis of considerations of surface structure complexity which parse will most likely be offered as the first choice by a native speaker. Second, and most important, parsing in computer languages differs from that in natural languages in that a computer parser is allowed an essentially unrestricted memory. For example, in the case of a parser for an LR(k) grammar, it is possible to look ahead *k* symbols and then decide that the appropriate action is to read in the next symbol. This can be done until the whole string is read into memory before the parse tree starts being built. On the other hand, there is considerable evidence that short-term memory (STM) in humans is quite restricted, and that a tree must be built over an input string constantly so that the initial parsed string may be cleared from STM. Principle Four below concerns what seems to be a fixed limit on linguistic STM. Because of the limitations on STM, the form of a parseable (acceptable) surface structure in natural language is quite restricted.

4. Six or seven principles of surface structure parsing

The following principles, although presented here as distinct, are closely linked and interact in various ways, as will be pointed out during the discussion. The central principle of the scheme discussed here is the last, and from it various of the others follow deductively.

4.1 *Principle One (Top-Down): Parsing in natural language proceeds according to a top-down algorithm*

The operation of such algorithms was outlined above in Section 3. The claim made is that parsing of natural languages is like that in LL(k) languages, with one variation noted below. Thus, the first node built in such parsing is the top S, while in bottom-up parsing this is the last node to be constructed. Before considering the consequences of such an assumption, let us consider it in operation in parsing a sentence like (7).

(7) That the boy and the girl left amazed us.

The first step upon hearing the initial terminal *that* is to build a tree down from S of the form (8a).

(8) a.

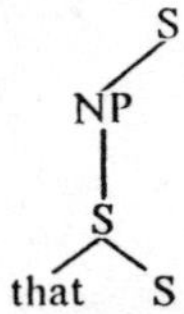

We may ask why such a tree is justified given that the initial terminal *that* could also have begun a sentence with a totally different structure, e.g. 'That is a nice flower'. To reduce the number of false starts, we may add the assumption that English is a look-ahead language. That is, the speaker is allowed to hear symbols subsequent to a given terminal before making up his mind as to the appropriate or most probable tree structure. I have no immediate empirical justification for this assumption, other than general considerations of simplicity and efficiency of parsing. The assumption is based on the conjecture that it may require less computation to hold a symbol in storage without tree attachment while one, but probably no more than two, subsequent symbols are scanned, than to build a tree only to have to return to alter it. (We will see later that sentences in which large tree changes are required during parsing are indeed perceptually complex.)

The reader will notice that complementizers are represented as Chomsky adjoined (*cf.* Kimball, 1972a, for a discussion of Chomsky adjunction) to the left of the appropriate notes, following Ross (1967). There is ample justification of this purely in terms of syntax. We will see later on that further justification arises from considering the role complementizers and other 'function' words play in perceptual routines.

Let us continue now to parse (7). Upon reading *the*, (8a) becomes (8b), which becomes (8c) after *boy* is read.

(8) b.

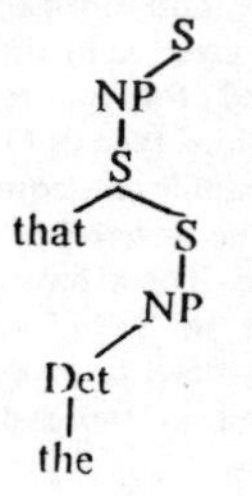

c.

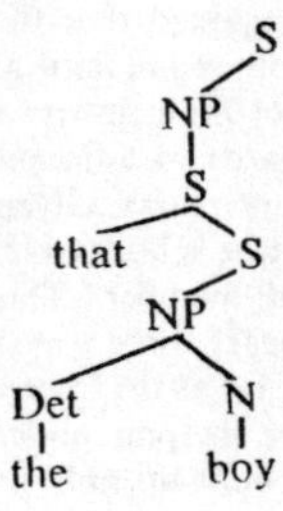

At this point *and* is read, signalling a conjunction, and three new NP nodes are constructed, shown in (8d).

(8) d.

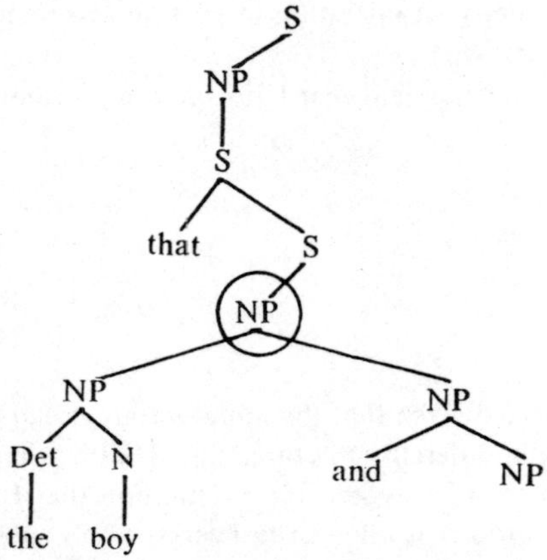

The *and* here is represented as Chomsky adjoined to the phrase following it. Again, justification for this can be found in Ross (1967). The circled NP is inserted between the lower S and the first NP. Insertion of such nodes I claim is possible, and such insertion is the only deviation from the general claim that trees are built top-down.[1] Furthermore, it seems to be the case that nodes that are so inserted are copies of already built nodes, so the S | NP structure is preserved.

Once the NP *the boy* is completed, all subsequent material is taken as belonging to a different phrase. If there were a relative clause construction, such as 'the boy who kissed the girl', as soon as *boy* is read, the NP is closed. Reading of *who* occasions building a new phrase, as shown in (9).

1. As this type of parsing differs from the usual top-down procedures, we may seek a new name for it. De Remer suggests over-the-top (OTT). In fact, it may be suggested that the mechanism of parsing in fact utilised in natural language is this: Trees are not built down to single terminals but with regards to adjacent pairs of terminals (discriminant pairs). Given an initial member of a pair, a tree is built over-the-top down to the second member. This could be done in one of at least three ways: (1) The tree is built up only as far as the lowest common dominating node for the pair under consideration; (2) the tree is built up only as far as the lowest common dominating S node for the pair, and then down to the second member; or (3) the tree is built all the way up to the highest S node, and down to the second member. As I have given it in the paper, the parsing hypothesized for natural language corresponds to this third type of OTT parsing.

There are some testable consequences which result from taking one or another of the positions outlined above. These have to do with what is maintained in STM, and what is cleared into the syntactic processor. This question is discussed in more detail under principle seven below.

(9)

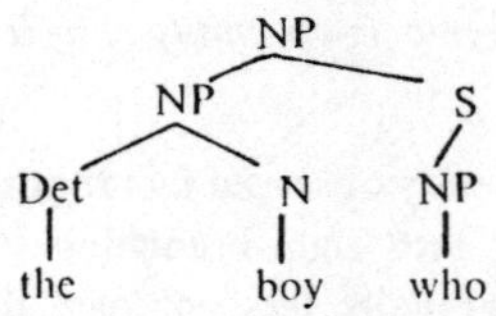

This observation concerning the closing a phrases is illustrative of a principle that operates generally in sentence parsing. This principle of closure will be discussed in detail below and is connected with another principle dealing with how semantic information is processed.

Returning once more to (7), after *the* and *girl* are read, the tree looks like (8e).

(8) e.

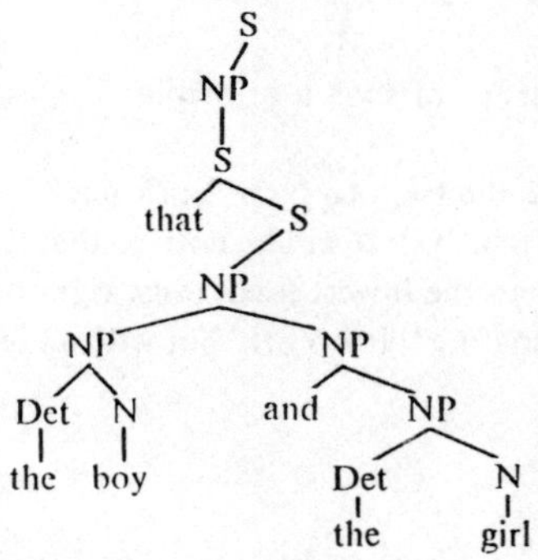

When *left* is read, the embedded sentence is closed, following the principle alluded to above, given look ahead to make sure no other possible parts of that sentence occurred (e.g., *early*, or *to NY*). When *amazed* is read a new VP in the main sentence is constructed, and an NP added as *us* is read. The final parse tree is shown in (8f).

(8) f.

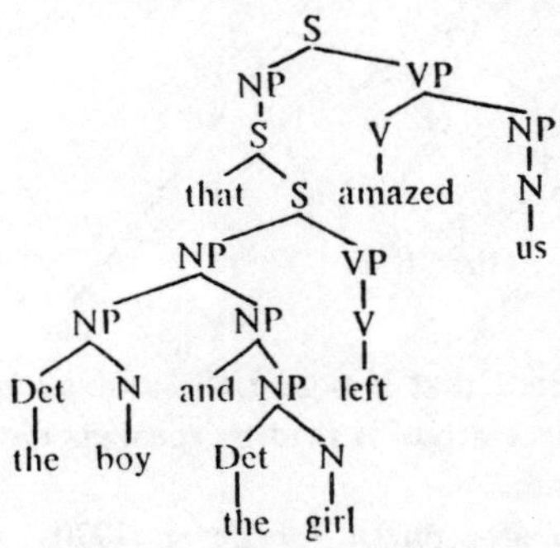

The 'top-down' principle interacts in an interesting way with the following principle.

4.2 *Principle Two (Right Association): Terminal symbols optimally associate to the lowest nonterminal node.*

This principle is designed in part to explain the frequently observed fact that sentences of natural language organize themselves generally into right-branching structures such as (10a), and that these structures are perceptually less complex than left-branching structures, such as (10b), or center embedded structures, such as (10c).

(10)
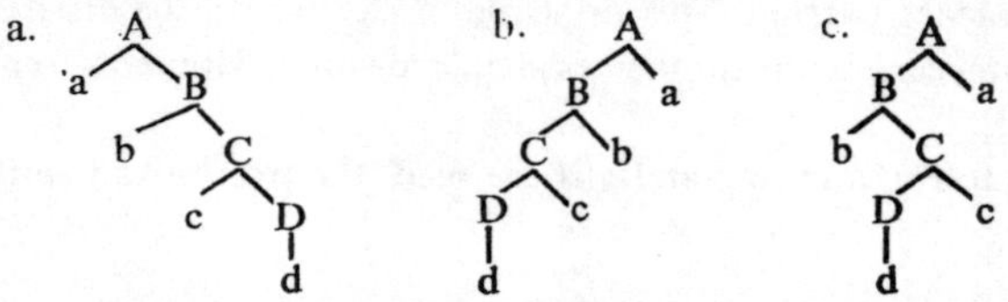

There is considerable evidence for the existence of such a principle.

First, consider a sentence such as (11).

(11) Joe figured that Susan wanted to take the train to New York out.

The surface structure of this sentence is shown in (12). It will be noticed that the particle *out* must be associated with a node other than the lowest (and, thus, rightmost). That is, 'out' is not associated with the node dominating 'New York' but with a higher node.

(12)
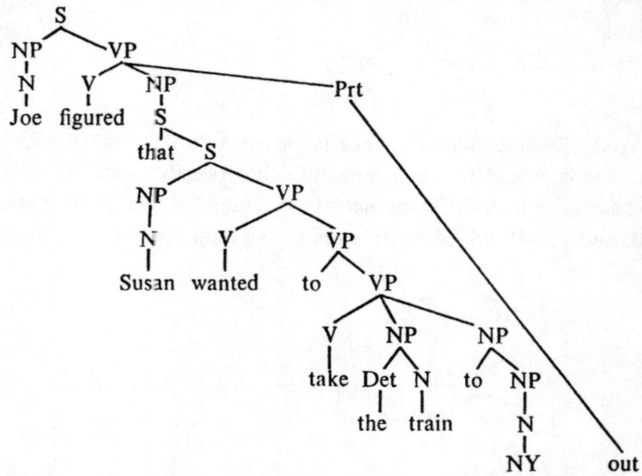

This principle also explains why a sentence that is potentially ambiguous, such as 'Joe figured that Susan wanted to take the cat out' is read by speakers naturally in a way such that 'take the cat out' is a phrase.

The status of such sentences constituted a puzzle for Bever (1970a) in that no known principle other than general 'memory limitation' would explain the difficulty of such sentences; he was unsure whether they should be marked ungrammatical or merely unacceptable. The sentence Bever gives is (37) (his numbering).

(37) I thought the request of the astronomer who was trying at the same time to count the constellations on his toes without taking his shoes off or looking at me over.

Such sentences should be counted as fully grammatical in that they are generated by general syntactic mechanisms. Their perceptual complexity is explained by Right Association, which is related to the principle of Closure discussed below.

Right Association accounts for the difficulty with phrases like 'the boy who Bill expected to leave's ball', or the preferred but incorrect interpretation of 'the boy who Sam gave the ball's book' (incorrect reading is that it's the ball's book; correct is that it is the boy's book). The reason is that the possessive *'s* optimally associates with the lowest constituent, instead of the higher NP dominating the whole phrase.

This principle also explains the preferred interpretation of a sentence like (13).

(13) The girl took the job that was attractive.

Which is that which is not synonymous with (14).

(14) The girl that was attractive took the job.

However, there is a general grammatical process that would form (13) out of (14), known as *Extraposition From NP*. This transformation maps a sentence like (15a) into (15b).

(15) a. The girl that was attractive went to NY.
b. The girl went to NY that was attractive.

The reason for the preferred interpretation of (13) can be seen from its parse tree (16).

(16)

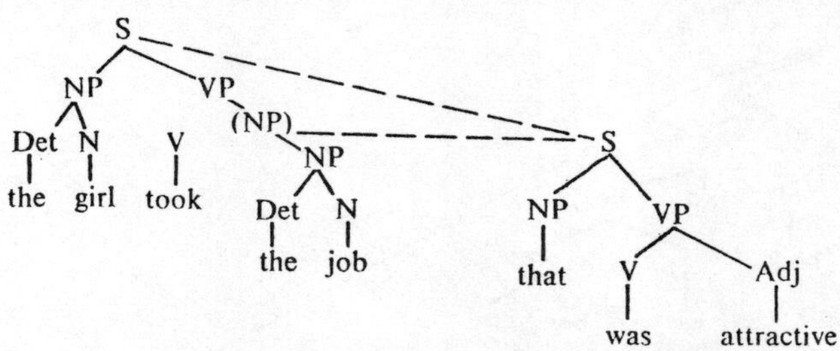

Principle Two predicts that the relative clause will be perceived as associated with the lowest, rightmost node, i.e., *the job* rather than as a daughter of the top S, where it would have to be interpreted as an extraposed relative.

The seeming unambiguity of sentences like (13) have been taken as evidence for the existence of a special and probably quite powerful form of grammatical mechanism known as transderivational constraints. Briefly, according to the hypothesizers of such constraints, a derivation in which (13) comes from (14) is to be blocked because there already exists a sentence of the same form with a different interpretation. The fact is that (13) can be read as (14) with a little effort; it's just that this reading is perceptually more difficult, due to Principle Two. Thus, this datum should be removed as evidence for the existence of transderivational constraints. Without going into the matter in detail, I conjecture that all putative evidence for these devices are of the form of that above, namely, they can be explained in terms of preferred interpretation on the basis of established principles of perception. If so, there is no reason to include transderivation constraints among the stock of possible grammatical mechanisms in the theory of universal grammar. (But *cf.* Hankammer, 1973.)

There seem to be grammatical mechanisms to avoid the generation of sentences that would be perceptually complex under Principle Two, i.e., which would involve assigning terminals to other than the lowest, rightmost non-terminal. A transformation known as *Heavy NP Shift* is a case in point. This transformation moves heavy NP's to the right hand side of sentences, where definition of 'heavy' is discussed in Ross (1967).

Thus, a sentence like (17a) would be mapped into (17b).

(17) a. Joe gave a book that was about the skinning of cats in Alberta between 1898 and 1901 to Berta.

b. Joe gave to Berta a book that was about the skinning of cats in Alberta between 1898 and 1901.

The perceptual complexity of (17a) can be seen in its surface parse tree (18).

(18)

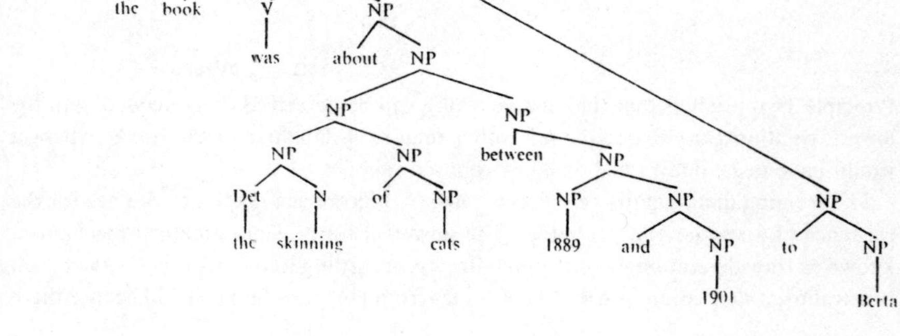

(It will be noticed that what are traditionally called prepositional phrases [e.g., *of cats*] are here represented as NPs with Chomsky adjoined prepositions. Justification for this may be found in Ross, 1967).

We can utilize the principle of Right Association in gaining a partial understanding of the complexity of sentences such as (1d) 'The girl the man the boy saw kissed left' which initiated the discussion of this paper. Part (but not all) of the difficulty resides in the fact that the verb *kissed* is optimally associated with the lowest, rightmost constituent of the tree. As this association is impossible on semantic grounds, it must receive association with a VP node in a higher sentence. The same goes for *left*. In this way (1d) violates Right Association.

Another confirmation of this principle comes from observing the natural association of adverbs. In this respect, consider a sentence like (19).

(19)

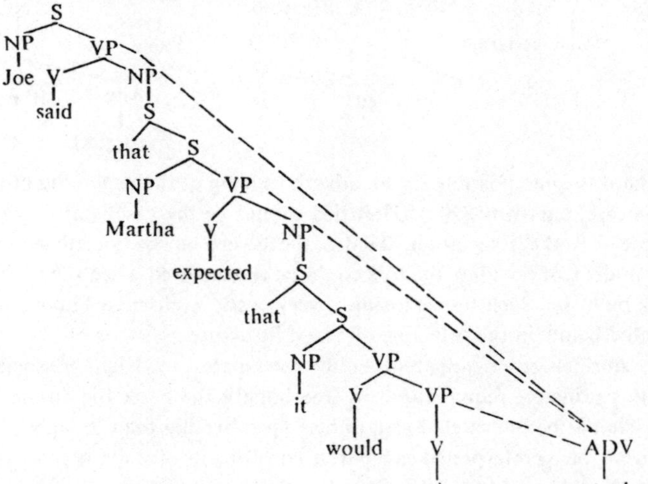

The dotted lines indicate the possible association of *yesterday*. Compare these with the natural associations of this adverb as the sentence is interpreted. The easiest reading is that in which the adverb is read as attached to the lowest VP, next easiest is that reading in which it hangs off the middle VP, and the hardest or least likely reading is that in which it is associated with *said*. This is exactly the prediction made by Right Association. In fact, we can define a metric such that the perceptual complexity of a sentence is proportional to the number of nodes above the lowest rightmost node to which a terminal is attached.

It is to be noted that there is a syntactic device available in English to disambiguate (19), as shown in (20a-c).

(20) a. Yesterday Joe said that Martha expected that it would rain.
b. Joe said that yesterday Martha expected it would rain.
c. Joe said that Martha expected that yesterday it would rain.

Notice that a sentence like (21) is read most naturally such that the adverb is associated with the higher clause.

(21) Haastiin knew yesterday it rained.

That is, the most natural structure imputed to (21) is that shown in (22a) rather than (22b),

(22) a.

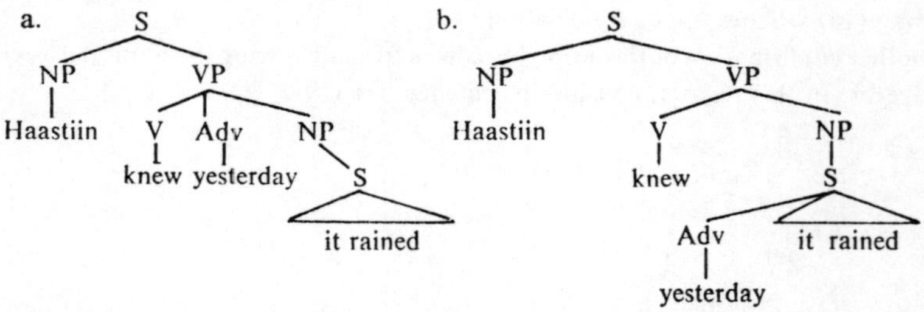

even though it is quite possible for an adverb to hang at the beginning of a sentence, as must be case, e.g., with (20a). That this should be the case is again predicted by the principle of Right Association. That is, the adverb must associate with the lowest, rightmost node. Conceivably, the tree could be restructured when the new embedded sentence is built, but such restructuring is very costly as discussed above and as will be elaborated below in the principle of Fixed Structure.

Let us consider some apparent counter-examples to Right Association. For example, this principle requires that in 'Joe bought the book for Susan', 'the book for Susan' should be interpreted as a phrase more readily than 'bought', 'the book', and 'for Susan' being interpreted as separate constituents of a VP, because the new NP 'for Susan' should preferably be assigned to the lowest completed node. That this is not the case is a function of the interaction of parsing with semantic information accessible to the speaker during the sentence scan. In particular, the verb 'buy' carries lexical information with it such that the speaker would expect to hear a 'for' phrase in its VP. This interaction of semantics with parsing will be discussed further below when the principle of *Processing* is presented.

In the same vein, notice that while 'Joe cooked the peas in the pot' is ambiguous, with either reading of equal complexity, 'Joe rode down the street in the car' does not carry the same ambiguity. That is, one does not read 'the street in the car' as a phrase because of semantics or knowledge about the world that streets usually aren't in cars. Again we see the role of outside information influencing parsing.

Thus, the above are not counter-examples to Right Association. Rather, this principle defines the optimal functioning of the parsing algorithm if no outside effects are relevant. Its operation may be superceded by mechanisms other than the parser.

The principle of Right Association operates in connection with another principle of perception, that of New Nodes. This principle is needed to account for the following observation: In processing a sentence, when the speaker has constructed a node A shown in (23a) and attached to it daughters a, b ..., upon reading the next terminal, *c*, Right Association demands that *c* be connected as shown in (23b). However, two other things may in fact happen. First, some new node, B, may appear and be subtended under A, as in (23c), or B may be Chomsky adjoined to the right of A, as in (23d).

(23)

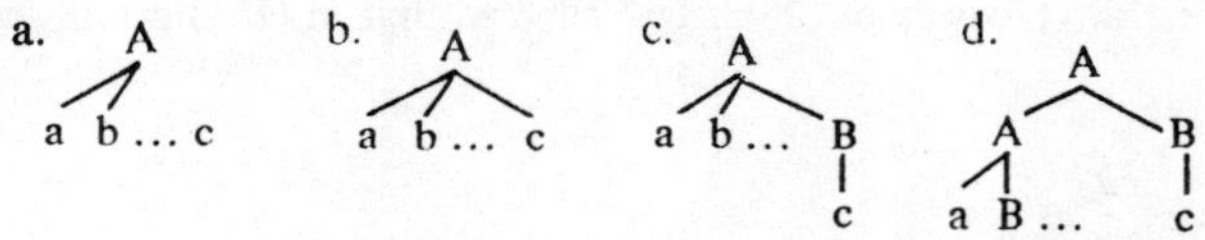

All three forms of assimilating the new terminal into the existing parse tree shown in (23b-d) are observed to occur in natural language. Right Association predicts that (23a) should become (23b). New Nodes is designed to predict when (23a) will become either (23c) or (23d), i.e., when the terminal *c* occasions the construction of a new phrase. This principle is stated as follows:

4.3 *Principle Three (New Nodes): The construction of a new node is signalled by the occurrence of a grammatical function word*

There is a traditional grammatical distinction in the discussion of the parts of speech between what are called content words (nouns, verbs, adjectives, etc.) and function words (prepositions, conjunctions, etc.). In the literature of transformational grammar, this distinction surfaces in terms of the difference between lexical formatives and grammatical formatives. For the time being I will focus on just prepositions, *wh*-words (e.g., *what, where, who, how, when, why*, etc.) conjunctions, and complementizers (*that, for-to*, and *pos-ing*). Later other categories will be examined as to whether they work like function words for purposes of perception.

There is syntactic evidence that grammatical formatives are Chomsky adjoined on surface structure (*cf.* Ross, 1967). (The assumption that this is the case is in fact not necessary to the correct operation of New Nodes, but I shall maintain the assumption in that which follows.) Thus, what is traditionally called a prepositional phrase is in fact a NP, as in (24a), and the complementizers and conjunctions appear on surface structure as in (24b,c).

(24) a. b. c.

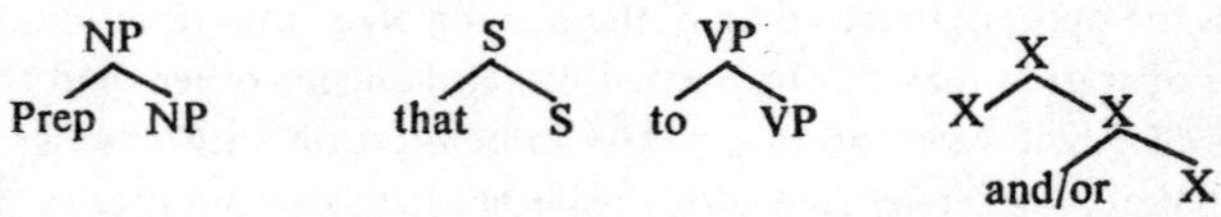

There is no direct proof that fronted *wh*-words, as in 'What did he say?' or 'The boy who you say' are Chomsky adjoined to the front of their clauses. As mentioned above, it makes no difference for New Nodes whether this is the case or not.

Let us examine how New Nodes operates to correctly predict the parsing of a sentence like (25).

(25) She asked him or she persuaded him to leave.

After reading the first three words of (25) a tree such as that in (26a) is constructed.

(26) a.

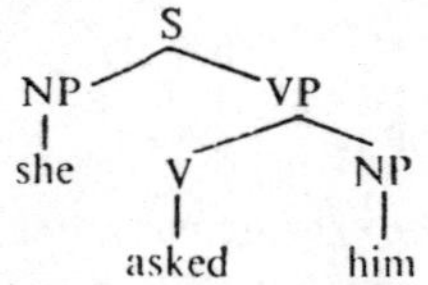

At this point a conjunction is reached, and the speaker must decide whether there is a conjunction of NPs (She asked him or her to leave), of VPs (She asked him or persuaded him to leave) or of Ss, as in (25). In this case a look ahead of one word reveals that the latter is the case, and (26b) is constructed, where the new node is Chomsky adjoined to the right of the top S.

(26) b.

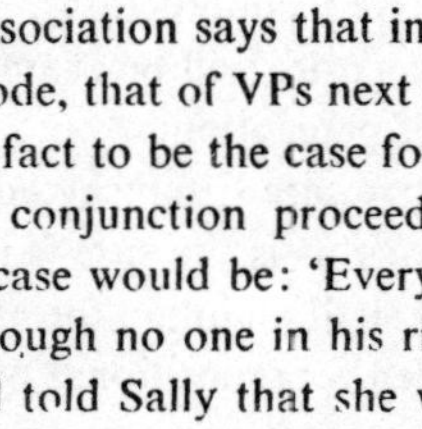

Right Association says that in this case the conjunction of NPs is easiest, as it is the lowest node, that of VPs next easy, and a conjunction of Ss hardest to perceive. This seems in fact to be the case for the sentences listed above. The deeper the node from which a conjunction proceeds, the more perceptually complex the sentence. An extreme case would be: 'Everyone said that Bill thought that Max believed that she was, although no one in his right mind who had been the movie would expect that Fred had told Sally that she was pregnant.' The perceptual complexity arises from the large internal constituent breaks, of which there are two in the above sentence, one before the conjunction *although*, and one before *pregnant*.

When the *to* of *to leave* in (25) is reached, this is a signal that a new node, in this case a VP, is to be formed. The structure of the conjunction forces this to be Chomsky adjoined to the top S, yielding a final parse tree (26c).

(26) c.

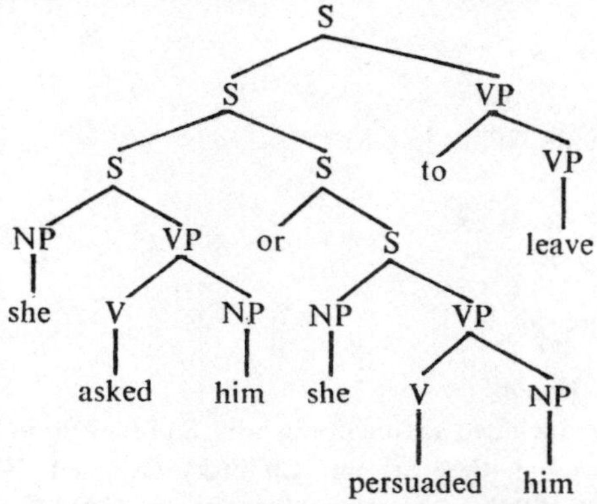

Thus, (25) illustrates how it is that both conjunctions and complementizers occasion the construction of a new node. New Nodes predicts that sentences in which the complementizers and relative pronouns have been deleted by optional rules will be perceptually more complex than those in which complementizers are present, i.e., that (27a) is more complex than (27b), and (28a) more complex than (28b).

(27) a. He knew the girl left.
     b. He knew that the girl left.

(28) a. The boy who the girl who the man saw kissed left.
     b. The boy the girl the man saw kissed left.

There is experimental evidence to support this contention. Hakes (1972) found that sentences with complementizers were processed faster than those without complementizers. He writes: 'When an optional cue to a sentence's underlying grammatical relations is deleted, the difficulty of comprehending is increased. These results taken together with the numerous results on relative pronoun deletion suggest that the cue deletion effect is general and not limited to a particular cue or structure' (pp. 283-284). Thus, New Nodes supplies a second piece in fitting together an explanation of the difficulty of a sentence like (1d). (The third, and perhaps crucial, piece comes from Principle Four below.)

The particular example considered above illustrates only the nodes formed by a conjunction and the complementizer *to*. In sentences like (27a) or (27b), the complementizer *that* signals the existence of an embedded sentence. The occurrence of this complementizer here introduces a structure with three new nodes, as shown in (27c).

(27) a. Joe knew that it was a duck.
b. That it was a duck annoyed Joe.
c.

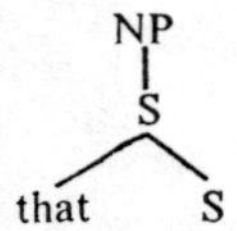

A preposition introduces the node NP, as in (28).

(28)

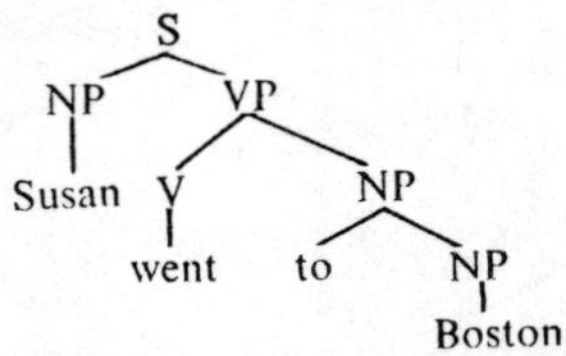

Traditionally articles (*a*, *the*) are included as function words, and they do in fact serve to introduce new phrases, although they are not Chomsky adjoined. We should include, perhaps, all words which fill the determiner slot in surface structure: *several*, *all*, *each*, *every*, *few*, etc. Finally, we may consider auxiliaries, which traditionally were counted as function words. There is a debate concerning the proper surface structure of the auxiliaries. Following the Chomsky (1957) analysis, it would be (29a); Ross' analysis (1967b) would predict (29b) as the correct structure.

(29) a. b.

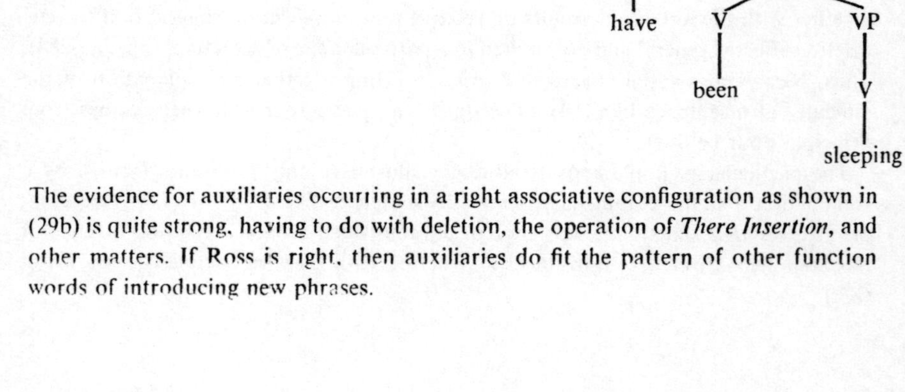

The evidence for auxiliaries occurring in a right associative configuration as shown in (29b) is quite strong, having to do with deletion, the operation of *There Insertion*, and other matters. If Ross is right, then auxiliaries do fit the pattern of other function words of introducing new phrases.

This statement of the role of function words in sentence perception as signallers of new phrases includes no hypothesis concerning their semantic role or their syntactic origin in deep structure. Function words themselves are among the things learned later in the process of acquisition, the first stage being that of telegraphic speech (Brown and Bellugi, 1964). Likewise, they are generally absent from 'pidgins'. In both cases, the grammatical structures may be conjectured to be not sufficiently complicated (say, in terms of occurrence of embedded sentences) to require cues to surface structure. We may hypothesize that there is a certain permissible complexity of surface structures which do not require indicators of constituent organization. In a free word order language, not so much of the surface tree is relevant to a determination of the underlying syntactic relations, and surface structures in such languages may be flat and relatively uncomplicated compared with a language such as English; thus, such a language may not need overt cues to surface parsings.

The operation of New Nodes in SOV languages needs further examination. In such languages grammatical formatives typically follow those constitutents to which they are attached (as pointed out to me by Jorge Hankammer). For cases where the constituent is a simple NP with a post-position, the principle could be operative, as this NP could be stored until a look-ahead to the post-position gave clue to its syntactic status. For large constitutents such as S's with following complementizers, New Nodes simply is inoperative. Note, however, that such cases are not counter examples; New Nodes has the logical form of a conditional: If a grammatical function word occurs, it signals construction of a new phrase.

It is possible now to consider a principle which is pervasive in application, and which is the first principle directly bearing on what short-term syntactic memory limitations are.

4.4 *Principle Four (Two Sentences): The constituents of no more than two sentences can be parsed at the same time*

The first pieces of supporting evidence for this principle come simply from considering pairs of sentences like (30a,b) and (31a,b) with respect to complexity.

(30) a. That Joe left bothered Susan.
 b. That that Joe left bothered Susan surprised Max.
 c. That for Joe to leave bothers Susan surprised Max.

(31) a. The boy the girl kissed slept.
 b. The boy the girl the man saw kissed slept.

In processing both (30b, 31b) at some point the constituents of three different sentences must be held in memory. E.g., when the second *that* of (30b) is heard and recognized as a complementizer, the imputed structure is (30), where three unfinished sentences are being processed at once.

(30) d.

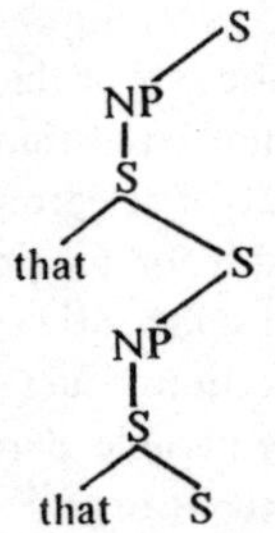

(30c) shows that repetition of 'that' does not here add noticeable complexity. Two Sentences provides the final principle explaining the difficulty of (1d). Part of the complexity of this sentence is due to the absence of *wh*-words to indicate the surface structure under the principle of New Nodes; part of the difficulty is that Right Association is violated; but the major difficulty here seems to be that the third sentence simply requires short-term memory space beyond the bounds of inherent capacity. Two Sentences is an attempt to state what that inherent capacity is. When the sentences in (30b) are nominatized, the result is much easier to parse, as in (32a).

(32) a.

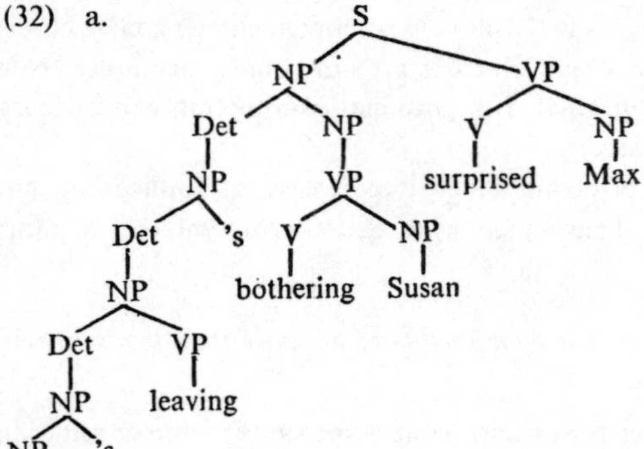

b.

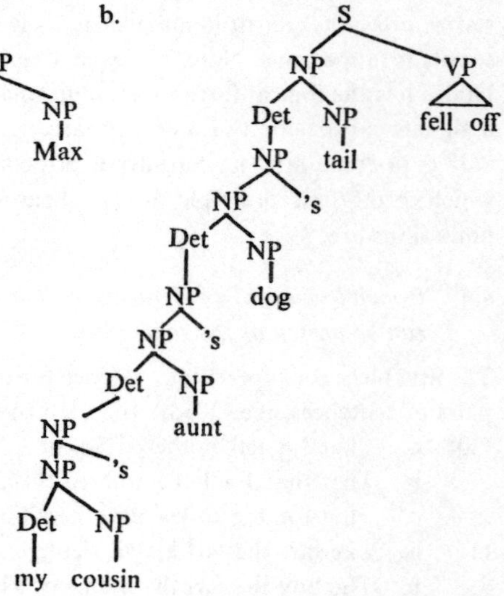

(32b) shows a large left-branching structure which is, nevertheless, not difficult to comprehend and which is within one S. On the other hand, when a fourth sentence is added to (30b in 33a), the result is totally incomprehensible, while the nominalized version, (33b), is not nearly as bad.

(33) a. That that that Joe left bothered Susan surprised Max annoyed no one.
b. Joe's leaving's bothering Susan's surprising Max annoyed no one.

One may conclude that the limitation is not on left branching, but rather on the number of Ss that must be processed at the same time. In a later discussion of semantic processing, I will discuss why this might be the case, and why it is permissible to string out relative clauses, embedded sentences, and prepositional phrases on the right of a sentence.

The Two Sentences principle accounts also for why right branching relative clause structures are permissible, while center embedded structures are not. Consider a structure like (34).

(34)

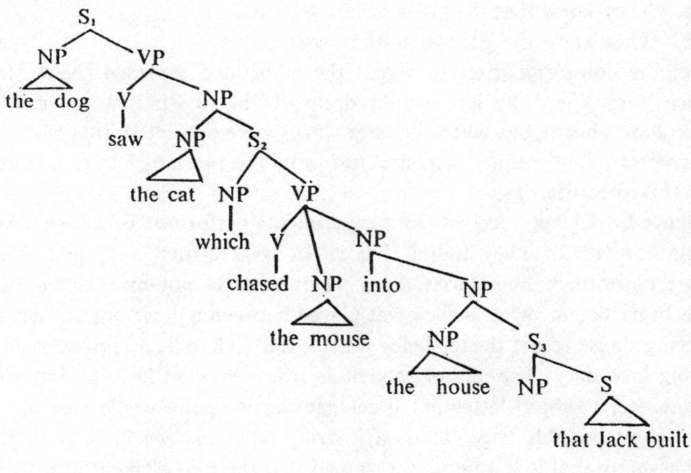

It may be thought that such a sentence violated Two Sentences because the top S is not finished until the last word of the bottom S is processed. To see why this is not so, let us say that we will consider a constituent to be 'closed' when the last immediately dominated rightmost daughter of that constituent is introduced in the process of parsing . In this sense, the top S is closed when the VP is reached, and the same for the second S. S_1 wouldn't be closed if, say, the sentential adverb *frankly* appeared sequentially after *built* and was to be attached to S_1. Why this definition of closure is appropriate and justified in that a phrase is through being parsed when it is closed will be discussed below under the principle of Processing. First, however, it is necessary to consider the principle of Closure.

4.5 *Principle Five (Closure): A phrase is closed as soon as possible, i.e., unless the next node parsed is an immediate constituent of that phrase*

Closure explains in part the complexity of sentences like (1c) 'The boat floated on the water sank'. In such sentences, as soon as the end of a potential S is reached, it is closed, unless the next phrase is also part of that S. Thus, at the end of 'the boat floated on the water', the assumption is that the S is closed. The remaining causes of the perceptual complexity of (1c) are accounted for below by Fixed Structure.

Also, the increased perceptual complexity of a sentence like (35b) over that of (35a) is explained by Closure, as well as New Nodes and Fixed Structure.

(35) a. They knew that the girl was in the closet.
b. They knew the girl was in the closet.

Without the complementizer to signal the embedded sentence (New Nodes), the sequence 'they knew the girl' would optimally be interpreted as an S (Closure) (modulo look-ahead), but when the later words were presented, that this assumption was incorrect would require a restructuring of the presumed tree, adding to complexity (Fixed Struture).

Evidence for Closure derives from experiments performed by Chapin, Smith, and Abrahamson (1972). They found that clicks were attracted to preceding surface structure constituent boundaries, even when they do not mark breaks of surface clauses. In particular, when a click was placed between a preceding surface break and a following clause break, the tendency was for the click to be perceived as closer to the preceding boundary. The authors conclude from this that in imposing a parse tree on a sentence, a subject 'attempts at each successive point to close off a constituent at the highest possible level. Thus, if a string of words *can* be a noun phrase, the subject assumes that it *is* a noun phrase and that the next element heard will be part of some subsequent constituent. Such a strategy would explain the strong preposing tendency observed in our experiment' (p. 171).

Closure has interacted closely with a number of the principles discussed above. In particular, it is not clear whether this principle is distinct from the principle of Right Association, for when the latter is violated, a terminal must be placed as a daughter of a node not the lowest, rightmost in the tree. Thus, this higher node must be 'reopened' to have a constituent added to it, contrary to the optimal situation described in Closure. That is, consider an abstract tree like (36).

(36)

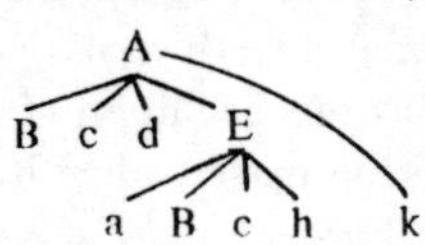

In building such a tree, as soon as E was being built, Closure would require that A be finished. However, when *k* is reached, A must be re-opened to receive the new constituent. In this sense, Closure operates the same as Right Association. However, I think that there are genuine cases where Closure operates which could not be covered by Right Association. For example, a phrase like (38a) should by Closure be interpreted more readily as (38b) rather than (38c) because of the tendency to close the phrase begun with *old*, even though Right Association predicts the opposite because by it the second phrase will be conjoined to the lowest NP available to it.

(38) a. old men who have small annual pensions and gardeners with thirty years of service

b.

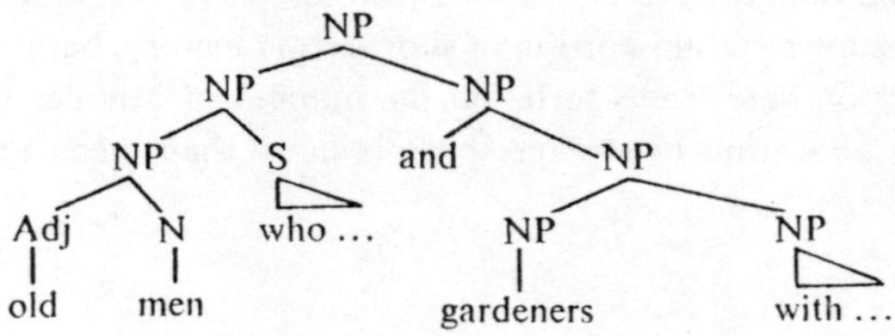

c.

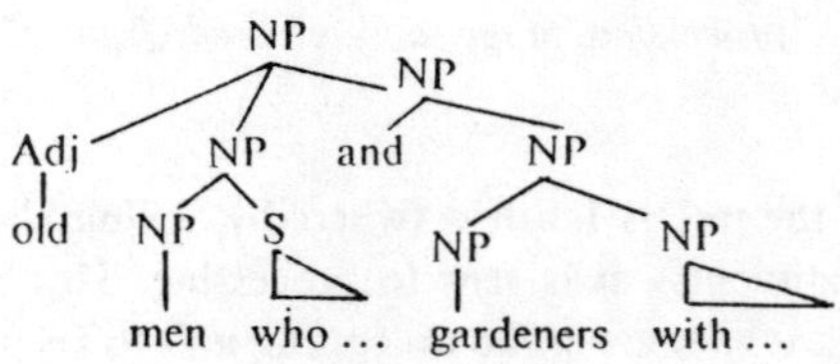

Bever (1970b) accounts for the difficulty in sentences like (1c) ('The boat Floated on the water sank') in terms of what he conjectures to be a general strategy of sentence perception. This principle (strategy B, p. 294) is that the first N ... V ... (N) sequence isolated in the course of parsing a sentence is interpreted as the main clause. This strategy is a particular case of Closure applied to sentences. Restated, it says that when an S node is 'opened' in the course of a parse, the first substring interpretable as an S (given some look ahead) will be so interpreted. In general, when a terminal string can be interpreted as an X-phrase, it will be.

With Closure established, we can now turn to Fixed Structure.

4.6 *Principle Six (Fixed Structure): When the last immediate constituent of a phrase has been formed, and the phrase E closed, it is costly in terms of perceptual complexity ever to have to go back to reorganize the constituents of that phrase*

This principle explains the complexity of sentences like (39a,b), as explained above.

(39) a. The horse raced past the barn fell.
b. The dog knew the cat disappeared, was rescued.

The principle is connected with the look-ahead capacities of the sentence analyzer. Part of the function of this assumed capacity is to prevent having to return to re-organize previously assigned constituents. For example, a sentence beginning with *that* could be continued in at least three different ways, in each of which *that* would be the initial constituent of different phrases ('That 2+2=4 is nice', 'that boy sang', 'that is a big camel'). Thus, the initial tree built down to *that* by Top-Down will not be determined until succeeding terminals have been scanned. From Fixed Structure we can conclude that English *is* a look-ahead language. The scanned but unconnected terminals occupy a certain portion of short-term memory, but not much, in that the biggest restriction here seems to be on the number of S nodes held and being processed, and the allocation of storage space is more than made up for by efficiency of parsing.

4.7 *Principle Seven (Processing): When a phrase is closed, it is pushed down into a syntactic (possibly semantic) processing stage and cleared from short-term memory*

By this principle when a chunk of the tree is finished (where by 'a chunk' is meant a node and all its immediate constituents), it is sent to processing. This principle requires the assumption that there are pointers in the processing unit to keep straight the original structure of the tree, but such devices are simple mechanisms of data organization and surely can be documented to occur in other kinds of data processing (for example, association) that humans perform. Under this assumption, consider how a sentence like (40) might be processed. At each stage, the contents of the processing unit (PU) are listed.

(40) Tom saw that the cow jumped over the moon.

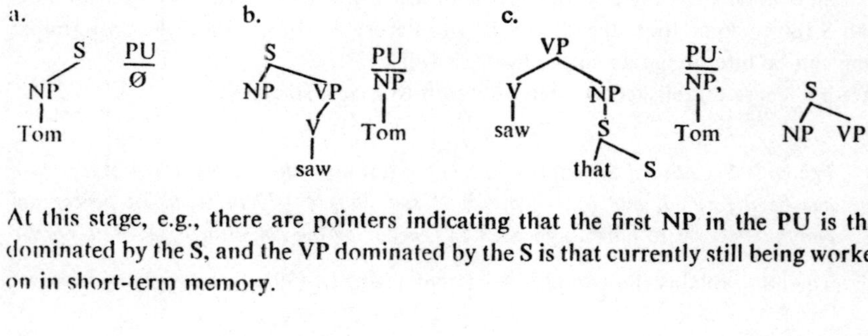

At this stage, e.g., there are pointers indicating that the first NP in the PU is that dominated by the S, and the VP dominated by the S is that currently still being worked on in short-term memory.

d.

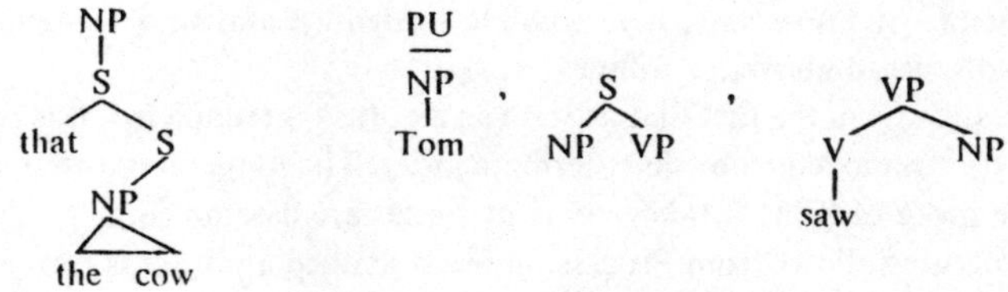

e.

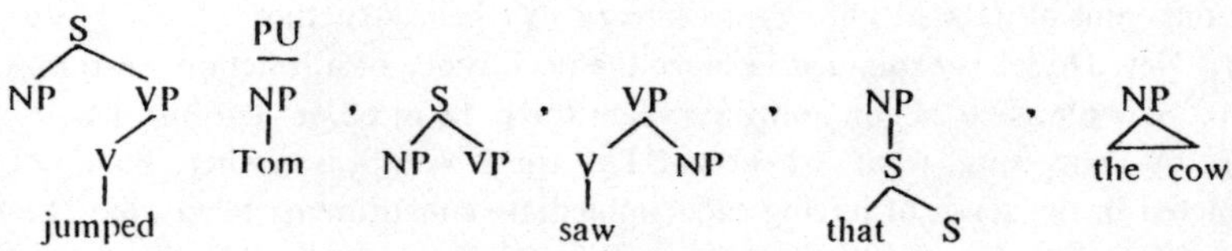

f.

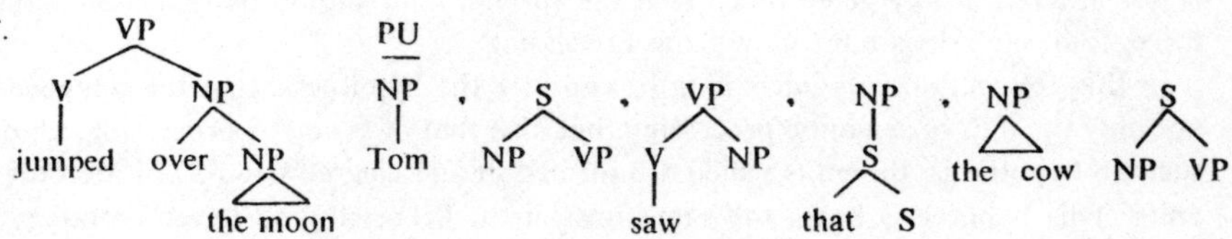

Again, pointers keep straight the relations among these tree chunks in the PU. E.g., there are two chunks of the form 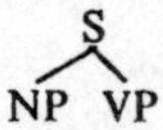 , and by the pointers it is kept straight which is matrix and which subordinate. Notice, in fact, that in a right branching structure the matrix sentence will always appear to the left in the PU of the embedded sentence. This suggests that the form of the tree on surface structure is relevant for the ease of its process in the PU.

I assume, further, that at any point semantic information in the PU is available for current decisions being made in constructing the parse tree, as in the different possibilities for parsing in 'They cooked the peas in the pot' absent in 'They rode the street in the car'.

Notice that at any given moment during the parse, not much more than a single phrase of one or at most two levels is held in short-term memory, a result of the fact that we have chosen for an example a sentence with a simple right branching structure. Sentences with center embedding require that a great deal more structure be held, simply because the higher phrases are not closed until the lower phrases are closed. Left branching structures don't present a problem, as each chunk of the tree is snipped off and placed in the PU as it is completed.

From the principle of Processing, it is possible to deduce and thus explain some of the principles discussed above, as follows:

(a) Closure results from the fact that as soon as a phrase is completed, it is pushed into the PU and thus removed from short-term memory. The longer a phrase remains uncompleted, the more of STM it takes up as its pieces are assembled.

(b) Fixed Structure follows from Processing because once a phrase is formed and pushed into the PU, then it should be difficult to reach down into PU, pull the phrase back out (plus all related phrases) to rework its (their) structure.

(c) New Nodes is explained because the occurrence of a function word indicated when a new phrase is begun and thus when the old can be pushed into PU.

(d) By Processing, what is held as STM are those phrases where have not been completed in the sense of having their immediate constitutents filled out. The statement of Two Sentences is thus made possible, for Processing establishes what is and is not in STM at any given time. That the specific limit should be two (versus one, three, four, etc.) does not follow from Processing.

In this connection it is interesting to consider the hypothesis that the sentence is not only the unit of semantic processing, but also that of syntactic processing. Under such an hypothesis, the units which are formed in and cleared from STM are only S units. This hypothesis bears the same relation to Processing as Bever's strategy B bears to Closure – namely, it is a specific case. At this time, I see no reason to prefer the restricted form over the mere general. I conjecture, then, that *syntactically* all phrases are treated alike in the process of establishing the surface tree. Semantically of course, S phrases occupy a special place. There is some evidence that the general Closure principle is correct – this same evidence seems to support Processing over the hypothesis considered above as Closure follows from Processing. If correct, this means that syntactically the unit of perception is the phrase – semantically the unit of perception is the sentence. This question deserves empirical investigation.

It is interesting to consider which among the seven principles adduced above are universal and which particular to English. I would conjecture that Processing and all those principles that follow from it deductively are universal, that it is common to all perceptual routines for parsing surface structures that phrases are formed, closed, and pushed into a processing unit, and that the semantic manipulations occur in that unit. Likewise, the condition on memory limitations stated in Two Sentences is probably universal. Thus, except for Top Down, which is an assumption for which it would be difficult to accrue evidence, it seems at first glance that all the principles above are universal. On the other hand, it would be most productive to look at a language like Japanese, which is 'backwards' from English in its order of constituents in the base, to determine which sentences are perceptually complex and why. Notice, again, that none of the principles predicts that left branching structures per se, as are found, e.g., in Japanese relative clauses, are difficult to parse. Center embeddings are difficult to process, as noted by everyone who has worked on the problem, and the principles above explain why.

Notice that the explanation of complexity of surface structures offered above does not refer to the transformational 'distance' of a surface structure from its deep structure. That is, these principles refer only to the tree pattern and not to how closely this resembles the tree pattern that represents the semantic relations in deep structure. In this way, these principles go against some of the earliest work on sentence recognition that sought to explain the complexity of a surface form in terms of its transformational history, which has also influenced later writings. For example, Foss and Cairns (1970) write: 'It seems reasonable to assume that the more the surface structure of a sentence distorts the grammatical relations in the structures underlying it, the more complex is the comprehension process and, hence, the more STS (short term storage) required to complete the task of understanding' (p. 541). If the principles above are correct, then distance between deep and surface structure bears no relation to sentence perception. It may bear on sentence comprehension and the nature of the computing process in PU, but this is a different matter.

So far I have said nothing concerning the internal structure of the PU, other than that its data input file consists of various tree chunks, plus an indication of how the pieces fit together. One model of the computations of the PU is that the surface tree is therein reformed, and transformations are applied 'backwards' to reconstruct a deep structure, which is then mapped into the meaning (under the identity mapping for a generative semanticist). I would conjecture that the sentence units of surface structure are reconstructed, being the basic units of comprehension. That is, in, say, 'John was believed by Bill to have been seen by Susan', part of the meaning is that John was believed and that John was seen.

5. Transformations and perceptual routines

Having a model of perceptual mechanism, it is now possible to discuss how various transformations and classes of transformations arrange the form of surface structure with respect to optimal perception. That is, one traditional explanation for the existence of transformations in natural language (Chomsky, 1965, chapter 1) is that they serve to arrange perceptually complex deep structures into perceptually simple surface structures. Given the definition of perceptual simplicity adduced above, it will now be possible to examine this claim in detail.

Research in transformational grammar has resulted in the accumulation of an inventory of transformations. The transformations which have been discovered, however, seem to fall into two distinct classes, the cyclic versus the last-cyclic transformations, with respect to a number of properties (Kimball, 1972c). That this should be the case is in no way predicted by the general theory of transformations. The properties of transformations in these two classes are listed below.

	Cyclic	*Last-cyclic*
(1)	Preserve form of input structure	Derange input structure
(2)	Make no essential use of variables	May make essential use of variables
(3)	May have lexical exceptions	No lexical exceptions
(4)	Several may apply within one S	Only one global transformation per S
(5)	Seem not to introduce structural ambiguities	May introduce structural ambiguities
(6)	Apply working upwards in tree	Apply only on top S

The last-cyclic transformations themselves can be divided into two groups, according to whether a transformation is global, moving constituents over a variable, or local. For example, a global transformation would be *Wh-Fronting* which moves question words to the front of sentences from arbitrarily far to the right. Thus, (41a) becomes (41b) by this transformation.

(41) a. He told you to ask Jill to go to the store to find *wh*-book?
b. *Wh*-book he told you to ask Jill to go to the store to find?
c. What book did he tell you to ask Jill to go to the store to find?

(41b) becomes (41c) by *Subject-Verb Inversion*, which is an example of a local last-cyclic transformation. As it turns out, the local transformations are essentially ordered with respect to some global transformations.

Perhaps the most interesting of the properties differentiating cyclic from last-cyclic transformations is the first, that cyclic transformations preserve the form of the input structures, while last-cyclic transformations tend to distort structure. For example, *Passive* operates on a structure of the form NP V NP, and produces an output structure of essentially the same form. *Dative* maps the structure V NP NP into one of the same form. An operation like *Equi NP Deletion* deletes an NP in an embedded sentence, but the sentence in the cycle on which it operates is unchanged. Likewise *Subject Raising* operating on a sentence embedded in subject position of a matrix sentence results in extraposing the VP of the embedded sentence to the end of the VP of the matrix, as shown in (42a,b).

(42)

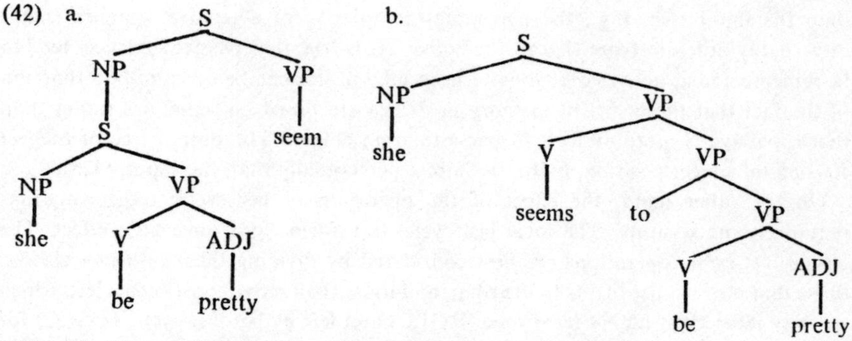

(The *to* is the remains of a *for-to* complementizer.)

Notice, however, that the form of the top S remains unchanged. Thus, the input structures to cyclic transformations are preserved across the operation of these transformations.

On the other hand, global as well as local last-cyclic transformations result in the production of structures unlike the input structures, structures which are also quite unlike those produced by the base rules of the grammar. For example, *Wh-Fronting* produces a structure shown in (43), leaving a hole where the *wh*-NP was extracted.

(43)

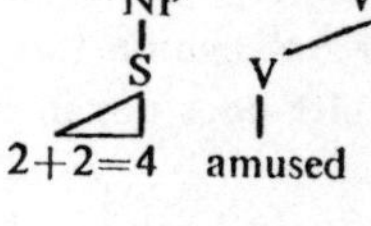

Extraposition moves sentences to the right, producing a structure like (44b) from (44a).

(44)

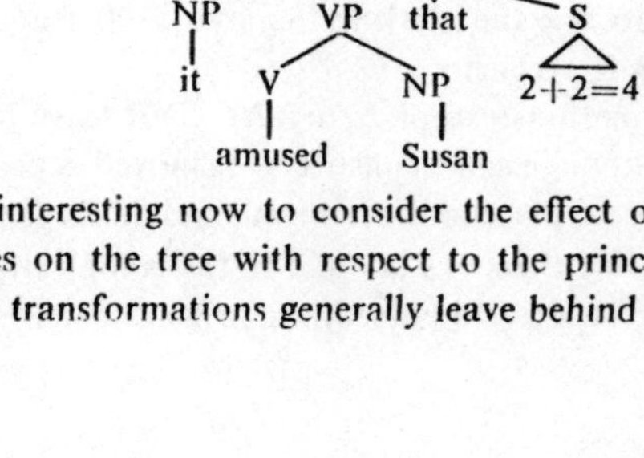

It is interesting now to consider the effect of the transformations of these various classes on the tree with respect to the principles of surface structure parsing. The cyclic transformations generally leave behind a tree with the same or less complexity than the input tree. E.g., the perceptual complexity of a passive sentence is not discernably different from that of the active. (It is true that passive sentences tend to be remembered as actives over long-term recall, but this can be taken to be a function of the fact that these sort of memory processes are based on semantics rather than that a passive is more difficult to parse than an active.) The output tree of *Subject Raising* in subject position, (42b), is simpler perceptually than the input, (42a).

On the other hand, the effect of the operation of last-cyclic transformations requires some scrutiny. The local last-cyclic transformations have little effect. The global last-cyclic operations are best considered by dividing them into two classes, those that move constituents to the right, and those than move constituents left, which we may label right global last-cyclic (RGLC) and left global last-cyclic (LGLC) for convenience. Some RGLC's are listed below:

(45) a. *Extraposition from NP*
the boy who was tall left → the boy left who was tall
b. *Extraposition of PP*
a review of this book will be coming out → a review will be coming out of this book
c. *Extraposition* (44a) → (44b)
d. *Right Dislocation*
Joe gave the book that was about ducks to Susan → Joe gave it to Susan, the book that was about ducks
e. *Heavy NP Shift* (discussed in Section 4)
he asked the girl with the bright blouse to leave → he asked to leave the girl with the bright blouse

All of these transformations hang constituents out on right branches. They all may simplify the sentence in terms of principles like Right Association and Closure, for constituents internal to a sentence are made 'lighter', permitting closure to occur in general earlier. The relation of *Heavy NP Shift* to Right Association was discussed earlier. Note, however, that blind application of these rules does not in every case lead to a perceptually simpler sentence. E.g., (46a,b).

(46) a. He told the girl with the blonde eyelashes to go to the bank to ask the clerk to remove $100 from their account.
b. He told to go to the bank to ask the clerk to remove $100 from their account the girl with the blonde eyelashes.

Finally, notice that all RGLC transformations except *Heavy NP Shift* leave behind some place marker in the tree. That is, although some constituent is moved, a pronoun or some lexical material remains to mark the place of the removal, and this is generally not true of LGLCs. In terms of Processing, this does not add to the complexity. For by the time the extraposed material is reached and parsed, the original material will most likely be in PU, and a pointer can be attached to the extraposed constituent assigning it to that place in the PU.

Let us consider next some operations that move constituents to the left.

(47) a. *Topicalization* (moves an NP to the front of the main S)
Joe told Martha to ask Susan to test the bagel for Will → the bagel Joe told Martha to ask Susan to test for Will

b. *Wh-Fronting* (discussed above)

c. *Relative clause formation*
Joe spanked the child Bill had seen Betty kiss *wh*-child → Joe spanked the child which Bill had seen Betty kiss

d. *Left Dislocation* (Like *Topicalization*, except leaves a pronoun)
Joe gave the book to Sally → the book, Joe gave it to Sally.

Notice now that all these operations except (47d) leave no place marker to indicate the spot of removal. This may be an accident; however, from the point of view of the principles of perception none is required. How it is that the moved material is assigned the correct place for semantic analysis is a problem solved in the PU. One could imagine the moved constituents as being placed in a special category in PU awaiting the first possible pointer assignment to a spot in the tree as it is parsed. But the difficulty in discovering the surface constituents of the tree is not increased by these operations that move elements leftwards.

It could be conjectured that the RGLC rules leave a marker to indicate a place in the tree so that the extraposed material need not be assigned as a new constituent under some node already in PU. That is, with *Extraposition*, for example, a place is opened and held for the extraposed sentence by the *it*. Without some marker, to correctly appoint the encountered extraposed sentence to its place, a new NP would have to be entered under the S and a pointer assigned to that place. This would violate the principle of Fixed Structure, for some structure that had been placed in PU would have to be altered. Thus it is that it is possible to explain this property of RGLC rules with respect to the operation of rules of perception.

In summary, the cyclic transformations either effect no major change in structure from the point of view of perceptual complexity, or, in the case of *Subject Raising*, may operate to hang material on right branches. A right branching tree is not difficult to parse, as predicted by Right Association and Closure. As pointed out to me by DeRemer, a right associative structure is easier for a top-down mechanism which is predictive. On the other hand, left associative structure is much easier for a bottom-up parser. One may conjecture that languages such as Japanese with characteristic left branching employ a mixed strategy of bottom-up and top-down parsing. RGLC transformations hang material on right branches, simplifying the tree. The fact that these transformations leave markers behind in the tree to indicate the spot of removal is significant; an empty place remains in the tree for the extraposed material to be reassigned in the PU. Finally, LGLC transformations do not produce a tree that is perceptually more complex than the input structure. Their operation does require that the moved material be located back to the original place in the tree, but this is evidently performed in the PU, and so does not add to the *perceptual* complexity of the surface structure.

REFERENCES

Bever, T. G. (1970a) The influence of speech performance on linguistic structure. In G. B. F. d'Arcais and W. J. M. Levelt (Eds.) *Advances in psycholinguistics*. New York, American Elsevier.

Bever, T. G. (1970b) The cognitive basis for linguistic structures. In J. R. Hayes (Ed.) *Cognition and the development of language*. New York, John Wiley.

Brown, R. and Bellugi, U. (1964) Three processes in the child's acquisition of syntax. *Harv. educ. Rev.*, 34 (2), 133-151.

Chapin, P. G., Smith, T. S., and Abrahamson, A. A., (1972) Two factors in perceptual segmentation of speech. *J. verb. Learn. verb. Beh.*, 11, 164-173.

Chomsky, N. A. (1957) *Syntactic structures*. The Hague, Mouton.

Chomsky, N. A. (1965) *Aspects of the theory of syntax*. Cambridge, M.I.T. Press.

Chomsky, N. A. and Miller, G. (1963) Introduction to the formal analysis of natural languages. In R. D. Luce, R. R. Bush and E. Galanter (Eds.) *Handbook of mathematical psychology*, vol. 2. New York, Wiley.

De Remer, F. (1971) Simple LR(k) grammars. *Communications of the ACM*, 14, 453-460.

Fodor, J. A. and Garrett, M. (1967) Some syntactic determinants of sentential complexity. *Perc. Psychophy.*

Foss, D. J. and Cairns H. S., (1970) Som effects of memory limitation upon sentence comprehension and recall. *J. verb. Learn. verb. Beh.*, 9, 541-547.

Hakes, D. T. (1972) Effects of reducing complement constructions on sentence comprehension. *J. verb. Learn. verb. Beh.*, 11, 278-286.

Hankammer, J. (1973) Unacceptable ambiguity, *Ling. Inq.*, 4 (1).

Kimball, J. (1972a) *The formal theory of grammar*. New Jersey, Prentice-Hall.

Kimball, J. (1972b) The modality of conditions. In J. Kimball (Ed.) *Syntax and semantics*, vol. 1. New York, Seminar Press.

Kimball, J. (1972c) Cyclic and linear grammars. In J. Kimball (Ed.) *Syntax and semantics*, vol. 1. New York, Seminar Press.

Knuth, D. E. (1965) On the translation of languages from left to right. *Information and Control*, 8.

McKeeman, W. M., Horning, J. J., and Wortman, D. B. (1970) *A compiler generator*. New Jersey, Prentice Hall.

Ross, J. R. (1967) *Constraints on variables in syntax*. M.I.T. Dissertation.

Ross, J. R. (1967b) Auxiliaries as main verbs. Mimeo.

[3]

LOGIC AND CONVERSATION*

H. P. GRICE
University of California, Berkeley

It is a commonplace of philosophical logic that there are, or appear to be, divergences in meaning between, on the one hand, at least some of what I shall call the FORMAL devices—$\sim$, $\wedge$, $\vee$, $\supset$, (x), $\exists(x)$, $\int x$ (when these are given a standard two-valued interpretation)—and, on the other, what are taken to be their analogs or counterparts in natural language—such expressions as *not, and, or, if, all, some* (or *at least one*), *the.* Some logicians may at some time have wanted to claim that there are in fact no such divergences; but such claims, if made at all, have been somewhat rashly made, and those suspected of making them have been subjected to some pretty rough handling.

Those who concede that such divergences exist adhere, in the main, to one or the other of two rival groups, which for the purposes of this article I shall call the formalist and the informalist groups. An outline of a not uncharacteristic formalist position may be given as follows: Insofar as logicians are concerned with the formulation of very general patterns of valid inference, the formal devices possess a decisive advantage over their natural counterparts. For it will be possible to construct in terms of the formal devices a system of very general formulas, a considerable number of which can be regarded as, or are closely related to, patterns of inferences the expression of which involves some or all of the devices: Such a system may consist of a

* Used by permission of the author and publisher from H. Paul Grice's William James Lectures, delivered at Harvard University in 1967, and to be published by Harvard University Press.

certain set of simple formulas that must be acceptable if the devices have the meaning that has been assigned to them, and an indefinite number of further formulas, many of them less obviously acceptable, each of which can be shown to be acceptable if the members of the original set are acceptable. We have, thus, a way of handling dubiously acceptable patterns of inference, and if, as is sometimes possible, we can apply a decision procedure, we have an even better way. Furthermore, from a philosophical point of view, the possession by the natural counterparts of those elements in their meaning, which they do not share with the corresponding formal devices, is to be regarded as an imperfection of natural languages; the elements in question are undesirable excrescences. For the presence of these elements has the result that the concepts within which they appear cannot be precisely/clearly defined, and that at least some statements involving them cannot, in some circumstances, be assigned a definite truth value; and the indefiniteness of these concepts is not only objectionable in itself but leaves open the way to metaphysics—we cannot be certain that none of these natural language expressions is metaphysically 'loaded'. For these reasons, the expressions, as used in natural speech, cannot be regarded as finally acceptable, and may turn out to be, finally, not fully intelligible. The proper course is to conceive and begin to construct an ideal language, incorporating the formal devices, the sentences of which will be clear, determinate in truth value, and certifiably free from metaphysical implications; the foundations of science will now be philosophically secure, since the statements of the scientist will be expressible (though not necessarily actually expressed) within this ideal language. (I do not wish to suggest that all formalists would accept the whole of this outline, but I think that all would accept at least some part of it.)

To this, an informalist might reply in the following vein. The philosophical demand for an ideal language rests on certain assumptions that should not be conceded; these are, that the primary yardstick by which to judge the adequacy of a language is its ability to serve the needs of science, that an expression cannot be guaranteed as fully intelligible unless an explication or analysis of its meaning has been provided, and that every explication or analysis must take the form of a precise definition that is the expression/assertion of a logical equivalence. Language serves many important purposes besides those of scientific inquiry; we can know perfectly well what an expression means (and so a fortiori that it is intelligible) without knowing its analysis, and the provision of an analysis may (and usually does) consist in the specification, as generalized as possible, of the conditions that count for or against the applicability of the expression being analyzed. Moreover, while it is no doubt true that the formal devices are especially amenable to systematic treatment by the logician, it remains the case that there are very many inferences and arguments, expressed in natural language and not in terms of these devices, that are nevertheless recognizably valid. So there must be a place for an unsimplified, and so more or less unsystematic, logic of the natural counterparts of these devices; this logic may be aided and guided by the simplified logic of the formal devices but cannot be supplanted by it; indeed, not only do the two logics differ, but sometimes they come into conflict; rules that hold for a formal device may not hold for its natural counterpart.

Now, on the general question of the place in philosophy of the reformation of natural language, I shall, in this article, have nothing to say. I shall confine myself to the dispute in its relation to the alleged divergences mentioned at the outset. I have, moreover, no intention of entering the fray on behalf of either contestant. I wish, rather, to maintain that the common assumption of the contestants that the divergences do in fact exist is (broadly speaking) a common mistake, and that the mistake arises from an inadequate attention to the nature and importance of the conditions governing conversation. I shall, therefore, proceed at once to inquire into the general conditions that, in one way or another, apply to conversation as such, irrespective of its subject matter.

IMPLICATURE

Suppose that A and B are talking about a mutual friend, C, who is now working in a bank. A asks B how C is getting on in his job, and B replies, *Oh quite well, I think; he likes his colleagues, and he hasn't been to prison yet.* At this point, A might well inquire what B was implying, what he was suggesting, or even what he meant by saying that C had not yet been to prison. The answer might be any one of such things as that C is the sort of person likely to yield to the temptation provided by his occupation, that C's colleagues are really very unpleasant and treacherous people, and so forth. It might, of course, be quite unnecessary for A to make such an inquiry of B, the answer to it being, in the context, clear in advance. I think it is clear that whatever B implied, suggested, meant, etc., in this example, is distinct from what B said, which was simply that C had not been to prison yet. I wish to introduce, as terms of art, the verb *implicate* and the related nouns *implicature* (cf. *implying*) and *implicatum* (cf. *what is implied*). The point of this maneuver is to avoid having, on each occasion, to choose between this or that member of the family of verbs for which *implicate* is to do general duty. I shall, for the time being at least, have to assume to a considerable extent an intuitive understanding of the meaning of *say* in such contexts, and an ability to recognize particular verbs as members of the family with which *implicate* is associated. I can, however, make one or two remarks that may help to clarify the more problematic of these assumptions, namely, that connected with the meaning of the word *say*.

In the sense in which I am using the word *say*, I intend what someone has said to be closely related to the conventional meaning of the words (the sentence) he has uttered. Suppose someone to have uttered the sentence *He is in the grip of a vice.* Given a knowledge of the English language, but no knowledge of the circumstances of the utterance, one would know something about what the speaker had said, on the assumption that he was speaking standard English, and speaking literally. One would know that he had said, about some particular male person or animal *x*, that at the time of the utterance (whatever that was), either (1) *x* was unable to rid himself of a certain kind of bad character trait or (2) some part of *x*'s person was caught in a certain kind of tool or instrument (approximate account, of course). But for a full identification of what the speaker had said, one would need to know (a) the identity of *x*, (b) the time of utterance, and (c) the meaning, on the particular occasion of utterance, of the phrase *in the grip of a vice* [a decision between (1) and (2)]. This brief indication of my use of *say* leaves it open whether a man who says (today) *Harold Wilson is a great man* and another who says (also today) *The British Prime Minister is a great man* would, if each knew that the two singular terms had the same reference, have said the same thing. But whatever decision is made about this question, the apparatus that I am about to provide will be capable of accounting for any implicatures that might depend on the presence of one rather than another of these singular terms in the sentence uttered. Such implicatures would merely be related to different maxims.

In some cases the conventional meaning of the words used will determine what is implicated, besides helping to determine what is said. If I say (smugly), *He is an Englishman; he is, therefore, brave,* I have certainly committed myself, by virtue of the meaning of my words, to its being the case that his being brave is a consequence of (follows from) his being an Englishman. But while I have said that he is an Englishman, and said that he is brave, I do not want to say that I have SAID (in the favored sense) that it follows from his being an Englishman that he is brave, though I have certainly indicated, and so implicated, that this is so. I do not want to say that my utterance of this sentence would be, STRICTLY SPEAKING, false should the consequence in question fail to hold. So SOME implicatures are conventional, unlike the one with which I introduced this discussion of implicature.

I wish to represent a certain subclass of nonconventional implicatures, which I shall call CONVERSATIONAL implicatures, as being essentially connected with certain general features of discourse; so my next step is to try to say what these features are.

The following may provide a first approximation to a general principle. Our talk exchanges do not normally consist of a succession of disconnected remarks, and would not be rational if they did. They are characteristically, to some degree at least, cooperative efforts; and each participant recognizes in them, to some extent, a common purpose or set of purposes, or at least a mutually accepted direction. This purpose or direction may be fixed from the start (e.g., by an initial proposal of a question for discussion), or it may evolve during the exchange; it may be fairly definite, or it may be so indefinite as to leave very considerable latitude to the participants (as in a casual conversation). But at each stage, SOME possible conversational moves would be excluded as conversationally unsuitable. We might then formulate a rough general principle which participants will be expected (ceteris paribus) to observe, namely: Make your conversational contribution such as is required, at the stage at which it occurs, by the accepted purpose or direction of the talk exchange in which you are engaged. One might label this the COOPERATIVE PRINCIPLE.

On the assumption that some such general principle as this is acceptable, one may perhaps distinguish four categories under one or another of which will fall certain more specific maxims and submaxims, the following of which will, in general, yield results in accordance with the Cooperative Principle. Echoing Kant, I call these categories Quantity, Quality, Relation, and Manner. The category of QUANTITY relates to the quantity of information to be provided, and under it fall the following maxims:

1. Make your contribution as informative as is required (for the current purposes of the exchange).

2. Do not make your contribution more informative than is required.

(The second maxim is disputable; it might be said that to be overinformative is not a transgression of the CP but merely a waste of time. However, it might be answered that such overinformativeness may be confusing in that it is liable to raise side issues; and there may also be an indirect effect, in that the hearers may be misled as a result of thinking that there is some particular POINT in the provision of the excess of information. However this may be, there is perhaps a different reason for doubt about the admission of this second maxim, namely, that its effect will be secured by a later maxim, which concerns relevance.)

Under the category of QUALITY falls a supermaxim—'Try to make your contribution one that is true'—and two more specific maxims:

1. Do not say what you believe to be false.
2. Do not say that for which you lack adequate evidence.

Under the category of RELATION I place a single maxim, namely, 'Be relevant.' Though the maxim itself is terse, its formulation conceals a number of problems that exercise me a good deal: questions about what different kinds and focuses of relevance there may be, how these shift in the course of a talk exchange, how to allow for the fact that subjects of conversation are legitimately changed, and so on. I find the treatment of such questions exceedingly difficult, and I hope to revert to them in a later work.

Finally, under the category of MANNER, which I understand as relating not (like the previous categories) to what is said but, rather, to HOW what is said is to be said, I include the supermaxim—'Be perspicuous'—and various maxims such as:

1. Avoid obscurity of expression.
2. Avoid ambiguity.
3. Be brief (avoid unnecessary prolixity).
4. Be orderly.

And one might need others.

It is obvious that the observance of some of these maxims is a matter of less urgency than is the observance of others; a man who has expressed himself with undue prolixity would, in general, be open to milder comment than would a man who has said something he believes to be false. Indeed, it might be felt that the importance of at least the first maxim of Quality is such that it should not be included in a scheme of the kind I am constructing; other maxims come into operation only on the assumption that this maxim of Quality is satisfied. While this may be correct, so far as the generation of implicatures is concerned it seems to play a role not totally different from the other maxims, and it will be convenient, for the present at least, to treat it as a member of the list of maxims.

There are, of course, all sorts of other maxims (aesthetic, social, or moral in character), such as 'Be polite', that are also normally observed by participants in talk exchanges, and these may also generate nonconventional implicatures. The conversational maxims, however, and the conversational implicatures connected with them, are specially connected (I hope) with the particular purposes that talk (and so, talk exchange) is adapted to serve and is primarily employed to serve. I have stated my maxims as if this purpose were a maximally effective exchange of information; this specification is, of course, too narrow, and the scheme needs to be generalized to allow for such general purposes as influencing or directing the actions of others.

As one of my avowed aims is to see talking as a special case or variety of purposive, indeed rational, behavior, it may be worth noting that the specific expectations or presumptions connected with at least some of the foregoing maxims have their analogues in the sphere of transactions that are not talk exchanges. I list briefly one such analog for each conversational category.

1. **Quantity.** If you are assisting me to mend a car, I expect your contribution to be neither more nor less than is required; if, for example, at a particular stage I need four screws, I expect you to hand me four, rather than two or six.
2. **Quality.** I expect your contributions to be genuine and not spurious. If I need sugar as an ingredient in the cake you are assisting me to make, I do not expect you to hand me salt; if I need a spoon, I do not expect a trick spoon made of rubber.
3. **Relation.** I expect a partner's contribution to be appropriate to immediate needs at each stage of the transaction; if I am mixing ingredients for a cake, I do not expect to be handed a good book, or even an oven cloth (though this might be an appropriate contribution at a later stage).
4. **Manner.** I expect a partner to make it clear what contribution he is making, and to execute his performance with reasonable dispatch.

These analogies are relevant to what I regard as a fundamental question about the CP and its attendant maxims, namely, what the basis is for the assumption which we seem to make, and on which (I hope) it will appear that a great range of implicatures depend, that talkers will in general (ceteris paribus and in the absence of indications to the contrary) proceed in the manner that these principles prescribe. A dull but, no doubt at a certain level, adequate answer is that it is just a well-recognized empirical fact that people DO behave in these ways; they have learned to do so in childhood and not lost the habit of doing so; and, indeed, it would involve a good deal of effort to make a radical departure from the habit. It is much easier, for example, to tell the truth than to invent lies.

I am, however, enough of a rationalist to want to find a basis that underlies these facts, undeniable though they may be; I would like to be able to think of the standard type of conversational practice not merely as something that all or most do IN FACT follow but as something that it is REASONABLE for us to follow, that we SHOULD NOT abandon. For a time, I was attracted by the idea that observance of the CP and the maxims, in a talk exchange, could be thought of as a quasi-contractual matter, with parallels outside the realm of discourse. If you pass by when I am struggling with my stranded car, I no doubt have some degree of expectation that you will offer help, but once you join me in tinkering under the hood, my expectations become stronger and take more specific forms (in the absence of indications that you are merely an incompetent meddler); and talk exchanges seemed to me to exhibit, characteristically, certain features that jointly distinguish cooperative transactions:

1. The participants have some common immediate aim, like getting a car mended; their ultimate aims may, of course, be independent and even in conflict—each may want to get the car mended in order to drive off, leaving the other stranded. In characteristic talk exchanges, there is a common aim even if, as in an over-the-wall chat, it is a second-order one, namely, that each party should, for the time being, identify himself with the transitory conversational interests of the other.
2. The contributions of the participants should be dovetailed, mutually dependent.
3. There is some sort of understanding (which may be explicit but which is often tacit) that, other things being equal, the transaction should continue in appropriate style unless both parties are agreeable that it should terminate. You do not just shove off or start doing something else.

But while some such quasi-contractual basis as this may apply to some cases, there are too many types of exchange, like quarreling and letter writing, that it fails to fit comfortably. In any case, one feels that the talker who is irrelevant or obscure has primarily let down not his audience but himself. So I would like to be able to show that observance of the CP and maxims is reasonable (rational) along the following lines: that any one who cares about the goals that are central to conversation/communication (e.g., giving and receiving information, influencing and being influenced by others) must be expected to have an interest, given suitable circumstances, in participation in talk exchanges that will be profitable only on the assumption that they are conducted in general accordance with the CP and the maxims. Whether any such conclusion can be reached, I am uncertain; in any case, I am fairly sure that I cannot reach it until I am a good deal clearer about the nature of relevance and of the circumstances in which it is required.

It is now time to show the connection between the CP and maxims, on the one hand, and conversational implicature on the other.

A participant in a talk exchange may fail to fulfill a maxim in various ways, which include the following:

1. He may quietly and unostentatiously VIOLATE a maxim; if so, in some cases he will be liable to mislead.
2. He may OPT OUT from the operation both of the maxim and of the CP; he may say, indicate, or allow it to become plain that he is unwilling to cooperate in the way the maxim requires. He may say, for example, *I cannot say more; my lips are sealed.*
3. He may be faced by a CLASH: He may be unable, for example, to fulfill the first maxim of Quantity (Be as informative as is required) without violating the second maxim of Quality (Have adequate evidence for what you say).
4. He may FLOUT a maxim; that is, he may BLATANTLY fail to fulfill it. On the assumption that the speaker is able to fulfill the maxim and to do so without violating another maxim (because of a clash), is not opting out, and is not, in view of the blatancy of his performance, trying to mislead, the hearer is faced with a minor problem: How can his saying what he did say be reconciled with the supposition that he is observing the overall CP? This situation is one that characteristically gives rise to a conversational implicature; and when a conversational implicature is generated in this way, I shall say that a maxim is being EXPLOITED.

I am now in a position to characterize the notion of conversational implicature. A man who, by (in, when) saying (or making as if to say) that *p* has implicated that *q*, may be said to have conversationally implicated that *q*, PROVIDED THAT (1) he is to be presumed to be observing the conversational maxims, or at least the cooperative principle; (2) the supposition that he is aware that, or thinks that, *q* is required in order to make his saying or making as if to say *p* (or doing so in THOSE terms) consistent with this presumption; and (3) the speaker thinks (and would expect the hearer to think that the speaker thinks) that it is within the competence of the hearer to work out, or grasp intuitively, that the supposition mentioned in (2) IS required. Apply this to my initial example, to B's remark that C has not yet been to prison. In a suitable setting A might reason as follows: '(1) B has apparently violated the maxim 'Be relevant' and so may be regarded as having flouted one of the maxims conjoining perspicuity, yet I have no reason to suppose that he is opting out from the operation of the CP; (2) given the circumstances, I can regard his irrelevance as only apparent if, and only if, I suppose him to think that C is potentially dishonest; (3) B knows that I am capable of working out step (2). So B implicates that C is potentially dishonest.'

The presence of a conversational implicature must be capable of being worked out; for even if it can in fact be intuitively grasped, unless the intuition is replaceable by an argument, the implicature (if present at all) will not count as a CONVERSATIONAL implicature; it will be a CONVENTIONAL implicature. To work out that a particular conversational implicature is present, the hearer will reply on the following data: (1) the conventional meaning of the words used, together with the identity of any references that may be involved; (2) the CP and its maxims; (3) the context, linguistic or otherwise, of the utterance; (4) other items of background knowledge; and (5) the fact (or supposed fact) that all relevant items falling under the previous headings are available to both participants and both participants know or assume this to be the case. A general pattern for the working out of a conversational implicature might be given as follows: 'He has said that *p*; there is no reason to suppose that he is not observing the maxims, or at least the CP; he could not be doing this unless he thought that *q*; he knows (and knows that I know that he knows) that I can see that the supposition that he thinks that *q* IS required; he has done nothing to stop me thinking that *q*; he intends me to think, or is at least willing to allow me to think, that *q*; and so he has implicated that *q*.'

Examples

I shall now offer a number of examples, which I shall divide into three groups.

GROUP A: *Examples in which no maxim is violated, or at least in which it is not clear that any maxim is violated*

A is standing by an obviously immobilized car and is approached by B; the following exchange takes place:

(1) A: *I am out of petrol.*

B: *There is a garage round the corner.* (Gloss: B would be infringing the maxim 'Be relevant' unless he thinks, or thinks it possible, that the garage is open, and has petrol to sell; so he implicates that the garage is, or at least may be open, etc.)

In this example, unlike the case of the remark *He hasn't been to prison yet,* the unstated connection between B's remark and A's remark is so obvious that, even if one interprets the supermaxim of Manner, 'Be perspicuous,' as applying not only to the expression of what is said but also to the connection of what is said with adjacent remarks, there seems to be no case for regarding that supermaxim as infringed in this example. The next example is perhaps a little less clear in this respect:

(2) A: *Smith doesn't seem to have a girlfriend these days.*

B: *He has been paying a lot of visits to New York lately.*

B implicates that Smith has, or may have, a girlfriend in New York. (A gloss is unnecessary in view of that given for the previous example.)

In both examples, the speaker implicates that which he must be assumed to believe in order to preserve the assumption that he is observing the maxim of relation.

GROUP B: *An example in which a maxim is violated, but its violation is to be explained by the supposition of a clash with another maxim*

A is planning with B an itinerary for a holiday in France. Both know that A wants to see his friend C, if to do so would not involve too great a prolongation of his journey:

(3) A: *Where does C live?*

B: *Somewhere in the South of France.* (Gloss: There is no reason to suppose that B is opting out; his answer is, as he well knows, less informative than is required to meet A's needs. This infringement of the first maxim of Quantity can be explained only by the supposition that B is aware that to be more informative would be to say something that infringed the maxim of Quality, 'Don't say what you lack adequate evidence for', so B implicates that he does not know in which town C lives.)

GROUP C: *Examples that involve exploitation, that is, a procedure by which a maxim is flouted for the purpose of getting in a conversational implicature by means of something of the nature of a figure of speech*

In these examples, though some maxim is violated at the level of what is said, the hearer is entitled to assume that that maxim, or at least the overall Cooperative Principle, is observed at the level of what is implicated.

(1a) *A flouting of the first maxim of Quantity*

A is writing a testimonial about a pupil who is a candidate for a philosophy job, and his letter reads as follows: 'Dear Sir, Mr. X's command of English is excellent, and his attendance at tutorials has been regular. Yours, etc.' (Gloss: A cannot be opting out, since if he wished to be uncooperative, why write at all? He cannot be unable, through ignorance, to say more, since the man is his pupil; moreover, he knows that more information than this is wanted. He must, therefore, be wishing to impart information that he is reluctant to write down. This supposition is tenable only on the assumption that he thinks Mr. X is no good at philosophy. This, then, is what he is implicating.)

Extreme examples of a flouting of the first maxim of Quantity are provided by utterances of patent tautologies like *Women are women* and *War is war.* I would wish to maintain that at the level of what is said, in my favored sense, such remarks are totally noninformative and so, at that level, cannot but infringe the first maxim of Quantity in any conversational context. They are, of course, informative at the level of what is implicated, and the hearer's identification of their informative content at this level is dependent on his ability to explain the speaker's selection of this PARTICULAR patent tautology.

(3b) *An infringement of the second maxim of Quantity, 'Do not give more information than is required', on the assumption that the existence of such a maxim should be admitted*

A wants to know whether *p*, and B volunteers not only the information that *p*, but information to the effect that it is certain that *p*, and that the evidence for its being the case that *p* is so-and-so and such-and-such.

B's volubility may be undesigned, and if it is so regarded by A it may raise in A's mind a doubt as to whether B is as certain as he says he is ('Methinks the lady doth protest too much'). But if it is thought of as designed, it would be an oblique way of conveying that it is to some degree controversial whether or not *p*. It is, however, arguable that such an implicature could be explained by reference to the maxim of Relation without invoking an alleged second maxim of Quantity.

(2a) *Examples in which the first maxim of Quality is flouted*

1. *Irony.* X, with whom A has been on close terms until now, has betrayed a secret of A's to a business rival. A and his audience both know this. A says *'X is a fine friend'*. (Gloss: It is perfectly obvious to A and his audience that what A has said or has made as if to say is something he does not believe, and the audience knows that A knows that this is obvious to the audience. So, unless A's utterance is entirely pointless, A must be trying to get across some other proposition than the one he purports to be putting forward. This must be some obviously related proposition; the most obviously related proposition is the contradictory of the one he purports to be putting forward.)

2. *Metaphor.* Examples like *You are the cream in my coffee* characteristically involve categorial falsity, so the contradictory of what the speaker has made as it to say will, strictly speaking, be a truism; so it cannot be THAT that such a speaker is trying to get across. The most likely supposition is that the speaker is attributing to his audience some feature or features in respect of which the audience resembles (more or less fancifully) the mentioned substance.

It is possible to combine metaphor and irony by imposing on the hearer two stages of interpretation. I say *You are the cream in my coffee,* intending the hearer to reach first the metaphor interpretant 'You are my pride and joy' and then the irony interpretant 'You are my bane.'

3. *Meiosis.* Of a man known to have broken up all the furniture, one says *He was a little intoxicated.*

4. *Hyperbole.* Every nice girl loves a sailor.

(2b) Examples in which the second maxim of Quality, 'Do not say that for which you lack adequate evidence', is flouted are perhaps not easy to find, but the following seems to be a specimen. I say of X's wife, *She is probably deceiving him this evening.* In a suitable context, or with a suitable gesture or tone of voice, it may be clear that I have no adequate reason for supposing this to be the case. My partner, to preserve the assumption that the conversational game is still being played, assumes that I am getting at some related proposition for the acceptance of which I DO have a reasonable basis. The related proposition might well be that she is given to deceiving her husband, or possibly that she is the sort of person who would not stop short of such conduct.

(3) *Examples in which an implicature is achieved by real, as distinct from apparent, violation of the maxim of Relation* are perhaps rare, but the following seems to be a good candidate. At a genteel tea party, A says *Mrs. X is an old bag.* There is a moment of appalled silence, and then B says *The weather has been quite delightful this summer, hasn't it?* B has blatantly refused to make what HE says relevant to A's preceding remark. He thereby implicates that A's remark should not be discussed and, perhaps more specifically, that A has committed a social gaffe.

(4) *Examples in which various maxims falling under the supermaxim 'Be perspicuous' are flouted*

1. *Ambiguity.* We must remember that we are concerned only with ambiguity that is deliberate, and that the speaker intends or expects to be recognized by his hearer. The problem the hearer has to solve is why a speaker should, when still playing the conversational game, go out of his way to choose an ambiguous utterance. There are two types of cases:

(a) Examples in which there is no difference, or no striking difference, between two interpretations of an utterance with respect to straightforwardness; neither interpretation is notably more sophisticated, less standard, more recondite or more far-fetched than the other. We might consider Blake's lines: 'Never seek to tell thy love, Love that never told can be.' To avoid the complications introduced by the presence of the imperative mood, I shall consider the related sentence, *I sought to tell my love, love that never told can be.* There may be a double ambiguity here. *My love* may refer to either a state of emotion or an object of emotion, and *love that never told can be* may mean either 'Love that cannot be told' or 'love that if told cannot continue to exist.' Partly because of the sophistication of the poet and partly because of internal evidence (that the ambiguity is kept up), there seems to be no alternative to supposing that the ambiguities are deliberate and that the poet is conveying both what he would be saying if one interpretation were intended rather than the other, and vice versa; though no doubt the poet is not explicitly SAYING any one of these things but only conveying or suggesting them (cf. 'Since she [nature] pricked thee out of women's pleasure, mine be thy love, and thy love's use their treasure.)

(b) Examples in which one interpretation is notably less straightforward than another. Take the complex example of the British General who captured the town of Sind and sent back the message *Peccavi.* The ambiguity involved ('I have Sind'/'I have sinned') is phonemic, not morphemic; and the expression actually used is unambiguous, but since it is in a language foreign to speaker and hearer, translation is called for, and the ambiguity resides in the standard translation into native English.

Whether or not the straightforward interpretant ('I have sinned') is being conveyed, it seems that the nonstraightforward must be. There might be stylistic reasons for conveying by a sentence merely its nonstraightforward interpretant, but it would be pointless, and perhaps also stylistically objectionable, to go to the trouble of finding an expression that nonstraightforwardly conveys that *p*, thus imposing on an audience the effort involved in finding this interpretant, if this interpretant were otiose so far as communication was concerned. Whether the straightforward interpretant is also being conveyed seems to depend on whether such a supposition would conflict with other conversational requirements, for example, would it be relevant, would it be something the speaker could be supposed to accept, and so on. If such requirements are not satisfied, then the straightforward interpretant is not being conveyed. If they are, it is. If the author of *Peccavi* could naturally be supposed to think that he had committed some kind of transgression, for example, had disobeyed his orders in capturing Sind, and if reference to such a transgression would be relevant to the presumed interests of the audience, then he would have been conveying both interpretants; otherwise he would be conveying only the nonstraightforward one.

2. *Obscurity.* How do I exploit, for the purposes of communication, a deliberate and overt violation of the requirement that I should avoid obscurity? Obviously, if the Cooperative Principle is to operate, I must intend my partner to understand what I am saying despite the obscurity I import into my utterance. Suppose that A and B are having a conversation in the presence of a third party, for example, a child, then A might be deliberately obscure, though not too obscure, in the hope that B would understand and the third party not. Furthermore, if A expects B to see that A is being deliberately obscure, it seems reasonable to suppose that, in making his conversational contribution in this way, A is implicating that the contents of his communication should not be imparted to the third party.

3. *Failure to be brief or succinct.* Compare the remarks:

(a) *Miss X sang 'Home sweet home.'*

(b) *Miss X produced a series of sounds that corresponded closely with the score of 'Home sweet home'.*

Suppose that a reviewer has chosen to utter (b) rather than (a). (Gloss: Why has he selected that rigmarole in place of the concise and nearly synonymous *sang?* Presumably, to indicate some striking difference between Miss X's performance and those to which the word *singing* is usually applied. The most obvious supposition is that Miss X's performance suffered from some hideous defect. The reviewer knows that this supposition is what is likely to spring to mind, so that is what he is implicating.)

I have so far considered only cases of what I might call particularized conversational implicature—that is to say, cases in which an implicature is carried by saying that *p* on a particular occasion in virtue of special features of the context, cases in which there is no room for the idea that an implicature of this sort is NORMALLY carried by saying that *p*. But there are cases of generalized conversational implicature. Sometimes one can say that the use of a certain form of words in an utterance would normally (in the ABSENCE of special circumstances) carry such-and-such an implicature or type of implicature. Noncontroversial examples are perhaps hard to find, since it is all too easy to treat a generalized conversational implicature as if it were a conventional implicature. I offer an example that I hope may be fairly noncontroversial.

Anyone who uses a sentence of the form *X is meeting a woman this evening* would normally implicate that the person to be met was someone other than X's wife, mother, sister, or perhaps even close platonic friend. Similarly, if I were to say *X went into a house yesterday and found a tortoise inside the front door,* my hearer would normally be surprised if some time later I revealed that the house was X's own. I could produce similar linguistic phenomena involving the expressions *a garden, a car, a college,* and so on. Sometimes, however, there would normally be no such implicature ('I have been sitting in a car all morning'), and sometimes a reverse implicature ('I broke a finger yesterday'). I am inclined to think that one would not lend a sympathetic ear to a philosopher who suggested that there are three senses of the form of expression *an X:* one in which it means roughly 'something that satisfies the conditions defining the word *X*,' another in which it means approximately 'an X (in the first sense) that is only remotely related in a certain way to some person indicated by the context,' and yet another in which it means 'an X (in the first sense) that is closely related in a certain way to some person indicated by the context.' Would we not much prefer an account on the following lines (which, of course, may be incorrect in detail):

When someone, by using the form of expression *an X*, implicates that the X does not belong to or is not otherwise closely connected with some identifiable person, the implicature is present because the speaker has failed to be specific in a way in which he might have been expected to be specific, with the consequence that it is likely to be assumed that he is not in a position to be specific. This is a familiar implicature situation and is classifiable as a failure, for one reason or another, to fulfill the first maxim of Quantity. The only difficult question is why it should, in certain cases, be presumed, independently of information about particular contexts of utterance, that specification of the closeness or remoteness of the connection between a particular person or object and a further person who is mentioned or indicated by the utterance should be likely to be of interest. The answer must lie in the following region: Transactions between a person and other persons or things closely connected with him are liable to be very different as regards their concomitants and results from the same sort of transactions involving only remotely connected persons or things; the concomitants and results, for instance, of my finding a hole in MY roof are likely to be very different from the concomitants and results of my finding a hole in someone else's roof. Information, like money, is often given without the giver's knowing to just what use the recipient will want to put it. If someone to whom a transaction is mentioned gives it further consideration, he is likely to find himself wanting the answers to further questions that the speaker may not be able to identify in advance; if the appropriate specification will be likely to enable the hearer to answer a considerable variety of such questions for himself, then there is a presumption that the speaker should include it in his remark; if not, then there is no such presumption.

Finally, we can now show that, conversational implicature being what it is, it must possess certain features:

1. Since, to assume the presence of a conversational implicature, we have to assume that at least the Cooperative Principle is being observed, and since it is possible to opt out of the observation of this principle, it follows that a generalized conversational implicature can be canceled in a particular case. It may be explicitly canceled, by the addition of a clause that states or implies that the speaker has opted out, or it may be contextually canceled, if the form of utterance that usually carries it is used in a context that makes it clear that the speaker IS opting out.

2. Insofar as the calculation that a particular conversational implicature is present requires, besides contextual and background infor-

mation, only a knowledge of what has been said (or of the conventional commitment of the utterance), and insofar as the manner of expression plays no role in the calculation, it will not be possible to find another way of saying the same thing, which simply lacks the implicature in question, except where some special feature of the substituted version is itself relevant to the determination of an implicature (in virtue of one of the maxims of Manner). If we call this feature NONDETACHABILITY, one may expect a generalized conversational implicature that is carried by a familiar, nonspecial locution to have a high degree of nondetachability.

3. To speak approximately, since the calculation of the presence of a conversational implicature presupposes an initial knowledge of the conventional force of the expression the utterance of which carries the implicature, a conversational implicatum will be a condition that is not included in the original specification of the expression's conventional force. Though it may not be impossible for what starts life, so to speak, as a conversational implicature to become conventionalized, to suppose that this is so in a given case would require special justification. So, initially at least, conversational implicata are not part of the meaning of the expressions to the employment of which they attach.

4. Since the truth of a conversational implicatum is not required by the truth of what is said (what is said may be true—what is implicated may be false), the implicature is not carried by what is said, but only by the saying of what is said, or by 'putting it that way.'

5. Since, to calculate a conversational implicature is to calculate what has to be supposed in order to preserve the supposition that the Cooperative Principle is being observed, and since there may be various possible specific explanations, a list of which may be open, the conversational implicatum in such cases will be disjunction of such specific explanations; and if the list of these is open, the implicatum will have just the kind of indeterminacy that many actual implicata do in fact seem to possess.

[4]

A THEORY OF SYNTACTIC RECOGNITION FOR NATURAL LANGUAGE

MITCHELL P. MARCUS

The Determinism Hypothesis

All current natural language parsers that are adequate to cover a wide range of syntactic constructions operate by simulating nondeterministic machines, either by using backtracking or by pseudo-parallelism. On the face of it, this seems to be necessary, for a cursory examination of natural language reveals many phenomema that seem to demand nondeterministic solutions if we restrict our attention to parsers that operate left-to-right.

A typical natural language parser is conceptually nondeterministic in that it will parse a given input if there is some sequence of grammar rules (however such rules are expressed) whose application yields a coherent analysis of the input, even if other legal sequences of rule application do not lead to such analyses. Since all physically existing machines must be deterministic, such a nondeterministic machine must be simulated by causing a deterministic machine to make "guesses" about what the proper sequence of actions for a given input should be, coupled with some mechanism for aborting incorrect guesses. For many inputs, this necessarily leads to the creation of some syntactic substructures which are not constituents in whatever final syntactic analysis is assigned to a given input.

To see that such an approach seems to be necessary, consider the sentences in figure 1. While the first seven words of both of these sentences are identical, their structures, as shown, are very different. In one "have" is the main verb of an imperative sentence, with the rest of the sentence a subordinate clause. In the other sentence, a question, "have" is an auxiliary of the main verb "taken," and "the students" is the subject of the main clause.

It would seem that to analyze the structure of these sentences in a left-to-right manner, a parser must necessarily simulate a nondeterministic process. Not only is it impossible to determine what role the word "have" serves in either of the given sentences on first encounter, but the two structures are identical up to the end of the NP "the students who missed the exam."

1.1a Have the students who missed the exam take the makeup today.

1.1b [$_S$ [$_{VP}$ Have [$_S$ the students who ... take the makeup today]]]

1.2a Have the students who missed the exam taken the makeup today?

1.2b [$_S$ [$_{AUX}$ Have] [$_{NP}$ the students who ...] [$_{VP}$ taken the exam today]]

Figure 1. Some examples that motivate non-deterministic parsing.

There is no possible way to tell the two structures apart until the morphology of the verb can be examined.

While this and other such examples appear to be quite compelling, it is contendable that natural language need not be parsed by a mechanism that simulates nondeterminism, that the seeming necessity for such an approach is an illusion. This work is based upon this hypothesis; the central idea behind the research is that *the syntax of any natural language can be parsed by a mechanism which operates "strictly deterministically" in that it does not simulate a nondeterministic machine,* in a sense which will be made precise below. (Actually, I will discuss this hypothesis only for the particular case of English; nothing will be said about languages other than English in what follows below.) I begin by assuming that the hypothesis is true, and then show that it leads directly to a simple mechanism, a grammar interpreter, that has the following properties:

- The grammar interpreter allows simple rules to be written which elegantly capture the significant generalizations behind such constructions as passives, yes/no questions, imperatives, and sentences with existential "there." These rules are reminiscent of the sorts of rules proposed within the framework of the theory of generative grammar, despite the fact that the rules presented here must recover underlying structure given only the terminal string of the surface form of the sentence. The component of the grammar interpreter which allows such rules to be formulated follows directly from the Determinism Hypothesis.

- The structure of the grammar interpreter constrains its operation in such a way that only very complex, *ad hoc* grammar rules can parse sentences which violate several of the constraints on rules of grammar proposed within the last several years by Chomsky [1977]. (The syntactic structures that the parser creates are by and large Annotated Surface Structures of the type currently proposed by Chomsky, complete with traces, although each parse node is also annotated with a set of features *a la* Winograd [1971].) Most of the structural properties of the grammar interpreter on which this result depends are directly motivated by the Determinism Hypothesis.

- The grammar interpreter provides a simple explanation for the difficulty caused by "garden path" sentences, sentences like "The horse raced past the barn fell." In essence, the grammar interpreter allows special diagnostic rules to be written which can diagnose between the alternative cases presented in figure 1, but there is a limitation on the power of such rules which follows from a parameter of the mechanism. By appropriately setting this parameter, sentences like those in figure 1 can be diagnosed, but those which typically cause garden paths cannot. The component of the mechanism which this parameter affects, and which is crucial for this diagnostic process, is the same component that allows the formulation of the linguistic generalizations discussed above. (These garden path sentences clearly disconfirm one possible form of the Determinism Hypothesis which would say that all sentences which are grammatical according to a purely competence grammar can be parsed strictly deterministically, but the explanation for such sentences afforded by this model is consistent with a more "psychological" formulation of the hypothesis: that all sentences which people can parse without conscious difficulty can be parsed strictly deterministically.)

To the extent that these properties, all of which reflect deep properties of natural language, follow from the Determinism Hypothesis, and in this sense are explained by it, they provide indirect evidence for the truth of the hypothesis.

From the properties of the grammar interpreter, it should be clear that the theory presented here is an *explanatory* theory of language, rather than merely a *descriptive* theory. My central concern is not the particular grammar of English that has been developed in the course of this work, but rather the general properties of the grammar interpreter that enable a grammar of this form to be written. Thus, the central focus of what follows will not be on the particular rules of grammar presented as examples, but rather on the properties of the interpreter that allow rules like these to be written.

The Notion of "Strictly Deterministic"

The notion of "strictly deterministic" is central to the meaning of the Determinism Hypothesis; exactly what does it mean?

To avoid any possible misunderstanding, let me state explicitly that the Determinism Hypothesis cannot mean simply that language can be parsed by a deterministic machine. As noted, any computational mechanism that physically exists is deterministic, and thus any process which is specified by an algorithm for such a machine must be deterministic. From this it follows that any parser, whether it simulates a nondeterministic machine or not, must itself be deterministic. (note that "deterministic" does *not* mean non-probabilistic, nor does "nondeterministic" mean probabilistic. A nondeterministic machine, instead, can be conceptualized as a machine which has a magical oracle which tells it the right decision to make whenever its course of action is not strictly determined by the input and the state of the machine.)

Rather than attempting to formulate any rigorous, general explanation of what it means to "not simulate a nondeterministic machine," I will focus instead on three specific properties of the grammar interpreter. These properties are special in that they will prevent this interpreter from simulating nondeterminism by blocking the implementation of either backtracking or pseudo-parallelism. This discussion is not intended to be definitional, but rather to give a better grasp of the computational restrictions which will be embodied in the grammar interpreter, whose structure will be sketched in the next section. These three properties are:

- *All syntactic substructures created by the machine are permanent.* This eliminates the possibility of simulating determinism by "backtracking," i.e by undoing the actions that were done while pursuing a guess that turns out to be incorrect.

- *All syntactic substructures created by the machine for a given input must be output as part of the syntactic structure assigned to that input.* Since the structure the grammar interpreter will assign to a given input expresses exactly one syntactic analysis, this property eliminates the possibility of simulating nondeterminism via pseudo-parallelism. If a machine were to simulate nondeterminism in this manner, it would necessarily create structures for the alternative analyses that would result from each of the paths such a machine pursues in parallel. (The reader may wonder how such an interpreter can handle true global ambiguity. The answer is that while such a machine can only build one syntactic analysis for a given input, and thus can only represent one interpretation of the input, it can also observe that the input was ambiguous, and flag the output analysis to indicate that this analysis is only one of a range of coherent analyses. Some external mechanism will then be needed to force the interpreter to reparse the input, taking a different analysis path, if the other consistent analyses are desired.)

- *The internal state of the mechanism must be constrained in such a way that no temporary syntactic structures are encoded within the internal state of the machine.* While this does not mean that the machine's internal state must be limited to a finite state control (the grammar interpreter to be presented below uses a push-down stack, among other control structures), it must be limited -- at least in its use -- in such a way that structure is not hidden in the state of the machine.

One immediate implication of all this is that a grammar for any interpreter which embodies these properties must constrain that interpreter from ever making a mistake, since the interpreter can only correctly analyze a given input if it never creates any incorrect structure. This means that such a grammar must at least implicitly specify how to decide what the grammar interpreter should do next, that is, it can never leave the grammar interpreter with more than one alternative.

The Structure of the Grammar Interpreter

Taking the Determinism Hypothesis as a given, an examination of natural language leads to a further set of properties which any deterministic grammar interpreter must embody. (Henceforth, the word "deterministic" means "strictly deterministic" in the sense discussed.) Any such interpreter *must* have the following properties: it must be at least partially data driven; but it must be able to reflect expectations that follow from general grammatical properties of the partial structures built up during the parsing process; and it must have some sort of look-ahead facility, even if it is basically left-to-right.

To show that each of these properties is necessary, it suffices to show a pair of sentences of English that cannot be distinguished by a mechanism without the given property, but which speakers of English understand without difficulty. The sentences shown in figure 2 below provide crucial pairs for each of the properties.

The parser must:

Be partially data driven.

(1a) John went to the store.

(1b) How much is the doggie in the window?

Reflect expectations.

(2a) I called [$_{NP}$ John] [$_{S}$ to make Sue feel better].

(2b) I wanted [$_{S}$ John to make Sue feel better].

Have some sort of look-ahead.

(3a) Have [$_{S}$ the boys take the exam today].

(3b) Have [$_{NP}$ the boys] [$_{VP}$taken the exam today].

Figure 2. Some examples which motivate the structure of the parser.

Almost by definition, a hypothesis driven parser cannot be deterministic, and thus a deterministic parser must necessarily be at least partially data driven. The essence of the problem is that any parser which is purely hypothesis driven, that is, which is purely top-down, must hypothesize several nested levels of structure before positing any constituents which can be checked against the input string itself.

For example, a top-down parser, newly given an input, might begin by hypothesizing that the input is a sentence. It might then hypothesize that the input is a declarative, and therefore hypothesize that the input begins with a noun phrase. Assuming that the input begins with a noun phrase, it might finally hypothesize that the NP begins with a determiner, a hypothesis which is testable against the input string. At this point, the parser has created structures that correspond to the S and the NP, which will necessarily have to be discarded for at least some inputs. (These structures might be implicit in the state of the machine, but this is simply a matter of how the constituents are represented at this point in the parsing process.) To take a concrete example, even so different a pair of sentences as 2.1a and 2.1b cannot be deterministically analyzed by a hypothesis driven parser. The problem, of course, is simply that any hypothesis driven parser must either attempt to parse a given input as a declarative sentence, beginning, say, with an NP, before it attempts to parse it as a question, beginning with an auxiliary, or vice versa. Whatever order the parser imposes upon these two possibilities relative to each other, the clause type attempted first must be at least occasionally wrong. It is clear that if a parser is to be deterministic, it must look before it leaps.

A deterministic parser cannot be entirely bottom-up, however. Any parser that is purely bottom-up must initially misparse one of the two sentences given as 2.2a and 2.2b. The problem is that the string "John to make Sue feel better" can be analyzed in two different ways: as one constituent that is an infinitive complement, as in 2.2b, or as two unrelated constituents, as in 2.2a, with the NP "John" the object of the verb and the phrase "to make Sue feel better" an adverbial "purpose" clause. The difference in structure between 2.2a and 2.2b can be predicted, however, if the parser can note that "want" typically takes an infinitive complement, while "call" cannot take such a complement. Thus, a deterministic parser must have some capacity to use whatever information and expectations can be gleaned from an examination of the structures that have been built up at any given point in the parsing process. If a parser is to operate deterministically, it must use such information to constrain the analysis imposed on the remainder of the input.

Finally, if a deterministic parser is to correctly analyze such pairs of sentences as 2.3a and 2.3b above, it cannot operate in an entirely left-to-right manner. As was discussed, it is impossible to distinguish between this pair of sentences before examining the morphology of the verb following the NP "the boys." These sentences can be distinguished, however, if the parser has a large enough "window" on the clause to see this verb; if the verb ends in "en" (in the simple case presented here), then the clause is an yes/no question, otherwise it is an imperative. Thus, if a parser is to be deterministic, it must have some facility for look ahead. *It must be stressed, however, that this look-ahead ability must be constrained in some manner; otherwise the determinism claim is vacuous.*

We now turn to a grammar interpreter called *PARSIFAL*, whose structure is motivated by the three principles discussed above. This grammar interpreter maintains two major data structures: a push down stack of incomplete constituents called *the active node stack*, and a small three-place *constituent buffer* which contains constituents which are complete, but whose higher level grammatical function is as yet uncertain.

Figure 3 shows a snapshot of the parser's data structures taken while parsing the sentence "John should have scheduled the meeting." At the bottom of the stack is an auxiliary node labelled with the features *modal*, *past*, among others, which has as a daughter the modal "should." (This stack grows *downward*, so that the structure of the stack reflects the structure of the emerging parse tree.) Above the bottom of the stack is an S node with an NP as a daughter, dominating the word "John." There are two words in the buffer, the verb "have" in the first buffer cell and the word "scheduled" in the second. The two words "the meeting" have not yet come to the attention of the parser. (The structures of form "(PARSE-AUX CPOOL)" and the like will be explained below.)

The constituent buffer is really the heart of the grammar interpreter; it is the central feature that distinguishes this parser

```
The Active Node Stack
   S1 (S DECL MAJOR S) / (PARSE-AUX CPOOL)
            NP : (John)
   AUX1 (MODAL PAST VSPL AUX) / (BUILD-AUX)
            MODAL : (should)

The Buffer
1 :   WORD3 (*HAVE VERB TNSLESS AUXVERB PRES V-3S) : (have)
2 :   WORD4 (*SCHEDULE COMP-OBJ VERB INF-OBJ v-3s
                      ED=EN EN PART PAST ED) : (scheduled)

Yet unseen words:   the meeting .
```

Figure 3. PARSIFAL's two major data structures.

from all others. The words that make up the parser's input first come to its attention when they appear at the end of this buffer after morphological analysis. After the parser builds these words into some larger grammatical structure at the bottom of the active node stack, it may then pop the new constituent from the active node stack and insert this completed constituent into the first cell of the buffer if the grammatical role of this larger structure is as yet undetermined. The parser is free to examine the constituents in this buffer, to act upon them, and to otherwise use the buffer as a workspace.

In general, the parser uses the buffer in a first-in, first-out fashion. It typically decides what to do with the constituent in the leftmost buffer position after taking the opportunity to examine its immediate neighbors to the right. The availability of the buffer allows the parser to defer using a word or a larger constituent that fills a single buffer cell until it has a chance to examine some of the right context of the constituent in question. Thus, for example, the parser must often decide whether the word "have" at the beginning of a clause initiates a yes/no question, as in 1.2b, or an imperative, as in 1.2a. The parser can often correctly decide what sort of clause it has encountered, and thus how to use the initial verb, by allowing several constituents to "pile up" in the buffer. Consider for example, the snapshot of the buffer shown in figure 4. By waiting until the NP "the boys," NP25, is formed, filling the 2nd

```
 ________________________________________
| WORD32 |    NP25    | WORD37 |
|  HAVE  |  THE BOYS  |   DO   |
|________|____________|________|
```

Figure 4. The buffer allows the parser to examine local context.

buffer position, and WORD37, the verb "do," enters the buffer, filling the 3rd buffer position, the parser can see that the clause must be an imperative, and that "have" is therefore the main verb of the major clause.

Note that each cell in the buffer can hold a *grammatical constituent* of any type, where a constituent is any tree that the parser has constructed under a single root node. The size of the structure underneath the node is immaterial; both "that" and "that the big green cookie monster's toe got stubbed" are perfectly good constituents once the parser has constructed the latter phrase into a subordinate clause.

The constituent buffer and the active node stack are acted upon by a grammar which is made up of pattern/action rules; this grammar can be viewed as an augmented form of Newell and Simon's production systems [Newell and Simon 1972]. Each rule is made up of a pattern, which is matched against some subset of the constituents of the buffer and the accessible nodes in the active node stack (about which more will be said below), and an action, a sequence of operations which acts on these constituents. Each rule is assigned a numerical *priority*, which the grammar interpreter uses to arbitrate simultaneous matches.

The grammar as a whole is structured into *rule packets*, clumps of grammar rules which can be turned on and off as a group; the grammar interpreter only attempts to match rules in packets that have been activated by the grammar. At any given time during the parsing process, the grammar interpreter only attempts to match those rules which are in *active* packets. Any grammar rule can activate a packet by associating that packet with the constituent at the bottom of the active node stack. If a node at the bottom of the stack is pushed into the stack, the active packets remain associated with it, but are only active when that node is again at the bottom of the stack. For example, in figure 3, the packet BUILD-AUX is associated with the bottom of the stack, and is thus active, while the packet PARSE-AUX is associated with the S node above the auxiliary.

The grammar rules themselves are written in a language called PIDGIN, an English-like formal language that is translated into LISP by a simple translator based on the notion of top-down operator precedence [Pratt 1973]. Figure 5 gives a schematic overview of the organization of the grammar, and exhibits some of the rules that make up the packet PARSE-AUX.

The parser (that is the grammar interpreter interpreting some grammar) operates by attaching constituents which are in the buffer to the constituent at the bottom of the stack until that constituent is complete, at which time it is popped from the stack. If the constituents in the buffer provide clear evidence that a constituent of a given type should be initiated, a new node of that type can be created and pushed onto the stack; this new node can also be attached to the node currently at the bottom of the stack before the stack is pushed, if the grammatical function of the new constituent is clear at the time it is created. When popped, a constituent either remains attached to its parent, if it was attached to some larger constituent when it was created, or

```
         Priority        Pattern                    Action
                       Description of:
                 1st     2nd    3rd    The Stack
                            PACKET1
         5:      [   ] [   ] [   ]              --> ACTION1
         10:     [   ]               [   ]      --> ACTION2
         10:     [   ] [   ] [   ] [   ]        --> ACTION3
                            PACKET2
         10:     [   ] [   ]                    --> ACTION4
         15:     [   ]               [   ]      --> ACTION5
                            PACKET3
         5:      [   ] [   ]                    --> ACTION6
         15:     [   ] [   ] [   ]              --> ACTION7
```

(a) - The structure of the grammar.

```
{RULE START-AUX PRIORITY: 10. IN PARSE-AUX
[=verb] -->
Create a new aux node.
Label C with the meet of the features of 1st and pres,
        past, future, tnsless.
Activate build-aux.}
{RULE TO-INFINITIVE PRIORITY: 10. IN PARSE-AUX
[=*to, auxverb] [=tnsless] -->
Label a new aux node inf.
Attach 1st to C as to.
Activate build-aux.}
{RULE AUX-ATTACH PRIORITY: 10. IN PARSE-AUX
[=aux] -->
Attach 1st to C as aux.
Activate parse-vp. Deactivate parse-aux.}
```

(b) - Some sample grammar rules that initiate and attach auxiliaries.

Figure 5. The structure of the grammar and some example rules.

else it falls into the constituent buffer (which will cause an error if the buffer was already full).

This structure embodies the principles discussed above in the following ways:

- *A deterministic parser must be at least partially data driven.* A grammar for PARSIFAL is made up of pattern/action rules which are triggered, in part, when lexical items or unattached higher level constituents fulfilling specific descriptions appear in the buffer. Thus, the parser is directly responsive to the input it is given.

- *A deterministic parser must be able to reflect expectations that follow from the partial structures built up during the parsing process.* Since PARSIFAL only attempts to match rules that are in active packets, grammar rules can activate and deactivate packets of rules to reflect the properties of the constituents in the active node stack. Thus grammar rules can easily be written that are constrained by whatever structure the parser is attempting to complete.

- *A deterministic parser must have some sort of constrained look-ahead facility.* PARSIFAL's buffer provides this constrained look-ahead. Because the buffer can hold several constituents, a grammar rule can examine the context that follows the first constituent in the buffer before deciding what grammatical role it fills in a higher level structure. I argue that a buffer of quite limited length suffices to allow deterministic parsing. The key idea is that the size of the buffer can be sharply constrained if each location in the buffer can hold a single complete constituent, regardless of that constituent's size.

We now turn to the structure of individual grammar rules. Some example rules were given in figure 5; this section will explain the syntax of those rules.

The pattern of a grammar rule is made up of a list of partial descriptions of parse nodes which must all be fulfilled if the pattern is to match. There can be up to five partial descriptions in each pattern: up to three consecutive descriptions which are matched against the first, second, and third constituents in the buffer (in order), and two descriptions which match against the two nodes in the active node stack accessible to the parser. These two nodes are the bottom node on the stack, which will be referred to as the *current active node*, and the S or NP node closest to the bottom of the stack which will be called the *dominating cyclic node* or alternatively, if an S, the *current S node*. In figure 3, AUX1 is the current active node, and S1 is the current S node. As we shall see later, making the dominating cyclic node explicitly available for examination and modification seems to eliminate the need for any tree climbing operations that ascend tree structures.

The syntax of the grammar rules presented is self-explanatory, but a few comments on the grammar notation itself are in order. The general form of each grammar rule is:

{Rule <name> priority: <priority> in <packet>
<pattern> --> <action>}

Each pattern is of the form:

[<description of 1st buffer constituent>] [<2nd>] [<3rd>]

The symbol "=" used only in pattern descriptions, is to be read as "has the feature(s)." Features of the form "*<word>" mean "has the root <word>," for example "**have*" means "has the root "have"." The tokens "1st," "2nd," "3rd," and "C" (or "c") refer to the constituents in the 1st, 2nd, and 3rd buffer positions and the current active node (that is the bottom of the stack), respectively. (I will also use these tags in the text below as names for their respective constituents.) The symbol "t" used in a pattern description is a predicate that is true of any node; thus "[t]" is the simplest always true description. Pattern descriptions to be matched against the current active node and the current S are flagged by "**C" appearing at the beginning of an additional pattern description.

Each description is made up of a Boolean combination of tests for given grammatical features. Each description can also include Boolean feature tests on the daughters of the target node; the grammar language provides a tree walking notation for indicating specific daughters of a node. (While the parser can access daughter nodes, it cannot modify them.) While this richness of specification seems to be necessary, it should be noted that the majority of rules in even a moderately complex grammar have patterns which consist only of tests for the positive presence of given features.

The action of a grammar rule consists of a rudimentary program that does the actual work of building constituent structures. An action is built up of primitives that perform such actions as these:

- Creating a new parse node, pushing the newly created node onto the bottom of the active node stack. A new node is presumably created whenever the parser decides that the first constituent(s) in the buffer are the initial daughters of a constituent not yet created by the parser.

- Inserting a specific lexical item into a specific buffer cell, which causes the previous contents of that cell to be shifted one place to the right.

- Popping the current active node from the active node stack, causing it to be inserted into the first buffer cell if it has not been previously attached to another node, shifting the previous contents of that cell one place to the right.

- Attaching a newly created node or a node in the buffer to the current active node or the current cyclic node. After each grammar rule is executed, the grammar interpreter removes all newly attached nodes from the buffer, with nodes to the previous right of each deleted node shifting to the left. Note that a newly created node can either be attached to a parent node at the time of its creation, if its function in higher level grammatical structure is clear at that time, or it can be created without an attachment, in which case it will be dropped into the buffer when it is popped from the active node stack.

- Assigning features to a node in the buffer or one of the accessible nodes in the stack.

- Activating and deactivating packets of rules.

PIDGIN also provides primitives from which Boolean tests of the features of a node can be constructed, as mentioned above. These predicates can be used within conditional "if...then...else..." expressions to conditionally perform various operations.

With the exception of allowing conditional expressions, PIDGIN rules fall into the simple class of programs called *fixed-instruction programs.* The PIDGIN language imposes the following constraints on rule actions:

- There are no user-settable variables within the rule actions. The constituents in the first three buffer cells are available as the values of the parameters *1st, 2nd,* and *3rd* within each rule, but these parameters are given values by the grammar interpreter before each rule action is called, and are not resettable within an action. The values of the parameters *C* and *the current cyclic node* do change within a rule as nodes are pushed and popped from the active node stack, but their values cannot be set by a grammar rule.

- PIDGIN allows no user-defined functions; the only functions within actions are PIDGIN primitives. (There is a limited ability for a rule to circumvent the pattern-matching process by explicitly naming its successor, but this is formally only a device for rule abbreviation, since the specification of the successor rule could simply be replaced with the code for the action of the rule named.)

- While conditional "if...then....else..." expressions are allowed, there is no recursion or iteration within actions.

- The only structure building operations in PIDGIN are (a) attaching one node to another, and (b) adding features to a node's feature set. In particular, the list building primitives of LISP are *not* available in PIDGIN.

Parsing a Simple Declarative Sentence

Now I present a small grammar which is just sufficient to parse a very simple declarative sentence, and then trace through the process of parsing the sentence immediately below, given this grammar. The emphasis here will not be on the complexities of the grammar, but rather on the form of the grammar in broad outline and on the details of the workings of the grammar interpreter.

John has scheduled the meeting.

One simplification will be imposed on this example: it will be assumed that all NPs come into the buffer preparsed, that the structure of NPs is determined by a mechanism that is transparent to the "clause-level" grammar rules that will be discussed here. Such a mechanism is in fact presented in Marcus [1977]; I will say here only that this mechanism involves relatively slight extensions of the mechanism presented here.

As a convention, the parser begins every parse by calling the grammar rule named INITIAL-RULE. This rule creates an S node and activates the packet SS-START (Simple Sentence-START), that contains rules that decide on the type of simple sentences. The parser's state after this rule is executed is depicted in figure 6. At this point there is nothing in the buffer.

Constituents enter the buffer on demand -- the buffer mechanism will get the next constituent from the input word stream when a rule pattern must be matched against a buffer cell that is currently empty. Furthermore, before the grammar interpreter will attempt to match a rule of a given priority, all higher prioritied rules must explicitly fail to match. This means that throughout the examples, constituents will often enter the buffer for no apparent reason. These constituents were requested by rules that ultimately failed to match, leaving no trace of why each constituent entered the buffer.

The packet SS-START, some of whose rules are shown in figure 7, contains rules which determine the type of a major clause. If the clause begins with an NP followed by a verb, then the clause is labelled a declarative; if it begins with an auxiliary verb followed by an NP, it is labelled a yes/no question. If the clause begins with a tenseless verb, then not only is the clause labelled an imperative, but the word "you" is inserted into the buffer, where it is inserted at the beginning of the buffer by convention.

After INITIAL-RULE has been executed and packet SS-START has been activated, the rule MAJOR-DECL-S matches, with the pattern matching process pulling the NP "John" and the verb "has" into the buffer. The action of the rule is now run, labelling the clause a major declarative clause, deactivating the packet SS-START, and activating the packet PARSE-SUBJ. The result of this is shown in figure 8.

The packet PARSE-SUBJ contains rules whose job it is

The Active Node Stack (0. deep)

C: S16 (S) / (SS-START)

The Buffer

```
{RULE INITIAL-RULE IN NOWHERE
[t] -->
Create a new s node.
Activate ss-start.}
```

Figure 6. After INITIAL-RULE has been run.

```
{RULE MAJOR-DECL-S IN SS-START
[=np] [=verb] -->
Label c s, decl, major.
Deactivate ss-start. Activate parse-subj.}

{RULE YES-NO-Q IN SS-START
[=auxverb] [=np] -->
Label c s, quest, ynquest, major.
Deactivate ss-start. Activate parse-subj.}

{RULE IMPERATIVE IN SS-START
[=tnsless] -->
Label c s, imper, major.
Insert the word 'you' into the buffer.
Deactivate ss-start. Activate parse-subj.}
```

Figure 7. Some rules that determine sentence type.

```
            The Active Node Stack ( 0. deep)
C:          S16 (S DECL MAJOR S) / (PARSE-SUBJ)

            The Buffer
1 :         NP40 (NP NAME NS N3P) : (John)
2 :         WORD125 (*HAVE VERB AUXVERB PRES V3S) : (has)

Yet unseen words:   scheduled a meeting .
```

Figure 8. After the rule MAJOR-DECL-S is run.

to find and attach the subject of the clause under construction. It contains two major rules which are shown in figure 9. The rule UNMARKED-ORDER picks out the subject in clauses

```
{RULE UNMARKED-ORDER IN PARSE-SUBJ
[=np] [=verb] -->
Attach 1st to c as np.
Deactivate parse-subj.
Activate parse-aux.}

{RULE AUX-INVERSION IN PARSE-SUBJ
[=auxverb] [=np] -->
Attach 2nd to c as np.
Deactivate parse-subj. Activate parse-aux.}
```

Figure 9. Two subject-parsing rules.

where the subject appears before the verb in surface order; the rule AUX-INVERSION picks out the subject in clauses where an element of the auxiliary occurs before the subject. Though the relevant rules will not be discussed here, the rule UNMARKED-ORDER will pick up the subject of imperatives and WH-questions where the subject of the clause is questioned, while AUX-INVERSION will pick up the subject of WH-questions that question other than the subject of the clause.

At this point in the parse, the rule UNMARKED-ORDER now matches, and its action is run. This rule attaches 1st, NP40, the NP "John," to C, the node S16, as subject. It also activates the packet PARSE-AUX after deactivating PARSE-SUBJ. After this rule has been executed, the interpreter notes that the NP has been attached, and removes it from the buffer. Figure 10 shows the state of the parser after

---

```
              The Active Node Stack ( 0. deep)
C:            S16 (S DECL MAJOR S) / (PARSE-AUX)
                        NP : (John)

              The Buffer
1 :           WORD125 (*HAVE VERB AUXVERB PRES V3S) : (has)

Yet unseen words:   scheduled a meeting .
```

Figure 10. After UNMARKED-ORDER has been executed.

---

this rule is executed.

For brevity the process of parsing the auxiliary verbs in this example will not be discussed, although figure 11 provides a trace of the application of rules during the parsing of the auxiliary phrase, through the attachment of the auxiliary to the VP node. The rules referred to in the trace are included in figure 12.

It should be noted that the rules of packet BUILD-AUX,

```
About to run: STARTAUX
            The Active Node Stack ( 0. deep)
C:          S16 (S DECL MAJOR S) / (PARSE-AUX)
                        NP : (John)

            The Buffer
1 :         WORD125 (*HAVE VERB AUXVERB PRES V3S) : (has)
Yet unseen words:   scheduled a meeting .
```

```
About to run: PERFECTIVE
            The Active Node Stack ( 1. deep)
              S16 (S DECL MAJOR S) / (PARSE-AUX)
                        NP : (John)
C:          AUX14 (PRES V3S AUX) / (BUILD-AUX)

            The Buffer
1 :         WORD125 (*HAVE VERB AUXVERB PRES V3S) : (has)
2 :         WORD126 (*SCHEDULE COMP-OBJ VERB INF-OBJ V-3S ...) : (scheduled)
Yet unseen words:   a meeting .
```

```
About to run: AUX-COMPLETE
            The Active Node Stack ( 1. deep)
              S16 (S DECL MAJOR S) / (PARSE-AUX)
                        NP : (John)
C:          AUX14 (PERF PRES V3S AUX) / (BUILD-AUX)
                        PERF : (has)

            The Buffer
1 :         WORD126 (*SCHEDULE COMP-OBJ VERB INF-OBJ V-3S ...) : (scheduled)
Yet unseen words:   a meeting .
```

```
About to run: AUX-ATTACH
            The Active Node Stack ( 0. deep)
C:          S16 (S DECL MAJOR S) / (PARSE-AUX)
                    NP : (John)

            The Buffer
 1 :        AUX14 (PERF PRES V3S AUX) : (has)
 2 :        WORD126 (*SCHEDULE COMP-OBJ VERB INF-OBJ V-3S ...) : (scheduled)
Yet unseen words:   a meeting .
```

```
            The Active Node Stack ( 0. deep)
C:          S16 (S DECL MAJOR S) / (PARSE-VP)
                    NP : (John)
                    AUX : (has)

            The Buffer
 1 :        WORD126 (*SCHEDULE COMP-OBJ VERB INF-OBJ V-3S ...) : (scheduled)
Yet unseen words:   a meeting .
```

Figure 11. Parsing the auxiliary of (i).

```
{RULE START-AUX PRIORITY: 10. IN PARSE-AUX
[=verb] -->
Create a new aux node.
Label C with the meet of the features of 1st and vspl, v1s,
        v+13s, vpl+2s, v-3s, v3s.
%(The above features are "person/number codes", e.g. "vpl+2s"
means that this verb goes with any plural or 2nd person singular
np as subject. The verb "are" has this feature.)%
Label C with the meet of the features of 1st and pres,
        past, future, tnsless.
Activate build-aux.}

{RULE AUX-ATTACH PRIORITY: 10. IN PARSE-AUX
[=aux] -->
Attach 1st to c as aux.
Activate parse-vp. Deactivate parse-aux.}

{RULE PERFECTIVE PRIORITY: 10. IN BUILD-AUX
[=*have] [=en] --> Attach 1st to c as perf. Label c perf.}

{RULE PROGRESSIVE PRIORITY: 10. IN BUILD-AUX
[=*be] [=ing] --> Attach 1st to c as prog. Label c prog.}

{RULE PASSIVE-AUX PRIORITY: 10. IN BUILD-AUX
[=*be] [=en] --> Attach 1st to c as passive. Label c passive.}

{RULE AUX-COMPLETE PRIORITY: 15. IN BUILD-AUX
[t] --> Drop c into the buffer.}
```

Figure 12. Some rules which parse auxiliaries.

some of which are shown in figure 12, are the equivalent of the transformational rule of affix-hopping. Note that these rules concisely state the relation between each auxiliary verb and its related affix by taking advantage of the ability to buffer both each auxiliary verb and the following verb. It might seem that some patch is needed to these rules to handle question constructions in which, typically, the verb cluster is discontinuous, but this is not the case, as will be shown later.

The packet PARSE-VP is now active. This packet contains, among other rules, the rule MVB (Main VerB), which creates and attaches a VP node and then attaches the main verb to it. This rule now matches and is run. The rule itself, and the resulting state of the parser, is shown in figure 13.

At the time MVB is executed, the packet PARSE-VP is associated with S16, the current active node, as shown in the last frame of figure 11. The action of MVB first deactivates the packet PARSE-VP, then activates either SS-FINAL, if C is a major clause, or EMB-S-FINAL, if it is an embedded clause. These two packets both contain rules that parse clause-level prepositional phrases, adverbs, and the like. They differ in that the rules in EMB-S-FINAL must decide whether a given modifier should be attached to the current clause, or left in the buffer to be attached to a constituent higher up in the parse tree after the current active node is completed. Whatever packet is activated, the newly activated packet will be associated with C, S16, and thus the grammar interpreter will attempt to match the rules in it whenever S16 is the current active node.

The execution of the next line in the action of MVB results in a new VP node, VP14, being attached to S16 and then pushed onto the active node stack, becoming the current active node. Next the verb "scheduled", WORD126, is attached to the VP, and then the action of MVB activates the packet SUBJ-VERB. As is always the case with packet activation, this packet is associated with the node which is the current active node at the time of its activation, in this case VP14. Thus, this rule leaves the parser with the packet SS-FINAL associated with

```
{RULE MVB IN PARSE-VP
[=verb] -->
Deactivate parse-vp.
If c is major then activate ss-final else
If c is sec then activate emb-s-final.
Attach a new vp node to c as vp.
Attach 1st to c %which is now the vp% as verb.
Activate subj-verb.}
```

```
             The Active Node Stack ( 1. deep)
              S16 (S DECL MAJOR S) / (SS-FINAL)
                        NP : (John)
                        AUX : (has)
                        VP : ↓
C:           VP14 (VP) / (SUBJ-VERB)
                        VERB : (scheduled)

             The Buffer
1 :          WORD127 (*A NGSTART DET NS N3P ...) : (a)

Yet unseen words:  meeting .
```

Figure 13. The rule MVB and the parser's state after its execution.

S16, and the packet SUBJ-VERB associated with VP14. Since VP14 is the current active node, SUBJ-VERB is now active. Once VP14 is popped from the stack, leaving S16 the current active node, SS-FINAL will be active.

The packet SUBJ-VERB contains rules which involve the deep grammatical relation of the surface subject of the clause to the verb, and which set up the proper environment for parsing the objects of the verb. The major rule in this packet, the rule

```
{RULE SUBJ-VERB PRIORITY: 15. IN SUBJ-VERB
[t] -->
%Activate packets to parse objects and complements.%
If the verb of c is inf-obj then activate inf-comp.
If the verb of c is to-less-inf-obj then activate to-less-inf-comp.
If the verb of c is that-obj then activate that-comp.
If there is a wh-comp of the s above c
        and it is not utilized then activate wh-vp
        else If the s above c is major then activate ss-vp
        else activate embedded-s-vp.
Deactivate subj-verb.}
```

Figure 14. The rule SUBJ-VERB.

SUBJ-VERB is shown in figure 14. Note that this rule has a very low priority and a pattern that will always match; this rule is a *default rule*, a rule which will become active when no other rule can apply. While most of the code of the action of the rule SUBJ-VERB does not apply to our example, I have refrained from abbreviating this rule to give a feel for the power of PIDGIN.

The rule SUBJ-VERB is now the rule of highest priority that matches, so its action is now executed. The purpose of this rule is to activate the appropriate packets to parse the objects and complements of a clause; the activation of some of these packets depends upon the verb of the clause, while the activation of others depends upon more global properties of the clause. Thus, the next several lines of the action activate packets for various sorts of complement constructions that a verb might take: infinitive phrases in general (the packet INF-COMP), infinitive phrases that are not flagged by "to," (the packet TO-LESS-INF-COMP in addition to INF-COMP), and tensed complements (the packet THAT-COMP). The next long clause activates one of a number of packets which will attach the objects of the verb. The packet activated depends on the clause type: whether this clause still has a WH-head that needs to be utilized (WH-VP), whether this clause is secondary without a WH-head (EMBEDDED-S-VP), or whether this clause is a major clause (SS-VP).

This rule provides a good example of one difference between the packets used to organize this grammar and the states of an ATN. Packets do *not* simply correspond to ATN states, for several packets will typically be active at a time. For instance, if parsing the sentence "Who did John see Jane kiss?," this rule would activate three packets: INF-COMP, TO-LESS-INF-COMP, and WH-VP. In terms of the ATN model, one can think of this rule as dynamically tailoring the arcs out of a state of an ATN to exactly match various properties of the clause being parsed.

Of the complement-initiating packets, only the packet INF-COMP is activated in the example, since "schedule" can take infinitive complements, as in "Schedule John to give a lecture on Tuesday." The packet SS-VP is then activated to attach the verb's objects, the packet SUBJ-VERB is deactivated, and the rule SUBJ-VERB is through. This rule thus changes the state of the parser only by activating and deactivating packets; the packets now associated with the current active node are SS-VP and INF-COMP. Figure 15 shows the rules in the packet SS-VP that will come into play in the example. Note that the rule VP-DONE, like the rule SUBJ-VERB above, is a default rule.

The rule OBJECTS is now triggered by NP41, and attaches the NP to VP14. The state of the parser at this point is shown in figure 16.

Completing the parse is now a simple matter. The default rule in packet SS-VP, VP-DONE, now triggers, popping the VP node from the active node stack. Since VP14 is attached

```
{RULE OBJECTS IN SS-VP
[=np] -->
Attach 1st to c as np.}

{RULE VP-DONE PRIORITY: 20 IN SS-VP
[t] --> Drop c.}
```

Figure 15.

```
                    The Active Node Stack ( 1. deep)
                    S16 (S DECL MAJOR S) / (SS-FINAL)
                              NP : (John)
                              AUX : (has)
                              VP : ↓
          C:        VP14 (VP) / (SS-VP INF-COMP)
                              VERB : (scheduled)
                              NP : (a meeting)

                    The Buffer
          1 :       WORD133 (*. FINALPUNC PUNC) : (.)

     Yet unseen words: (none)
```

Figure 16. After the rule OBJECTS has been run.

to S16, the S node above it on the stack, it remains hanging from the S node. The S node is once again the current active node, and packet SS-FINAL is now active. The default rule in this packet, SS-DONE, shown in figure 17, immediately triggers. This

```
{RULE S-DONE IN SS-FINAL
[=finalpunc] -->
Attach 1st to c as finalpunc.
%The next line is really part of the cf mechanism.%
Finalize the cf of s.
Parse is finished.}
```

Figure 17. The rule S-DONE.

rule attaches the final punctuation mark to the clause, pops the S node from the stack, and signals the grammar interpreter that the parse is now complete, bringing our example to a close.

Implications of the Mechanism

This research began with the assumption that English could be parsed with greater ease than had previously been assumed. This assumption in turn leads to several principals which any such parser must embody, principals which PARSIFAL reflects rather directly. The grammar is made up of pattern/action rules, allowing the parser to be data directed. These rules themselves are clustered in packets, which can be activated and deactivated to reflect global expectations. The grammar interpreter's constituent buffer gives it a small window through which to view the input string, allowing restricted look-ahead.

In conclusion, I will briefly sketch how specific structural properties of the parser allow the range of linguistic generalizations indicated at the beginning of this report to be captured.

Of the structures that make up the grammar interpreter, I believe, it is the constituent buffer which is most central to these results. The most important of these results follow from the sorts of grammar operations that the buffer makes feasible; others follow from the limitations that its fixed length imposes. All however, hinge crucially upon the existence of the buffer. This data structure, I submit, is the primary source of the power of the parser.

For example:

- Because the buffer automatically compacts upon the attachment of the constituents that it contains, the parsing of a yes/no question and the related declarative will differ in one rule of grammar, with the key difference restricted to the rule patterns and one line of the rules' actions. The yes/no question rule explicitly states only that the NP in the second buffer cell should be attached as the subject of the clause. Because the buffer will then compact, auxiliary parsing rules that expect the terminals of the verb cluster to be contiguous will then apply without need for modification.

- Because the buffer provides a three-constituent window on the structure of a clause, diagnostic rules can be formulated that allow the differential diagnosis of pairs of constructions that would seem to be indistinguishable by a deterministic parser without such a buffer. Thus, evidence that would seem to imply that natural language must be parsed nondeterministically can be explained away.

- Because the buffer can only contain a limited number of constituents, the power of these diagnostic rules is limited, leading to a principled distinction between sentences which are perceived as garden paths and those which are not. This explanation for why certain sentences cause garden paths is consistent with a series of informal experiments conducted during this research. Furthermore, this theory led to the counter-intuitive prediction that as simple a sentence as "Have the packages delivered tomorrow." would cause a garden path, a prediction which was confirmed by informal experiment (20 out of 40 people questioned were "led down the garden path" by this sentence).

In short, this one mechanism not only allows the formulation of rules of syntax that elegantly capture linguistic generalizations, it also provides the underpinnings for a psychological theory that seems to have high initial plausibility.

Another important source of power is the use of Chomsky's notion of a trace, a dummy constituent that allows the parser to indicate the "deep" underlying positions of constituents in an annotated surface structure. Especially important is the fact that a trace can be dropped into the buffer, thereby indicating its underlying position in a factorization of the terminal string without specifying its position in the underlying tree.

From this follows a simple formulation of passive which accounts for the phenomenon of "raising." The essence of the passive rule -- create a trace, bind it to the subject of the current S, drop it into the buffer -- is noteworthy in its simplicity. Again, the availability of the buffer yields a very simple solution to a seemingly complex linguistic phenomenon.

Moreover, we also have an explanation for the phenomena which underlie Chomsky's Specified Subject Constraint and Subjacency Principle. It is most interesting that the simple formulation of Passive presented here -- perhaps the simplest possible formulation of this rule within the framework of PARSIFAL -- behaves exactly as if these constraints were true; that is, that the formulation presented here, by itself and with no extraneous stipulations, leads to the behavior that these constraints attempt to specify.

There are several important areas for which an account must be given if the model presented is to resemble a full model of the syntactic recognition of natural language. The following is a list of several topics which have yet to be adequately investigated:

- Much further investigation of phenomena which I believe require extensive semantic processing, such as *conjunction*, or *PP attachment*, is still required. I believe that they will require mechanisms outside of the basic grammar interpreter presented here. It should be noted that Woods does not handle these two phenomena within the framework of his ATN mechanism itself, but rather he posits special purpose mechanisms to deal with them. I share his intuitions.

- I have not yet investigated *lexical ambiguity*, but rather have focussed on *structural ambiguity*. An ambiguity is structural when two different structures can be built up out of smaller constituents of the same given structure and type; an ambiguity is lexical when one word can serve as various parts of speech. While part of the solution to this very pervasive problem seems to involve notions of *discourse context*, I believe that much of the problem can be solved using the techniques presented in this paper, but have not yet investigated this problem.

- I have yet to provide an account in terms of this model for the psychological difficulty caused by *center embedded sentences* such as "The rat the cat the dog chased bit ate the cheese." For a recent theory of this phenomena which is consistent with the model presented here, see Cowper [1976]. Cowper's results have yet to be integrated into this model, however.

References

N. Chomsky, "On Wh-Movement," in A. Akmajian, P. Culicover, and T. Wasow (eds.), *Formal Syntax*, Academic Press,

1977.

E. A. Cowper, *Constraints on Sentence Complexity: A Model for Syntactic Processing*, PhD thesis, Brown University, 1976.

M. P. Marcus, *A Theory of Syntactic Recognition for Natural Language*, PhD thesis ,1977. To be published by MIT Press, 1978.

A. Newell and H. A. Simon, *Human Problem Solving*, Prentice-Hall, 1972.

V. R. Pratt, "Top-Down Operator Precedence," in the Proceedings of *The SIGACT/SIGPLAN Symposium on Principles of Programming Languages*, 1973.

T. Winograd, *Procedures as a Representation for Data in a Computer Program for Understanding Natural Language*, MIT Laboratory for Computer Science TR-84, 1971.

W. A. Woods, "Transition Network Grammars for Natural Language Analysis," *CACM*, Vol. 13, p. 591, 1970.

Cognition, 7 (1979) 217–283

Formal models of language learning*

STEVEN PINKER**
Harvard University

Abstract

Research is reviewed that addresses itself to human language learning by developing precise, mechanistic models that are capable in principle of acquiring languages on the basis of exposure to linguistic data. Such research includes theorems on language learnability from mathematical linguistics, computer models of language acquisition from cognitive simulation and artificial intelligence, and models of transformational grammar acquisition from theoretical linguistics. It is argued that such research bears strongly on major issues in developmental psycholinguistics, in particular, nativism and empiricism, the role of semantics and pragmatics in language learning, cognitive development, and the importance of the simplified speech addressed to children.

I. Introduction

How children learn to speak is one of the most important problems in the cognitive sciences, a problem both inherently interesting and scientifically promising. It is *interesting* because it is a species of the puzzle of induction: how humans are capable of forming valid generalizations on the basis of a finite number of observations. In this case, the generalizations are those that allow one to speak and understand the language of one's community, and are based on a finite amount of speech heard in the first few years of life. And language acquisition can claim to be a particularly *promising* example of this

*I am grateful to John Anderson, Roger Brown, Michael Cohen, Martha Danly, Jill de Villiers, Nancy Etcoff, Kenji Hakuta, Reid Hastie, Stephen Kosslyn, Peter Kugel, John Macnamara, Robert Matthews, Laurence Miller, Dan Slobin, and an anonymous reviewer for their helpful comments on earlier drafts of this paper. Preparation of this paper was supported in part by funds from the Department of Psychology and Social Relations, Harvard University; the author was supported by NRC and NSERC Canada Postgraduate Scholarships and by a Frank Knox Memorial Fellowship.
**Reprints may be obtained from the author, who is now at the Center for Cognitive Science, Massachusetts Institute of Technology, Cambridge, MA 02139.

puzzle, promising to the extent that empirical constraints on theory construction promote scientific progress in a given domain. This is because any plausible theory of language learning will have to meet an unusually rich set of empirical conditions. The theory will have to account for the fact that all normal children succeed at learning language, and will have to be consistent with our knowledge of what language is and of which stages the child passes through in learning it.

It is instructive to spell out these conditions one by one and examine the progress that has been made in meeting them. First, since all normal children learn the language of their community, a viable theory will have to posit mechanisms powerful enough to acquire a natural language. This criterion is doubly stringent: though the rules of language are beyond doubt highly intricate and abstract, children uniformly *succeed* at learning them nonetheless, unlike chess, calculus, and other complex cognitive skills. Let us say that a theory that can account for the fact that languages can be learned in the first place has met the *Learnability Condition.* Second, the theory should not account for the child's success by positing mechanisms narrowly adapted to the acquisition of a particular language. For example, a theory positing an innate grammar for English would fail to meet this criterion, which can be called the *Equipotentiality Condition.* Third, the mechanisms of a viable theory must allow the child to learn his language within the time span normally taken by children, which is in the order of three years for the basic components of language skill. Fourth, the mechanisms must not require as input types of information or amounts of information that are unavailable to the child. Let us call these the *Time* and *Input Conditions*, respectively. Fifth, the theory should make predictions about the intermediate stages of acquisition that agree with empirical findings in the study of child language. Sixth, the mechanisms described by the theory should not be wildly inconsistent with what is known about the cognitive faculties of the child, such as the perceptual discriminations he can make, his conceptual abilities, his memory, attention, and so forth. These can be called the *Developmental* and *Cognitive Conditions,* respectively.

It should come as no surprise that no current theory of language learning satisfies, or even addresses itself to, all six conditions. Research in psychology has by and large focused on the last three, the Input, Developmental, and Cognitive Conditions, with much of the research directed toward further specifying or articulating the conditions themselves. For example, there has been research on the nature of the speech available to children learning language (see Snow and Ferguson, 1977), on the nature of children's early word combinations (e.g., Braine, 1963), and on similarities between linguistic and cognitive abilities at various ages (e.g., Sinclair-de Zwart, 1969). Less often, there have been attempts to construct theoretical accounts for one or more of such findings, such as the usefulness of parental speech to children (e.g., Newport, Gleitman, and Gleitman, 1977), the reasons that words are put together the way they are in the first sentences (e.g., Brown, 1973; Schlesinger, 1971), and the ways that cognitive development interacts with linguistic development (e.g., Slobin, 1973). Research in linguistics that has addressed itself to language learning at all has articulated the Equipotentiality Condition, trying to distinguish the kinds of properties that are universal from those that are found only in particular languages (e.g., Chomsky, 1965, 1973).

In contrast, the attempts to account for the acquisition of language itself (the Learnability Condition) have been disappointingly vague. Language Acquisition has been attributed to everything from "innate schematisms" to "general multipurpose learning strategies"; it has been described as a mere by-product of cognitive development, of perceptual development, of motor development, or of social development; it has been said to draw on "input regularities", "semantic relations", "perceived intentions", "formal causality", "pragmatic knowledge", "action schemas", and so on. Whether the mechanisms implicated by a particular theory are adequate to the task of learning human languages is usually left unanswered.

There are, however, several bodies of research that address themselves to the Learnability criterion. These theories try to specify which learning mechanisms will succeed in which ways, for which types of languages, and with which types of input. A body of research called *Grammatical Induction*, which has grown out of mathematical linguistics and the theory of computation, treats languages as formal objects and tries to prove theorems about when it is possible, in principle, to learn a language on the basis of a set of sentences of the language. A second body of research, which has grown out of artificial intelligence and cognitive simulation, consists of attempts to program computers to acquire languages and/or to simulate human language acquisition. In a third research effort, which has grown out of transformational linguistics, a learning model capable of acquiring a certain class of transformational grammars has been described. However, these bodies of research are seldom cited in the psychological literature, and researchers in developmental psycholinguistics for the most part do not seem to be familiar with them. The present paper is an attempt to remedy this situation. I will try to give a critical review of these formal models of language acquisition, focusing on their relevance to human language learning.

There are two reasons why formal models of language learning are likely to contribute to our understanding of how children learn to speak, even if none of the models I will discuss satisfies all of our six criteria. First of all, a theory that is powerful enough to account for the *fact* of language acquisition may be a more promising first approximation of an ultimately viable theory than one that is able to describe the *course* of language acquisition, which has been the traditional focus of developmental psycholinguistics. As the reader shall see, the Learnability criterion is extraordinarily stringent, and it becomes quite obvious when a theory cannot pass it. On the other hand, theories concerning the mechanisms responsible for child language per se are notoriously underdetermined by the child's observable linguistic behavior. This is because the child's knowledge, motivation, memory, and perceptual, motor, and social skills are developing at the same time that he is learning the language of his community.

The second potential benefit of formal models is the explicitness that they force on the theorist, which in turn can clarify many conceptual and substantive issues that have preoccupied the field. Despite over a decade and a half of vigorous debates, we still do not know that sort of a priori knowledge, if any, is necessary to learn a natural language; nor whether different sorts of input to a language learner can make his task easy or difficult, possible or impossible; nor how semantic information affects the learning of the syntax of a language. In part this is because we know so little about the mechanisms of language learning, and so do not know how to translate vague terms such as "semantic information" into the information structures that play a causal role in the acquisition process. Developing explicit, mechanistic theories of language learning may be the only way that these issues can be stated clearly enough to evaluate. It seems to be the consensus in other areas of cognitive psychology that mechanistic theories have engendered enormous conceptual advances in the understanding of mental faculties, such as long-term memory (Anderson and Bower, 1973), visual imagery (Kosslyn and Schwartz, 1977), and problem solving (Newell and Simon, 1973).

The rest of the paper is organized into eight sections. In Section II, I will introduce the vocabulary and concepts of mathematical linguistics, which serve as the foundation for research on language learnability. Sections III and IV present E. Gold's seminal theorems on language learnability, and the subsequent research they inspired. Section V describes the so-called "heuristic" language learning models, several of which have been implemented as computer simulations of human language acquisition. Sections VI and VII discuss the rationale for the "semantic" or "cognitive" approach to language learning, focusing on John R. Anderson's computer simulation of a semantics-based learner. Section VIII describes a model developed by Henry Hamburger, Kenneth Wexler, and Peter Culicover that is capable of learning transformational grammars for languages. Finally, in Section IX, I discuss the implications of this research for developmental psycholinguistics.

II. Formal Models of Language

In this section I define the elementary concepts of mathematical linguistics found in discussions of language learnability. More thorough accounts can be found in Gross (1972) and in Hopcroft and Ullman (1969).

Languages and Grammars

To describe a language in mathematical terms, one begins with a finite set of *symbols*, or a *vocabulary*. In the case of English, the symbols would be English words or morphemes. Any finite sequence of these symbols is called a *string*, and any finite or infinite collection of strings is called a *language*. Those strings in the language are called *sentences;* the strings not in the language are called *non-sentences*.

Languages with a finite number of sentences can be exhaustively described simply by listing the sentences. However, it is a celebrated observation that natural and computer languages are infinite, even though they are used by beings with finite memory. Therefore the languages must have some finite characterization, such as a recipe or program for specifying which sentences are in a given language. A *grammar*, a set of rules that generates all the sentences in a language, but no non-sentences, is one such characterization. Any language that can be generated by a set of rules (that is, any language that is not completely arbitrary) is called a *recursively enumerable* language.

A grammar has four parts. First of all, there is the vocabulary, which will now be called the *terminal vocabulary* to distinguish it from the second component of the grammar, called the *auxiliary vocabulary*. The auxiliary vocabulary consists of another finite set of symbols, which may not appear in sentences themselves, but which may act as stand-ins for groups of symbols, such as the English "noun", "verb", and "prepositional phrase". The third component of the grammar is the finite set of *rewrite rules*, each of which replaces one sequence of symbols, whenever it occurs, by another sequence. For example, one rewrite rule in the grammar for English replaces the symbol "noun phrase" by the symbols "article noun"; another replaces the symbol "verb" by the symbol "grow". Finally, there is a special symbol, called the *start symbol*, usually denoted S, which initiates the sequence of rule operations that generate a sentence. If one of the rewrite rules can rewrite the "S" as another string of symbols it does so; then if any rule can replace part or all of that new string by yet another string, it follows suit. This procedure continues, one rule taking over from where another left off, until no auxiliary symbols remain, at which point a sentence has been generated. The language is simply the set of all strings that can be generated in this way.

Classes of Languages

There is a natural way to subdivide grammars and the languages they generate into classes. First, the grammars of different sorts of languages make use of different types of rewrite rules. Second, these different types of languages require different sorts of computational machinery to produce or recognize their sentences, using various amounts of working memory and various ways of accessing it. Finally, the theorems one can prove about language and grammars tend to apply to entire classes of languages, delineated in these ways. In particular, theorems on language learnability refer to such classes, so I will discuss them briefly.

These classes fall into a hierarchy (sometimes called the *Chomsky hierarchy*), each class properly containing the languages in the classes below it. I have already mentioned the largest class, the recursively enumerable languages, those that have grammars that generate all their member sentences. However, not all of these languages have a *decision procedure,* that is, a means of determining whether or not a given string of symbols is a sentence in the language. Those that have decision procedures are called *decidable* or *recursive* languages. Unfortunately, there is no general way of knowing whether a recursively enumerable language will turn out to be decidable or not. However, there is a very large subset of the decidable languages, called the *primitive recursive* languages, whose decidability *is* known. It is possible to *enumerate* this class of languages, that is, there exists a finite procedure called a *grammar-grammar* capable of listing each grammar in the class, one at a time, without including any grammar not in the class. (It is not hard to see why this is impossible for the class of decidable languages: one can never be sure whether a given language is decidable or not.)

The primitive recursive languages can be further broken down by restricting the form of the rewrite rules that the grammars are permitted to use. *Context-sensitive* grammars contain rules that replace a single auxiliary symbol by a string of symbols whenever that symbol is flanked by certain neighboring symbols. *Context-free* grammars have rules that replace a single auxiliary symbol by a string of symbols regardless of where that symbol occurs. The rules of *finite state* grammars may replace a single auxiliary symbol only by another auxiliary symbol plus a terminal symbol; these auxiliary symbols are often called *states* in discussions of the corresponding sentence-producing machines. Finally, there are grammars that have no auxiliary symbols, and hence these grammars can generate only a finite number of strings altogether. Thus they are called *finite cardinality* grammars. This hierarchy is summarized in Table 1, which lists the classes of languages from most to least inclusive.

Table 1. *Classes of Languages*

Class	Learnable from an informant?	Learnable from a text?	Contains natural languages?
Recursively Enumerable	no	no	yes*
Decidable (Recursive)	no	no	?
Primitive Recursive	yes	no	?
Context-Sensitive	yes	no	?
Context-Free	yes	no	no
Finite State	yes	no	no
Finite Cardinality	yes	yes	no

*by assumption.

Natural Languages

Almost all theorems on language learnability, and much of the research on computer simulations of language learning, make reference to classes in the Chomsky hierarchy. However, unless we know where natural languages fall in the classification, it is obviously of little psychological interest. Clearly, natural languages are not of finite cardinality; one can always produce a new sentence by adding, say, "he insists that" to the beginning of an old sentence. It is also not very difficult to show that natural languages are not finite state: as Chomsky (1957) has demonstrated, finite state grammars cannot generate sentences with an arbitrary number of embeddings, which natural languages permit (e.g., "he works", "either he works or he plays", "if either he works or he plays, then he tires", "since if either he...", etc.). It is more difficult, though not impossible, to show that natural languages are not context-free (Gross, 1972; Postal, 1964). Unfortunately, it is not clear how much higher in the hierarchy one must go to accomodate natural languages. Chomsky and most other linguists (including his opponents of the "generative semantics" school) use *transformational* grammars of various sorts to describe natural languages. These grammars generate bracketed strings called *deep structures,* usually by means of a context-free grammar, and then, by means of rewrite rules called *transformations,* permute, delete, or copy elements of the deep structures to produce sentences. Since transformational grammars are constructed and evaluated by a variety of criteria, and not just by the ability to generate the sentences of a language, their place in the hierarchy is uncertain. Although the matter is by no means settled, Peters and Ritchie (1973) have persuasively argued that the species of transformational grammar necessary for generating natural languages can be placed in the context-sensitive class, as Chomsky conjectured earlier (1965, p. 61). Accordingly, in the sections following, I will treat the set of all existing and possible human languages as a subset of the context-sensitive class.

III. Grammatical Induction: Gold's Theorems

Language Learning as Grammatical Induction

Since people presumably do not consult an internal list of the sentences of their language when they speak, knowing a particular language corresponds to knowing a particular set of rules of some sort capable of producing and recognizing the sentences of that language. Therefore learning a language consists of inducing that set of rules, using the language behavior of the community as evidence of what the rules must be. In the paragraphs following I will treat such a set of rules as a grammar. This should not imply the belief that humans mentally execute rewrite rules one by one before uttering a sentence. Since every grammar can be translated into a left-to-right sentence producer or recognizer, "inducing a grammar" can be taken as shorthand for acquiring the ability to produce and recognize just those sentences that the grammar generates. The advantage of talking about the grammar is that it allows us to focus on the process by which a particular language is learned (i.e., as opposed to some other language), requiring no commitment as to the detailed nature of the production or comprehension process in general (i.e., the features common to producers or recognizers for *all* languages).

The most straightforward solution to this induction problem would be to find some algorithm that produces a grammar for a language given a sample of its sentences, and then to attribute some version of this algorithm to the child. This would also be the most *general* conceivable solution. It would not be necessary to attribute to the child any a priori knowledge about the particular type of language that he is to learn (except perhaps that it falls into one of the classes in the Chomsky hierarchy, which could correspond to some putative memory or processing limitation). We would not even have to attribute to the child a special language acquisition faculty. Since a grammar is simply one way of talking about a computational procedure or set of rules, an algorithm that could produce a grammar for a language from a sample of sentences could also presumably produce a set of rules for a different sort of data (appropriately encoded), such as rules that correctly classify the exemplars and non-exemplars in a laboratory concept attainment task. In that case it could be argued that the child learned language via a general induction procedure, one that simply "captured regularity" in the form of computational rules from the environment.

Unfortunately, the algorithm that we need does not exist. An elementary theorem of mathematical linguistics states that there are an infinite number of different grammars that can generate any finite set of strings. Each grammar will make different predictions about the strings not in the set. Consider the sample consisting of the single sentence "the dog barks". It could have been taken from the language consisting of: 1) all three-word strings; 2) all article-noun-verb sequences; 3) all sentences with a noun phrase; 4) that sentence alone; 5) that sentence plus all those in the July 4, 1976 edition of the New York Times; as well as 6) all English sentences. When the sample consists of more than one sentence, the class of possible languages is reduced but is still infinitely large, as long as the number of sentences in the sample is finite. Therefore it is impossible for *any* learner to observe a finite sample of sentences of a language and always produce a correct grammar for the language.

Language Identification in the Limit

Gold (1967) solved this problem with a paradigm he called *language identification in the limit.* The paradigm works as follows: time is divided into discrete trials with a definite starting point. The teacher or environment "chooses" a language (called the *target language*) from a predetermined class in the hierarchy. At each trial, the learner has access to a single string. In one version of the paradigm, the learner has access sooner or later to all the sentences in the language. This sample can be called a *text*, or *positive information presentation.* Alternately, the learner can have access to both grammatical sentences and ungrammatical strings, each appropriately labelled. Because this is equivalent to allowing the learner to receive feedback from a native informant as to whether or not a given string is an acceptable sentence, it can be called *informant* or *complete information presentation*. Each time the learner views a string, he must guess what the target grammar is. This process continues forever, with the learner allowed to change his mind at any time. If, after a finite amount of time, the learner always guesses the same grammar, and if that grammar correctly generates the target language, he is said to have *identified the language in the limit.* Is is noteworthy that by this definition the learner can never know when or even whether he has succeeded. This is because he can never be sure that future strings will not force him to change his mind.

Gold, in effect, asked: How well can a completely general learner do in this situation? That is, are there any classes of languages in the hierarchy whose members can all be identified in the limit? He was able to prove that language learnability depends on the information available: if both sentences and non-sentences are available to a learner (informant presentation), the class of primitive recursive languages, and all its subclasses (which include the natural languages) are learnable. But if only sentences are available (text presentation), *no* class of languages other than the finite cardinality languages is learnable.

The proofs of these theorems are straightforward. The learner can use a maximally general strategy: he enumerates every grammar of the class, one at a time, rejecting one grammar and moving on to the next whenever the grammar is inconsistent with any of the sample strings (see Figure 1). With informant presentation, any incorrect grammar will eventually be rejected when it is unable to generate a sentence in the language, or when it generates a string that the informant indicates is not in the language. Since the correct grammar, whatever it is, has a definite position in the enumeration of grammars, it will be hypothesized after a finite amount of time and there will never again be any reason to change the hypothesis. The class of primitive recursive languages is the highest learnable class because it is the highest class whose languages are decidable, and whose grammars and decision procedures can be enumerated, both necessary properties for the procedure to work.

The situation is different under text presentation. Here, finite cardinality languages are trivially learnable – the learner can simply guess that the language is the set of sentences that have appeared in the sample so far, and when every sentence in the language has appeared at least once, the learner will be correct. But say the class contains all finite languages and at least one infinite language (as do classes higher than finite cardinality). If the learner guesses that the language is just the set of sentences in the sample, then when the target language is infinite the learner will have to change his mind an infinite number of times. But if the learner guesses only infinite languages, then when the target language is finite he will guess an incorrect language and will never be forced to change his mind. If non-sentences were also available, any overgeneral grammar would have been rejected when a sentence that it was capable of generating appeared, marked as a non-sentence. As Gold put it, "the problem with text is that if you guess too large a language, the sample will never tell you you're wrong".

Implication of Gold's theorems

Do children learn from a text or an informant? What evidence we have strongly suggests that children are not usually corrected when they speak ungrammatically, and when they are corrected they take little notice (Braine, 1971; Brown and Hanlon, 1970; McNeill, 1966). Nor does the child seem to have access to more indirect evidence about what is not a sentence. Brown and Hanlon (1970) were unable to discern any differences in how parents responded to the grammatical versus the ungrammatical sentences of their children. Thus the child seems to be in a text situation, in which Gold's

Figure 1. *A flowchart for Gold's enumeration procedure. Note that there is no "stop" symbol; the learner samples strings and guesses grammars forever. If the learner at some point enters loop "A" and never leaves it, he has identified the language in the limit.*

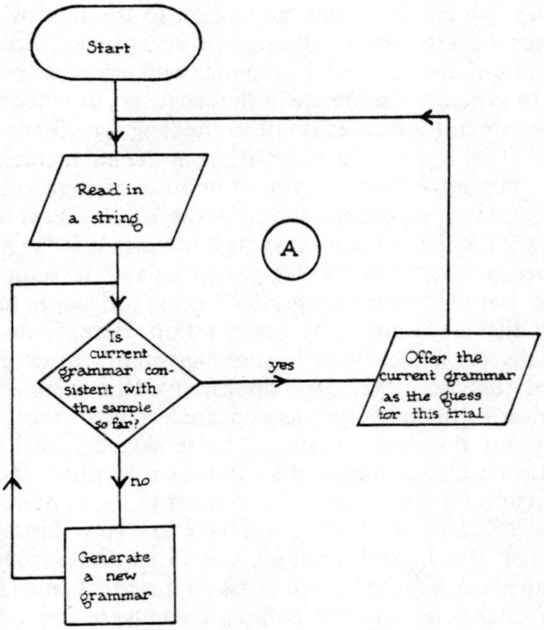

learner must fail. However, all other models must fail in this situation as well – there can be no learning procedure more powerful than the one that enumerates all the grammars in a class.

An even more depressing result is the astronomical amount of time that the learning of most languages would take. The enumeration procedure, which gives the learner maximum generality, exacts its price: the learner must test astronomically large numbers of grammars before he is likely to hit upon the correct one. For example, in considering all the finite state grammars that use seven terminal symbols and seven auxiliary symbols (states), which the learner must do before going on to more complex grammars, he must test over a googol (10^{100}) candidates. The learner's predicament is reminiscent of Jorge Luis Borges's "librarians of Babel", who search a vast library containing books with all possible combinations of alphabetic characters for the book that clarifies the basic mysteries of humanity. Nevertheless, Gold has proved that *no* general procedure is uniformly faster than his learner's enumeration procedure. This is a consequence of the fact that an infinite number of grammars is consistent with any finite sample. Imagine a rival procedure of any sort that correctly guessed a certain language at an earlier trial than did the enumeration procedure. In that case the enumeration procedure must have guessed a different language at that point. But the sample of sentences up to that point could have been produced by many different grammars, including the one that the enumeration procedure mistakenly guessed. If the target language had happened to be that other language, then at that time the enumeration procedure would have been correct, and its rival incorrect. Therefore, for every language that a rival procedure identifies faster than the enumeration procedure, there is a language for which the reverse is true. A corollary is that every form of enumeration procedure (i.e., every order of enumeration) is, on the whole, equivalent in speed to every other one.

Gold's model can be seen as an attempt to construct some model, any model, that can meet the Learnability Condition. But Gold has shown that even if a model is unhindered by psychological considerations (i.e., the Developmental, Cognitive, and Time Conditions), learnability cannot be established (that is, unless one flagrantly violates the Input Condition by requiring that the learner receive negative information). What's more, no model can do better than Gold's, whether or not it is designed to model the child. However, since children presumably do have a procedure whereby they learn the language of their community, there must be some feature of Gold's learning paradigm itself that precludes learnability, such as the criterion for success or access to information. In Section IV, I will review research inspired by Gold's theorems that tries to establish under what conditions language learnability from a sample of sentences is possible.

IV. Grammatical Induction: Other Results

Grammatical Induction from a Text

This section will describe four ways in which languages can be learned from samples of sentences. One can either restrict the order of presentation of the sample sentences, relax the success criterion, define a statistical distribution over the sample sentences, or constrain the learner's hypotheses.

Order of sentence presentation

In Section III it was assumed that the sample strings could be presented to the learner in any order whatsoever. Gold (1967) proved that if it can be known that the sample sentences are ordered in some way as a function of time, then all recursively enumerable languages are learnable from a positive sample. Specifically, it is assumed that the "teacher" selects the sentence to be presented at time *t* by consulting a *primitive recursive function* that accepts a value of *t* as input and produces a sentence as output. Primitive recursive functions in this case refer to primitive recursive grammars that associate each sentence in the language with a unique natural number. Like primitive recursive grammars, they can be enumerated and tested, and the learner merely has to identify in the limit which function the teacher is using, in the same way that the learner discussed in Section III (and illustrated in Figure 1) identified primitive recursive grammars. This is sufficient to generate the sentences in the target language (although not necessarily sufficient to recognize them). Although it is hard to believe that every sentence the child hears is uniquely determined by the time that has elapsed since the onset of learning, we shall see in Section VI how a similar learning procedure allows the child to profit from semantic information.

Another useful type of sequencing is called *effective approximate ordering* (Feldman, 1972). Suppose that there was a point in time by which every grammatical sentence of a given length or less had appeared in the sample. Suppose further that the learner can calculate, for any length of sentence, what that time is. Then, at that point, the learner can compute all the strings of that length or less that are *not* in the language, namely, the strings that have not yet appeared. This is equivalent to having access to non-sentences; thus learning can occur. Although it is generally true that children are exposed to longer and longer sentences as language learning proceeds (see Snow and Ferguson, 1977), it would be difficult to see how they could take advantage of this procedure, since there is never a point at which short sentences are excluded altogether. More generally, though, it is possible that the fairly systematic changes in the speech directed to the developing child (see Snow and Ferguson, 1977) contain information that is useful to the task of inducing a grammar, as Clark (1973) and Levelt (1973) have suggested. For example, if it were true that sentences early in the sample were always generated by fewer rules or needed fewer derivational steps than sentences later in the sample, perhaps a learner could reject any candidate grammar that used more rules or steps for the earlier sentences than for the later ones. However, the attempts to discern such an ordering in parental speech have been disappointing (see Newport *et al.*, 1977) and it remains to be seen whether the speech directed to the child is sufficiently well-ordered with respect to this or any other syntactic dimension for an order-exploiting strategy to be effective. I will discuss this issue in greater depth in Section IX.

Relaxing the success criterion

Perhaps the learner should not be required to identify the target language exactly. We can, for example, simply demand that the learner *approach* the target language, defining approachability as follows (Biermann and Feldman, 1972; Feldman, 1972): 1) Every sentence in the sample is eventually included in the language guessed by the learner; 2) any incorrect grammar will at some point be permanently rejected; and 3) the correct grammar will be guessed an infinite number of times (this last condition defining *strong approachability*). The difference between strong approachability and identifiability is that, in the former case, we do not require the learner to stick to the correct grammar once he has guessed it. Feldman has shown that the class of primitive recursive languages is approachable in the limit from a sample of sentences.

The success criterion can also be weakened so as to allow the learner to identify a language that is an *approximation* of the target language. Wharton (1974) proposes a way to define a *metric* on the set of languages that use a given terminal vocabulary, which would allow one to measure the degree of similarity between any two languages. What happens, then, if the learner is required to identify any language whatsoever that is of a given degree of similarity to the target language? Wharton shows that a learner can approximate any primitive recursive language to any degree of accuracy using only a text. Furthermore, there is always a degree of accuracy that can be imposed on the learner that will have the effect of making him choose the target language exactly. However, there is no way of knowing how high that level of accuracy must be (if there were, Gold's theorem would be false). Since it is unlikely that the child ever duplicates exactly the language of his community, Wharton and Feldman have shown that a Gold-type learner *can* meet the Learnability condition if it is suitably redefined.

There is a third way that we can relax the success criterion. Instead of asking for the *only* grammar that fits the sample, we can ask for the *simplest* grammar from among the infinity of candidates. Feldman (1972) defines the *complexity* of a grammar, given a sample, as a joint function (say, the sum) of the *intrinsic complexity* of the grammar (say, the number of rewrite rules) and the *derivational complexity* of the grammar with respect to the sample (say, the average number of steps needed to generate the sample sentences). He then describes a procedure which enumerates grammars in order of increasing intrinsic complexity, thereby finding the simplest grammar that is consistent with a positive sample. However it is important to point out that such a procedure will *not* identify or even strongly approach the target language when it considers larger and larger samples. It is easy to see why not. There is a grammar of finite complexity that will generate every possible string from a given vocabulary. If the target language is more complex than this *universal grammar*, it will never even be considered, because the universal grammar will always be consistent with the text and occurs earlier in the enumeration than the target grammar (Gold, 1967). Thus equipping the child with Occam's Razor will not help him learn languages.

Bayesian grammar induction

If a grammar specifies the probabilities with which its rules are to be used, it is called a *stochastic* grammar, and it will generate a sample of sentences with a predictable statistical distribution. This constitutes an additional source of information that a learner can exploit in attempting to identify a language.

Horning (1969) considers grammars whose rewrite rules are applied with fixed probabilities. It is possible to calculate the *probability of a sentence given a grammar* by multiplying together the probabilities of the rewrite rules used to generate the sentence. One can calculate the *probability of a sample of sentences* with respect to the grammar in the same way. In Horning's paradigm, the learner also knows the *a priori probability* that any grammar will have been selected as the target grammar. The learner enumerates grammars in approximate order of decreasing a priori probability, and calculates the probability of the sample with respect to each grammar. He then can use the equivalent of Bayes's Theorem to determine the *a posteriori probability* of a grammar given the sample. The learner always guesses the grammar with the highest a posteriori probability. Horning shows how an algorithm of this sort can converge on the most probable correct grammar for any text.

Constraining the hypothesis space

In its use of a priori knowledge concerning the likelihood that certain types of languages will be faced, Horning's procedure is like a stochastic version of Chomsky's (1965) abstract description of a language acquisition device. Chomsky, citing the infinity of grammars consistent with any finite sample, proposes that there is a *weighting function* that represents the child's selection of hypothesis grammars in the face of a finite sample. The weighting function assigns a "scattered" distribution of probabilities to grammars, so that the candidate grammars that incorporate the basic properties of natural languages are assigned high values, while those (equally correct) grammars that are not of this form are assigned extremely low or zero values. In weighting grammars in this way, the child is making assumptions about the probability that he will be faced with a particular type of language, namely, a natural language. If his weighting function is so constructed that only one highly-weighted grammar will be consistent with the sample once it has grown to a certain size, then learnability from a text is possible. To take an artificial example, if the child gave high values only to a set of languages with completely disjoint vocabularies (e.g., Hindi, Yiddish, Swahili, etc.), then even a single sentence would be sufficient evidence to learn a language. However, in Gold's paradigm, a learner that assigned weights of zero to some languages would fail to learn those languages should they be chosen as targets. But in the case of the child, this need not be a concern. We need only show how the child is able to learn human languages; it would not be surprising if the child was thereby rendered unable to learn various gerrymandered or exotic languages.

There are two points to be made about escaping Gold's conclusions by constraining the learner's hypothesis set. First, we lose the ability to talk about a general rule-inducing strategy constrained only by the computation-theoretic "lines of fracture" separating classes of languages. Instead, we are committed to at least a weak form of nativism, according to which "the child approaches the data with the presumption that they are drawn from a language of an antecedently well-defined type" (Chomsky, 1965, p. 27). Second, we are begging the question of whether the required weighting function exists, and what form it should take. It is not sufficient simply to constrain the learner's hypotheses, even severely. Consider Figure 2, a Venn diagram representing the set of languages assigned high a priori values (Circle A) and the set of languages that are consistent with the sample at a given point in the learning process (Circle B). To ensure learnability, the set of languages in the intersection between the two circles must shrink to a single member as more and more of the sample is considered. Circle B must not encompass Circle A completely, nor coincide with it, nor overlap with it to a large degree (a priori set too broad); nor can it be disjoint from it (a priori set too narrow). Specifying an a priori class of languages with these properties corresponds to the *explanatory adequacy* requirement in transformational linguistics. In Section VIII I shall examine an attempt to prove learnability in this way.

We have seen several ways to achieve learnability, within the constraint that only grammatical sentences be available to the learner. However, in

Figure 2. *Achieving learnability by constraining the learner's hypothesis set.*

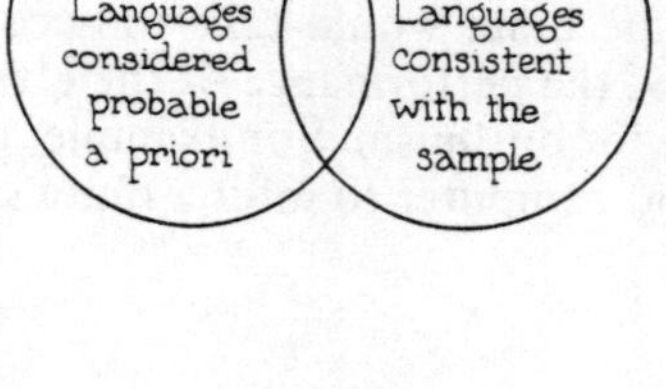

severing one head of this hydra, we see that two more have grown in its place. The learning procedures discussed in this section still require astronomical amounts of time. They also proceed in an implausible manner, violating both the Developmental and the Cognitive criteria. First, children do not adopt and jettison grammars in one piece; they seem to add, replace, and modify individual rules (see Brown, 1973). Second, it is unreasonable to suppose that children can remember every sentence they have heard, which they must do to test a grammar against "the sample". In the next paragraphs I will review some proposals addressed to the Time Condition, and in Section V, research addressed more directly to the Developmental and Cognitive Conditions.

Reducing Learning Time

Efficient enumeration

The learners we have considered generate grammars rather blindly, by using a grammar-grammar that creates rules out of all possible combinations of symbols. This process will yield many grammars that can be shown to be undesirable even before they are tested against the sample. For example, grammars could be completely equivalent to other grammars except for the names of their auxiliary symbols; they could have some rules that grind to a halt without producing a sentence, and others that spin freely without affecting the sentence that the other rules produce; they could be redundant or ambiguous, or lack altogether a certain word known to appear in the language. Perhaps our estimate of the enormous time required by an enumeration procedure is artificially inflated by including various sorts of silly or bad grammars in the enumeration. Wharton (1977) has shown that if a learner had a "quality control inspector" that rejected these bad grammars before testing them against the sample, he could save a great deal of testing time. Furthermore, if the learner could reject not one but an entire set of grammars every time a single grammar failed a quality control test or was incompatible with the sample, he could save even more time, a second trick sometimes called *grammatical covering* (Biermann and Feldman, 1972; Horning, 1969; Wharton, 1977; Van der Mude and Walker, 1978). Horning and Wharton have implemented various enumeration techniques as computer programs in order to estimate their efficiency, and have found that these "quality control" and "covering" strategies are faster than blind enumeration by many orders of magnitude. Of course, there is no simple way to compare computation time in a digital computer with the time the brain would take to accomplish an analogous computation, but somehow, the performance of the efficient enumeration algorithms leaves little cause for optimism. For example, these techniques in one case allowed an IBM 360 computer to infer a finite state grammar with two auxiliary symbols and two terminal symbols after several minutes of computation. However natural languages have on the order of 10–100 auxiliary symbols, and in general the number of grammars using n auxiliary symbols grown as 2^{n^3}. Clearly, stronger medicine is needed.

Ordering by a priori probability

The use of an a priori probability metric over the space of hypothesis grammars, which allowed Horning's procedure to learn a language without an informant, also reduces the average time needed for identification. Since Horning's learner must enumerate grammars in approximate order of decreasing a priori probability, the grammars most likely to have been chosen as targets are also the ones first hypothesized. Thus countless unlikely grammars need never be considered. Similarly, if the learner could enumerate the "natural grammars" before the "unnatural" ones, he would learn more quickly than he would if the enumeration order was arbitrary. Unfortunately, still not quickly enough. Despite its approximate ordering by a priori probability, Horning's procedure requires vast amounts of computation in learning even the simplest grammars; as he puts it, "although the enumeration procedure... is formally optimal, its Achilles's heal is efficiency". Similarly, the set of natural languages is presumably enormous, and more or less equiprobable as far as the neonate is concerned; thus even enumerating only the natural languages would not be a shortcut to learning. In general, the problem of learning by enumeration within a reasonable time bound is likely to be intractable. In the following section I describe the alternative to enumeration procedures.

V. Heuristic Grammar Construction

Algorithms and Heuristics for Language Learning

Like many other computational problems, language learning can be attempted by *algorithmic* or *heuristic* techniques (see Newell and Simon, 1973). The enumerative procedures we have been discussing are algorithmic in that they guarantee a solution in those cases where one exists.[1] Unfortunately they are also prohibitively time-consuming and wildly implausible as models of children. Heuristic language learning procedures, on the other hand, may hold greater promise in these regards. They differ from the enumerative procedures in two respects. First, the grammars are not acquired and discarded whole, but are built up rule by rule as learning proceeds. Second, the input sentences

[1] Strictly speaking, they are not "algorithms" in the usual sense of effective procedures, since they do not compute a solution and then halt, but compute an infinite series of guesses.

do not just contribute to the binary decision of whether or not a grammar is consistent with the sample, but some property possessed by sample sentences is used as a hint, guiding the process of rule construction. Thus heuristic language learning procedures are prima facie candidates for theories of human language acquisition. They acquire language piecemeal, as children do (Brown, 1973), and they have the potential for doing so in a reasonable amount of time, drawing their power from the exploitation of detailed properties of the sample sentences instead of the exhaustive enumeration of a class of grammars.

Many heuristic procedures for acquiring rules of finite state and context-free grammars have been proposed (for examples see Biermann and Feldman, 1972; Fu and Booth, 1975; and Knobe and Knobe, 1977). The following example should give the reader the flavor of these procedures. Solomonoff (1964) suggested a heuristic for inferring recursive context-free rules from a sample, in this case with the aid of an informant to provide negative information. *Recursive rules* (not to be confused with the "recursive grammars" discussed earlier) rewrite a symbol as a string containing the original symbol, i.e., rules of the form A → BAC. They are important because they can be successively applied an infinite number of times, giving the grammar the power to generate an infinite number of sentences. An English example might rewrite the symbol for an adjective "A" as the sequence "very A". Solomonoff's learner would delete flanking substrings from an acceptable sample string, and ascertain whether the remaining string was grammatical. If so, he would sandwich that string repetitively with the substrings that were initially deleted, testing each multi-layered string for grammaticality. If they were all grammatical, a recursive rule would be constructed. For example, given the string XYZ in the original sample, the learner would test Y, then if successful, XXYZZ, XXXYZZZ, and so on. If a number of these were acceptable, the rules A → XAZ and A → Y would be coined.

Caveats concerning heuristic methods

Several points must be made about heuristic methods, lest it appear that in trading enumerative procedures for heuristic ones one gets something for nothing. First, as I have mentioned, no procedure can do better than Gold's, either in overall success or in speed, when the set of target languages consists of one of the classes in the Chomsky hierarchy. If the heuristic procedures succeed in learning some languages in a reasonable amount of time, they must take large amounts of time or fail altogether for many other ones. Thus we must again abandon the notion of a general rule learner who is constrained only by the sorts of processing or memory limits that implicitly define classes of computational procedures. Second, heuristic procedures commit the learner to assumptions not only about the target languages, but about the sentences that find their way into the sample. That is, the procedures could be fooled by using unusual or unrepresentative sets of sentences as the basis for rule construction. Consider Solomonoff's heuristic. If the target language permitted no more than three levels of embedding, the learner would have erred by constructing a rule that permitted an infinite number of embeddings. On the other hand, if the sample was a text lacking the multiply-embedded sentences that in Solomonoff's case were provided by the informant, the learner would have erred by constructing the overly-narrow rule which simply generates the original string XYZ. In the natural language case, of course, these problems are less worrisome. Not only will the child do well by "assuming" that the target language is a member of a relatively constrained set (viz., the natural languages), but he will do well in "assuming" that his sample will be a well-defined subset of the target language, not some capricious collection of sentences. Whatever its exact function may turn out to be, the dialect of speech addressed to children learning language has been found to have indisputably consistent properties across different cultures and learning environments (see Snow and Ferguson, 1977).

However, one difference between algorithmic and heuristic procedures advises caution. Whereas enumeration procedures guarantee success in learning an entire language, each heuristic at best gives hope for success in acquiring some piece of the grammar. But one can never be sure that a large collection of heuristics will be sufficient to acquire all or even a significant portion of the language. Nor can one know whether a heuristic that works well for simple constructions or small samples (e.g., the research on the construction of context-free and finite state rules cited earlier) will continue to be successful when applied to more complex, and hence more realistic tasks. In other words, in striving to meet the Developmental, Cognitive, or Time Conditions, we may be sacrificing our original goal, Learnability. The research to be discussed in the remainder of this section illustrates this tradeoff.

The computer simulation of heuristic language acquisition

Since one cannot prove whether or not a set of heuristics will succeed in learning a language, several investigators have implemented heuristic strategies as computer programs in order to observe how effective the heuristics turn out to be when they are set to the task of acquiring rules from some sample. Constructing a learning model in the form of a computer program also gives the designer the freedom to tailor various aspects of the program to certain characteristics of human language learners, known or hypothesized. Thus the theorist can try to meet several of our conditions, and is in a better position to submit the model as a *theory* of human language acquisition.

Kelley's Program
Kalon Kelley (1967) wrote the first computer simulation of language acquisition. His priority was to meet the Developmental criterion, so his program was designed to mimic the very early stages of the child's linguistic development.

Kelley's program uses a heuristic that we may call *word-class position learning.* It assumes that the words of a language fall into classes, and that each class can be associated with an absolute or relative ordinal position in the sentence. At the time that Kelley wrote the program, an influential theory ("pivot grammar", Braine, 1963) asserted that early child language could be characterized in this way. As an example of how the heuristic works, consider the following sentences:

1. (a) He smokes grass.
   (b) He mows grass.
   (c) She smokes grass.
   (d) She smokes tobacco.

A learner using the word-class position heuristic would infer that "he" and "she" belong to one word class, because they both occur as the first word of the sentence (or perhaps because they both precede the word "smokes"); similarly, "smokes" and "mows" can be placed in another word class, and "grass" and "tobacco" can be placed into a third. The learner can also infer that a sentence can be composed of a word from the first class, followed by a word from the second class, followed by a word from the third class. A learner who uses this heuristic can now produce or recognize eight sentences after having heard only four.

Kelley's program is equipped with three sets of hypotheses, corresponding to the periods in which the child uses one-, two-, and three-word utterances, respectively. The program advances from one stage to the next at arbitrary moments designated by the programmer. Its first strategy is to count the number of occurrences of various "content" words in the sample sentences; these words are explicitly tagged as content words by the "adult". It retains the most frequent ones, and can produce them as one-word sentences. In its second stage, it looks for two word classes, called "things" and "actions". Kelley assumes that children can tell whether a word refers to a thing or an action by the non-linguistic context in which it was uttered. To model this assumption, his program guesses arbitrarily that a particular word is in one or the other class, and has access to its "correct" classification. If the guess is correct, it is strengthened as a hypothesis; if incorrect, it is weakened. At the same time, the program tabulates the frequency with which the word classes precede or follow each other, thereby hypothesizing rules that generate the frequent sequences of word classes (e.g., S → thing action; S → thing thing). Like the hypotheses that assign words to classes, these rules increase or decrease in strength according to how frequently they are consistent with the input sentences. In its third state, the program retains its two word classes, and adds a class consisting of two-item sequences (e.g., thing-action) from the previous stage. As before, it accumulates evidence regarding which of these classes can occur in which sentence positions relative to one another, thereby hypothesizing rules that generate frequent sequences of classes (e.g., S → thing-action thing). A separate feature of the program is its ability to learn the "functions" of the individual sentence constituents, such as which is the subject and which is the predicate. As before, the program learns these by making rather arbitrary guesses and checking them against the "correct" answer, to which it has access.

An evaluation

Though Kelley's program was a brave first attempt, it is unsatisfactory on many counts. For one thing, children seem unaffected by the frequency of syntactic forms in adult speech (Brown, 1973), whereas frequency of input forms is the very life-blood of Kelley's learning procedure. Second, the role of the "correct" structural descriptions of sentences given to the program is puzzling. Kelley intends them to be analogous to the child's perception that a word uttered in the context of some action is an "action" word, that a part of a sentence denoting an object being attended to is the "subject" of the sentence, and so on. But in the context of the program, this is reduced to the trivial process of guessing the class or function of a word, and being told whether or not the guess is correct. I will review more systematic attempts to simulate perceptual and pragmatic clues in Sections VI-VIII. Finally, the heuristics that the program uses are inadequate to advance beyond the three-word stage since, as we shall see, natural languages cannot be characterized by sequences of word classes. In any case, one must question whether there is really any point in doing simulations that address themselves only to the Developmental Condition. The early stages of language development can easily be accounted for by all sorts of ad hoc models; it is the acquisition of the full adult grammar that is the mystery.

The Distributional Analysis Heuristic

The problem with the word-class position heuristic when it is applied to learning natural languages is that it analyzes sentences at too microscopic a level. It is practically impossible to state natural language regularities in terms of contiguous word classes in sentences. Consider the following sentences:

2. (a) That dog bothers me.
   (b) What she wears botners me.
   (c) Cheese that is smelly bothers me.
   (d) Singing loudly bothers me.
   (e) The religion she belongs to bothers me.

In the different sentences, the word "bothers" is preceded by a noun, a verb, an adjective, an adverb, and a preposition. Clearly there is a generalization here that an astute learner should make: in all the sentences, "bothers" is preceded by a noun phrase. But noting that certain word classes precede "bothers" will not capture that generalization, and will only lead to errors (e.g., "Loudly bothers me").

A more general heuristic should look for more flexible contexts than either ordinal position in a sentence or position relative to an adjacent item, and should define classes more broadly, so that each class can consist of strings of words or subclasses instead of single words. Kelley's program moved in this direction in its third stage. Heuristics of this sort are often called *distributional analysis* procedures (see Harris, 1964), and exploit the fact that in context-free languages, the different instantiations of a grammatical class are interchangeable in the same linguistic context. Thus it is often a good bet that the different strings of words that all precede (or follow, or are embedded in) the same string of words all fall into the same class, and that if one member of such a class is found in another context, the other members of that class can be inserted there, too. Thus in sentences 2(a-e), a distributional analysis learner would recognize that all strings that preceed "bothers me" fall into a class, and that a member of that class followed by the phrase "bothers me" constitutes a sentence. If the learner then encounters the sentence "That dog scares me", he can place "scares me" and "bothers me" into a class, and "scares" and "bothers" into a subclass. If he were to encounter "Sol hates that dog", he could place all the noun phrases in the first class after the phrase "Sol hates". By this process, the learner could build up categories at different levels of abstraction, and catalogue the different ways of combining them in sentences.

Problems with distributional analysis

There are several hurdles in the way of using distributional analysis to learn a natural language. First, it requires a great many sets of minimally-contrasting sentences as input. We know that American children often do hear closely-spaced sets of sentences with common constituents (e.g., Brown, Cazden, and Bellugi, 1969; Snow, 1972; see Snow and Ferguson, 1977), but we do not know whether this pattern is universal, nor whether it occurs with enough grammatical constituents to determine uniquely every rule that the child can master. Second, a distributional analysis of a sample of a natural language is fraught with the possibility for serious error, because many words belong to more than one word class, and because virtually any subsequence of words in a sentence could have been generated by many different rules. For example, sentences 3(a-d)

3. (a) Hottentots must survive.
   (b) Hottentots must fish.
   (c) Hottentots eat fish.
   (d) Hottentots eat rabbits.

would seduce a distributional analysis learner into combining heterogeneous words such as "must" and "eat" into a single class, leading to the production of "Hottentots must rabbits", "Hottentots eat survive", and other monstrosities.

Finally, there is a combinatorial explosion of possibilities for defining the context for a given item. Given n words in a sentence other than the item of interest, there are $2^n - 1$ different ways of defining the "context" for that item – it could be the word on the immediate right, the two words on the immediate left, the two flanking words, and so on. In combination with the multiple possibilities for focusing on an item to be generalized, and with the multiple ways of comparing items and contexts across large sets of sentences, these tasks could swamp the learner. However by restricting the types of contexts that a learner may consider, one can trade off the first and third problems against the second. An extremely conservative learner would combine two words in different sentences into the same class only if all the remaining words in the two sentences were identical. This would eliminate the explosion of hypotheses, and sharply reduce the chances of making overgeneralization errors, but would require a highly overlapping sample of sentences to prevent undergeneralization errors (for example, considering every sentence to have been generated by a separate rule). Siklóssy (1971, 1972) developed a model that relies on this strategy. On the other hand, a bolder learner could exploit more tenuous similarities between sentences, making fewer demands on the sample but risking more blunders, and possibly having to test for more similarities. It is difficult to see whether there is an "ideal" point along this continuum. In any case no one has reported a successful formalization or computer implementation of a "pure" distributional analysis learner. Instead, researchers have been forced to bolster a distributional analysis learner with various back-up techniques.

An "Automated Linguist"
Klein and Kuppin (1970) have devised what they call "an automatic linguistic fieldworker intended to duplicate the functions of a human fieldworker in learning a grammar through interaction with a live human informant". Though never intended as a model of a child, "Autoling", as they call it, was the most ambitious implementation of a heuristic language learner, and served as a prototype for later efforts at modelling the child's language learning (e.g., Anderson, 1974; Klein, 1976).

Use of distributional analysis
The program is at heart a distributional analysis learner. As it reads in a sentence, it tries to parse it using the grammar it has developed up until that point. At first each rule simply generates a single sentence, but as new sentences begin to overlap with old ones, the distributional heuristics begin to combine words and word strings into classes, and define rules that generate sequences of classes and words. Out of the many ways of detecting similar contexts across sentences, Autoling relies most heavily on two: identical strings of words to the left of different items, and alternating matching and mismatching items.

Generalizing rules
Autoling also has heuristics for generalizing rules once they have been coined. For example, if one rule generates a string containing a substring that is already generated by a second rule (e.g., X → ABCD and Y → BC), the first rule is restated so as to mention the left-hand symbol of the second rule instead of the substring (i.e., X → AYD; note that this is a version of Solomonoff's heuristic). Or, if a rule generates a string composed of identical substrings (e.g., X → ABCABC), it will be converted to a recursive pair of rules (i.e., X → ABC; X → XABC). Each such generalization increases the range of sentences accepted by the grammar.

Taming generalizations
In constructing rules in these ways, Autoling is generalizing beyond the data willy-nilly, and if left unchecked, would soon accept or generate vast numbers of bad strings. Autoling has three mechanisms to circumvent this tendency. First, whenever it coins a rule, it uses it to generate a test string, and asks the informant whether or not that string is grammatical. If not, the rule is discarded and Autoling tries again, deploying its heuristics in a slightly different way. If this fails repeatedly, Autoling tries its second option: creating a transformational rule. It asks its informant now for a correct version of the malformed string, and then aligns the two strings, trying to analyze the cor-

rect string into constituents similar to those of the malformed string. It then generates a rule that transforms the malformed into the correct string, permuting or deleting the most inclusive common constituents. As before, it uses the new transformation to generate a test string, and asks the informant for a verdict on its grammaticality, discarding the rule and trying again if the verdict is negative. Finally, if nothing succeeds, the entire grammar self-destructs, and the heuristics begin again from scratch on the entire collection of acceptable sentences, which have been retained since the beginning of the learning session.

An evaluation

Autoling was not meant to be a model of the child, and needless to say, it is far from one. Unlike children, it scans back and forth over sentences, makes extensive use of negative feedback and corrections from an informant (cf., Brown *et al.*, 1969), tests each new rule methodically, remembers every sentence it hears, and gives up and restarts from scratch when in serious trouble. But it is important as a vivid illustration of the pitfalls of building a language learning model around a collection of heuristics. It is bad enough that Autoling resembles one of Rube Goldberg's creations, with its battery of heuristics (only a few of which I have mentioned), its periodic checkings and recheckings for overlapping, redundant, or idle rules, its various cleanup routines, its counters tabulating its various unsuccessful attempts, and so on. But even with all these mechanisms, Autoling's success as a language learner is very much in doubt. Klein and Kuppin do present records of the program successfully inducing grammars for artificial languages such as a set of well-formed arithmetic expressions. But as an illustration of its ability to learn a natural language, they present a rather unparsimonious grammar, constructed on its second attempt, which generates a finite fragment of English together with a variety of gibberish such as "need she" and "the want take he". Klein and Kuppin are simply unable to specify in any way what Autoling can or cannot learn. Thus Autoling – and, I would argue, any other attempt to model grammar acquisition via a large set of ad hoc heuristics – does not seem a promising start for an adequate theory of language learning. Not only does it violate the Developmental, Cognitive, and Input Conditions, but it does not even come close to meeting the Learnability Condition – the chief motivation for designing learning simulations in the first place.

VI. Semantics and Language Learning

I have postponed discussing the role of semantics in language learning for as long as possible, so as to push the purely syntactic models as far as they can go. But the implausibility of both the enumerative and the heuristic learners seems to indicate that the time has come.

The "Cognitive Theory" of Language Learning
The semantic approach to language learning is based on two premises. First, when children learn a language, they do not just learn a set of admissible sentences; they also learn how to express meanings in sentences. Second, children do not hear sentences in isolation; they hear them in contexts in which they can often make out the intended meanings of sentences by non-linguistic means. That is, they can see what objects and actions are being referred to in the sentences they hear, and they can discern what their parents are trying to communicate as they speak. (Kelley incorporated a version of this assumption into his model.) An extremely influential theory in developmental psycholinguistics (often called the "Cognitive Theory") asserts that children learn syntax by inferring the meanings of sentences from their non-linguistic contexts, then finding rules to convert the meanings into sentences and vice-versa (Macnamara, 1972; Schlesinger, 1971). Several considerations favor the Cognitive Theory. The first (though rarely cited) consideration is that semantic information can substitute for information about non-sentences to make classes of languages formally learnable. The second is that there is some empirical evidence that both children and adults use semantic information when they learn syntactic rules. The third consideration is that this task is thought to be "easier" than inferring a grammar from a set of strings alone, because the mental representations corresponding to sentence meanings are thought to resemble the syntactic structures of sentences. I will discuss each justification for the semantic approach in turn.

Learnability with Semantic Information
John Anderson (1974, 1975, 1976) has described a semantic version of Gold's language acquisition scenario, formalizing an earlier speculation by Clark (1973). First, he assumes that whatever "sentence meanings" are, they can be expressed in a formal symbolic notation, and thus can be put into one-to-one correspondence with the set of natural numbers by the mathematical technique known as "Gödelization". Second, he assumes that a natural language is a function that maps sentences onto their meanings, or equivalently, well-formed strings onto natural numbers, and vice-versa. (In contrast, we have been assuming that natural languages are functions that map strings onto the judgments "grammatical" and "non-grammatical", or equivalently, "1" and "0".) Third, he assumes that children have access to a series of pairs consisting of a sentence and its meaning, inferred from the non-linguistic context.

The child's task is to identify in the limit a function which maps sentences onto their meanings.

Recall that Gold (1967) proved that the class of primitive recursive functions, which map strings onto numbers, is learnable provided that the learner has eventual access to all number-string pairs. For Gold, the numbers represented the trial number or time since the start of learning, but in Anderson's model, the numbers correspond to sentence meanings. The learner enumerates the primitive recursive functions, testing each one against the sample of sentence-meaning pairs, retaining a function if it is consistent with the sample (see Figure 1). In this way the learner will identify the function (and hence the language) in the limit, since all incorrect functions will be rejected when they pair a meaning with a different string than the one in the sample.

Although in this version the learner can be proved to succeed without requiring information as to what is not a sentence, all of Gold's other conclusions remain in force. It will take the learner an astronomical amount of time until he arrives at the correct function, but there is no quicker or more successful method, on the whole, than enumerating functions one by one. By suitably restricting the learner's hypothesis space, learning time can be reduced, and by using heuristic procedures that exploit properties of individual meaning-sentence pairs, it can be reduced even further. But once again the learner ceases to be a multipurpose rule learner – he makes tacit assumptions about the syntax of the target language, about the way that meanings are mapped onto strings, and about the representativeness of the meaning-sentence pairs in the sample at a given time. He will fail to learn any language that violates these assumptions. As Chomsky (1965) has noted, the hypothesis that the child uses semantics in learning syntax is in some senses stronger, not weaker, than the hypothesis that sentences alone are used.

Evidence for the Cognitive Theory

Cognitive development and language acquisition

Two sorts of evidence have been martialled in support of the view that humans base their learning of syntax upon their conceptualization or perception of the meanings of sentences. The first consists of various correlations between language development and cognitive development, which are thought to imply that the non-linguistic mental representations available to the child constrain the linguistic hypotheses that he will entertain. For example, the early two- and three-word utterances of children seem to reflect closely certain semantic relations such as agent-action, possessor-possessed, etc. (Bowerman, 1973; Brown, 1973; Schlesinger, 1971). As well, the "cognitive complexity" of the semantic functions underlying various grammatical rules has been shown to predict in a rough way the order of the child's mastery of those rules (Brown, 1973). Similarly, it has been found that some syntactically simple rules (such as the conditional in Russian) are not acquired until the underlying semantic functions (in this case, implication) have been mastered (Slobin, 1973).

Semantics and artificial language learning

The second sort of evidence comes from a set of experiments in which adult subjects are required to learn artificial languages, that is, they must learn to discriminate grammatical from ungrammatical test strings as defined by a grammar concocted by the experimenter. In early experiments of this type (e.g., Miller, 1967), where subjects saw various strings of nonsense syllables, even the simplest grammars were extremely difficult for the subjects to learn. However, in a famous set of experiments, Moeser and Bregman (1972, 1973) presented some subjects with a sample of strings, and other subjects with a sample in which each string was paired with a picture of geometric forms such that the shapes, colors, and spatial relations of the forms corresponded to the words and syntactic relations in the sentences (that is, the pictures were intended to serve as the semantic referents of the strings). After more than 3000 strings had been presented, the subjects who saw only strings failed utterly to discriminate grammatical from ungrammatical test strings, while those who saw strings and pictures had no trouble making the discrimination. This finding has led many theorists to conclude that it is intrinsically easier for humans to learn syntactic rules if they use semantic information in addition to sentences.

However Anderson (1974, 1975) has pointed out that semantics-based learners, including the subjects in Moeser and Bregman's studies, learn by virtue of specific assumptions they make about the way the target language uses syntactic structures to express semantic relations. For example, he notes that natural languages require an adjective to predicate something about the referent of the noun in its own noun phrase, never a noun in another noun phrase in the sentence. That is, in no natural language could a phrase such as "the blue stripes and the red rectangle" refer to an American flag, even though the sentences of such a language might be identical to the sentences of (say) English, and the semantic relations expressible in that language might be identical to those expressible in (say) English. Anderson performed an experiment in which subjects saw strings of English words (referring to shapes, colors, and spatial relations) generated by an artificial grammar. A second group saw the same strings paired with pictures in such a way that each adjective in the sentence modified the noun in its phrase; a third group saw the same strings and pictures, but they were paired in such a way that each adjective modified a noun in a different phrase (like our example with the flag).

Only the second group of subjects, with the "natural semantics", were later able to discriminate grammatical from ungrammatical test strings. Thus, Anderson argues, it is not the availability of semantic information per se that facilitates syntax learning in humans, but semantic information that corresponds to the syntactic structures in the target language in some assumed way.[2] These correspondences will be explained in the next section, in which semantics-based learning heuristics are discussed.

Heuristics that use Semantics

The most important fact about the natural language acquisition task is that the units composing linguistic rules are abstract, and cannot be derived from sample strings in any simple way. The problem with distributional analysis was that these units or "constituents" do not uniquely reveal themselves in the patterns of sentence overlappings in a sample. However, if the semantic representation of a sentence corresponds in a fairly direct way to the syntactic description of that sentence, semantic information can serve the same purpose as distributional regularities. The syntactic structure of a sentence in a context-free or context-sensitive language can be depicted as a tree, with each node representing a constituent, and the set of branches emanating from a node representing the application of a rule rewriting that constituent as a sequence of lower-order constituents. Similarly, the mental representational structures corresponding to percepts and sentence meanings are also often represented as trees or similar graph structures (e.g., Anderson and Bower, 1973; Norman and Rumelhart, 1975; Winston, 1975). The top nodes of such trees usually correspond to logical propositions, and the branches of these trees correspond to the breakdown of propositions into their subjects and predicates, and to the successive breakdown of the subject and predicate into concepts and relations, or into further propositions. If the tree representing a sentence meaning is partially isomorphic to the constituent structure of the sentence, presumably there is a way that a child can use the meaning structure, which by assumption he has, to discern the constituent structure of the sentence, which he does not have. Anderson (1974, 1975, 1977) has demonstrated precisely how such heuristics could work. In the following paragraphs I shall explain the operation of these heuristics; then, in Section VII, I shall show how Anderson has embodied these heuristics in a computer model of the language learner.

[2] Of course, in this particular case the assumption about semantics and syntax need not have been innate, since the subjects' tacit knowledge of English could have constrained their hypotheses.

Using semantics to delineate constituents: the Tree-fitting heuristic

This heuristic begins with the assumption that the child knows the meaning of all the "content" words in the sentence, that is, he knows to which concept node in the meaning structure each word corresponds. The learner matches the concepts in the meaning structure to the words in the sentence, and attempts to fit the tree structure for the meaning onto the sentence, spatially rearranging the nodes and branches as necessary but preserving all links between nodes. The learner now has a tree-structure for the sentence, and can deduce what the constituents are and how the rules of the grammar rewrite the major constituents as sequences of minor ones.

An example will make this heuristic clearer. Say the child saw a white cat eating a mouse. His perceptual system might construct the propositions "X is a CAT", "X is WHITE", "Y is a MOUSE", and "X EATS Y", which can be depicted as a single tree-structure like the one in Figure 3(a). Say the child simultaneously heard the string of words "the white cat eats a mouse". By matching the word "white" onto the concept "WHITE" (and so on for the other words), reversing the order of the respective links to "CAT" and to "MOUSE", and straightening out continuous series of links, the child can arrive at the tree-structure for the sentence which is depicted in Figure 3(c). He can then hypothesize rules specifying that a sentence can be broken down into two constituents, that one constituent can be broken down into a class containing the word "white" and another containing the word "cat", and that the second main constituent can be broken down into the word "eats" and a constituent containing a class containing the word "mouse". Furthermore, the child can construct rules translating syntactic constituents into semantic propositions and vice-versa. In this example, he could hypothesize that the first major constituent of a sentence refers to some individual that is the subject of an underlying proposition, the first word class in this constituent refers to some property predicated of that individual, and so on.

The problem with this heuristic is that there are usually many ways to fit a semantic structure onto a string of words, only one of which will correspond to the correct breakdown of the sentence into its syntactic constituents. For example, nothing would have prevented the child in our example from constructing the syntactic trees depicted in Figures 3(d) and (e) instead of the one in Figure 3(c). Anderson has proposed two mechanisms by which the heuristic could "know" the best way to fit the semantic tree onto the string. First, the learner must know which node of the semantic tree should be highest in the syntactic tree, in order to distinguish between the possibilities represented in Figures 3(c) and (d). This corresponds to knowing the main proposition of the sentence, that is, what is the major topic of the sentence and what is the major thing being asserted of it. Anderson suggests that

Figure 3. *Semantic structure (a) to be fitted onto the string (b) in various ways by the Tree-fitting heuristic. In this formalism for semantic structures (HAM; Anderson and Bower, 1973), S = subject, P = predicate, R = relation, O = object, X and Y represent individuals, and capitalized terms are concepts, which correspond to words.*

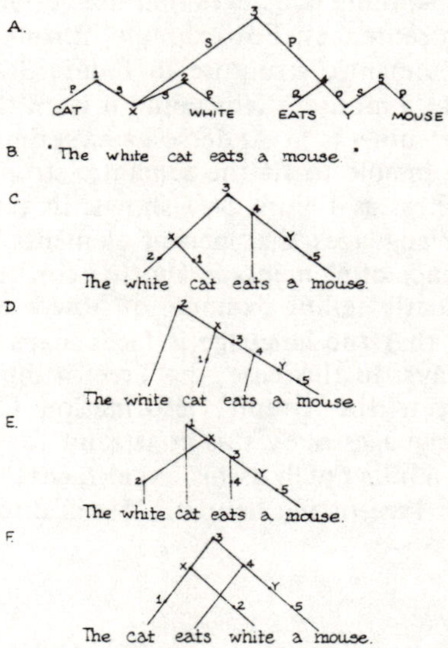

this pragmatic information is communicated to the child during his normal interactions with adults; in other words, the social and communicative context in which a sentence is uttered makes it clear what the adult intends to assert about what (see Bruner, 1975, for supporting arguments and evidence). For the tree-fitting heuristic, this means that one of the propositions in the semantic structure is tagged as the "principal" one, and its node will be highest when the semantic tree is fitted onto the string of words. The nodes connected to this "root" node by one link are placed one level lower, followed by the nodes connected to the root by two links, and so on. Thus if the main proposition concerns what the cat did to the mouse, the heuristic will fit the tree depicted in Figure 3(c) onto the string. On the other hand, if it is the whiteness of the mouse-eating cat that is being asserted (e.g., "white is the cat that eats the mouse"), the heuristic will fit the tree depicted in Figure 3(d) onto the string.

The second constraint on the heuristic is that no branches be allowed to cross. Thus the heuristic would be prohibited from fitting the tree depicted in Figure 3(e) onto the string. No set of context-free rules can generate a tree like this, and in fact what the constraint does is prevent the heuristic from constructing trees from which no context-free rules can possibly be derived. Thus this constraint, which Anderson calls the *Graph Deformation Condition*, will prevent the learner from learning languages that use certain rules to transform meaning structures into sentences. For example, it cannot learn a language that could express the semantic structure in Figure 3(a) by the string of words "the cat eats white a mouse". Nor could it learn the "unnatural semantics" language that the subjects in Anderson's experiment failed to learn. In each case it would be unable to fit the semantic structure onto the string without crossing branches, as Figure 3(f) shows. In general, the heuristic is incapable of learning languages that permit elements from one constituent to interrupt the sequence of elements in another constituent. As Anderson argues, this is a particularly telling example of how a semantics-based heuristic in effect assumes that the language it faces maps meanings onto sentences only in certain ways. In this case, the Tree-fitting heuristic "assumes" that the language meets the Graph Deformation Condition. Anderson believes that natural languages obey this constraint for the most part, and that both children and adults (such as his experimental subjects) tacitly assume so as they use the Tree-fitting heuristic. I will discuss these claims in Section VII.

Using semantics to generalize rules

Once the learner has broken down sentences into their constituents and hypothesized the corresponding rewrite rules, he must combine rules that have been derived from different sentences – otherwise he is left with one set of rules for each sentence, not much better than a learner who simply memorized the sentences whole. Rule-merging is a particularly rocky step for distributional analysis heuristics (as sentences 3(a-d) showed), since sentences from natural languages provide countless temptations to merge dissimilar constituents owing to the syntactic ambiguity of most short substrings. Klein and Kuppin's program tentatively merged rules with overlapping constituents, used the newly-merged rules to generate a sentence, and submitted the sentence to the informant for approval before it would declare the merger permanent. But this is an unrealistic way to keep overgeneralizations in check. Not only do children not have access to such an informant, but even if they did, it is unlikely that the Autoling strategy would work as required. A merged rule can usually generate many sentences, sometimes an infinite number, so the knowledge that one string is acceptable does not mean that all the strings generated by the rule will be acceptable.

However information in the semantic representation might be used instead to decide whether rules can safely be merged. First, Anderson suggests that words in the same positions in different sentences whose concepts have identical roles in the semantic structure can be merged into one class. For example, say the learner, after processing the meaning–sentence pair in Figure 3(c), encountered the sentence "The green snail nibbles the leaf", together with its semantic structure, as shown in Figure 4(a) and (b). After fitting the semantic tree onto the string (see Figure 4(c)) and deriving the corresponding rules, the learner can use the similarities between the semantic representations in Figures 3(a) and 4(a) to merge the two sets of rules. For example, "EATS" in Figure 3(a) corresponds to the "relation" branch of the predicate of the main proposition, and so does "NIBBLES" in Figure 4(a). The learner can then merge the corresponding words into one class, and by similar means can merge "white" and "green", "eat" and "snail", and so on.

Now the learner must recognize that the higher constituents in the two sentences can also be merged, such as the ones embracing "the white cat" and "the green snail". Anderson suggests a double criterion for when to merge higher-order constituents: they must decompose into identical sub-constituents, and they must serve the same semantic role. In this example, both are satisfied: the word classes in the two constituents have already been merged, and both constituents serve as the subject of their respective main propositions. Once all the parallel constituents in the two sentences have been merged, the learner will end up with a grammar that generates sixteen different sentences: "the green cat eats a leaf", "the white snail nibbles a mouse",

Figure 4. *Semantic structure* (A), *string* (B), *and tree* (C) *which, in conjunction with Figure 3, illustrate the Semantics-Induced Equivalence Heuristic.*

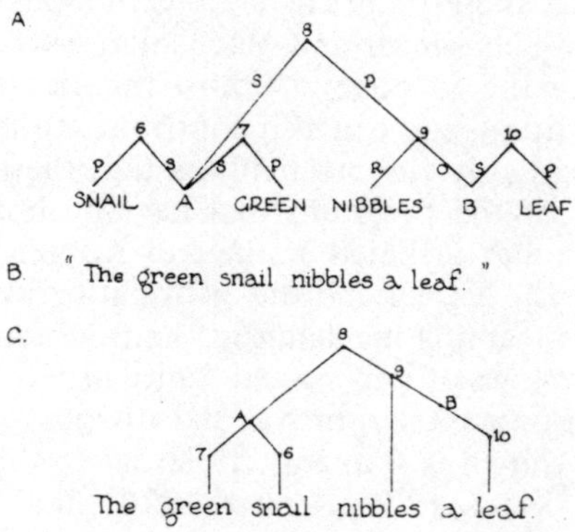

and so on. Anderson calls this heuristic (and the putative property of natural languages that it exploits) *Semantics-Induced Equivalence of Syntax*. He asserts that the heuristic exploits the tendency of natural languages always to use the same syntactic construction to express a particular semantic relation within a given higher-order constituent. Whether or not this claim is true of English will be discussed in Section VII.

It is interesting to note that the Semantics-Induced Equivalence of Syntax heuristic is neither more nor less conservative, on the whole, than Distributional Analysis. Each will try to merge in situations where the other would not. Thus the Distributional Analysis heuristic would make no generalization embracing the sentences in Figures 3 and 4, since they share no content words. Instead it would have to wait until some sentence like "the green snail eats a mouse" appeared. On the other hand, the Semantics-Induced Equivalence heuristic, upon encountering the sentence "the white cat eats slowly", would not merge "slowly" with "the mouse" (as would Distributional Analysis), since "MOUSE" and "SLOWLY" would not have the same roles in their semantic structures. It should be clear from these examples that the Semantics-Induced Equivalence heuristic will, in general, make the wiser generalization.

VII. Anderson's Language Acquisition System

The Computer Simulation of Semantics-based Heuristic Language Acquisition
Heuristics that exploit syntax-semantics correlations, like those that exploit properties of sentences alone, are often implemented as computer programs (Anderson, 1974, 1975, 1977; Fabens and Smith, 1975; Klein, 1976; Klein and Rozencvejg, 1974; McMaster, Sampson, and King, 1976; Reeker, 1976; Siklóssy, 1971, 1972). In a sense, these programs are incarnations of the informal Cognitive Theories of the Schlesinger and Macnamara sort. As such, they serve as a testing ground for the adequacy of those theories, especially at meeting the Learnability Condition, and can also contribute to the goal of specifying more precisely and explicitly the mechanisms that these theories implicate. Unfortunately, many of the programs that have been developed succumb to the same syndrome that afflicted Klein and Kuppin's model: unreasonable assumptions about the learner and the information available to him, ad hoc and unparsimonious learning mechanisms, and dubious success at learning. For example, the program of Fabens and Smith (1975) modifies its rules in accordance with environmental approval and disapproval, which Brown and Hanlon (1970) have shown is probably irrelevant to the learning of syntax. Other programs (e.g., Klein, 1976; Reeker, 1976; Siklóssy, 1971,

1972) avoid this device but only learn to produce meager, ill-defined fragments of natural languages, often generating many non-sentences at the same time. The exception among these efforts is Anderson's Language Acquisition System (LAS; 1974, 1975, 1977). As we have seen, Anderson has carefully defined certain heuristics that his program employs and the properties of natural languages that make these heuristics useful. As well, the program can acquire well-defined infinite subsets of natural languages, its semantic representations have an independent theoretical motivation, and it avoids for the most part psychologically unrealistic strategies. For these reasons, I will discuss only Anderson's simulation from among the many that have been reported (which in any case rely on heuristics remarkably similar to the ones Anderson uses).

How LAS works

General architecture

LAS uses a formalism for semantic representations that Anderson has used elsewhere as a theory of information representation in long term memory (the Human Associative Memory system (HAM) of Anderson and Bower, 1973). Its grammar is in the form of an Augmented Transition Network (ATN), which is held by many to be a plausible model of human language processing (see Kaplan, 1975). The ATN that LAS uses corresponds rule-for-rule to a context-free grammar, but can be incorporated more easily into a left-to-right sentence recognizer or producer. LAS has a subroutine corresponding to sentence production, which uses the ATN to convert a semantic structure into a sentence. It also has a subroutine that corresponds to sentence comprehension, which uses the ATN to convert a sentence into its semantic structure. Finally, it has a learning program that uses pairs consisting of semantic structures and sentences to build the ATN piece-by-piece. The latter program is the one of interest here.

Like Kelley's and Klein and Kuppin's programs, LAS is driven by the comprehension process. It tries to interpret a sentence from left-to-right with its current grammar, and alters parts of the grammar if it fails. If a particular rule gets the learner part way in interpreting a sentence, it is the one that will be expanded. LAS also forgets the exact sentences that it hears, so that a sentence contributes to grammatical development only in the way that it alters the grammar as it is being understood. These features give LAS a psychologically realistic flavor compared to other models I have discussed.

Use of the Tree-fitting heuristic

When LAS receives its first sentence-meaning pair, there is no grammar to interpret it yet, so it must build the first pieces of the grammar relying entirely on the Tree-fitting heuristic. But in general, the HAM structure representing the learner's perception of the situation in which the sentence has been uttered is not really suitable for fitting onto the string right away. It contains too many sentence-irrelevant propositions, and has no way of indicating the proposition corresponding to the principle assertion of the sentence (see Section VI). Thus the program is forced to compute an intermediate representation, called the *Prototype Structure,* which omits propositions whose concepts have no counterparts among the words of the sentence, and highlights the principle proposition (in line with supposed pragmatic cues). It is this Prototype structure, not the meaning structure itself, that the Tree-fitting heuristic tries to fit onto the string of words. Once an acceptable tree has been derived by the heuristic, LAS constructs ATN *arcs,* each one corresponding to a left-to-right sequence of constituents composing a higher constituent, and the corresponding rules that map these syntactic constituents onto their semantic counterparts.

Use of the semantics-based equivalence heuristic

When subsequent pairs come in, LAS tries to interpret the strings using all its rules simultaneously. Using the Semantics-Induced Equivalence heuristic, it unites into a single class words whose concepts serve the same role in their respective HAM structures. Similarly, it merges any two arcs (i.e., higher-order constituents) that simultaneously assign the same semantic role to their respective sentence constituents. These mechanisms were discussed in Section VI. In addition, LAS will merge two arcs if one is a proper subsequence of another, as long as they both specify the same semantic role. For example, assume that LAS has induced an arc that will parse sequences like "the mouse" in Figure 3, and that it is forced by a subsequent sentence to construct an arc that will parse "the mouse that nibbles the house". Then the old arc will be swallowed into the new one automatically (with the last four words marked as "optional"). In this way, LAS can construct recursive rules, allowing it to generate infinite languages. In the present example, it would construct a low-level arc to parse "the house"; however, this substring can already be parsed with the higher-level arc built to parse "the mouse that nibbles the house" (since "mouse" and "house" would presumably be merged, and the last four words are marked as optional). Consequently it would merge the two arcs, ending up with the recursive arc corresponding to the rule "*noun phrase* → the *noun* that nibbles *noun phrase*". Now it can generate "the mouse that nibbles the cat that eats the mouse that nibbles the house" and so on.

Finally, LAS has a special heuristic with which it handles the so-called "grammatical morphemes" such as articles, auxiliaries, relative pronouns, and so on, which have no direct counterparts in the semantic representations. This heuristic will be discussed in a later paragraph.

Learning powers of LAS

How well does LAS do? Anderson presents several examples in which LAS is faced with artificial languages or fragments of natural languages, all context-free, which can be used to describe arrangements of two-dimensional shapes of various colors and sizes. In all cases LAS succeeded in acquiring a grammar for the language, including infinitely large subsets of English and French, after taking in 10-15 meaning-sentence pairs. For example, it could handle sentences like "the large blue square which is below the triangle is above the red circle which is small", and other sentences using these grammatical constructions. Anderson conjectures that LAS could learn any context-free language with a semantic system that respected the Graph Deformation Condition and the Semantics-Induced Equivalence of Syntax Condition.

An Evaluation of LAS

LAS is unquestionably an impressive effort. Anderson is alone in showing how a learner with semantics-based heuristics can succeed in learning chunks of natural languages in a plausible manner. Furthermore, there are possibilities for extending the powers of LAS. If LAS were built like Winograd's (1972) program to converse with another speaker instead of receiving sentences passively, it would have representational structures that conceivably could be useful in acquiring rules for interrogatives, conditionals, imperatives, and so on. And if it had a more childlike semantic representational system, which categorized the world into actors, actions, and recipients of actions, possessors and possessed, objects and locations, and so on, its linguistic abilities might even resemble those of young children (cf., Brown, 1973). By enriching the semantic system gradually, it might even be possible to generate a sequence of stages parallel to the child's linguistic development, which would be a unique accomplishment among formal models of language learning (outside of Kelley's limited attempts). Of course, all of this remains to be shown.

In any case, rather than spelling out the various ways that LAS can be extended, I shall focus in this section on the *limits* of LAS's abilities, on Anderson's claim that "the weakness of LAS... is sufficiently minor that I am of the opinion that LAS-like learning mechanisms, with the addition of some correcting procedures, could serve as the basis for language learning" (1977, p. 155-156). Since LAS is an incarnation of the currently popular Cognitive Theory of language learning, Anderson's claim is an important one. If true, it would support the contention that the child's perceptual and cognitive representations are sufficiently rich data structures to support language acquisition (e.g., Bowerman, 1973; Sinclair-de Zwart, 1969; Schlesinger, 1971), obviating the need for innate language-specific data structures (e.g.,

Chomsky, 1965; Fodor, 1966; McNeill, 1966). On this view, the innate constraints on the learner derive only from his cognitive representational structures and, as Anderson points out, his tacit assumptions about how these correspond to syntactic structures. For this reason I will examine LAS's abilities in some detail. In particular, I shall scrutinize Anderson's central claim, that most syntactic rules can be derived from distinctions made at the semantic level, while the rest can be derived with the help of a few miscellaneous heuristics.

Do natural languages obey the Graph Deformation Condition?

This condition, on which the Tree-fitting heuristic depends, decrees in effect that natural languages must be context-free, a conclusion that Anderson explicitly supports (despite its near-universal rejection by linguists). There are a number of natural language constructions which cross branches, and Anderson must find reason to dismiss them as counter-examples to the omnipotence of the Tree-fitting heuristic. One example is the "respectively" construction. As Figures 5(a) and (b) show, the semantic structures for these sentences cannot be fitted onto the strings without branches crossing. A second example can be found in languages that indicate semantic roles by case markers instead of by word order (e.g., Russian, Latin, Wolbiri). In these languages it is possible for an element that belongs to one phrase to interrupt a sequence of elements in a second phrase, provided that the intruding element is suitably marked as belonging to its phrase. Anderson cites both these counter-examples, and argues that they are atypical constructions, possibly acquired by special problem-solving strategies outside the normal language induction mechanisms. While the rarity of constructions of the "respectively" sort make this conclusion tenable for these constructions, it is less easy to forgive the paucity of mechanisms in LAS for acquiring case-inflection rules, prevalent in languages other than English, which naturally give rise to constructions with crossing branches.

A second class of counter-examples consists of discontinuous elements, which give rise to crossing syntactic dependencies in a sentence. For example, in the sentence "Irving threw the meat out that had green spots", the phrase "the meat" is part of a constituent that includes "that had green spots", whereas the word "threw" is part of a constituent that includes the word "out". Figure 5(c) and (d) show how these branches must cross (similar crossing dependencies can occur with auxiliary and tense morphemes under certain analyses, see Gross, 1972). Anderson exempts the Tree-fitting heuristic from having to deal with such constructions on the grounds that they involve "non-meaning bearing morphemes" which are outside its province. But this is not quite true – the morpheme "out" in the sentence in Figure

Figure 5. *Violations of the Graph Deformation Condition.*

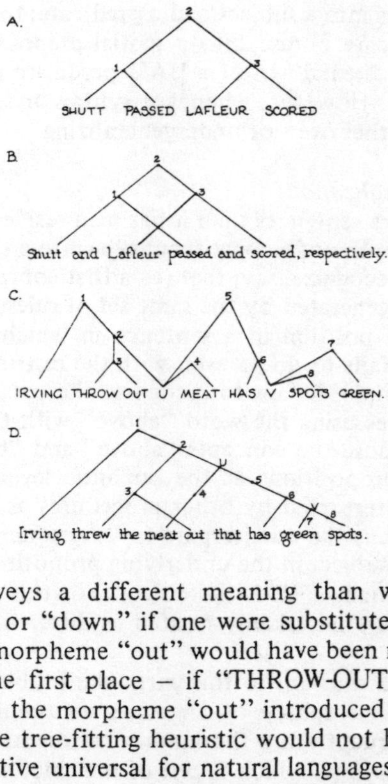

5(d) conveys a different meaning than would the morphemes "up" or "around" or "down" if one were substituted in its place. But it is not clear how the morpheme "out" would have been mapped onto the semantic structure in the first place – if "THROW-OUT" were represented as a unitary node, and the morpheme "out" introduced into the sentence by some other means, the tree-fitting heuristic would not have to deal with the morpheme. As a putative universal for natural languages, the Graph Deformation Condition can be criticized in that the HAM structures representing the meanings of various sentence types are not specified a priori, but seem to be made up as they are needed. For this reason it is hard to disconfirm the Condition with the present examples, though.

Do natural languages permit Semantics-Induced Generalizations?

The Tree-fitting heuristic has a function other than giving a tree-structure to the sentence. The heuristic attaches semantic labels to the branches of the tree, and the Semantics-Induced Equivalence of Syntax heuristic uses these labels as criteria for merging rules derived from different sentences. These heuristics serve LAS well, but only because the subset of English grammar and the subset of HAM structures that Anderson has chosen correspond almost feature for feature. For example, the grammatical rule that specifies that sentences consist of a noun phrase and a verb phrase corresponds to the breakdown of HAM propositions into a subject and a predicate; the grammatical rule that breaks the predicate phrase into a spatial preposition and a noun phrase corresponds to the breakdown of a HAM predicate into a relation and an object, and so on. However, whenever syntax and semantics diverge, I will show, LAS errs, either over- or undergeneralizing.

Semantics-induced undergeneralizations

LAS's powers to generate more sentences than it has seen reside in its abilities to merge the different exemplars of a constituent type into a single class. Thus one would want LAS to recognize, say, that (to a first approximation) all noun phrases in English are generated by the same set of rules, regardless of the type of sentence or the position in a sentence in which the noun phrase is found. However LAS fails to do so even with the restricted subset of English it is given. For example, it fails to recognize the equivalence of subject noun phrases in sentences using the word "above" with those using the word "below". This is because the concepts "above" and "below" are represented identically in the propositions at the semantic level, with the subject of such a proposition interpreted by other procedures as the higher of the two objects in space. Thus the counterpart to "the square" in "the square is above the circle" is the subject in the underlying proposition, whereas in "the square is below the triangle" it is the object. For this reason the two occurrences of the phrase are mistakenly treated as different syntactic units.

Although Anderson suggests a solution to this particular problem, related problems will pop up when different subsets of languages are attempted. This is because natural languages frequently use the same constituents to express different underlying logical functions (which is one of the chief motivations for developing transformational grammars, with their distinction between deep and surface structures). Thus the Semantics-Induced Equivalence heuristic would never realize that the different tokens of the phrase "the cop" in 4(a-e)

4. (a) The cop frightens the thief.
   (b) The cop is frightened by the thief.
   (c) The cop tends to like thieves.
   (d) The cop who arrests thieves...
   (e) The cop who thieves frighten...

are examples of the same type of sentence constituent, since in the different sentences and phrases it functions variously as subject or object of the underlying proposition, or as part of the principal proposition or one of the secondary propositions. LAS would develop ad hoc rules for the different types of sentences, and would be unable to conclude that a subject noun phrase in an active sentence can also appear as the subject of a passive sentence, a "tend"-type sentence, and so on.

One interesting way to remedy this problem would be to posit distinct mental predicates corresponding to the different syntactic constructions that a verb can enter into. Thus there would be mental predicates for "FRIGHTEN", "IS-FRIGHTENED-BY", "TENDS-TO-FRIGHTEN", "IS-EASY-TO-FRIGHTEN", and so on (which is similar to a proposal Anderson has made elsewhere in discussing memory for sentences, see Anderson and Bower, 1973). Since the subjects of the sentences with all these constructions are also the subjects of their underlying propositions at the semantic level, LAS would have grounds to merge them. Unfortunately, this raises the problem of how the learner could tell when to encode a situation using one type of mental predicate rather than another. For example, how would the learner know to use the "FRIGHTEN" predicate just when hearing "It is easy to frighten the cat", but the "IS-EASY-TO-FRIGHTEN" predicate when hearing "That cat is easy to frighten"? This "encoding problem" and its possible solutions will be discussed further in Section IX.

Semantics-induced overgeneralizations

In relying on semantic criteria, LAS also generalizes in cases where it should not. For example, the proposition asserting that an entity is square-shaped can appear in a sentence either as "the square" or "the square thing", but the proposition asserting that an entity is colored red can appear only as "the red thing". Nonetheless, since the two propositions have the same format, LAS overgeneralizes and accepts "the red". Anderson solves this problem by providing LAS with an innate schema for noun phrases, including the stipulation that a noun phrase must contain at least one noun. If indeed the general form of the noun phrase is innate, it cannot have the format Anderson proposes, however, since many noun phrases lack nouns – consider the subject noun phrases in the sentences "Jogging exhausts me", "It is a total bore", and "That he chortles is irritating".

A similar overgeneralization problem follows from the fact that verbs with similar semantic representations have different case structures, that is, require different numbers and arrangements of noun phrases in the sentences in which they appear. Thus "give" might appear in a semantic structure with a subject and two objects, corresponding to the giver, gift, and recipient. Using this structure, LAS could build rules that parsed "Rockefeller gave Brown a million dollars", with two consecutive noun phrases after the verb; it would also construct rules to parse "Rockefeller gave a million dollars to Brown", with a noun phrase and a prepositional phrase. However when LAS later encounters sentences like "The IMF transferred a billion dollars to Ghana", or "Rockefeller donated a Wyeth to the museum", it would merge "give", "transfer", and "donate" into a single class, since they would have similar roles in their semantic representations, and would mistakenly produce "Rockefeller donated the museum a Wyeth", "The IMF transferred Ghana a billion dollars", and so on. Anderson does suggest a heuristic that might help LAS in learning the case structures of verbs: placing all the concepts that are causally related to the verb at the same level of embedding in the Prototype structure. This would not help for the present examples, however, since the different verbs have the same causal relation to the noun phrases, but have different case structures nonetheless. Many similar examples can be found in English: "throw" versus "propel", "show" versus "display", "teach" versus "instruct", and so on.

Learning grammatical morphemes

The class of grammatical morphemes (e.g., articles, inflections, conjunctions, relative pronouns, etc.) poses special problems for LAS, since they have no counterparts in its semantic structures. Anderson argues that learning the rules for ordering these terms in the absence of semantic information is *not* problematic, at least in a formal sense. Since grammatical morphemes occur in sub-sequences of finite length, they constitute a finite cardinality language, which, according to Gold's theorems, can be learned with neither an informant nor a semantic referent. This argument is misleading, however, because whether or not a string of grammatical morphemes is acceptable will depend on its context. Since the relevant context can be indefinitely long, there would be an infinite number of cases for the finite cardinality learner to memorize. Thus what the learner faces is *not* a finite cardinality language after all. For example, the occurrence of the string "to which" in sentence 5(a) is grammatical only because the verb "give", which can take a prepositional phrase beginning with "to", appears later in the sentence (compare the same sentence with "spent" in place of "gave"). But as sentences 5(b-d) show, that verb can be an arbitrary distance away, resulting in an infinite number of contexts to learn.

5. (a) The museum to which he gave a million dollars is in Chicago.
   (b) The museum to which it is obvious he gave a million dollars is in Chicago.
   (c) The museum to which I think it is obvious he gave a million dollars is in Chicago.

(d) The museum to which I think without any justification whatsoever it is obvious he gave a million dollars is in Chicago.

Thus learning rules for these classes of items is formally far from a trivial matter, and it is worth examining the heuristic solutions to the problem that Anderson proposes.

To begin with, it should be noted that LAS faces a language with few grammatical morphemes: only the articles "the" and "a", the copula "is", and the relative pronoun "which". The spatial prepositions such as "above" are treated as content words, since they correspond directly to nodes in the semantic representation, and to simplify matters even further, the expression "to the left of" has been collapsed into the single word "left-of". With this simple language, LAS can survive with a single heuristic: when it encounters one or more grammatical morphemes, it brackets them with the content word immediately to the right, creating a new constituent.

Problems with the grammatical morpheme heuristic

Although this heuristic works well enough to prevent LAS from making any gross errors, it prevents it from making important generalizations as well. For example, LAS cannot recognize the equivalence in its grammar of predicate phrases in the main clause of a sentence and predicate phrases in relative clauses, because the latter have the word "which" grafted onto them. This also seems to be the reason that LAS fails to merge its class for prenominal adjectives ("*red* square") with its identical class for predicate adjectives ("the square *is red*"). In any case, the heuristic clearly would not work for larger subsets of natural languages. As Anderson notes, in sentences like

6. The woman that he ran after is nimble.

LAS would create the nonsense constituent "after is nimble", leading to many possibilities for error (e.g., "The woman he loved after is nimble").

"Correcting procedures" for handling grammatical morphemes

Anderson does suggest remedies for some of these problems. For example, the first problem could be solved by allowing LAS to merge arcs with identical subconstituents, whether or not one arc is wholly contained in the other. However this procedure would still not make the required generalization in the general case – it would not help detect the similarities between main clauses and other sorts of relative clauses, such as those in which the objects have been deleted. For example, in 7(b), there is no constituent corresponding to the "the monster devoured" in 7(a), as the brackets indicate. Nonetheless one would want a learner to be able to generalize that whatever can be expressed in a main clause like 7(b)

7. (a) The cookie that the monster devoured is huge.
   (b) (The monster) (devoured (the cookie))

can also be expressed in a relative clause like the one in 7(a).

Anderson also suggests that redundant word classes, such as the predicate and prenominal adjective classes in our example, should be merged if they have enough members in common. But this would only lead to trouble. In natural languages, many if not most nouns can also serve as verbs and adjectives, but it would be disastrous to merge those classes outright, since many adjectives and verbs *cannot* serve as nouns.

Finally, Anderson suggests that the incorrect parse of sentences like 6 could be avoided if the learner would exploit the pause often found after the preposition in spoken speech as a cue to the correct location of the constituent boundary. However, natural speech is full of pauses that do *not* signal phrase boundaries (see Rochester, 1973), so such a heuristic would not, in general, do much good.

Conclusion

In sum, careful scrutiny of the learning mechanisms of LAS does not bear out Anderson's claim that such mechanisms are sufficient to learn natural languages. We have seen a number of cases in which the semantics-based heuristics are inadequate in principle to learn important features of English. This would not be a serious criticism if there were principled ways of extending LAS to handle these features. But virtually all of Anderson's proposals for extending LAS would at best work for the particular glitches they were designed to fix, and would be ineffective if applied to larger subsets of natural languages.

None of this diminishes the importance of Anderson's contribution. In the traditional psycholinguistic literature, the "Cognitive" theory of language learning is usually discussed in such vague terms that it is impossible to evaluate. In embodying this theory in a computer program, Anderson has shown what assumptions the theory rests on, which aspects of language learning the theory can account for, and which aspects are beyond its reach. In Section IX, I will discuss further the implications of LAS and other models for theories of human language learning.

VIII. A Theory of Learning Transformational Grammars

The features of natural language that give LAS the most trouble are precisely those features that cannot easily be handled by context-free grammars, and that motivated the development of transformational grammars (Chomsky,

1957, 1965). Examples are discontinuous constituents, "respectively"-type constructions, case- and complement structures of various verbs, the divergence of semantic roles and syntactic constituent structures, the placement of "grammatical morphemes", and generalizations that hold across related syntactic constructions. An adequate theory of language learning will have to account for the acquisition of languages with these sorts of properties. Henry Hamburger, Kenneth Wexler, and Peter Culicover have taken a large step in this direction by constructing a mathematical model which incorporates some reasonable assumptions about the language learner, and which they prove is capable of learning transformational grammars of a certain type (Hamburger and Wexler, 1975; Wexler, Culicover, and Hamburger, 1975; Culicover and Wexler, 1977).

Central to Hamburger, Wexler, and Culicover's theory is the assumption that the learner is innately constrained to entertain hypotheses of a certain sort, and is therefore capable of acquiring only certain types of languages. As I have mentioned, this assumption could conceivably enable an enumerative language learner to learn a language with access only to a sample of sentences. The assumption is also implicit in a weak form in the heuristic approach to language learning, and is explicitly embraced by Anderson when he claims that the learner "assumes" that the target language conforms to the Graph Deformation Condition and to the Semantics-Induced Equivalence of Syntax Condition. But Hamburger *et al.*, take the strongest view, originally proposed by Chomsky (1962, 1965), that innate, language-specific constraints cause the child to consider only a very narrowly-defined class of transformational grammars. Hamburger, Wexler, and Culicover's feat was to define these constraints in a precise way, show why they contribute to learnability, and make the case that natural languages fall into the class they define.

Hamburger *et al.*, begin with a version of Chomsky's transformational grammar, in which a set of context-free *base rules* generates a *deep structure* tree which *transformations* operate upon to produce a sentence. The base rules can generate arbitrarily large deep structures only by rewriting sentences within sentences, that is, by repetitively applying one of the rules that rewrites the "S" symbol. Each occurrence of an "S" delineates a *level* in the deep structure. Transformational rules are applied first at the lowest level (i.e., the most deeply embedded subsentence), then to the second lowest level, and so on.

Learnability of transformational grammars from a text

Wexler and Hamburger (1973) first attempted to prove that a constrained class of transformational grammars was identifiable in the limit from a sample of sentences (see the section on "Constraining the Hypothesis Space" in

Section IV). They made the assumption, known to be overly strong, that all languages have identical base rules and differ only in their transformational rules. Thus they made the base rules innate, and required the learner to identify in the limit a set of transformations that generated the target language. This they proved to be impossible. Therefore, in their next attempts (Hamburger and Wexler, 1975; Wexler, Culicover, and Hamburger, 1975) they assumed, with Anderson and the "Cognitive" theorists, that the child has simultaneous access to a string and its meaning, and must learn rules that translate one into the other.

Semantic representations and the Invariance Principle
In Hamburger *et al.*'s model, a sentence meaning is represented by a tree structure that has the same hierarchical breakdown of constituents as the deep structure of the sentence, but with no particular left-to-right ordering of the constituents (such a structure is similar to Anderson's "Prototype structure"). Since deep structure constituents are ordered differently in different languages, the first task for the learner is to learn the base rules which define the orderings his language uses. Wexler and Culicover note that this can be accomplished in a number of simple ways (in fact, Anderson's Tree-fitting heuristic is one such way). Like Anderson, they point out that this assumes that in all natural languages the deep structures *will* preserve the hierarchical connectivity of nodes in semantic structures, differing only in their linear order (i.e., branches may not cross, nor may links be severed and re-attached elsewhere). They justify this *Invariance Condition* (similar, of course, to Anderson's Graph Deformation Condition) by showing that out of all the combinatorial possibilities for ordering constituents of a certain type in deep structures, only those that respect the Invariance Condition are found in natural languages (over 200 of which they examine, Culicover and Wexler, 1974).

The learning procedure
From then on the learner must hypothesize a set of transformations, or a *transformational component*, that in combination with the base rules generates the target language. The procedure is simple. The learner undergoes an infinite series of trials in which he is presented with a meaning-sentence pair and is required to guess a grammar. For each pair, the learner applies his current transformational rules to the deep structure (which he computes from the meaning structure), and compares the result against the input string. If they match, the learner leaves his grammar untouched and proceeds to the next pair. If they do not match, the learner randomly decides between two courses of action. He can discard, at random, any of the transformations he

used to derive the incorrect string; or, he can hypothesize a set consisting of all the transformations capable of transforming the deep structure to the input string in conjunction with the rest of the grammar, and select one of these transformations at random for inclusion in the grammar. Hamburger *et al.*, prove that with suitable constraints on the transformations used by the target language (and hypothesized by the learner), the learner will converge on a correct grammar for the language (i.e., the probability that the learner will have guessed a correct grammar becomes arbitrarily close to 1 as time passes). The proof is long and complex and will not be outlined here. Instead I will summarize how the constraints that Hamburger *et al.*, propose function to guarantee learnability. This, of course, is the crux of the Chomskian claim that learnability considerations favor a strongly nativist theory of language acquisition.

Proving learnability

As I have mentioned in Section IV, restricting the learner's hypothesis space only yields learnability if the intersection between the grammars in the hypothesis space and the grammars consistent with the sample becomes smaller and smaller as learning proceeds (see Figure 2). Hamburger *et al.* must show that when the learner has guessed an incorrect transformational component, he need not wait an arbitrarily long time before discovering his error, that is, encountering a semantic structure that the Component does not properly transform into the corresponding sentence. This in turn implies that the learner must not have to wait until an arbitrarily *complex* meaning-sentence pair appears in the sample before knowing that his transformational component is incorrect, since by the laws of probability he would have to wait an arbitrarily long time for an arbitrarily complex pair. In other words, if the learner has an incorrect transformational component, that component must make an error on a sentence-meaning pair that is no more complex than a certain bound (where complexity is measured by the number of S-nodes or levels in the deep structure).

This condition is not satisfied for unconstrained transformational grammars. In transformational grammars, each transformation is triggered by a particular configuration of symbols in a deep structure, or *structural description.* If a structural description can be arbitrarily complex for a transformation in the grammar, then the learner would have to wait until a meaning-sentence pair of that (arbitrary) complexity appeared in the sample before having occasion to hypothesize such a transformation. It would then be impossible to prove that the probability of the learner having hypothesized a complete, correct grammar approaches unity with increasing exposure to the sample. So Hamburger *et al.*, proposed the following constraint on transformations: no transformation may have a structural description that refers to symbols in more than two adjacent levels in the deep structure. Consider the deep structure-sentence pair in Figure 6 (the example has been simplified drastically from Chomsky, 1973). Assuming that the learner's transformational component does not yet correctly map one onto the other, the learner could hypothesize something like the following transformation (assuming that other transformations place the grammatical morphemes properly):

NP VP NP VP NP VP what → what NP VP NP VP NP VP.

However this transformation would be forbidden by Hamburger *et al.*'s constraint, because the symbols on the left hand side span across three levels in the deep structure. Instead, the learner could hypothesize something like the following:

NP VP what → what NP VP

which, applied successively from the deepest level upward, would produce the same string. (It is interesting to note that in this example the learner would not even have had to wait until encountering a pair this complex to hypothesize the transformation – an interrogative sentence with one level would have sufficed.) Hamburger *et al.*, argue that virtually all transformations in English and other languages conform to this condition, which they call the *Binary Principle.* Although they proposed the principle because, without it, they could not have proved learnability, they point out that Chomsky (1973) independently proposed an identical constraint, the *Subjacency Condition*, which he justified on descriptive grounds. That is, there seems to be independently motivated evidence that the Binary Principle is true of natural languages.

Figure 6. *Deep structure* (A) *and string* (B) *illustrating the Binary Principle.*

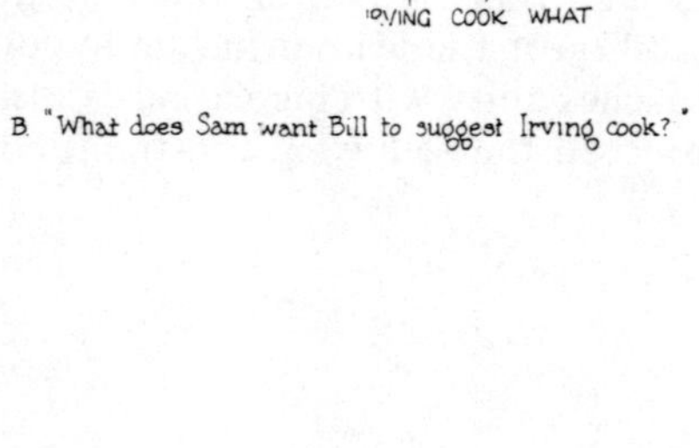

The Freezing Principle

The Binary Principle is not sufficient, however, to guarantee that an incorrect transformational component will make a telltale error on a meaning-sentence pair less complex than a certain bound. The base rules of a grammar can generate only a finite number of structures within a single level, by definition. Together with the Binary Principle, this would seem to ensure that input data of bounded complexity would suffice to exhaust all the structural descriptions that could trigger transformations. Unfortunately, whenever a transformation is applied at one level, it can alter the configuration of symbols within another level, creating new potential structural descriptions for transformations. Thus a series of transformations starting arbitrarily far down in a deep structure can alter the configuration of symbols within another level (as, in fact, the example in Figure 6 showed), creating new potential structural descriptions for transformations. A learner whose transformational component was in error only when applied to this altered configuration would never discover the error until coming across this arbitrarily complex structure. To remedy this situation, Culicover, Wexler, and Hamburger (1975) proposed a new constraint, the *Freezing Principle*, which forbids a transformation to apply to a configuration of symbols that could only have been created by the previous application of another transformation. The artificial example in Figure 7 shows how the constraint works. Say the learner must transform the deep structure 7(a) into the string 7(c), and already has a transformation that reverses the two morphemes C and B, as shown in 7(b). Now he must coin a transformation that reverses the morphemes A and B. The following transformation, for example, would accomplish this reversal:

ABDC → BADC.

However, the Freezing Principle forbids this hypothesis, since it refers to the symbol sequence DC, which was not generated by a base rule but was created by another transformation. Instead, the learner can hypothesize the following transformation:[3]

AB → BA.

With the Binary and Freezing Principles, Hamburger, Wexler, and Culicover not only prove that the learner will converge on a correct grammar, but that

[3] As the example suggest, the Binary and Freezing Principles tend to reduce the context-sensitivity of rules in grammars by preventing large parts of tree structures from entering into the structural descriptions of transformations. This is not a coincidence, since in general context-free rules are more easily learnable than context-sensitive rules. See also Kaplan (1978) who argues that the reduction of context-sensitivity afforded by the Subjacency (i.e., Binary) Principle contributes to efficient sentence parsing.

Figure 7. *Deep structure* (A) *and string* (C) *illustrating the Freezing Principle.*

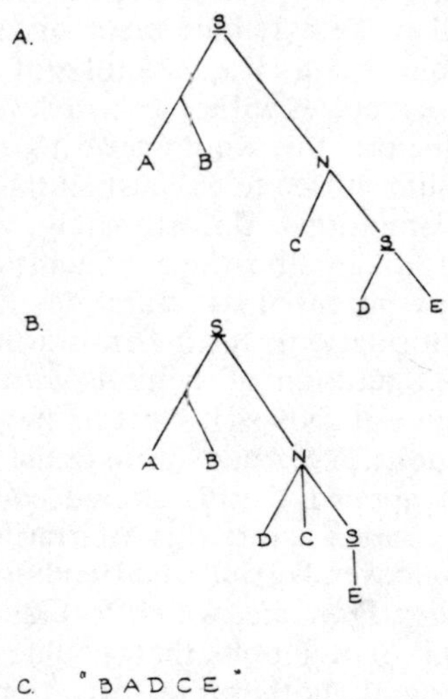

he can do so without even having to consider any structure with more than two levels of embedded sentences (i.e., three S-nodes).

Of course, Hamburger, Wexler, and Culicover must show that their constraint does not prevent their learner from acquiring any natural language. In Wexler, Culicover, and Hamburger (1975) and Culicover and Wexler (1977), examples of many sorts of English constructions are adduced to support the contention that natural languages obey the Freezing Principle. Moreover, Wexler *et al.* argue that in some cases the Freezing Principle does a better job than other constraints proposed in the linguistics literature at explaining why certain types of sentences are judged ungrammatical, and that in other cases, it mandates a choice between competing, hitherto equally plausible theories.

An evaluation

In evaluating the Hamburger *et al.*, model, it is important to note that I have changed the emphasis somewhat from their presentation. Their chief goal was to develop an "explanatorily adequate" linguistic theory (see Chomsky,

1965), which not only accounts for various linguistic phenomena, but *explains* why they must be one way and not another. Thus Hamburger, Wexler, and Culicover claim that the reason why natural languages conform to the Invariance, Binary, and Freezing Principles is that if they did not, they would not be learnable. Their model of a language learner was their means of justifying the claim.

Secondarily, they present their learning model as a first step toward an adequate theory of language learning (which is what I have been emphasizing). As such, they can claim no more than that their model is (at least) "minimally plausible". It requires no information about non-sentences, does not have to remember the entire sample, requires sentences no more complex than those with two levels of subsentences, employs semantic information in learning, processes sentences one at a time, and changes its gramma rule by rule. In other words, it does not flagrantly contradict some of the obvious facts of human language development. However, since the model is only a specification of the boundary conditions of a theory of language learning (i.e., they are claiming that the child's hypotheses must be *no less constrained* than those of the model), many features would have to be fleshed out before it could be considered any more than "minimally plausible". First, there is no indication at present that the learner would converge in a time-span comparable to a human childhood. It seems inefficient and implausible to have the child enumerating sets of transformations and mentally rolling dice to decide which to keep or discard. What is needed is a theory showing how the child's hypotheses are guided in a more direct way by the meaning-sentence pair under consideration, and how these hypotheses are computed during the left-to-right processing of a sentence. Third, unordered deep structures are questionable candidates for a theory of the child's representational system (although this will be discussed further in Section IX). Finally, we are left with few suggestions as to how the transformational component, once acquired, is used in producing and comprehending sentences.

In any case the Hamburger, Wexler, and Culicover model is a unique and extremely impressive achievement. Theirs is the only model that is capable of learning natural languages in all their complexity and that at the same time is not blatantly counter to what we know about the child and his learning environment. They also have clarified and justified, more clearly than anyone else has, two central tenets of transformational linguistics: that considerations of language learnability can dictate a choice between rival linguistic theories, and that learnability considerations imply strong innate constraints of a certain type on the child's language learning faculties. As they put it, "The bridge that Chomsky has re-erected between psychology and linguistics bears two-way traffic" (Hamburger and Wexler, 1975).

IX. Implications for Developmental Psycholinguistics

Toward a Theory of Language Learning

Among the models of language learning that I have considered, two seem worthy upon examination to serve as prototypes for a *theory* of human language acquisition. Anderson's LAS program roughly meets the Cognitive, Input, and Time Conditions, while faring less well with the Learnability and Equipotentiality Conditions. Hamburger, Wexler, and Culicover's transformational model meets the Learnability and Equipotentiality Conditions (clearly), and the Input Condition (perhaps), while faring less well with the Cognitive and Time Conditions. I hope it is not too banal to suggest that we need a theory that combines the best features of both models. It must incorporate a psychologically realistic comprehension process, like Anderson's system, since language acquisition is most plausibly thought of as being driven by the comprehension process. But at the same time, the model's semantic structures must be rich enough, and the hypothesization procedure constrained enough, that any natural language can be shown to be learnable (like the Hamburger *et al.*, model), so that the model does not become buried under a pile of ad hoc, semi-successful heuristics when it is extended to more and more linguistic domains. Of course, developing such a theory has been hampered by the lack of a suitable theory of language itself, one that both gives a principled explanation for linguistic phenomena in various domains and languages, and that can be incorporated in a reasonable way into a comprehension model (see Bresnan, 1978, for a step in this direction). Of course, here is not the place to attempt to present a new theory synthesizing the best features of previous efforts. Instead, I will attempt to point out the implications that the formal study of language learning has for current issues in developmental psycholinguistics.

Developmental Psycholinguistics and Language Acquisition Devices

Current attitudes toward language acquisition models among developmental psycholinguists have been strongly influenced by the fate of a research framework adopted during the 1960's that went under the name of the Language Acquisition Device, or LAD. There were in fact two different meanings to the expression Language Acquisition Device, and I think it is important to distinguish them. In one formulation (Chomsky, 1962), the child was idealized as an abstract device that constructed rules for an unknown language on the basis of a sample of sentences from that language; characterizing the workings of that "device" was proposed as a goal for linguistics and psychology. As an analogy, we could think of a physiologist interested in electrolyte regulation who idealized the brain as "a bag of salt water", proceeding then to study the structure of the membrane, concentration of ions, and so on. Of course, in this sense, I have been talking about language acquisition devices throughout the present paper. However there is a second, stronger sense in which LAD is taken to describe a specific theory of language acquisition (e.g., Clark, 1973; Levelt, 1973). In this sense (Fodor, 1966; McNeill, 1966), the child is said to possess an innate mental faculty containing highly specific knowledge about transformational grammars, which extracts deep structures from the speech around him and adopts transformational rules, one at a time, culminating in a transformational grammar for the language. Pursuing the analogy with physiology, LAD would correspond in this sense to our physiologist proposing that the brain accomplished electrolyte regulation by means of a special purpose structure, "a bag of salt water", with various properties. In support of this theory, it was claimed that the child based his learning on a sample of speech composed largely of fragments and complex, semi-grammatical expressions (Chomsky, 1965), that the early utterances of the child displayed mastery of highly abstract syntactic relations (McNeill, 1966), and that the linguistic progress of the child seemed to reflect the accretion of transformations (e.g., Brown and Hanlon, 1970). However the entire approach quickly fell into disfavor when it was found that the speech directed to children was well-formed and structurally simple (Snow, 1972), that the child might exploit semantic information in addition to sentences themselves (e.g., Macnamara, 1972), that the early speech of children might be better broken down into "cognitive" or semantic relations than into abstract syntactic ones (e.g., Bowerman, 1973; Brown, 1973), and that in many cases children learned transformationally complex constructions before they learned their simpler counterparts (e.g., Maratsos, 1978). As a result, LAD has been abandoned by developmental psycholinguists as a *theory*, and in its place I think there has developed a rough consensus that semantic and pragmatic information, together with the simplified speech of parents, allows children to learn language by using general cognitive skills rather than a special language-specific faculty. However, LAD has also been rejected in its more general sense as a *problem* to be addressed, and it also seems to me that most debates in developmental psycholinguistics are, unfortunately, no longer carried out with an eye toward ultimately specifying the mechanisms of syntax acquisition. When specific proposals concerning such mechanisms are considered, I shall argue, the substance of many of these debates can change significantly.

Nativism versus Empiricism: Two Extreme Proposals

Formal results from the study of language learnability give us grounds for dismissing quite decisively two general proposals concerning what sort of mechanisms are necessary and sufficient for language learning, one empiricist, one nativist.

The extreme empiricist proposal is that there are no language-specific a priori constraints on the types of rules that humans can acquire. In this vein, it is argued that once a sufficient number of sentences has been observed, languages can be learned by "general multipurpose learning strategies" (Putnam, 1971), by "discovery procedures" (Braine, 1971), or by "learning algorithms" like a "discretizer-plus-generalizer" that "extracts regularity from the environment" (Derwing, 1973). As I have mentioned, Gold's enumeration procedure is the most powerful imaginable realization of a general learning algorithm. Nevertheless, even this procedure is inadequate in principle to acquire rules on the basis of a sample of sentences. And if the criterion for "acquisition" is weakened (by requiring only approachability, approximations to the target language, etc.), then learning is possible, but not within a human lifespan.

At the other extreme is the proposal that innate knowledge of the properties of natural languages, especially those of deep structures, allows the child to learn a language from a sample of sentences (e.g., Fodor, 1966; McNeill, 1966). In one of Hamburger and Wexler's early models (Wexler and Hamburger, 1973), they imposed constraints on the learner's hypotheses that were known to be unrealistically stringent (e.g., that all languages share identical deep structure rules). Nevertheless they proved that this class of languages is unlearnable on the basis of a sample of sentences, and therefore, that the same must be true of classes that are specified more weakly (and hence more realistically). Of course, it is still possible that a different sort of innate constraint might guarantee learnability, but this will remain a matter of speculation until someone puts forth such a proposal.

Problems for the Cognitive Theory of Language Acquisition

The inability of these procedures to induce grammars from samples of sentences suggests strongly that semantic and pragmatic information is used in language learning. The moderate success of the models of Anderson and of Hamburger *et al.*, also lends credence to this conclusion. However, despite the great popularity of the Cognitive Theory among developmental psycholinguists, there has been little discussion of what I believe to be the foundation of the theory: the precise nature of the child's internal representations. The Cognitive Theory requires that children have available to them a system of representational structures similar enough in format to syntactic structures to promote language learning, and at the same time, flexible and general enough to be computable by children's cognitive and perceptual faculties on the basis of nonlinguistic information. Until we have a theory of the child's mental representations that meets these conditions, the Cognitive Theory will remain an unsupported hypothesis. Unfortunately, designing a representational system with the desired properties will be far from a simple task. The two main problems, which I call the "encoding problem" and the "format problem", pit the Cognitive Condition against the Learnability and Equipotentiality Conditions.

The encoding problem

This problem is a consequence of the fact that languages can describe a situation in a number of ways, and that humans can perceive a situation in a number of ways. One might plausibly attribute many different representational structures to a child perceiving a given situation, but only one of these structures will be the appropriate one to try to convert into the sentence being heard simultaneously. Barring telepathy, how does the child manage to encode a situation into just the structure that underlies the sentence that the adult is uttering?

Consider an earlier example. Anderson assumes that when a child sees, say, a white cat eating a mouse, his mind constructs a structure something like the one in Figure 3(a). This is fortunate for the child (and for the model-builder), since in the example the sentence arriving concurrently happens to be "The white cat eats a mouse", whose meaning corresponds to that structure. But what if the sentence were "The mouse is being eaten by the cat", "That's the second mouse that the cat has eaten", "Some cats don't eat mice", "What's that white cat doing with the mouse?", and so on? To put it differently, assuming that the original sentence was the one uttered, what if the child were to have constructed a cognitive structure containing propositions asserting that the mouse was 'all gone', or that the cat and mouse were playing, or that the mouse looked easy for the cat to eat, and so on? In any of these cases, the child would face the task of trying to map a meaning structure onto a string with which it has only a tenuous connection. Thus the semantic representation would offer few clues, or misleading ones, about how to hypothesize new rules.[4]

[4]Dan Slobin (1978; personal communication) has pointed out that the child faces a similar problem in learning the morphology of his language. Natural languages dictate that certain semantic features of the sentence referent (e.g. number, person, gender, definiteness, animacy, nearness to the speaker, completedness, and so on) must be signalled in prefixes, suffices, alternate vowel forms, and other means. However, these features are by no means all that a child *could* encode about an event: the color, absolute position, and texture of an object, the time of day, the temperature, and so on, though certainly perceptible to the child, are ignored by the morphology of languages, and hence should not be encoded as part of the semantic structure that the child must learn to map onto the string. To make matters worse, the morphological rules of different languages select different subsets of these features

(Continued overleaf)

I have already mentioned that Anderson would face this problem if he were to multiply the number of available mental predicates that correspond to a given verb, in order to foster certain generalizations. Hamburger *et al.* face a similar problem. In their model, the structures underlying synonymous sentences, such as actives and passives, are presumably identical except for a marker triggering a transformation in cases like the passive (since each transformation is obligatorily triggered by some deep structure configuration). Again, it is not clear how the child knows when to insert into his semantic structure the markers that signal the transformations that the adult happens to have applied.

Possible solutions to the encoding problem

I see three partial solutions to the encoding problem that together would serve to reduce the uncertainty associated with typical language learning situations, ensuring that the child will encode situations into unique representations appropriate to the sentences the adult is uttering. The first relies on the hypothesis that the representational system of the child is less powerful and flexible than that of the adult, and is capable of representing a given situation in only a small number of ways. Thus in the preceding example, the child is unlikely to encode the scene as propositions asserting that the mouse was not eating the cat, that all cats eat mice, etc. As the child develops, presumably his representational powers increase gradually, and so does the range of syntactic constructions addressed to him by his parents. If, as is often suggested (e.g., Cross, 1977), parents "fine-tune" their speech to the cognitive abilities of their children, that is, they use syntactic constructions whose semantics correspond to the representations most likely to be used by the child at a given moment, then the correspondence between the adult's sentence meaning and the child's encoding of the situation would be closer than we have supposed.

The second solution would posit that the child's social perception is acute enough to detect all the pragmatic or communicative differences that are concurrently signaled by syntactic means in different sentences (see Bruner, 1975). That is, the child knows from the conversational context what the adult is presupposing, what he or she is calling attention to, what is being asserted of what, and so on. For example, the child must not only *see* that

---

to signal obligatorily, and disagree further over which features should be mapped one-to-one onto morphological markers, and which sets of features should be conflated in a many-to-one fashion in particular markers. Thus there has to be some mechanism in the child's rule-hypothesization faculty whereby his possible conceptualizations of an event are narrowed down to only those semantic features that languages signal, and ultimately, down to only those semantic features that his target language signals.

the cat is eating the mouse, but must know that the adult is asserting of the cat that it is eating a mouse, instead of asserting of the mouse that it is disappearing into the cat, or many other possibilities. (As mentioned earlier, Anderson used this rationale in developing LAS, when he marked one of the propositions in each semantic structure as the intended "main proposition" of the sentence.) If this line of reasoning is correct, strong conditions are imposed both on the language and on the learner. The syntax of languages must not allow synonymy, in a strict sense: any two "base" structures (i.e., Anderson's Prototype structure or Hamburger *et al.*'s deep structure) that do not differ semantically (i.e., instantiate the same propositions) must differ pragmatically in some way. Conversely, the pragmatic and perceptual faculties of the child must be capable of discriminating the types of situations that occasion the use of different syntactic devices.

The third solution would equip the child with a strategy that exploited some simple property of the sentence to narrow down the possible interpretations of what the adult is asserting. Anderson implicated a strategy of this sort when LAS examined the set of words in a sentence and retained only the propositions in its meaning structure whose concepts corresponded to those words. In the present example, the child might always construct a proposition whose subject corresponds to the first noun in the sentence, and then choose (or, if necessary, create) some mental predicate that both corresponds to the verb and is consistent with his perception of the scene. Thus, when hearing an active sentence, the child would construct a proposition with the cat as the subject and "EATS" as part of the predicate; when hearing the passive version, the proposition would have the mouse as the subject and "IS-EATEN-BY" as part of the predicate.[5] One can even speculate that such a strategy is responsible for Bever's (1970) classic finding that children of a certain age interpret the referent of the first noun of both active and passive sentences as the agent of the action designated by the verb. The children may have set up the concept corresponding to the first noun as the subject of a proposition, but, lacking mental predicates like "IS-EATEN-BY" at that stage in their development, they may have mistakenly chosen predicates like "EATS" by default.

I hope to have shown how consideration of the requirements and implications of formal theories of language learning (in this case, those of Anderson and of Hamburger *et al.*) lead one to assign more precise roles to several phenomena studied intensively by developmental psycholinguists. Specific-

[5] This example follows the Anderson model with the "multiple predicate" modification I suggested. In the Hamburger *et al.* model, the child could insert a "transformation marker" into his deep structure whenever the subject of the deep structure proposition was not the first noun in the sentence.

ally, I suggest that the primary role in syntax learning of cognitive development, "fine-tuning" of adult speech to children learning language, knowledge of the pragmatics of a situation, and perceptual strategies is to ensure that the child encodes a situation into the same representational structure that underlies the sentence that the adult is uttering concurrently (cf. Bruner, 1975; Bever, 1970; Sinclair de-Zwart, 1969; and Snow, 1972; for different interpretations of the respective phenomena).

The Format Problem

Once we are satisfied that the child has encoded the situation into a unique representation, corresponding to the meaning of the adult's sentence, we must ensure that that representation is of the appropriate *format* to support the structural analyses and generalizations required by the learning process.

To take an extreme example of the problem, imagine that the study of perceptual and cognitive development forced us to conclude that the internal representations of the child were simply lists of perceptual features. Using a semantics-based generalization heuristic, the learner would have no trouble merging words like "cat" and "mouse", since both are objects, furry, animate, four-legged, etc. But the learner would be unable to admit into this class nouns like "flutter" or "clang", which have no perceptual features in common with "cat", nor "fallacy" or "realization", which have no perceptual features at all. The difficulties would intensify with more abstract syntactic structures, since there are no conjunctions of perceptual features that correspond to noun phrases, relative clauses, and so on. The problem with this representational format is that even if it were adequate for perception, it is not adaptable to syntax learning. It does not provide the units that indicate how to break a sentence into its correct units, and to generalize to similar units across different sentences.

In other words, what is needed is a theory of representations whose elements correspond more closely to the elements of a grammar. In Anderson's theory, for example, a representation is composed of a "subject" and a "predicate", which in turn is composed of a "relation" and an "object". These correspond nicely to the syntactic rules that break down a sentence into a noun phrase and a verb phrase, then the verb phrase into a verb and another noun phrase. Furthermore, propositions encoded for different situations in which syntactically similar sentences would be uttered would all have the same format, regardless of whether they represent furry things, square things, events, actions, abstract mathematical concepts, or other propositions. Hamburger *et al.*, posit a cognitive representation with a format even more suitable to language learning: unordered deep structures. This is one of the reasons why their model is more successful at acquiring syntactic rules than

LAS is. In sum, these theorists posit that the syntax of the language of thought is similar to the syntax of natural languages.

However, this solution might create problems of its own. It is possible for theorists to use "cognitive" representations with a format so suitable to syntactic rule learning that the representations may no longer be plausible in a theory of perception or cognition. To take a hypothetical example, in standard transformational grammars a coordinated sentence such as "Jim put mustard and relish on his hot dog" is derived from a two-part deep structure, with trees corresponding to the propositions "Jim put mustard on his hot dog" and "Jim put relish on his hot dog". However a theory of cognitive or perceptual representations based on independent evidence (e.g., reaction times, recall probabilities, etc.), when applied to this situation, might not call for two separate propositions, but for a single proposition in which one of the arguments was divided into two parts, corresponding to the two conjoined nouns (which is the way it is done in Anderson and Bower, 1973, for example). Cases like this, if widespread and convincing, would undermine Hamburger *et al.*'s premise that unordered deep structures are plausible as cognitive representations.

In this vein, it is noteworthy that even though Anderson's semantic structures were lifted from his theory of long term memory, they too are more similar to linguistic deep structures than those of any other theory of memory representation, incorporating features like a binary subject-predicate division, distinct labels for each proposition, and a hierarchical arrangement of nodes (cf., Norman and Rumelhart, 1975; Winston, 1975). In fact, many of these features are not particularly well-supported by empirical evidence (see Anderson, 1976), and others may be deficient on other grounds (see Woods, 1975). Concerning other computer models in which "the designer feeds in what he thinks are the semantic representations of utterances", McMaster *et al.* (1976, p. 377) remark that "the risk is that [the designer] will define semantics in such a way that it is hardly different from syntax. He is actually providing high-level syntactic information. This gives the grammar-inferrer an easy task, but makes the process less realistic...".[6]

Implications of the format problem

Faced with possibly conflicting demands on a theory of the form of mental representation from the study of language learning and the study of other

[6]This discussion has assumed that the language-specific structures posited as cognitive representations are specific to languages in general, not to particular languages. If the representations are tailored to *one* language (e.g., when predicates in LAS's propositions take the same number of arguments as the verb they correspond to, even though the same verbs in different languages take different numbers of arguments), a second and equally serious problem results.

cognitive processes, we have two options. One is to assert that, all other considerations notwithstanding, the format of mental representations *must* be similar to syntactic structures, in order to make language learning possible. Fodor (1976), for example, has put forth this argument.[7] The second is to posit at least two representational formats, one that is optimally suited for perception and cognition, and one that is optimally suited for language learning, together with a conversion procedure that transforms a representation from the former to the latter format during language learning. Anderson and Hamburger *et al.*, already incorporate a version of this hypothesis. In LAS, the semantic structures are not entirely suitable for rule learning, so there is a procedure that converts them into the "prototype structures". And in the Hamburger *et al.*, model, the deep structures are not entirely suitable as cognitive representations (being too specific to particular languages), so there is a procedure whereby they are derived from "semantic structures". Ultimately the Cognitive Theory of language learning must posit one or more representational formats appropriate to cognition in general and to language learning in particular, and, if necessary, the procedures that transform one sort of representation into the other.

Nativism and empiricism revisited

It is often supposed that if children indeed base their rule learning on cognitive representational structures, the traditional case for nativism has been weakened (e.g., Schlesinger, 1971; Sinclair de-Zwart, 1969). According to this reasoning, cognitive structures already exist for other purposes, such as perception, reasoning, memory, and so forth, so there is no need to claim that humans possess an innate set of mental structures specific to language. However, this conclusion is at best premature. It is far from obvious that the type of representational structures motivated by a theory of perception or memory is suitably adaptable to the task of syntactic rule learning. For if the foregoing discussion is correct, the requirements of language learning dictate that cognitive structures are either language-like themselves, or an innate procedure transforms them into structures that are language-like. When one considers as well the proposed innate constraints on how these structures enter into the rule hypothesization process (i.e., Anderson's Graph Deformation and Semantics-Induced Equivalence Principles, and Hamburger *et al.*'s Binary and Freezing Principles), one must conclude that the Cognitive Theory

[7]Incidentally, it is ironic that Anderson, in a different context, fails to mention this argument when he examines the case for propositional theories of mental representation in general (Anderson, 1978).

of language learning, in its most successful implementations, vindicates Chomsky's innateness hypothesis if it bears on it at all.[8]

Language learning and other forms of learning

It might be conjectured that if one were to build models of other instances of human induction (e.g., visual concept learning, observational learning of behavior patterns, or scientific induction), one would be forced to propose innate constraints identical to those proposed by the designers of language learning models. If so, it could be argued that the constraints on language learning are necessitated by the requirements of *induction* in general, and not natural language induction in particular. While it is still too early to evaluate this claim, the computer models of other types of induction that have appeared thus far do not seem to support it. In each case, the representational structures in which data and hypotheses are couched are innately tailored to the requirements of the *particular domain of rules being induced.* Consider Winston's (1975) famous program, which was designed to induce classes of block-structures, such as arches and tables, upon observing exemplars and non-exemplars of the classes. The units of the program's propositional structures can designate either individual blocks, blocks of triangular or rectangular shape, or any block whatsoever; the connecting terms can refer to a few spatial relations (e.g., adjacency, support, contact) and a few logical relations (e.g., part-whole, subset-superset). The program literally cannot conceive of distance, angle, color, number, other shapes, disjunction, or implication. This removes the danger of the program entertaining hypotheses other than the ones the programmer is trying to teach it. Similarly. Soloway and Riseman's (1977) program for inducing the rules of baseball upon observing sample plays is fitted with innate knowledge of the kind of rules and activities found in competitive sports in general. Langley's (1977) program for inducing physical laws upon observing the behavior of moving bodies is confined to considering assertions about the values of parameters for the positions, velocities, and accelerations of bodies, and is deliberately fed only those attributes of bodies that are significant in the particular mock universe in which it is "placed" for a given run. These restrictions are not just adventitious shortcuts, of course. Induction has been called "scandalous" because any finite set of observations supports an intractably large number of gener-

[8] One could contest this conclusion by pointing out that it has only been shown that the various nativist assumptions are *sufficient* for learnability, not that they are *necessary*. But as Hamburger and Wexler put it (1975), "anyone who thinks the assumption[s are] not necessary is welcome to try to devise proofs corresponding to ours without depending on [those] assumptions".

alizations. Constraining the type of generalizations that the inducer is allowed to consider in a particular task is one way to defuse the scandal.

Parental Speech to Children
Frequently it is argued that the special properties of parents' speech to children learning language reduces the need for innate constraints on the learning process (e.g., Snow, 1972). Since these claims have not been accompanied by discussions of specific learning mechanisms that benefit from the special speech, they seem to be based on the assumption that something in the formal properties of the language learning task makes short, simple, grammatical, redundant sentences optimal for rule learning. However a glance at the models considered in the present paper belies this assumption: the different models in fact impose very different requirements on their input.

Consider the effects of interspersing a few ungrammatical strings among the sample sentences. Gold's enumeration learner would fail miserably if a malformed string appeared in the sample – it would jettison its correct hypothesis, never to recover it, and would proceed to change its mind an infinite number of times. On the other hand, Horning's Bayesian learner can easily tolerate a noisy sample, because here the sample does not mandate the wholesale acceptance or rejection of grammars, but a selection from among them of the one with the highest posterior probability. The Hamburger *et al.*, model would also converge despite the occasional incorrect input datum, since at any point in the learning process at which it has an incorrect grammar (e.g., if it were led astray by a bad string), there is a nonzero probability that it will hypothesize a correct grammar within a certain number of trials (assuming, of course, that it does not encounter another bad string before converging).

Similarly, it is doubtful that the length or complexity of sentences has a uniform effect on different models. Feldman described a procedure requiring that the sample sentences be ordered approximately by increasing length, whereas Gold's procedure is completely indifferent to length. In the Hamburger *et al*, model, contrary to the intuition of some, learning is facilitated by *complex* sentences – not only will the learner fail to converge if he does not receive sentences with at least two levels of embedded sentences, but he will converge faster with increasingly complex sentences, since in a complex sentence there are more opportunities for incorrect transformations or the absence of correct transformations to manifest themselves by generating the wrong string. Nevertheless, short and simple sentences may indeed facilitate learning in humans, but for a different reason. Since children have limited attention and memory spans, they are more likely to retain a short string of words for sufficient time to process it than they would a long string of words. Similarly, they are more likely to encode successfully a simple conceptualization of an event than a complex one. Thus short, simple sentences may set the stage for rule hypothesization while playing no role (or a detrimental role) in the hypothesization process itself.

Other models are sensitive to other features of the input. Since Klein and Kuppin's Autoling relies on distributional analysis, it thrives on sets of minimally-contrasting sentences. Since Anderson's LAS merges constituents with the same semantic counterparts, it progresses with sets of sentences with similar or overlapping propositional structures.

In sum, the utility of various aspects of the input available to a language learner depends entirely on the learning procedure he uses. A claim that some feature of parental speech facilitates rule learning is completely groundless unless its proponent specifies some learning mechanism.

Conclusions

In an address called "Word from the Language Acquisition Front", Roger Brown (1977) has cautioned:

"Developmental psycholinguistics has enjoyed an enormous growth in research popularity... which, strange to say, may come to nothing. There have been greater research enthusiasms than this in psychology: Clark Hull's principles of behavior, the study of the Authoritarian personality, and, of course, Dissonance Theory. And in all these cases, very little advance in knowledge took place. ...A danger in great research activity which we have not yet surmounted, but which we may surmount, is that a large quantity of frequently conflicting theory and data can become cognitively ugly and so repellent as to be swiftly deserted, its issues unresolved."

It is my belief that one way to surmount this danger is to frame issues in the context of precise models of the language learning process, following the lead of other branches of the cognitive sciences. I hope to have shown in this section why it may be necessary to find out how language learning *could* work in order for the developmental data to tell us how it *does* work.

References

Anderson, J. (1974) Language acquisition by computer and child. (Human Performance Center Technical Report No. 55.) Ann Arbor, University of Michigan.

Anderson, J. (1975) Computer simulation of a Language Acquisition System: A first report. In R. Solso (ed.), *Information processing and cognition: The Loyola Symposium.* Washington, Erlbaum.

Anderson, J. (1976) *Language, Memory, and Thought*. Hillsdale, N.J.: Erlbaum.
Anderson, J. (1977) Induction of augmented transition networks. *Cog. Sci., 1*, 125–157.
Anderson, J. (1978) Arguments concerning representations for mental imagery. *Psychol. Rev., 85*, 249–277.
Anderson, J. and G. Bower (1973) *Human Associative Memory*. Washi· on, Winston.
Bever, T. (1970) The cognitive basis for linguistic structures. In J. ..ayes (ed.), *Cognition and the Development of Language*. New York, Wiley.
Biermann, A. and J. Feldman (1972) A survey of results in grammatical inference. In S. Watanabe (ed.), *Frontiers in Pattern Recognition*. New York, Academic Press.
Bowerman, M. (1973) *Learning to talk: A Cross-sectional Study of Early Syntactic Development, with Special Reference to Finnish*. Cambridge, U.K., Cambridge University Press.
Braine, M. (1963) The ontogeny of English phrase structure: The first phrase. *Lang., 39*, 1–14.
Braine, M. (1971) On two models of the internalization of grammars. In D. Slobin (ed.), *The Ontogenesis of Grammar*. New York, Academic Press.
Bresnan, J. (1978) A realistic transformational grammar. In G. Miller, J. Bresnan and M. Halle (eds.), *Linguistic Theory and Psychological Reality*. Cambridge, Mass., MIT Press.
Brown, R. (1973) *A First Language: The Early Stages*. Cambridge, Mass., Harvard University Press.
Brown, R. (1977) Word from the language acquisition front. Invited address at the meeting of the Eastern Psychological Association, Boston.
Brown, R, C. Cazden and U. Bellugi (1969) The child's grammar from I to III. In J. Hill (ed.), *Minnesota Symposium on Child Psychology, Vol. II*. Minneapolis, University of Minnesota Press.
Brown, R. and C. Hanlon (1970) Derivational complexity and order of acquisition in child speech. In J. Hayes (ed.), *Cognition and the Development of Language*. New York, Wiley.
Bruner, J. (1975) The ontogenesis of speech acts. *J. child Lang., 2*, 1–19.
Chomsky, N. (1957) *Syntactic Structures*. The Hague, Mouton.
Chomsky, N. (1962) Explanatory models in linguistics. In E. Nagel and P. Suppes (eds.), *Logic, Methodology, and Philosophy of Science*. Stanford, Stanford University Press.
Chomsky, N. (1965) *Aspects of the Theory of Syntax*. Cambridge, Mass., MIT Press.
Chomsky, N. (1973) Conditions on transformations. In S. Anderson and P. Kiparsky (eds.), *A Festschrift for Morris Halle*. New York, Holt, Rinehart and Winston.
Clark, E. (1973) What should LAD look like? Some comments on Levelt. In *The Role of Grammar in Interdisciplinary Linguistic Research*. Colloquium at the University of Bielefeld, Bielefeld, W. Germany.
Cross, T. (1977) Mothers' speech adjustments: The contribution of selected child listener variables. In C. Snow and C. Ferguson (eds.), *Talking to Children: Input and Acquisition*. New York, Cambridge University Press.
Culicover, P. and K. Wexler (1974) The Invariance Principle and universals of grammar. (Social Science Working Paper No. 55.) Irvine, Cal., University of California.
Culicover, P. and K. Wexler (1977) Some syntactic implications of a theory of language learnability. In P. Culicover, T. Wasow, and A. Akmajian (eds.), *Formal Syntax*. New York, Academic Press.
Derwing, B. (1973) *Transformational Grammar as a Theory of Language Acquisition*. Cambridge, UK, Cambridge University Press.
Fabens, W. and D. Smith (1975) A model of language acquisition using a conceptual base. (Technical Report CBM-TR-55, Department of Computer Science.) New Brunswick, N.J.) Rutgers – The State University.
Feldman, J. (1972) Some decidability results on grammatical inference and complexity. *Information and Control, 20*, 244–262.
Fodor, J. (1966) How to learn to talk: Some simple ways. In F. Smith and G. Miller (eds.), *The Genesis of Language*. Cambridge, Mass., MIT Press.
Fodor, J. (1975) *The Language of Thought*. New York, Thomas Crowell.
Fu, K. and T. Booth (1975) Grammatical inference: Introduction and survey. *IEEE Transactions on Systems, Man, and Cybernetics, SMC-5(1)*, 95–111; *SMC–5(4)*, 409–423.
Gold, E. (1967) Language identification in the limit. *Information and Control, 16*, 447–474.
Gross, M. (1972) *Mathematical models in linguistics*. Englewood Cliffs, N.J., Prentice-Hall.

Hamburger, H. and K. Wexler (1975) A mathematical theory of learning transformational grammar. *J. Math. Psychol., 12,* 137–177.

Harris, Z. (1964) Distributional structure. In J. Fodor and J. Katz (eds.), *The Structure of Language.* Englewood Cliffs, N.J., Prentice Hall.

Hopcroft, J. and J. Ullman (1969) *Formal languages and their relation to automata.* Reading, Mass., Addison Wesley.

Horning, J. (1969) A study of grammatical inference. (Technical Report No. CS 139, Computer Science Dept.) Stanford, Stanford University.

Kaplan, R. (1975) On process models for sentence analysis. In D. Norman and D. Rumelhart (eds.), *Explorations in cognition.* San Francisco, W. H. Freeman.

Kaplan, R. (1978) Computational resources and linguistic theory. Paper presented at the Second Theoretical Issues in Natural Language Processing Conference, Urbana, Ill.

Kelley, K. (1967) Early syntactic acquisition. (Report No. P-3719.) Santa Monica, Cal., The Rand Corporation.

Klein, S. (1976) Automatic inference of semantic deep structure rules in generative semantic grammars. In A. Zampoli (ed.), *Computational and Mathematical Linguistics: Proceedings of 1973 International Conference on Computational Linguistics, Pisa.* Florence, Italy, Olschki.

Klein, S. and M. Kuppin (1970) An interactive program for learning transformational grammars. *Computer Studies in the Humanities and Verbal Behavior, III,* 144–162.

Klein, S. and V. Rozencvejg (1974) A computer model for the ontogeny of pidgin and creole languages. (Tehcnical Report No. 238, Computer Science Dept.) Madison: University of Wisconsin.

Knobe, B. and K. Knobe (1977) A method for inferring context-free grammars. *Information and Control, 31,* 129–146.

Kosslyn, S. and S. Schwartz (1977) A simulation of visual imagery. *Cog. Sci., 1,* 265–296.

Langley, P. (1977) BACON: A production system that discovers empirical laws. (CIP Working Paper No. 360.) Pittsburg, Carnegie Mellon University.

Levelt, W (1973) Grammatical inference and theories of language acquisition. In *The role of Grammar in Interdisciplinary Linguistic Research.,* Colloquium at the University of Bielefeld, Bielefeld, W. Germany.

Macnamara, J. (1972) Cognitive basis for language learning in infants. *Psychol. Rev., 79,* 1–13.

Maratsos, M. (1978) New models in linguistics and language acquisition. In G. Miller, J. Bresnan and M. Halle (eds.), *Linguistic Theory and Psychological Reality.* Cambridge, Mass., MIT Press.

McMaster, I., J. Sampson and J. King (1976) Computer acquisition of natural language: A review and prospectus. *Intern. J. Man-Machine Studies, 8,* 367–396.

McNeill, D. (1966) Developmental psycholinguistics. In F. Smith and G. Miller (eds.), *The genesis of language.* Cambridge, Mass., MIT Press.

Miller, G. (1967) Project Grammarama. In *The Psychology of Communication.* Hammonsworth, NY: Basic Books.

Moeser, S. and A. Bregman (1972) The role of reference in the acquisition of a miniature artificial language. *J. verb. Learn. verb. Behav., 12,* 91–98.

Moeser, S. and A. Bregman (1973) Imagery and language acquisition. *J. verb. Learn. verb. Behav., 12,* 91–98.

Newell, A. and H. Simon (1973) *Human problem solving.* Englewood Cliffs, N.J., Prentice Hall.

Newport, E., H. Gleitman and L. Gleitman (1977) Mother, I'd rather do it myself: Some effects and non-effects of maternal speech style. In C. Snow and C. Ferguson (eds.), *Talking to Children: Input and Acquisition.* New York, Cambridge University Press.

Norman, D. and D. Rumelhart (1975) *Explorations in Cognition.* San Francisco, W. H. Freeman.

Peters, S. and R. Ritchie (1973) On the generative power of transformational grammars. *Infor. Sci., 6,* 49–83.

Postal, P. (1964) Limitations of phrase structure grammars. In J. Fodor and J. Katz (eds.), *The Structure of Language.* Englewood Cliffs, N.J., Prentice Hall.

Putnam, H. (1971) The "Innateness Hypothesis" and explanatory models in linguistics. In J. Searle (ed.), *The Philosophy of Language.* London, Oxford University Press.

Reeker, L. (1976) The computational study of language acquisition. In M. Yovits and M. Rubinoff (eds.), *Advances in Computers, Vol. 15.* New York, Academic Press.

Rochester, S. (1973) The significance of pauses in spontaneous speech. *J. Psycholing. Res., 2,* 51–81.

Schlesinger, I. (1971) Production of utterances and language acquisition. In D. Slobin (ed.), *The Ontogenesis of Grammar.* New York, Academic Press.

Siklóssy, L. (1971) A language learning heuristic program. *Cog. Psychol., 2,* 279–295.

Siklóssy, L. (1972) Natural language learning by computer. In H. Simon and L. Siklóssy (eds.), *Representation and Meaning: Experiments with Information-processing Systems.* Englewood Cliffs, N.J., Prentice Hall.

Sinclair de-Zwart, H. (1969) Developmental psycholinguistics. In D. Elkind and J. Flavell (eds.), *Studies in Cognitive Development: Essays in Honor of Jean Piaget.* New York, Oxford University Press.

Slobin, D. (1973) Cognitive prerequisites for the development of grammar. In C. Ferguson and D. Slobin (eds.), *Studies in Child Language Development.* New York, Holt, Rinehart and Winston.

Slobin, D. (1978) Universal and particular in the acquisition of language. In *Language Acquisition: State of the Art.* Conference at the University of Pennsylvania, Philadelphia, May 1978.

Snow, C. (1972) Mothers' speech to children learning language. *Child Devel., 43,* 549–565.

Snow, C. and C. Ferguson (1977) *Talking to children: Language Input and Acquisition.* New York: Cambridge University Press.

Solomonoff, R. (1964) A formal theory of inductive inference. *Infor. Control, 7,* 1–22; 224–254.

Soloway, E. and E. Riseman (1977) Levels of pattern description in learning. (COINS Technical Report 77-5), Computer and Information Science Dept., Amherst, Mass., University of Massachusett.

Van der Mude, A. and A. Walker (1978) On the inference of stochastic regular grammars. *Infor. Control, 38,* 310–329.

Wexler, K., P. Culicover and H. Hamburger (1975) Learning-theoretic foundations of linguistic universals. *Theoret. Ling., 2,* 215–253.

Wexler, K. and H. Hamburger (1973) On the insufficiency of surface data for the learning of transformational languages. In K. Hintikka, J. Moravcsik and P. Suppes (eds.), *Approaches to Natural Languages.* Dordrecht, Netherlands: Reidel.

Wharton, R. (1974) Approximate language identification. *Infor. Control, 26,* 236–255.

Wharton, R. (1977) Grammar enumeration and inference. *Infor. Control, 33,* 253–272.

Winograd, T. (1972) A program for understanding natural languages. *Cog. Psychol., 3,* 1–191.

Winston, P. (1975) Learning structural descriptions from examples. In P. Winston (ed.), *The Psychology of Computer Vision.* New York, McGraw-Hill.

Woods, W. (1975) What's in a link: Foundations of semantic networks. In D. Bobrow and A. Collins (eds.), *Representation and Understanding: Studies in Cognitive Science.* New York, Academic Press.

[6]

COGNITIVE SCIENCE 4, 209–241 (1980)

What Does It Mean to Understand Language?

TERRY WINOGRAD
Stanford University

INTRODUCTION

In its earliest drafts, this paper was a structured argument, presenting a comprehensive view of cognitive science, criticizing prevailing approaches to the study of language and thought and advocating a new way of looking at things. Although I strongly believed in the approach it outlined, somehow it didn't have the convincingness on paper that it had in my own reflection. After some discouraging attempts at reorganization and rewriting, I realized that there was a mismatch between the nature of what I wanted to say and the form in which I was trying to communicate.

The understanding on which it was based does not have the form of a carefully structured framework into which all of cognitive science can be placed. It is more an orientation—a way of approaching the phenomena—that has grown out of many different experiences and influences and that bears the marks of its history. I found myself wanting to describe a path rather than justify its destination, finding that in the flow, the ideas came across more clearly. Since this collection was envisioned as a panorama of contrasting individual views, I have taken the liberty of making this chapter explicitly personal and describing the evolution of my own understanding.

My interests have centered around natural language. I have been engaged in the design of computer programs that in some sense could be said to "understand language," and this has led to looking at many aspects of the problems, including theories of meaning, representation formalisms, and the design and construction of complex computer systems. There has been a continuous evolution in my understanding of just what it means to say that a person or computer "understands," and this story[1] can be read as recounting that evolution. It is

[1]This is a "story" because like all histories it is made up. In an attempt to make sense of the chaos of past events one imposes more of a sense of orderliness than they deserve. Things didn't actually happen exactly in this order, and the events contain inconsistencies, throwbacks, and other misfortunes that would make it much harder to tell.

long, because it is still too early to look back and say "What I was *really* getting at for all those years was the one basic idea that . . ." I am too close and too involved in its continuation to see beyond the twists and turns. The last sections of the paper describe a viewpoint that differs in significant ways from most current approaches, and that offers new possibilities for a deeper understanding of language and a grasp on some previously intractable or unrecognized problems. I hope that it will give some sense of where the path is headed.

2. UP THROUGH SHRDLU

The Background

In the mid 1960s, natural language research with computers proceeded in the wake of widespread disillusionment caused by the failure of the highly touted and heavily funded machine translation projects. There was a feeling that researchers had failed to make good on their early confident claims, and that computers might not be able to deal with the complexities of human language at all. In AI research laboratories there were attempts to develop a new approach, going beyond the syntactic word-shuffling that dominated machine translation and other approaches based on key word search or statistical analysis. It was clear that for effective machine processing of language—whether for translation, question answering, or sophisticated information retrieval—an analysis of the syntactic structures and identification of the lexical items was not sufficient. Programs had to deal somehow with what the words and sentences meant.

There were a number of programs in this new vein described in the early collections of AI papers.[2] Each program worked in some very limited domain (baseball scores, family trees, algebra word problems, etc.) within which it was possible to set up a formal representational structure corresponding to the underlying meaning of sentences. This structure could be used in a systematic reasoning process as part of the overall language comprehension system. The model of language understanding that was implicit in those programs and in many AI programs since then is illustrated in Figure 1.

This model rests on some basic assumptions about language and representation:

1. Sentences in a natural language correspond to facts about the world.
2. It is possible to create a formal representation system such that
   (a) For any relevant fact about the world there is a corresponding structure in the representation system;

[2]Green et al. and Lindsay in Feigenbaum and Feldman (1963); Black, Bobrow, Qullian, and Raphael in Minsky (1967).

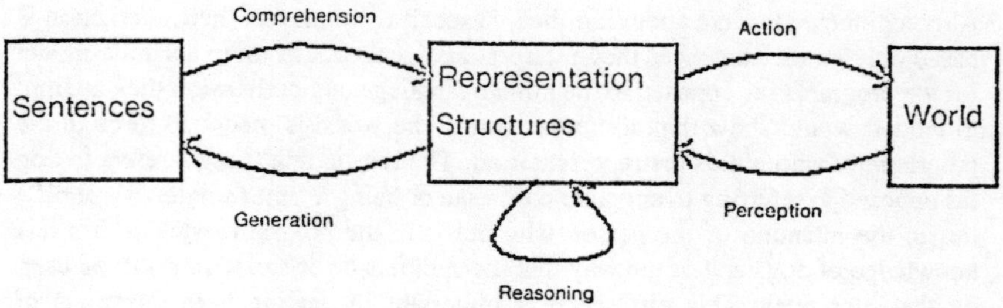

Figure 1. Basic AI model of language understanding.

(b) There is a systematic way of correlating sentences in natural language with the structure in the representation system that correspond to the same facts about the world; and
(c) Systematic formal operations can be specified that operate on the representation structures to do "reasoning." Given structures corresponding to facts about the world, these operations will generate structures corresponding to other facts, without introducing falsehoods.

This somewhat simplistic formulation needs some elaboration to be comprehensive. It is clear, for example, that a question or command does not "correspond to facts" in the same manner as a statement, and that it is unlikely that any actual reasoning system will be error-free. We will discuss some of these elaborations later, but for a first understanding, they do not play a major role.

The critical element in this model that distinguishes it from the pre-AI programs for language is the explicit manipulation of a formal representation. Operations carried out on the representation structures are justified not by facts about language, but by the correspondence between the representation and the world being described. This is the sense in which such programs were said to "understand" the words and sentences they dealt with where the earlier machine translation programs had "manipulated them without understanding."

This general model was not a new idea. It corresponds quite closely to the model of language and meaning developed by philosophers of language like Frege, drawing on ideas back to Aristotle and beyond. There has been a good deal of flag waving at times about the ways in which the "artificial intelligence paradigm" is new and superior to the older philosophical ideas. In large degree (some exceptions are discussed later) this has been a rather empty claim. As Fodor (1978) has pointed out, to the extent they are clearly defined, AI models are generally equivalent to older philosophical ones. A formal logical system can play the role of a representation system as described in the figure, without being explicit about the nature of the processing activity by which reasoning is done.

In fact, AI programs dealing with language do not really fit the model of Figure 1, since they have no modes of perception or action in a real world.

Although they converse about families, baseball or whatever, their interaction is based only on the sentences they interpret and generate. A more accurate model for the programs (as opposed to the human language comprehension they attempt to model) would show that all connection to the world is mediated through the programmer who builds the representation. The reason that ''dog'' refers to dog (as opposed to referring to eggplant parmesan or being a ''meaningless symbol'') lies in the intention of the person who put it in the program, who in turn has knowledge of dogs and of the way that the symbols he or she writes will be used by the interpreter. This difference is important in dealing with questions of ''background'' discussed later.

SHRDLU

SHRDLU (Winograd, 1972) was a computer program for natural language conversation that I developed at MIT between 1968 and 1970.[3] The program carried on a dialog (via teletype) with a person concerning the activity of a simulated ''robot'' arm in a tabletop world of toy objects. It could answer questions, carry out commands, and incorporate new facts about its world. It displayed the simulated world on a CRT screen, showing the activities it carried out as it moved the objects around.

SHRDLU had a large impact, both inside and outside the field, and ten years later it is still one of the most frequently mentioned AI programs, especially in introductory texts and in the popular media. There are several reasons why so many people (including critics of AI, such as Lighthill (1973)) found the program appealing. One major factor was its comprehensiveness. In writing the program I attempted to deal seriously with all of the aspects of language comprehension illustrated in the model. Earlier programs had focussed on one or another aspect, ignoring or shortcutting others. Programs that analyzed complex syntax did not attempt reasoning. Programs that could do logical deduction used simple patterns for analyzing natural language inputs. SHRDLU combined a sophisticated syntax analysis with a fairly general deductive system, operating in a ''world'' with visible analogs of perception and action. It provided a framework in which to study the interactions between different aspects of language and emphasized the relevance of nonlinguistic knowledge to the understanding process.

Another factor was its relatively natural use of language. The fact that person and machine were engaged in a visible activity in a (pseudo-)physical world gave the dialog a kind of vitality that was absent in the question-answer or problem-solution interactions of earlier systems. Further naturalness came from the substantial body of programs dealing with linguistic phenomena of conversation and context, such as pronouns (''it,'' ''that,'' ''then,'' etc.), substitute

[3]Winograd (1971) was the original dissertation. Winograd (1972) is a rewritten version that owes much to the editing and encouragement of Walter Reitman. Winograd (1973) is a shortened account, which also appears in various reworkings in several later publications.

nouns ("a green *one*"), and ellipsis (e.g., answering the one-word question "Why?"). Dialog can be carried on without these devices, but it is stilted. SHRDLU incorporated mechanisms to deal with these phenomena in enough cases (both in comprehension and generation) to make the sample dialogs feel different from the stereotype of mechanical computer conversations.

In the technical dimension, it incorporated a number of ideas. Among them were:

1. Use of a reasoning formalism (MicroPlanner) based on the "procedural embedding of knowledge." Specific facts about the world were encoded directly as procedures that operate on the representation structures, instead of as structures to be used by a more general deductive process. The idea of "procedural embedding of knowledge" grew out of early AI work and had been promoted by Hewitt (1971). SHRDLU was the first implementation and use of his Planner language. The difference between "procedural" and "declarative" knowledge has subsequently been the source of much debate (and confusion) in AI.[4] Although procedural embedding in its simplistic form has many disadvantages, more sophisticated versions appear in most current representation systems.
2. An emphasis on how language triggers action. The meaning of a sentence was represented not as a fact about the world, but as a command for the program to do something. A question was a command to generate an answer, and even a statement like "I own the biggest red pyramid" was represented as a program for adding information to a data base. This view that meaning is based on "imperative" rather than "declarative" force is related to some of the speech act theories discussed below.
3. A representation of lexical meaning (the meaning of individual words and idioms) based on procedures that operate in the building of representation structures. This contrasted with earlier approaches in which the lexical items simply provided (through a dictionary lookup) chunks to be incorporated into the representation structures by a general "semantic analysis" program. This was one of the things that made it possible to deal with conversational phenomena such as pronominalization. Some equivalent device is present in most current natural language programs, and there is a formal analog in the generality of function application in Montague Grammar formalizations of word meaning.
4. An explicit representation of the cognitive context. In order to decide what a phrase like "the red block" refers to, it is not sufficient to consider facts about the world being described. There may be several red blocks, one of which is more in focus than the others because of having been mentioned or acted on recently. In order to translate this phrase into the appropriate representation structure, reasoning must be done using representation structures corresponding to facts about the text preceding the phrase, and structures corresponding to facts about which objects are "in focus."

The attempt to deal with conversational phenomena called for an extension to the model of language understanding, as illustrated in Figure 2. It includes additional structures (as part of the overall representation in the language understander) labelled "model of the text" and "model of the speaker/hearer." The label

[4]See Winograd, (1975) for discussion.

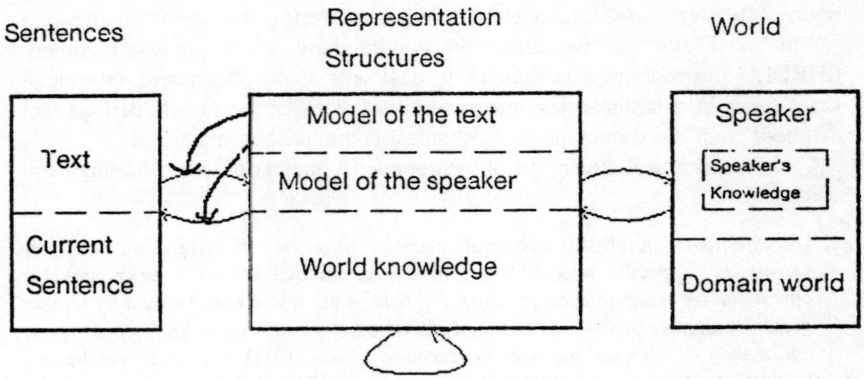

Figure 2. Extended A1 model of language understanding.

"model of the speaker" was chosen to reflect the particular approach taken to the problem. It is assumed that inferences about which objects are in focus (and other related properties) can be made on the basis of facts about the knowledge and current internal state (presumably corresponding to representation structures) of the other participant in the conversation. The question "could I use this phrase to refer to object X?" is treated as equivalent to "if I used this phrase would the hearer be able to identify it as naming object X?" On the other side, "what does he mean by this phrase?" is treated as "what object in his mind would he be most likely to choose the phrase for?"

In addition to reasoning about the domain world (the world of toy blocks), the system reasons about the structure of the conversation and about the hypothesized internal structure and state of the other participant. In SHRDLU, this aspect of reasoning was not done using the same representation formalism as for the domain world, but in an *ad hoc* style within the programs. Nevertheless, in essence it was no different from any other reasoning process carried out on representation structures.[5]

3. SEEING SOME SHORTCOMINGS

SHRDLU demonstrated that for a carefully constrained dialog in a limited domain it was possible to deal with meaning in a fairly comprehensive way, and to achieve apparently natural communication. However, there were some obvious problems with the approach, summarized here and discussed below:

[5]For a more elaborate version of this model, along with many examples of conversational phenomena not handled by SHRDLU or any other existing computer system, see Winograd (1977a). Winograd (in preparation) presents an overview of syntactic and semantic structures within a viewpoint drawn from this model.

1. The explicit representation of speaker/hearer internal structure was *ad hoc,* and there was no principled way to evaluate extensions.
2. The notion of word definition by program, even though it opened up possibilities beyond more traditional logical forms of definition, was still inadequate.
3. It took rather strained reasoning to maintain that the meaning of every utterance could be structured as a command to carry out some procedure.
4. The representation and reasoning operations seemed inadequate for dealing with common-sense knowledge and thought reflected in language.

The Internal Structure

In building a simpler system as illustrated in Figure 1, the programmer is creating a model of the language comprehension process. In creating the representation structures corresponding to facts about the domain, he or she is guided by an idea of what is true in the domain world—in representing facts about blocks, one can draw on common-sense knowledge about physical objects. On the other hand, in trying to create structures constituting the model of the speaker/hearer as in Figure 2, there is no such practical guide. In essence, this model is a psychological theory, purporting to describe structures that exist in the mind. This model is then used in a reasoning process, as part of a program whose overall structure itself can be thought of as a hypothesis about the psychological structure of a language understander.

Experimental psychology provides some suggestive concepts, but little else of direct use. A language comprehension system depends on models of memory, attention, and inference, all dealing with meaningful material, not the well-controlled stimuli of the typical laboratory experiment. Research in cognitive psychology has focussed on tasks that do not clearly generalize to these more complex activities. In fact, much current psychological research on how people deal with meaningful material has been guided by AI concepts rather than the other way around.

The problem is hard to delimit, since it touches on broad issues of understanding. In SHRDLU, for example, the program for determining the referent of a definite noun phrase such as "the block" made use of a list of previously mentioned objects. The most recently mentioned thing fitting the description was assumed to be the referent. Although this approach covers a large number of cases, and there are extensions in the same spirit which cover even more, there is a more general phenomenon that must be dealt with. Winograd (1974a) discusses the text "Tommy had just been given a new set of blocks. He was opening the box when he saw Jimmy coming in."

> There is no mention of what is in the box—no clue as to what box it is at all. But a person reading the text makes the immediate assumption that it is the box which contains the set of blocks. We can do this because we know that new items often come in boxes, and that opening the box is a usual thing to do. Most important, we assume that we are receiving a connected message. There is no reason why the box has to be

connected with the blocks, but if it weren't, it couldn't be mentioned without further introduction. (Winograd, 1974a)

Important differences in meaning can hinge on subtle aspects of the speaker/hearer model. For example, in the first sentence below, it is appropriate to assume that the refrigerator has only one door, while in the second it can be concluded that it has more than one. On the other hand, we cannot conclude from the third sentence that the house has only one door.

When our new refrigerator arrived, the door was broken.
When our new refrigerator arrived, a door was broken.
When we got home from our vacation, we discovered that the door had been broken open.

The problem, then, is to model the ways in which these connections are made. In general this has led to an introspective/pragmatic approach. Things get added to the representation of the speaker/hearer because the programmer feels they will be relevant. They are kept because with them the system is perceived as performing better in some way than it does without them. There have been some interesting ideas for what should be included in the model of the speaker/hearer, and how some of it might be organized[6] but the overall feeling is of undirected and untested speculation, rather than of persuasive evidence or of convergence towards a model that would give a satisfactory account of a broad range of language phenomena.

Word Definition

The difficulty of formulating appropriate word definitions was apparent even in the simple vocabulary of the blocks world and becomes more serious as the domain expands. In SHRDLU, for example, the word "big" was translated into a representation structure corresponding to "having X,Y, and Z coordinates summing to more than 600 units (in the dimensions used for display on the screen)." This was clearly an *ad hoc* stopgap, which avoided dealing with the fact that the meaning of words like "big" is always relative to an expected set. The statement "They were expecting a big crowd" could refer to twenty or twenty thousand, depending on the context. By having word definitions as programs, it was theoretically possible to take an arbitrary number of contextual factors into account, and this constituted a major departure from more standard "compositional" semantics in which the meaning of any unit can depend only on the independent meanings of its parts. However, the mere possibility did not provide a guide for just what it meant to consider context, and what kind of formal structures were needed.

On looking more closely, it became apparent that this problem was not a special issue for comparative adjectives like "big," but was a fundamental part

[6]See for example Schank and Abelson (1975), Hobbs (1978). and Grosz (1980).

of the meaning of most words. Linguists have pointed out that a natural language categorization cannot be equated with a finite set of logical predicates to be applied to the situation.[7] The applicability of a word depends on an understanding of the purposes of speaker and hearer. Winograd (1976) discusses the meaning of "bachelor." In classical discussions of semantics, "bachelor" has been used as an example of a word with a clear paraphrase in more elementary terms—"unmarried adult human male."[8] But if someone refers to a person as a "bachelor" in normal conversation, much more is meant. It is inaccurate if used in describing the Pope or a member of a monogamous homosexual couple, and might well be used in describing an independent career woman.

The issue is not that the definition of bachelor is complex and involves more terms than usually accounted for. There is no coherent checklist of any length such that objects meeting all of its conditions will consistently be called "bachelors" and those failing one or more of them will not. The question "Is X a bachelor?" cannot be answered without asking "Why do you want to know?". Of course, it is possible to create artificial strict definitions, but these do not account for the normal use of language. When we move to a larger vocabulary, the problem becomes even more obvious. Each of the nouns in the sentence "The administration's dishonesty provoked a crisis of confidence in government" raises a significant problem of definition, and it is clear that purpose and context play a major role in determining what will be called a "crisis," "dishonesty," or even in deciding just what constitutes an "administration."

The problem of considering the contextual background in the meaning of words is not solved simply by moving from a "declarative" form of definition to a "procedural" one, as in SHRDLU. The activity of a program in any bounded amount of time can be described in purely logical terms, and a logical formula can be written which can be proved true in exactly those cases where the program would return the answer "yes" and false when it would say "no."

Meaning as Command

SHRDLU was based on a formalism in which the meaning of a sentence was represented as a command to carry out some action. A question is a command to generate a sentence satisfying a set of constraints, and a statement is a command to add a formula to the data base. This shift of viewpoint from meaning-as-statement to meaning-as-command provided some interesting ways of talking about sentences, but in its naive form it is clearly unworkable.

The good part of the idea was the view of an utterance as triggering some kind of activity in the hearer. The bad part of the idea was the analogy with

[7]See, for example, Fillmore (1975) and Labov (1973).

[8]This is, of course, only one of its definitions. Others, as pointed out by Katz and Fodor (1964), relate to fur seals and chivalry.

computer programming languages, in which there is a direct correspondence between the linguistic form and the sequence of activities to be carried out. In the case of natural language, much more is going on. Both speaker and hearer are engaged in ongoing processes of trying to make sense of their conversation and the world they inhabit. The interpretation of utterances is only one of many activities, and interacts with perception, reasoning, memory, and all the other aspects of cognition mentioned by Norman (this volume). When I utter a sentence, I have no way of anticipating in any detail the processing it will invoke in the hearer. It clearly includes much more than simply obeying or storing away a fact. The following simple examples illustrate some of what goes on.

1. Tom has never failed a student in Linguistics 265.
2. I'm sorry I missed the meeting yesterday. My car had a flat tire.
3. There's an animal over there in the bushes.

Sentence 1 is true in many circumstances, including the one in which Tom has never taught Linguistics 265. However, in ordinary conversation, the hearer makes the additional implication that Tom has taught the course, and is justified in accusing the speaker of bad faith if the implication is not warranted. Similarly in sentence 2, the hearer assumes that there is a coherence to the events being described. If the second sentence were "There are fifteen million people in Mexico City" the hearer would be puzzled, and if the flat tire had nothing to do with missing the meeting (even though it actually did happen), the speaker is practicing deception.

Sentence 3 is a more subtle case. If the hearer looks over and sees a dog in the bushes, and finds out that the speaker knew it was a dog, he or she will feel that the statement was inappropriate, and might say "If you knew it was a dog, why didn't you say so?". On the other hand, the statement "There's a dog over there in the bushes" is perfectly appropriate even if both speaker and hearer know that it is a beagle,[9] and sentence 3 would be fine for a dog if it were a response to something like "There are no animals anywhere around here."

The common element in all of these is that the "meaning" for a hearer is the result of a complex process of trying to understand what the speaker is saying and why. In effect, every statement is the answer to a question, which may be implicit in the context. Its meaning depends as much on the question as on the form of the answer. There is an important germ of truth in saying that the meaning of a sentence "is" the process it invokes, but this view is not compatible with a formal compositional semantics of the kind that has generally interested linguists and philosophers. What is needed is an understanding of meaning-as-triggering, which deals with the interaction between the utterance and the full range of cognitive processes going on in the language user.

[9]These phenomena are related to the properties of human categorization systems studied by Rosch (1975).

[10]See the discussion in Fodor (1978).

Natural Reasoning

In looking at any significant sample of natural language, it becomes quickly apparent that only a small fraction of human "reasoning" fits the mold of deductive logic. One is often presented with a fragmentary description of some object or situation, and on the basis of knowledge about what is "typical" jumps to a number of conclusions that are not justifiable as logical deductions, and may at times be false. Most AI systems have been based (either explicitly or unwittingly) on a notion of deduction that does not account for this style of reasoning. In the "opening the box" example quoted above, it was assumed that "the box" was the one in which the blocks arrived even though this is not rigorously deducible from the text, or even from knowledge of the world. This kind of inference is a predominant aspect of reasoning and one which calls for formal systems having very different properties from deductive logic.[11]

4. KRL: TOWARD A NEW FORMALISM

My recognition of these problems with the approach underlying SHRDLU came at the time (1972–73) I participated in a series of discussions in the Artificial Intelligence Lab at MIT about the problems of natural reasoning. These discussions led among other things to Minksy's (1975) formulation of "frames" and to my own attempts to make these intuitions more precise (1974b, 1975). The next major step toward clarifying them was in the specification and construction of KRL (Knowledge Representation Language), a project done jointly with Daniel Bobrow at Xerox PARC.[12] The project has included two rounds of design and implementation, with a limited amount of testing in small applications. Each implementation has covered only a fraction of the overall intended features of the language, and the discussion below (except where otherwise indicated) deals with the basic concepts, not the current implentations.

The Formalism

Viewed as a language for use in building systems of the kind described in Figures 1 and 2, KRL was intended to have the following properties:

1. *Resource-limited processing as a basis for reasoning.* In any act of interpretation or reasoning, a system (biological or computer) has a finite quantity of processing resources to expend. The nature of these resources will be affected by the details of the processor, its environment, and previous history. The outcome of the process is determined by the interaction between the structure of the task and the allocation of process-

[11]Winograd (1980) and several other papers in the same issue of *Artificial Intelligence* deal with problems of "nonmonotonic logic" related to this problem.

[12]See Bobrow and Winograd (1977).

ing. The ability to deal with partial or imprecise information comes from the ability to do a finite amount of processing, then jump to a conclusion on the basis of what has happened so far.

2. *Separation of logical form from memory form*. In order to build a theory of resource use in processing, it is necessary to consider the details of how memory processes operate. Two pieces of information may be logically equivalent but differ in how easily they can be accessed or stored. Most representation systems have either ignored memory properties (as in most theories of formal logic), or assumed that they followed from the logical properties of the information. In KRL, stored knowledge has separable dimensions of logical content and memory "chunking." The memory structure depends not just on the logical content, but also potentially the particular history of the way it was entered into the structures.
3. *The integration of meta-knowledge*. In order to use information about the knowledge structures in a systematic way, the system needs a capacity of partial self-description. Formal structures can be used to represent objects in an "outside world" such as blocks, other structures of the same form can be used to represent those structures in turn, and so on indefinitely. This makes it possible to formulate algorithms for resource-limited processing in the same language that is used for other representation.
4. *An overall control structure based on matching*. The basic organization of KRL provides a framework for an extended "matching" process in which the choice of what to do at each point is driven by an attempt to match stored "prototypes" against the current inputs. This matching can involve complex deductions.

The development of KRL led to deeper thinking about the nature of representation formalisms and the reasoning operations done with them. Looking from one point of view, a KRL system can be thought of as a purely logical formal system—it operates with precise rules on well-defined structures as does any computer program. At the same time, there is another view from which the system can be seen as carrying out a kind of informal reasoning—one that comes to conclusions based on partial evidence, makes assumptions about what is to be expected in typical cases, and leaves open the possibility of mistake and contradiction. A KRL system (or any system using resource-limited reasoning) can draw some conclusion, then reverse it on the basis of further reasoning.

The key to this paradox lies in the system's use of formal rules that make reference to the structure of the system of itself. In coming to a conclusion about some world being represented (such as a world of toy blocks,) a KRL system can come to conclusions on the basis not only of statements about the world, but also on the basis of the form of its structures (for example, concluding something is false because it is normally false, and its truth in this case does not emerge in some bounded amount of reasoning). There is a fundamental philosophical and mathematical difference between truth-based systems of logic, and process-based systems like KRL.[13] One of the major theoretical directions of the KRL effort is to make this difference clear and understand its consequences.

[13]These issues are discussed at length in Winograd (1980).

Reasoning and Language

KRL was designed to make possible a different formal basis for the definition of words in natural language. In a standard deductive system, the definition of a word is a formula specifying a set of necessary and sufficient conditions for its applicability. Some terms are designated as *primitive,* and the rest are defined in terms of them. Although nonprimitive terms can appear in definitions, there can ultimately be no *circular definitions*—every nonprimitive term can be expanded through its definition (and the definitions of the terms appearing in it, recursively) to a formula containing only connectives and primitives. In deciding whether a word applies to an object, this expansion is carried out and the primitive terms are checked against the case at hand. If (and only if) they all check out, the word applies.

Formalisms like KRL provide another way of looking at this.[14] A word is associated with a "prototype" in the representation formalism. This prototype (like a definition) includes a description of the object in other terms. However, unlike a definition, this further description is not taken to be sufficient or necessary for determining the applicability of the prototype. It can include things that are typical (but not always the case) or that are relevant only in some contexts. In deciding whether a word applies to an object, the reasoning system compares these further descriptions to what is known about the object. It does so in a resource-dependent way, possibly looking at only some of the description, and choosing which to do on the basis of context. After a limited amount of processing, it makes a decision (as to whether the word applies) on the basis of what has been examined so far.

A process of this type has the potential to treat word meanings in the open-ended way discussed for the "bachelor" example above. Although there is a "checklist" of further descriptions, the process of checking is context dependent and limited. It is possible to imagine strategies for deciding which items on the list to examine, depending on the current purposes and background. The KRL framework does not specify how this is to be done in detail, but through the combination of resource-limited reasoning and meta-description it provides tools that can be used.

Similarly, these tools can be applied to the other problems raised in the critique of SHRDLU above. The whole concept of resource-limited reasoning grew out of an analysis of the kinds of "natural reasoning" that cannot be done with a straightforward deductive system. The idea of "self-description" was partly an attempt to provide an explicit way of allowing the system to have a model of itself and, by reflection, of a hypothesized dialog partner. The matching framework provides a way to treat inputs to the system as elements of an overall

[14]The problem of word definition, and the notions of "prototype" and "primitive" are discussed at length in Winograd (1976, 1978) and Bobrow and Winograd (1979).

pattern being matched, rather than as items to be stored or commands to be carried out.

With all of these KRL provided only a framework. It was designed to facilitate exploration of these problems, not to embody solutions to them. In that sense it is a kind of "calculus" of natural reasoning, just as the predicate calculus provides a formal basis for systems of deduction and the differential calculus serves as language for describing physical systems. Having this calculus is not the same as having a solution to the specific problems, or a theory of language understanding; but it is a potentially valuable tool. The experience in using it so far is difficult to evaluate. No substantial project has been done that uses its capabilities to really address the kinds of problems that motivated its creation. In trying to use it more fully, technical difficulties (including everything from slow execution speed to user interfaces) are intertwined with conceptual difficulties (which are hard to isolate clearly). As discussed below, there are deep questions of just what could be expected if KRL were to "succeed," but in any event it is clear that it cannot realistically be evaluated in its current implementations.

5. THE HERMENEUTIC CONNECTION

At the same time that KRL was being developed I took part in a series of informal discussions in Berkeley about the nature of language and cognition. These discussions included philosophers, linguists, and computer scientists and ranged from the narrowest technical details to the broadest philosophical concerns. They raised questions about what it was that KRL (and all other computer representations) were claimed to achieve. Among the works discussed were those of Maturana (1977) on the biological foundations of cognition, and the literature on hermeneutics.[15] This paper is not the place for a detailed exposition of these ideas,[16] but there were some elements of both that were applicable to the problem of building programs that interacted in natural language. The following paragraphs attempt to lay them out in outline, but for a more thorough understanding, it is necessary to go to the original sources.

The Nervous System as a Closed, Plastic, Structure-determined System

Maturana proposes an understanding of the nervous system that is not built around the usual notions of input, output, memory, perception, etc. He adopts instead an orientation toward it as a system of components whose activities

[15]Gadamer (1976) was an important source, and Palmer (1969) is an excellent overall introduction.

[16]Flores and Winograd (in preparation) does so at some length.

trigger further activity within the system. The system is "structure determined" in that its activity at any moment is fully determined by the structure (or state) at that moment. It is "plastic" in that its structure can be changed by the activity, so that its structure at any moment is a product of the entire previous history of activity and changing structure. It is "closed," in the sense that the system can do only that which is determined by its own structure and activity—its action cannot be understood as a reflection of an external world it perceives.

At first reading, for most people this approach seems bizarre, apparently denying the obvious fact that we see, hear, and generally perceive a world outside of our own nervous system. It is not a denial, but a change of stance. Instead of looking at vision as a mapping of external reality onto mental structures, we can look at it biologically as a change to the structure of the nervous system, in particular to the chemical and electrical properties of various cells in the retina. The subjective introspection is that we "see something," but from a neurophysiological standpoint, there is a structure-determined causal network in which "perturbations" to the structure of the system lead to patterns of activity that are different from those that would have happened with different perturbations. The focus is shifted away from the structure of the phenomena that led to the perturbations toward the structure of changes in the ongoing activity of the system as it is perturbed.

This view meshed well with the "triggering" view of language understanding described above. An utterance is neither a description of something in the world, nor a command specifying what the hearer is to carry out. It is a "perturbation" to an active ongoing cognitive system that is trying to make sense of things. The central questions to be asked of an utterance are those dealing with the changes in activity that it triggers, not with its correspondence to a world it describes.

The Hermeneutic Circle

Hermeneutics is the study of interpretation. It began as a discipline with the problem of interpreting sacred texts, but has come to encompass not only the interpretation of language, but also the larger understanding of how we interpret the world in which we live. One of the fundamental insights of hermeneutics is the importance of *pre-understanding*. In any situation where we are interpreting language, we begin with a system of understanding that provides a basis within which to generate an interpretation. This pre-understanding in turn arises and evolves through the acts of interpretation. This circle, in which understanding is necessary for interpretation, which in turn creates understanding, is called the *hermeneutic circle*.

> But there is a contradiction here. How can a text be understood, when the condition for its understanding is already to have understood what it is about? The answer is that somehow, by a dialectical process, a partial understanding is used to understand still further, like using pieces of a puzzle to figure out what is missing . . . A certain

pre-understanding of the subject is necessary or no communication will happen, yet that understanding must be altered in the act of understanding. (Palmer, 1969)

The parallel with the issues of reasoning with frame representations should be obvious. The set of stored schemas in a system is its ''pre-understanding.'' The use of these schemas affects the interpretation of what is said:

> In comprehension, the set of stored schemas is actively used in a process of ''pattern recognition.'' The hearer assumes that the discourse is made up of instances of known discourse and reasoning patterns. . . . Some feature of an utterance, together with the current context, can trigger a hypothesis that an instance of some particular schema is being conveyed. This hypothesis is tested by attempting to fit other parts of the utterance in as pieces of the hypothesized schema. As a result the way in which the input is analyzed can be controlled (or biased) by the fact that it is being processed as part of looking for an instance of a specific hypothesized schema. (Winograd, 1977a)

Although little has been said within AI and cognitive science about how new schemas are generated, it is clear that it must be the result of a history of previous interactions, each of which is mediated through other schemas in a hermeneutic circle.[17] In emphasizing the relationship between an act of interpretation and the history of previous interpretations by the system, hermeneutics raises some of the same questions as Maturana's approach to plastic, structure-determined systems.

6. FOUNDERING ON THE OPPOSITE SHOALS

My course to this point was a gradual steering away from the logical-deductive model and its view of language understanding based on objective truth. Starting with the essentially conventional semantic underpinnings of SHRDLU, I had become more and more concerned with the properties of language and thought that could not easily be embodied in traditional logical systems. Maturana and Gadamer provided some philosophical wind for the rudimentary KRL sails, and this new direction seemed ripe for exploration. Structure-dependent resource-limited reasoning offered a way of formalizing natural kinds of inference, and its dependence on the specific past history of the language understander opened up room for systematically dealing with the ''nonlogical'' phenomena of language.

In many ways, this direction is still open for development and there are many areas to be explored. But early *en route* I was exposed to some troubling questions about the rocks we faced on the other shore. Two basic issues stood out: the problem of subjective relativism and the problem of representation. In this section, we will lay out these problems without proposing solutions. The directions described in later sections were guided in part by an attempt to solve them.

[17]The development of schemas is of growing concern in AI, as indicated by Minsky (this volume) and Schank (this volume.)

Subjective Relativism

The first issue can be described in simplistic terms as a dispute about two different starting points for understanding language:

> *Objectivity:* an utterance has meaning by virtue of corresponding to a state of affairs. We approach the study of language by analyzing how the structures of utterances correspond systematically to the states of affairs they describe.
>
> *Subjectivity:* an utterance has meaning by virtue of triggering processes within a hearer whose cognitive structure depends on prior history and current processing activity. We approach the study of language by analyzing the nature of those cognitive structures and activities.

The examples given above suggest that a subject-dependent view must be taken even in such seemingly objective issues as the appropriate use of "John is a bachelor." If we are seriously interested in understanding the regularities in the use of real language in real situations we will be misled by persisting with idealizations of objective truth. As Lakoff (this volume) points out, ordinary language leans more to metaphor than to mathematics.

But there is a problem with unbridled relativism. If the "meaning" of an utterance can only be described in terms of its effects on a particular understander with a particular history, how do we talk about inter-subjective meaning at all? Since no two people have identical histories, and since any aspect of cognitive structure can potentially have an effect on the processing triggered by a particular utterance, there is a different meaning for every hearer. There is no objective "right meaning"—only a meaning for a particular person at a particular moment in a particular situation. Carried to an extreme, if you interpret my use of the word "dog" as referring to eggplant parmesan, what allows me to argue that you are wrong? We want to understand meaning in a way that makes sense of the fact that you and I may not have identical (or even mutually consistent) understandings of "democracy" or "system," but we cannot ignore the common sense intuition that there are broad areas of obvious agreement.

The most obvious fix is to take some kind of behaviorist or operationalist criterion. There is a long history in discussions of AI of using these criteria in arguing about whether a program can be said to "think" or "understand." The Turing test (Turing, 1950) is the most often cited form of an operational criterion, and many defenses of AI naively take for granted that it is the only sensible (scientific) way to deal with the philosophical issues. Even though no two individuals have the same internal structure, we can talk about equivalent classes of behavior and say that a person "understands" something if he behaves in the appropriate ways upon hearing it. But this is only sweeping the problem under a different rug. In trying to define what constitutes the class of "appropriate behavior on hearing the word 'dog' " we are stuck with a problem no easier than that of defining the meaning of "dog." If we posit some objective standards by which the behavior can be tested, we are making the same kinds of assumptions as in setting objective definitions for words. If we don't, then the question "appropriate behavior according to whom in what situation" is just as troubling as "meaning according to whom in what situation." We cannot avoid relativism by converting it to objective propositions about behavior, and thus must deal in some other way with the problem.

The Problem of Representation

The other problem lies in the assumption that we can build formal structures that represent the knowledge and state of a language understander. In going from a standard deductive system to a structure-dependent formalism like KRL, we still maintain the basic idea of representing the relevant knowledge in formal structures that can be set down according to our understanding of linguistics, psychology, and computation. The question of whether human knowledge can be represented in formal structures has been a major concern for a number of philosophers. Dreyfus (1979) has been the most active critic of artificial intelligence from this perspective, drawing on the philosophy of Heidegger (1962).[18]

For a long time I found the arguments rather incomprehensible, seeing the position that knowledge was not representable as equivalent to the belief that the human nervous system could not operate according to deterministic principles. But Maturana, starting by viewing the nervous system as a mechanistic system, argued in ways that were disturbingly similar. One of the most challenging of Maturana's views is his dogmatic insistence that cognition is not based on the manipulation of mental models or representations of the world. For someone trained in AI (or in cognitive science generally, as illustrated by the other papers in this volume) it is hard to understand what other kind of explanation there could be.

Maturana sees much of the discussion of representation as exhibiting a serious error of confusing "phenomenic domains." Anything we choose to describe as a system can be described in different domains,[19] each with its relevant pheomena. For example, we can look at a TV screen and see an array of luminescent dots excited by a moving electron beam, or we can see a comedian telling jokes. We can talk coherently about what we see in either domain, but cannot combine them meaningfully. Maturana argues that in describing cognition we often fail to carefully distinguish the relevant domains. The error takes the form:

1. A scientist observes some recurrent pattern of interactions of an organism.
2. He or she devises some formal representation (for example a set of generative rules or a "schema") that characterizes the regularities.

[18] For a discussion of the issues from within AI, see Barr (1980).

[19] This use of the word "domain" follows Maturana. It is somewhat different from the more common use in AI, where the "domain" of a particular program is something like "blocks," "airline reservations," or "blood infections."

3. The organism is assumed to "have" the representation, in order to be able to exhibit the regularities.
4. (Depending on the particular sub-field) The scientist looks for experiments that will demonstrate the presence of the representation, or designs a computer program using it to see whether the behavior can be generated by the program.

The error is in the reification of the representation at step 3. Working from basic biological examples, Maturana points to many phenomena that *for an observer* can be described in terms of representation, but that can also be understood as the activity of a structure-determined system with no mechanism corresponding to a representation. As a simple example, we might watch a baby successfully getting milk from its mother's nipple and argue that it has a "representation" of the relevant anatomy, or of the activity of feeding. On the other hand, we might note that there is a reflex that causes it to react to a touch on the cheek by turning its head in that direction, and another that triggers sucking when something touches its mouth. From the viewpoint of effective behavior, it has a "correct representation," but it would be fruitless to look for neurophysiological mechanisms that correspond to reasoning that uses facts about breasts or milk.[20]

Maturana argues that there is a "domain of description" in which it is appropriate to talk about the correspondence between effective behavior and the structure of the environment or "medium" in which it takes place, but that we must not confuse this kind of description with the description of the causal mechanisms operating to produce it. In saying that a representation is "present in the nervous system," we are indulging in misplaced concreteness and can easily be led into fruitless quests for the corresponding mechanisms. Whereas the point is obvious for reflexive behavior (which can certainly be quite complex, as pointed out by the animal ethologists), he sees it as central to our understanding of all behavior, including complex cognitive and linguistic activities.

After an initial scepticism (of the "How could it be anything but . . ." variety), I thought about how this view might be directly applied to the problems of language understanding. There is a good deal of confusion of domains apparent in the work on "schemas," "scripts," and "frames." Some kind of regularity is observed in text patterns, or the ability to answer certain kinds of questions given text. The cognitive researcher builds a formal representation of this pattern, and often builds some kind of program that uses it to produce minor variants on the observed behavior. The resulting claim is that a person must "have" the script or schema and use it explicitly (perhaps not consciously) in carrying out the process.

[20]Geschwind (this volume) exhibits this kind of reification in most direct form: ". . . the cat who has never seen a mouse attacked will bite through the nape of the neck, thus adopting the "best strategy" for immobilizing the prey. These findings suggest the surprising conclusion that a "model" of the nape must be present in the nervous system . . ."

My own discussion of discourse (Winograd, 1977a) carried this representational view to its logical extreme. It gives examples of many different dimensions of discourse patterning, referring to them as kinds of "schemas." "Each speaker of a language possesses a large and rather diverse set of schemas dealing with the process of natural language communication. The understanding of these schemas will form the core of a science of discourse." I still feel that the kinds of phenomena that were pointed out and categorized were interesting and important, but dressing up the observations in the language of schemas did little or nothing to sharpen or develop them. A direct implementation of the purported schemas as representation structures in a computer program would have been uninteresting. It would have the flavor of much of the work we see, in which a program "plays back" the schemas or scripts put into it, but in doing so it does not provide any insight into what happens in those cases that don't closely match exactly one of the hypothesized schemas. *The schemas correspond to classes of external behavior, which may not correlate in any straightforward way to the components of the internal mechanism (either physical or functional).*

It was also interesting that the same questions could be raised with respect to computers. In other work (Winograd 1979b) I have been concerned with the problem of languages and programming environments for developing complex computer systems. The appropriate level of description for explaining a system to another person often includes terms that do not correspond to any mechanism in the program. If I say of a program "It has the goal of minimizing the number of jobs on the waiting queue," there is unlikely to be a "goal structure" somewhere in memory or a "problem solving" mechanism that uses strategies to achieve specified goals. There may be dozens or even hundreds of places throughout the code where specific actions are taken, the net effect of which is being described. In the case of computer systems (as opposed to naturally evolving biological systems) the goal can be more appropriately thought of as existing in the specification that was given to the people who programmed it. The situation would be more parallel to that of a living mind if the system had simply "survived" because a collection of changes led to a particular behavior pattern without conscious intention by programmers. But in any case, there is an important lesson in the fact that there are important regularities in the "descriptive domain" that *do not* correspond to mechanisms in the program.

Returning to the problems of language understanding, we can see what the appropriate domain is for talking about context. In traditional linguistics, the context of a sentence is made up of the sentences preceding and following it. In the AI model, as described in Figure 2 above, we assume that context can be understood as a set of cognitive structures within the speaker and hearer. Some of these structures are records (or analyses) of other sentences, but others are things like a "topic structure" representing what the discourse is about, a "script" that is being applied, and a "focus list" of things recently mentioned or thought about. Taking Maturana's examples seriously, we are led to question whether these descriptive notions can appropriately be used as a guide for building or analyzing mechanisms. Perhaps some of the difficulties mentioned in section 3 result from an attempt to characterize regularities in the wrong domain. We need to ask, then, just what the relevant domain might be.

7. UNDERSTANDING IN A DOMAIN OF ACTION

Four Domains for Understanding Language

In getting to this point we have described language in three different phenomenic domains. The assumptions central to a given domain form a major element in the generation of a "paradigm" in the sense of Kuhn (1962). However, it is not a simple matter of choosing the "right" domain. For any field of inquiry, there can be several relevant domains of regularity. In focussing on one domain as central we are led to ask certain questions and pay attention to certain phenomena. For any choice, there are some phenomena that become more easily describable, and others that become more obscure. The three domains we have discussed so far are:

> *The domain of linguistic structure*. This is the domain of traditional linguistics. One looks for regularities in the patterning of structural elements (phonemes, words, phrases, sentences, etc.) in utterances and text. As mentioned above, most of the work on the larger-scale structure of discourse is in this domain, even when it is reformulated in terms of "schemas."
>
> *The domain of correspondence between linguistic structures and the world*. In this domain, one is concerned with regularities in the correspondence between the structures of linguistic objects and the states of affairs in the world that those objects describe. Much of the current work in the philosophy of language is an attempt to formalize this correspondence. Much of the AI work on natural language has had this orientation as well.
>
> *The domain of cognitive processes*.[21] In this domain the relevant regularities are not in the linguistic structures themselves, or their correspondence to a world, but in the cognitive structures and processes of a person (or machine) that generates or interprets them. This is the domain explored in much of cognitive psychology and artificial intelligence.

My current work is moving in the direction of a fourth domain for understanding language:

> *The domain of human action and interaction*. In this domain the relevant regularities are in the network of actions and interactions within a human society. An utterance is a linguistic *act* that has consequences for the participants, leading to other immediate actions and to commitments for future action.

[21]This view is used as a basis for a comprehensive description in a textbook entitled *Language as a Cognitive Process* (Winograd, forthcoming) and is contrasted with the first two views in Winograd (1977b).

This domain has been explored under the rubric of *speech acts*. The work of Austin (1962) and Searle (1970, 1975) has shown that utterances can be understood as acts rather than as representations. In giving a command or making a promise, a person is not uttering a sentence whose meaning lies in whether it does or does not correspond truthfully to the world. The speaker is entering into an interaction pattern, playing a certain role, committing both speaker and hearer to future actions, some linguistic and others not. In this domain the relevant question about an utterance is "What is the speaker doing?" Understanding is connected with the ability to recognize what the speaker is doing and to participate in the appropriate pattern of actions.

My first reaction on being urged to look at this domain was "Oh, just speech acts." I had been familiar with the basic ideas of speech acts for quite a while and viewed them as a kind of add-on to the central theory of language. In the paper on discourse discussed above they were described, true to form, as a collection of "speech act schemas" that were among the cognitive structures of a speaker or hearer. The challenge, though, was to see what it meant to look at language as a whole from the perspective of speech acts—as action rather than structure or the result of a cognitive process.

This was a difficult shift of view, but one with interesting consequences. It is perhaps easiest if we begin by looking at one of the most often described speech acts, that of *promising*. If someone asks me to come to a meeting tomorrow and I respond "I'll be there," I am performing a speech act[22] that is quite different from describing a state of affairs in the world or making a prediction about future states of affairs. By virtue of the utterance, I am entering into a commitment. It is not relevant to ask whether the promise is "true" or "false," but rather whether it is appropriate, or, to use Austin's term, "felicitous."

An essential thing about speech acts is that they always occur in a social context, with a background implied by that context. If I find out tomorrow that the meeting has been moved to Katmandu and don't show up, I can justifiably argue with you that I haven't broken my promise. What I really meant was "Assuming it is held as scheduled. . . ." On the other hand, if the meeting is moved to an adjacent room, and I know it but don't show up, you are justified in arguing that I have broken my promise, and that the "Katmandu excuse" doesn't apply. This kind of argumentation can be pursued back and forth indefinitely, and forms a part of the network of potential actions related to a promise. The legal system provides a regular mechanism for exactly this kind of interaction.

[22]As Searle points out, we can talk about a speech act as being a promise even though it does not use any explicit illocutionary verb, such as "promise" or "assure." Technically, it is said that the utterance has "promise" as its "illocutionary force."

Statements as Speech Acts Initiating Commitment

In the basic works on speech acts, there is a separation between the propositional content of an utterance and its illocutionary force. The fact that my utterance is a promise is its illocutionary force. The fact that it involves my attendance at a particular meeting at a particular time is its propositional content. In further pursuing the connection between meaning and speech acts, it is possible to view more of meaning (including what has been seen as propositional content) in the domain of action, rather than the domain of correspondence with the world.

Consider the following dialog:

A: I'm thirsty.

B: There's some water in the refrigerator.

A: Where? I don't see it.

B: In the cells of the eggplant.

A claims that B's first response was a lie (or "misleading"), whereas B contends that everything he said was literally true. Most work in semantics (including artificial intelligence) can be seen as providing formal grounds to support B. But there is an important sense in which a theory of "meaning" needs to deal with the grounds for A's complaint. In making the statement "There's some water in the refrigerator" B is doing something more than stating an abstract objective fact.

At first, it seems like it might be possible to expand on the definition of "water." Perhaps there is a "sense" of the word that means "water in its liquid phase in a sufficient quantity to act as a fluid," and the statement about water is ambiguous in whether it refers to this sense or to a sense dealing purely with chemical composition. But this doesn't help us in dealing with some other possible responses of B to the initial request:

B: There's no water in the refrigerator, but there's somc lemonade.

B: There's a bottle of water in the refrigerator, with a little lemon in it to cover up the taste of the rust from the pipes.

In the first case, the presence of lemon in the water is taken as making it not "water." In the second, the lemon (perhaps the same quantity) is considered irrelevant. The difference lies in a background of assumptions the speaker has about the hearer's purposes and experience. After any amount of fiddling with the definition, one can always come up with a new context (e.g., what if the person were doing a science experiment or checking for sources of humidity), in which the acceptability of the statement "There is water" would not be accounted for by the definition. Every speech act occurs in a context, with a background understood by speaker and hearer. There are "felicity conditions" that depend on mutual knowledge and intentions. The speaker is responsible for things he can anticipate that the hearer will infer from what he says, not just its abstract correspondence with the state of affairs.

What happens, then, if we try to understand the problem of "truth" in the terms of social action and commitment? In making a statement I am doing something like making a promise—committing myself to acting in appropriate ways in the future. In this case, there is a different kind of satisfaction condition. There is no specific action that I am bound to, but there is a structure of potential dialog that we could enter into in the face of a "breakdown." If I say "There is water in the refrigerator" and you can't find any, I am committed to give you an account. Either we agree that I was wrong, or we discuss the assumed background ("I assumed we were talking about something to drink." "I assumed we were talking about chemical composition").

There are several reasons why this shift of viewpoint is potentially advantageous:

It lets us deal with what happens when we actually make statements. Formal approaches to meaning often take as their model the language of mathematics, in which it is generally assumed that the truth of a statement can be determined without reference to outside context or situation.[23] But in real language, we rarely if ever make a statement that could not be construed as having a literal meaning we don't intend. If I say "snow is white" you can point to the murky grey polluted stuff at our feet. I reply "I meant pure snow," and you respond "You didn't say so, and anyway no snow is absolutely pure." It is an edifying exercise to look at the statements we make both in our writing and our everyday conversation and see how few of them can even apparently be judged true or false without an appeal to an unstated background.

It shifts us out of the objective/subjective dichotomy. In the previous section we saw a dilemma arising from trying to identify the "meaning" of a word or utterance. By assuming it had an objective meaning independent of a particular speaker/hearer situation, we banished many aspects of meaning that play a central role in language. But in assuming that meaning is defined in terms of effect on a particular individual in a situation, we lose the sense that meaning can be the same across individuals and situations. In moving to the domain of interactions (rather than that of objective truth correspondence or cognitive process), we are directing attention to the interactional situation in which something is uttered. We draw generalizations across these situations (and their potential for continued conversation) rather than across objective correspondence or mental states.

It places central emphasis on the potential for further articulation of unstated background. By looking at statements as language acts analogous to promises, we bring into prominence the fact that human action always occurs in an unarticulated background. When I promise to do something, it goes without saying that the commitment is relative to a large number of assumptions about the rest of our world continuing as expected. The same properties carry over to language acts. Sociologists working in "ethnomethodology"[24] have explored the problems in recognizing our background

[23]In fact, as pointed out by Lakatos (1976), this is not really the case even in mathematics.
[24]For example, Garfinkel (1967).

assumptions. We can never make the background fully explicit, but we can study the nature of the dialog by which people come to a consensus about things that were previously in the background.

8. WHITHER FROM HERE?

Having turned to a different "phenomenic domain" we must raise the question of methodology. How does one proceed in day to day work? My own answer to this question includes both constructive and interpretive activities:

Developing Theories and Formalisms

Just as the structural formalisms of linguistics and the deductive formalisms of predicate calculus were developed to provide a language for description in their respective domains, we need an appropriate "calculus of language acts" if we want to develop detailed theories of language interaction. There will be several parts of such a theory:

1. *Illocutionary logic*. The basic groundwork of speech act theory includes an analysis of the different kinds of illocutionary points and the structure of the felicity conditions associated with them. Searle (1975) proposes a universal taxonomy of five basic illocutionary points, further developed by Searle and vanDerVeken (in press). This analysis can serve as a starting point for understanding the structure of larger composite patterns made up of the basic acts. For example an "offer-negotiation-acceptance" sequence is a standardly observed pattern made up of individual "commissives" and "requests." The formal tools for describing the "syntax" of such patterns may be quite different from those used in traditional linguistics, since they must take into account the passage of time (e.g., not overtly responding to an invitation constitutes a kind of response).

2. *Taxonomy of linguistic grounding*. In order to carry out the suggested program of viewing the truthfulness of statements in the domain of interaction, we need a "logic of argument," where "argument" stands for the kind of elucidation of background assumptions discussed above. When I make a statement, I am making a commitment to provide some kind of "grounding" in case of a "breakdown." This grounding is in the form of another speech act (also in a situational context) that will satisfy the hearer that the objection is met. There appear to be three basic kinds of grounding: experiential, formal, and social.

> *Experiential:* If asked to justify the statement "Snow is white" I may give a set of instructions ("Go outside and look!") such that a person who follows them will be led to concur on the basis of experience. The methodology of science is designed to provide this kind of grounding for all empirical statements. Maturana (in press) points out that the "objectivity" of science derives from the assumption that for any observation, one can provide instructions that if followed by a "standard observer" will lead him or her to the same conclusion. This does not necessarily mean that the result is observer free, simply that it is anticipated to be uniform for all potential observers.

Formal: Deductive logic is based on the playing of a kind of "language game" in which a set of basic rules are taken for granted, and argument proceeds as a series of moves constrained by those rules. For example, if I expect you to believe that all Swedes are blonde and that Sven is a redhead, then I can use a particular series of moves to provide grounding for the statement that Sven is not Swedish. Of course, this depends on the grounding of the statements used in the process—one can recursively demand grounding for each of them. Under this category fall most of the issues that have been discussed in formal semantics,[25] but with a different emphasis. The focus is not on their coherence as a mathematical abstraction, but on the way they play a role in the logic of conversation.

Social: Much of what we say in conversation is based neither on experience nor logic, but on other conversations. We believe that water is H_2O and that Napoleon was the Emperor of France not because we have relevant experience but because someone told us. One possible form of grounding is to "pass the buck"—to argue that whoever made the statement could have provided grounding. This is also recursive, but we assume that the buck stops somewhere. Of course this need not be so (as illustrated by the development of ideologies in societies), but this is not relevant to its role in the dynamics of conversation.

Just as one can develop taxonomies and structural analyses of illocutionary points, it is important to develop a precise analysis of these structures of argumentation. There are many ways in which such a logic will parallel standard formal logic, and others in which it will not. In particular, it seems that the role of analogy and metaphor will be much more central when the focus is on patterns of argumentation between individuals with a shared background rather than on deductive inference from axioms.

In thinking about these problems it is interesting to look at the structure of reasoning done by lawyers rather than mathematicians. The lawyer is not guided by a formal theory of proof that can be used mechanically to establish the validity of a conclusion. Instead, he or she attempts to anticipate the objections that could be raised (by another person in a particular context) and to prepare justifications that can overcome the objections. Those justifications are in turn based on statements that may themselves need other justifications, *ad infinitum*. Superficially, there is an analogy to the hierarchical structure of theorems based on simpler theorems in mathematics, but there is a deep fundamental difference. The legal brief does not "ground out" on unquestionable axions. Instead it rests on suppositions that the lawyer assumes are part of the basic background (experiential, social, etc.) and hence will not be called into question. Of course this is only an anticipation, and there are mechanisms in court for going deeper into the logic of any statement in the form of an adversary dialog.

3. *Theory of reasoning as triggered by breaking-down*. Going one step further, we need to inquire into the relationship between argumentation and reasoning. The classical understanding is that reasoning proceeds by logical

[25]Hintikka (1976) uses the notion of games in his development of deductive logic, including modal logic.

deductive steps, and argumentation is a laying-out of the steps for someone else to follow. The orientation provided here is almost the reverse—reasoning is a form of arguing with yourself, and the justification for a step is that it (temporarily, at least) quiets objections to some statement. The motivation for going farther is that something that previously seemed obvious (either on examination or because it had never been examined) has broken down. Starting with the analyses of philosophers like Heidegger (1962) we can look for a "logic of breaking down" that applies not only to linguistic conversations, but to our non-verbal interactions and understanding.

There are interesting ways in which its characteristics (of breakdowns leading to questions to a certain depth, still within an implicit background) are parallel to those of systems using resource dependent reasoning. Perhaps the mechanisms of systems like KRL can be usefully reinterpreted as representing a logic of argument, rather than a logic of cognitive processing. The issues raised in these systems (such as default knowledge, meta-description, and partial matching) have intriguing connections to problems of background and representation. It would be presumptuous to see the existing formalism as an "answer," but it provides a starting point for exploring the interaction between implicit background, formal structures, and patterns of conversation and justification.

Understanding Existing Systems and Designing New Ones

Computer Systems. Much of cognitive science has been both stimulated by and directed towards the construction of computer programs that behave "intelligently." Hundreds of books and articles have been written on how computer systems will soon become prevalent in every arena of life. The question asked of the cognitive scientist is: "What kind of theories do we need in order to build intelligent systems we can use?"

The prevalent view is that in AI we design "expert systems" that can stand as surrogates for a person doing some job. From a viewpoint of human interaction we see the computer's role differently It is not a surrogate expert, but an intermediary—a sophisticated medium of communication. A group of people (typically including both computer specialists and experts in the subject domain) build a program incorporating a formal representation of their beliefs. The computer communicates their statements to users of the system, typically doing some combinations and rearrangements along the way. The fact that these combinations may involve complex deductive logic, heuristic rule application or statistical analysis does not alter the basic structure of communicative acts.

A person writing a program (or contributing to its "knowledge base") does so within a background of assumptions about how the program will be used and who will be interpreting its responses. Part of this can be made explicit in documentation, but part is an implicit background of what can be "normally understood." Except for systems operating within strongly constrained domains, there inevitably comes a time when the system "breaks down" because it is being used in a way that does not fit the assumptions. This is true not only of "expert systems," but of computer programs in all areas. Many of the biggest failures of mundane computer systems (management systems, inventory systems, etc.) have come not because the system failed to do what the designers specified, but because the assumptions underlying that specification were not appropriate for the situation in which the program was used. This will become even more the case as we build systems that are more flexible—that allow the user to develop new modes of interaction in the course of using the program, rather than staying within a fixed set of alternatives.

Of course, to the degree that system builders can anticipate areas of potential breakdown, they can make explicit statements in advance, which the computer can convey (again perhaps with complex recombination) through an "explanation system."[26] Some existing programs incorporate explanation facilities that move in this direction, but they are able to deal with only a limited range of the potential dialogs of explanation. There is always a limit set by what has been made explicit, and always the potential of breakdowns that call for moving beyond this limit.

If we see the machine as an intermediary, it is clear that the commitment (in the sense discussed above in viewing truth in the context of speech acts) is made by those who produce the system. A dialog must be carried on with the people who performed the original speech acts, or those to whom they have delegated the responsibility. In the absence of this perspective it becomes easy to make the dangerous mistake of interpreting the machine as making commitments, and losing the sense that some person or group of people has responsibility for what it does. We need theories that can help in understanding the properties of dialogs in which a person tries to elucidate the background assumptions that may have led to a breakdown. In designing the computer system the focus is shifted from the problem of "How do we make sure it always gives the right answer?" to "What is the best kind of organization for figuring out when and where things are going wrong?"

Medical diagnosis programs provide a good example. Imagine a program written by a team of computer specialists, working with a group of medical experts. It is installed by the administration of a hospital and used by a member of the medical house staff in choosing a treatment. If the diagnosis was wrong and the patient is harmed, who is responsible? The problem may not be one of wrong medical knowledge, but rather one of background assumptions. An answer that is correct in one context may be inappropriate in another. For example, if the program was written with ambulatory patients in mind, it might not be

[26]Winograd (1979a) discusses the importance of developing systems of this type.

appropriate for a chronic bedridden invalid who shows very different symptom patterns. How can the user of the system systematically find out what the relevant assumptions are? If we do not develop theories that help us to cope with these issues, the systems will be useless or worse.

I also believe that there are potential new directions in the design of computer languages in which these issues will play a prominent role. The writing of a program is a communication between one programmer and another. The understanding of existing programs is one of the most important and difficult things we do, and future systems will need to be oriented primarily towards aiding in this understanding. The issues of language and understanding that have been at the heart of this paper will be at the heart of those systems.[27]

Management. Another practical area that looks to cognitive science for a theoretical base is management. There is a significant overlap of concepts, as illustrated by terms like "problem solving," "decision making," and "heuristic strategy." Of course this is no accident. Herbert Simon, one of the founders of cognitive science, is best known for his work on organizational decision making. The management and decision metaphor has played a large role in shaping artificial intelligence from the earliest days.

My own experience is not in this area, but my work with Fernando Flores (who has been a cabinet minister in the Chilean government) has pointed out concerns that are parallel to those arising in the study of language. The view of management as problem solving and decision making suffers from the same problems of background and context. In looking at the role of a manager as optimizing some value by choosing among alternatives for action, we are in the same position as in looking at the understanding of a word as choosing among a set of formal definitions. The hard part is understanding how the alternatives relevant to a given context come into being. The critical act of problem solving in the recognition of the problem.

Again, this is not a new insight. Management specialists (including Simon) have at times pointed out the essential role of "problem acquisition" as a prerequisite to "problem solving." The practical wisdom of management recognizes the importance of the role a manager plays in continually creating the "problem space" in which he or she operates. But the theories provided by cognitive science and artificial intelligence have little to say about this process of creation. To a large extent, a problem is created by the linguistic acts in which it is identified and categorized. Of course, some situation exists previous to the formulation, but its structure as a problem (which constrains the space of possible solutions) is generated by the commitment of those who talk about it.

Once again we cannot look to simple notions of truth and deduction. The "energy crisis" was not created by specific acts of the oil companies, the Arabs,

[27]Winograd (1979b) discusses these issues in a more technical vein.

or the American consumer, but by those with the power to create consensus who looked at a long-term situation and determined it to be a crisis. The relevant question is not whether it is "true" or "false" that there is a crisis, but what commitments are entailed (for speaker and hearer) by the speech acts that created it.

Education. A third area where cognitive theories play a central role is in education. To a large extent our current educational practice has grown from experience with no theory at all, or with a hodgepodge of theories borrowed from behaviorism, structuralism, and various kinds of interpersonal psychology. There have been attempts to attract educators to a theory based on work in artificial intelligence, emphasizing how human thought can be described in terms of concepts such as "algorithm," "heuristic," and "bug." Researchers like Papert (1972) point out that in teaching children subject material, we are also "teaching children thinking." In providing instructions for how to go about particular tasks we are also providing models for how to go about tasks in general. By being conscious of this "meta-teaching" and by applying the best theories we can of thought and language, we can provide a more consistent and effective educational environment.

Again, the questions we choose to look at depend on the domain we see as central. If we concentrate on cognitive processing, students will learn to think in that domain. Faced, for example, with a breakdown in some kind of interpersonal situation, they will ask "What is the bug in my model of the other person?" Starting from an orientation toward the structure of interaction, the question becomes "What do we do to make explicit the assumptions that led to the breakdown?" The emphasis is on the nature of the continuing interaction, rather than on the cognitive structure of the participants. No one orientation is good for everything, and the purpose of an educational system should be to give students experience with the broadest possible range. There is a significant place for understanding language and thought in terms of social interaction.

9. CONCLUSION

Those who are familiar with the philosophical discussion about artificial intelligence will have noticed that many of the ideas and sources discussed here (such as the incompleteness of formal representation, Heidegger's notion of background, etc.) are among those cited by critics like Dreyfus (1979), who deny the possibility of developing any formalization of human thought and knowledge. A conclusion one might draw is that having brought these questions to light, we can only proceed by abandoning formal cognitive science and our attempts to program computers to do things we consider "intelligent."

It should be clear from the previous section that this is not my conclusion. I am not advocating (or planning) the abandonment of the scientific study of cognition, but trying to better understand what we are doing, and refocusing my efforts in the light of that understanding. However, it could be argued that the path described in this paper is one leading away from the paradigm of cognitive science. Even granting that an orientation towards language as social action is interesting and useful, it is arguable that it is not "cognitive science"—that it represents a turning away from the domain of "cognitive processing" (or, as it is often called, "information processing"). In some ways this observation is valid, but in others it is misleading.

It is important to recognize what we are doing we apply words like "cognitive" or "science" to a particular enterprise or approach. In our writing, teaching, and interactions with people (both in the field and outside), we are performing speech acts that give those words meaning. Different orientations lead to different practical suggestions, to different ways of interpreting and acting. As has been pointed out in the philosophy of science, the importance of a paradigm may not lie so much in the answers it provides as in the questions it leads one to consider, and a paradigm (like a crisis) is created by a web of interlinked speech acts.

Some day in the future there may be a field called "cognitive science" whose boundaries are defined by a narrow common approach and domain, just as there are fields of "anatomy," "physiology," and "pathology" dealing with the physical structures and functioning of the human body in their own domains. This narrowly defined cognitive science would deal with those aspects of language, thought, and action that are best understood in the domain of information processing. At the moment though, this is not the case. As indicated by this volume and the nascent professional society it represents, "cognitive science" is a broad rubric, intended to include anyone who is concerned with phenomena related to mind. I believe that the kinds of issues I have been led to in looking at language are relevant to a broad segment of cognitive science as generally interpreted, and a serious consideration of them may call into question many of the assumptions at the heart of our current understanding.

ACKNOWLEDGMENTS

My thinking about language has been developed through interactions with a number of people, as chronicled in the paper. I am particularly indebted to Fernando Flores for continually introducing new vistas. It was in conversations with Danny Bobrow and Brian Smith that the ideas of KRL emerged, and work on KRL has involved many other people. Coversations with Don Norman led to some of the original thinking about memory structure. Henry Thompson, Rich Fikes, David Levy, Mitch Model, Paul Martin, Jonathan King, and Wendy Lehnert have all contributed to the ideas and implementations. In addition to Bobrow, Flores, Levy, and Thompson, the Berkeley discussions mentioned at the beginning of section 5 included Hubert Dreyfus, John Searle, and Stephen White. Danny Bobrow, Fernando Flores, David Levy, and David Lowe provided insightful comments on an earlier draft of this paper.

REFERENCES

Austin, J. L. *How to do things with words*. Cambridge, Mass.: Harvard University Press, 1962.

Barr, A. The representation hypothesis. (Stanford HPP-Memo, HPP-80-1, Dept. of Computer Science 1980).

Bobrow, D. G., & Winograd, T. An overview of KRL, a knowledge representation language. *Cognitive Science*, 1977, *1*, 3–46.

Bobrow, D. G., & Winograd, T. KRL: Another perspective. *Cognitive Science*, 1979, *3*, pp. 29–42.

Dreyfus, H. *What computers can't do: A critique of artificial reason*. (2nd Ed.) San Francisco: Freeman, 1979.

Feigenbaum, E., & Feldman, J. *Computers and thought*. New York: McGraw Hill, 1963.

Fillmore, C. An alternative to checklist theories of meaning. In Cogen et al. (Eds.), *Proceedings of the First Annual Meeting of the Berkeley Linguistics Society*. University of California, Berkeley, 1975.

Flores, C. F., & Winograd, T. *Understanding cognition as understanding*. Norwood, N.J.: Ablex Publishing Corporation, in press.

Fodor, J. A. Tom Swift and his procedural grandmother. *Cognition*, 1978, *6*, pp. 229–247.

Fodor, J. A. Methodological solipsism as a research strategy in psychology. *Brain and Behavioral Sciences*, forthcoming.

Gadamer, H-G. *Philosophical hermeneutics*. Translated by David E. Linge. Berkeley: University of California Press, 1976.

Garfinkel, H. What is ethnomethodology? In H. Garfinkel (Ed.), *Studies in Ethnomethodology*. Englewood Cliffs, N.J.: Prentice–Hall, 1967.

Geschwind, N. Neurological knowledge and complex behaviors. *Cognitive Science*, 1980, *4*, pp. 185–193.

Grosz, B. J. Focusing and description in natural language dialogues. In A. K. Joshi, I. A. Sag, & B. L. Webber (Eds.), *Elements of discourse understanding*. Cambridge: Cambridge University Press, 1980.

Heidegger, M. *Being and time*. New York: Harper & Row, 1962.

Hintikka, J. Quantifiers in logic and quantifiers in natural languages. In S. Korner (Ed.), *Philosophy of Logic*. Berkeley, Calif: University of California Press, 1976.

Hobbs, J. R. Resolving pronoun references. *Lingua*, 1978, *44*, 311–338.

Katz, J. J., & Fodor, J. A. The structure of a semantic theory. In J. Fodor & J. Katz, (Eds.), *The Structure of Language*. Englewood Cliffs, N.J.: Prentice–Hall, 1964.

Kuhn, T. *The structure of scientific revolutions*. Chicago: Ill.: University of Chicago Press, 1962.

Labov, W. The boundaries of words and their meanings. In C–J. N. Bailey & R. Shuy (Eds.), *New ways of analyzing variation in English*. Washington, D.C.: Georgetown University, 1973.

Lakatos, I. *Proofs and refutations*. Cambridge, Mass.: Cambridge University Press, 1976.

Lakoff, G. & Johnson, M. The metaphorical structure of the human conceptual system. *Cognitive Science*, 1980, *4*, pp. 195–208.

Lighthill, Sir. J. *Artificial intelligence: A general survey*. London: Science Research Council, 1973.

Maturana, H. R. Biology of language. In R. W. Rieber (Ed.), *The neuropsychology of language*. New York: Plenum Press, 1977.

Minsky, M. *Semantic information processing*. Cambridge, Mass.: MIT Press, 1967.

Minsky, M. A framework for representing knowledge. In P. Winston, (Ed.), *The psychology of computer vision*. New York: McGraw–Hill, 1975.

Minsky, M. K-Lines: A theory of memory. *Cognitive Science*, 1980, *4*, pp. 117–133.

Norman, D. A. Twelve issues for Cognitive Science. *Cognitive Science*, 1980, *4*, pp. 1–32.

Palmer, R. E. *Hermeneutics: Interpretation theory in Schleiermacher, Dilthey, Heidegger and Gadamer*. Evanston, Ill.: Northwestern University Press, 1969.

Papert, S. Teaching children thinking. *Mathematics Teaching: The Bulletin of the Association of Mathematics* No. 58, Spring, 1972.

Rosch, E. Cognitive representations of semantic categories. *Journal of Experimental Psychology: General*, 1975, *104*, pp. 192–233.

Schank, R. & Abelson, R. *Scripts, plans, goals, and understanding*. Hillsdale, N.J.: Lawrence Erlbaum Associates, 1975.

Schank, R. C. Language and memory. *Cognitive Science*, 1980, *4*, pp. 243–284.

Searle, J. *Speech Arts*. Cambridge, Mass.: Cambridge University Press, 1970.

Searle, J. A taxonomy of illocutionary acts. In K. Gunderson (Ed.), *Language, mind, and knowledge: Minnesota studies in the philosophy of science*, Vol. XI. Minneapolis, Minn.: University of Minnesota Press, 1975.

Searle, J. R. The intentionality of intention and action. *Cognitive Science*, 1980, *4*, 47–70.

Turing, A. M. Computing machinery and the mind. *Mind: A Quarterly Review of Psychology and Philosophy*, 1950.

Winograd, T. *Procedures as a Representation for Data in a Computer Program for Understanding Natural Language*. MAC-TR-84, MIT Project MAC, 1971.

Winograd, T. *Understanding natural language*. New York: Academic Press, 1972.

Winograd, T. A process model of language understanding. In R. C. Schank & K. M. Colby (Eds.), *Computer models of thought and language*. San Francisco, Calif.: W. H. Freeman, 1973.

Winograd, T. When will computers understand people? *Psychology Today*, 1974, *7*, pp. 73–79. (a)

Winograd, T. Five lectures on artificial intelligence, PIPS-R. No. 5, Electrotechnical Laboratory, Tokyo, Japan, 1974. Reprinted in A. Zampolli (Ed.), *Linguistic structures processing*. North Holland, 1977. (b)

Winograd, T. Frame representations and the procedural–declarative controversy. In D. Bobrow & A. Collins (Eds.), *Representation and understanding: Studies in Cognitive Science*. New York: Academic Press, 1975.

Winograd, T. Towards a procedural understanding of semantics. *Revue Internationale de Philosophie*, 1976, *3*, pp. 117–118.

Winograd, T. A framework for understanding discourse. In M. Just & P. Carpenter (Eds.), *Cognitive processes in comprehension*. Hillsdale, N.J.: Lawrence Erlbaum Associates, 1977. (a)

Winograd, T. On some contested suppositions of generative linguistics about the scientific study of language. *Cognition*, 1977, *5*, pp. 151–179. (b)

Winograd, T. On primitives, prototypes, and other semantic anomalies. *Proceedings from Theoretical Issues in Natural Language Processing II*. University of Illinois at Champaign–Urbana, 25–32, 1978.

Winograd, T. Towards convivial computing. In M. L. Dertouzos & J. Moses (Eds.), *The computer age: A twenty-year view*. Cambridge, Mass.: MIT Press, 1979. (a)

Winograd, T. Beyond Programming languages. *Communications of the ACM*, 1979, *22*, 391–401. (b)

Winograd, T. Extended inference modes in computer reasoning systems. *Artificial Intelligence*.

Winograd, T. *Language as a cognitive process*. Reading, Mass.: Addison–Wesley, (in preparation).

[7]

LANGUAGE AND SPEECH, Vol. 23, Part 1, 1980

SPEECH PERCEPTION

MICHAEL STUDDERT-KENNEDY, REPORTER
Queens College, City University of New York
and
Haskins Laboratories

The paper reviews selected studies in speech perception, most of them published in the past five years. Topics include the contributions of prosody to segmental perception, the problems of segmentation and invariance, categorical perception of speech and non-speech, the role of feature detectors, the scaling of speech sounds to an auditory-articulatory space, acoustic phonetic dependencies within the syllable, the contributions of higher order (non-phonetic) factors to the comprehension of fluent speech, and cerebral specialization. The bias of the paper is toward viewing phonetic segments as abstract processes that link sound and articulation, and that become available to the listener through specialized sensorimotor mechanisms.

The past few years of research in speech perception have been very active. The old questions are still there – What are the units? How do we segment? Where are the invariants? – but some old answers have turned out to be wrong and some new ones are beginning to emerge. The intricate articulatory and acoustic structure of the syllable is still at the center of the maze, but other sources of information for the listener – prosody, syntax, semantics – have begun to receive experimental attention. Studies of fluent speech are taking their place beside the established methods of syllable analysis and synthesis. Theory has dropped into the background (or perhaps the back room) and no one seems very eager to argue the merits of analysis-by-synthesis or the "motor theory" any more. Certainly, theory continues to guide research, but a refreshing atheoretical breeze has been blowing in from artifical speech understanding research (Klatt, 1977, 1978) and from developmental psychology (Aslin and Pisoni, 1978). In the latter regard, I shall not have much to say directly about infant speech perception, but much of what I have to say will bear on it. The infant is a listener, a very attentive one, because by learning to listen it learns to speak. In my opinion, only by carefully tracking the infant through its first two years of life shall we come to understand adult speech perception and, in particular, how speaking and listening establish their links at the base of the language system. This said, let us begin, as infants do, with prosody.

PROSODY

Prosody means the melody, rhythm, rate, amplitude, quality and temporal organization of speech. There has been an upsurge of interest in these factors in recent years, partly because they seem to hold a key to improved speech synthesis, partly because prosodic contributions to speech perception have been unjustly neglected (Cohen and Nooteboom, 1975; Nooteboom, Brokx and de Rooij, 1976). To say that prosody "contributes" to speech perception may seem to imply that speech perception is confined to segmental processes, of which prosody is a mere subsidiary, conveying no distinctive information of its own. This, of course, is false. Prosody carries much of that important indexical information (Abercrombie, 1967) without which, if it is dark, you don't know who is talking to you or whether he means what he says. However, it is with the adjutant function – contributions to segmental perception – that I am concerned here.

One prosodic function is to maintain a coherent auditory signal. Darwin (1975) asked listeners to shadow a sentence in one ear, while a competing sentence was led into the other. At some arbitrary point, prosodic contours were suddently switched across ears, while syntactic and semantic sequences were maintained. Prosodic continuity then often overrode syntax, semantics and ear of entry, leading to the intrusion of words from the supposedly unattended ear. Evidently, listeners were tracking the prosodic contour, a process that Nooteboom *et al.* (1976) suggest may be necessary to maintain "perceptual integrity."

What physical dimensions of the signal sustain this integrity? Rate is probably not important, because quite sharp rate variations are regularly used to convey syntactic information (e.g., Klatt, 1976). Of course, rate *can* affect segmental classification (Ainsworth, 1972), but listeners adjust rapidly, within less than a second (Fujisaki, Nakamura and Imoto, 1975; Summerfield, 1975; Nooteboom *et al.*, 1976). Amplitude changes, within limits, are also probably of little importance (Darwin and Bethell-Fox, 1977). In fact, the principal determinants of prosodic continuity seem to be fundamental frequency (F_0) and spectrum: Nooteboom *et al.* (1976) showed that when pitches, alternating over a 2–6 Hz range, are imposed on a sequence of three vowels, repeated at intervals of less than 150 msec, the vowels split into two streams, as though from two speakers. The effect is reduced, if the vowels are granted a degree of spectral continuity by being placed into consonantal context. This work, taken with similar studies by Dorman, Cutting and Raphael (1975) and by Darwin and Bethell-Fox (1977), leads to the conclusion that continuity of both formant structure and F_0 underlies the perceptual integrity of running speech.

A second prosodic function is to facilitate phrasal grouping. Here the main variables seem to be F_0 and segment duration. Several studies have documented syntactic control of timing and segment duration in production (e.g., Cooper, 1976; Klatt, 1976). Klatt and Cooper (1975) show, further, that listeners expect segment duration to vary with the syntactic position of a word in a sentence. For example, they judge lengthened syllables to be more natural at the end of a clause than at the beginning or middle. Similarly, Nooteboom *et al.* (1976) report that listeners judge a vowel of a particular length to be shorter if it occurs at the end of a word than if it occurs at the beginning. Presumably, such observations reflect listeners' habitual use of final lengthening as an aid to parsing.

The role of F_0 has been more extensively studied. For example, Collier and 't Hart (1975) constructed synthetic utterances consisting of 13 or 15 200-msec steady-state, vowel-like "syllables," separated by 50-msec silent intervals. They imposed ten theoretically derived F_0 contours ('t Hart and Cohen, 1973) on these syllables, deploying characteristic "continuation rises" and "non-final falls" to delimit the ends and the beginnings, respectively, of possible syntactic constituents. Finally, following Svensson (1974) and Kozhevnikov and Chistovich (1965), they asked listeners to write down syntactically acceptable sentences to match each contour in number of syllables, location of stresses and overall intonation. Of the resulting sentences, 72% matched the predicted syntactic structures. Since two hypotheses were under test here – both the correctness of the theoretically derived contours and the listeners' capacity to infer syntactic structure from intonation – this is a remarkably high score.

Finally, a third perceptual function of prosody has aroused a great deal of interest in recent years. This is the function – nobody knows what it is – supposedly fulfilled by rhythm. Martin (1972) wrote a persuasive paper in which he argued that speaking involves more than a simple concatenation of motor elements: like other motor behaviors speech is compelled, by natural constraints on the relative timing of components, to be rhythmic. Moreover, some components (syllables) are "accented," and these are predictable: accent level (or stress) covaries with timing and the main accents are equidistant (i.e., isochronous). Finally, since ". . . speaking and listening are dynamically coupled rhythmic activities . . ." (p. 489), listeners can predict the main stresses and can use that fact to "cycle" their attention, saving it, as it were, for the more important words.

There is, in fact, evidence from phoneme-monitoring experiments that reaction time (RT) is shorter to initial phonemes in stressed words than in unstressed words (Shields, McHugh and Martin, 1974). This is apparently not due to the greater energy of the stressed words, since, if the words are presented in isolation, no RT difference appears (Shields *et al.*, 1974). Moreover, Cutler (1976) has found that the RT difference holds, even if stress, or the lack of it, is merely "predicted" by prior prosodic contour and if the actual target is acoustically identical in both conditions. Cutler and Foss (1977) demonstrate, further, that RT advantage is not due to syntactic form class, since it is found for stressed function words as well as for stressed content words. They conclude that the reduced reaction time may reflect heightened attention to the semantic focus of a sentence, and they cite (p. 10) unpublished evidence from Allen and O'Shaugnessy that ". . . reliable correlates of semantic focus are to be found in the fundamental frequency contour."

By this last point Cutler and Foss seem to be cutting themselves free from Martin's (1972) claim for isochrony, whether wisely or not remains to be seen. Lehiste (1977) has recently reopened the isochrony issue in a paper summarizing much of her research on the topic. She concludes that although isochrony is "primarily a perceptual phenomenon" (p. 253), it does have some basis in production and is therefore available for communicative uses. Lehiste shows that English interstress intervals are often lengthened to signal a syntactic boundary.

Isochrony has also come under experimental scrutiny. Morton, Marcus and Frankish (1976), recording a list of spoken digits for experimental use, discovered that acoustically (onset to onset) isochronous sequences sounded anisochronous. Moreover, listeners, asked to adjust a sequence to perceptual isochrony, made it acoustically anisochronous. Morton *et al.* (1976) coined the term "perceptual centers" ("P-centers") to refer to those points in a sequence of words that are equidistant when the words sound isochronous.

But they were unable to locate the points or specify their acoustic correlates. Surprisingly, the P-center does not correspond to any obvious acoustic marker, such as sound onset, vowel onset or syllable peak. However, Fowler (1979, p. 375) has recently discovered that "... when asked to *produce* isochronous sequences, talkers generate precisely the acoustic anisochronies that listeners require in order to *hear* a sequence as isochronous." The acoustic anisochronies apparently arise because the articulatory onsets of words beginning with sounds from different manner classes have acoustic consequences at different relative points in time. From a review of her own and related studies (e.g., Allen, 1972; Lindblom and Rapp, 1973), Fowler (1979, p. 375) concludes that "... listeners judge isochrony based on acoustic information about *articulatory* timing rather than on some articulation-free acoustic basis." Finally, although this work seems to be a thread that might unravel isochrony, Fowler is cautious in her claims. Most of the relevant experimental studies have used monosyllables and artificially repetitive utterances. What inroads this approach can make into the apparent isochrony of phonetically heterogeneous running speech remains to be seen.

SEGMENTATION AND INVARIANCE

We turn now from the broad questions of prosody to the narrower puzzle of the syllable on which the prosody is carried. In what follows, I assume (together with most other investigators) that our task is to understand the process by which phonemes or features are extracted from the signal. Let us begin with a question raised by Myers, Zhukova, Chistovich and Mushnikov (1975): Is segmentation an auditory process, preceding phonetic classification, or an automatic consequence of classification itself? Several studies from the Pavlov Institute in Leningrad speak to the question. Chistovich, Fyodorova, Lissenko and Zhukova (1975) showed that a sudden amplitude drop, roughly in the middle of a 460-msec steady-state vowel, caused listeners to hear either two vowels or a VCV sequence, depending on the magnitude and rate of the amplitude decrease. Subsequently, Myers *et al.* (1975) used an ingenious dichotic technique to suggest that such amplitude decreases are registered by the peripheral auditory system; they inferred that, since classification is presumably central, segmentation must precede classification. Finally, Zhukov, Zhukova and Chistovich (1974) reported on the use of a similar technique to study the effects of spectral variation at segment boundaries. The investigators presented a time-varying value of the second formant (roughly 2200 to 800 Hz over 200 msec), to one ear, steady-state values of F1 and F3 to the other. The latter were interrupted by a 12–15 msec pause, of which the position could be set by the subject so as to vary the fused percept from hard to soft [r], that is, from [iru] to [iŗu]. Subjects reliably set the pause so that its endpoint coincided with an F2 value of roughly 1600 Hz. Since this value is close to that of the hard-soft boundary previously determined for the steady-state isolated consonants [s] and [ş], the authors infer that listeners were also judging the soft consonant [ŗ] by its F2 value at onset. They conclude (p. 237) that "the auditory system interprets the acoustic flow as a sequence of time segments between instants of variation," and that it derives consonantal information by sampling formant frequencies at these instants.

However, this conclusion does not seem to be forced by the data. On the one hand, the presumed peripheral segment boundary, determined by a sharp amplitude drop, seems to have something in common with the boundary proposed by certain automatic recognition procedures for isolating syllables rather than phonemes (e.g., Mermelstein, 1975). On the other hand, an invariant formant onset is not incompatible with the use of formant movement into the following vowel as a consonantal cue (see Dorman, Studdert-Kennedy and Raphael, 1977). My inclination, therefore, is to suppose that the preliminary auditory segmentation (if any) is syllabic rather than phonemic, and that within-syllable segmentation may often be synonymous with classification. I will return to this point below.

The view of the perceptual process, proposed by the Russian group, as a succession of brief time slices (rather than as the active continuous tracking suggested by studies of prosody), is close to that currently being explored by K.N. Stevens. In a succession of publications over recent years, Stevens (e.g., 1975) has elaborated on the "quantal nature of speech." He points out that, although the vocal apparatus is capable of producing a wide variety of sounds, relatively few are actually used in the languages of the world. He attributes this restriction to a nonlinear relation between articulatory and acoustic parameters: some articulatory configurations are acoustically stable, in the sense that small changes in articulation have little acoustic effect; others are unstable in the sense that equally small changes have a substantial effect. The universal set of phonetic features is drawn from those articulatory configurations that generate acoustically stable, invariant "properties." The properties, it should be stressed, are higher order spectral configurations, rather than isolated cues such as F2 onset frequency. To define these configurations, Stevens has largely relied on computations from a vocal tract model. Finally, to assure quantal (or categorical) perception of the invariant properties and to afford the human infant a mechanism for netting them in the speech stream, Stevens postulates a matching set of innate "property detectors."

Empirical tests of the quantal theory have been few. But a recent study of English stops (Blumstein and Stevens, 1979; see also Stevens and Blumstein, 1978) is a good illustration of the approach, since it deals with a notoriously context-dependent set of sounds. The goal was to demonstrate the presence of invariant properties in the acoustic signal, sufficient for recognition by fixed templates. The first step was to record two male speakers reading random lists of the voiced stops [b d g], followed by each of five vowels [i e a o u]. Short-time spectra were then determined, integrated over a 26-msec window at onset. The spectra were used to construct, by trial and error, a template fitted to each place of articulation, such that it either correctly accepted or correctly rejected the majority of utterances. Descriptions of the templates ("diffuse-rising" for alveolar, "diffuse-falling" for labial and "compact" for velar) recall the terminology of distinctive feature theory.

In the second part of the study, a corpus of utterances was collected for classification by the templates. Six subjects (4 males, 2 females) recorded five repetitions each of the voiced and voiceless stop consonants [b d g p t k], followed by each of the vowels [æ e i o u], or preceded by each of the vowels [i ɛ æ ʌ u]. The resulting 1800 utterances were then analyzed spectrally in the same way as the original utterances, and compared with the templates. The results were at least 80% (and often higher) correct rejection and correct acceptance for initial stops and a slightly lower performance for released final stops, although for some unreleased final stops scores dropped as low as 40%. Analysis of variance revealed significant differences in template-matching performance as a function of vowel context, but performance was significantly above chance in every case. Quite similar results have been reported by Searle, Jacobson and Rayment (1979) using a very much longer time slice (100-200 msec) and deriving their invariant patterns from a running sequence of spectra.

Where then does this leave us? A score of 80% or better is good, although, as A.M. Liberman has suggested to me, we might do almost as well with the binary recipe proposed by Cooper, Delattre, Liberman, Borst and Gerstman (1952); high burst, falling F2 transition for alveolar; low burst, falling F2 transition for velar; low burst, rising F2 transition for labial.

The question, of course, is: Is this really the way that humans do it? Dorman *et al.* (1977), modeling their study on the work of Fischer-Jørgensen (1972), edited release bursts and/or formant transitions out of English voiced stop consonants ([b d g]), spoken before nine different vowels. Acoustic analysis of the bursts for a given place of articulation showed them to be largely invariant (cf. Zue, 1976). However, the bursts were not invariant in their effect: for the most part, listeners only perceived the bursts correctly if their main spectral weight lay close to the main formant of the following vowel, as Stevens himself has remarked (1975, pp. 312-313). Kuhn (1975) has shown that the main vowel formant varies with the length of the cavity in front of the point of maximum tongue constriction. Since front cavity length is a function of place of articulation, an estimate of front cavity resonance is tantamount to an estimate of place of articulation. Thus, proximity on the frequency scale may facilitate perceptual integration of the burst with the vowel, enabling the listener to track the changing cavity shape characteristic of a particular place of articulation followed by a particular vowel.

Stevens (see especially, 1975) does not deny that contextually variable cues – such as formant transitions, voice onset time, vowel formant structure – can be used by the human listener. However, he regards them as "secondary," learned cues, acquired by repeated association with the "primary" invariant properties, and used only as safety devices when invariant cues fail. Given the many knotty questions concerning the possible mechanisms for extracting and interpreting these "secondary" context-dependent cues, one may wonder how an organism whose primary endowment is a set of passive templates learns to use them at all.

The question becomes even more pressing when one considers that there is no independent evidence for the existence of the hypothesized templates or property detectors. To understand this we must briefly review recent findings in the study of categorical perception.

Categorical Perception

As is well known, early work with speech synthesizers showed that a useful procedure for defining the acoustic properties of a phoneme was to construct tokens of opponent categories, distinguished on a single phonological feature, by varying a single acoustic parameter along a continuum (e.g., [ba] to [da], [da] to [ta], etc.). If listeners were asked to identify these tokens, they tended to identify any particular stimulus in the same way every time they heard it; there were few ambiguous tokens. Moreover, if they were asked to discriminate between neighboring tokens, they tended to do very badly if they assigned the two tokens to the same class, very well if they assigned them to different classes – even though the acoustic distance between tokens was identical in the two cases. This phenomenon was dubbed "categorical perception" (Liberman, Harris, Hoffman and Griffith, 1957). Although there were usually no grounds for supposing that the acoustic variations along synthetic continua mimicked the intrinsic allophonic variations of natural speech, categorical perception in the laboratory was taken to reflect a necessary aspect of normal speech perception, namely, the rapid transfer of speech sounds into a phonetic or phonological code. The phenomenon was also believed by some people, including myself, to be peculiar to speech (Studdert-Kennedy, Liberman, Cooper and Harris, 1970).

However, we now know that categorical perception, as observed in the laboratory, is neither peculiar nor necessary to speech. Demonstrations that it is not peculiar we owe to Cutting and Rosner (1974) (rise-time at the onset of sawtooth waves, analogous to a fricative-affricate series); to Miller, Wier, Pastore, Kelly and Dooling (1976) (noise-buzz sequences analogous to the aspiration-voice sequence of a voice onset time (VOT) series); to Pisoni (1977) (relative onset time of two tones); and to Pastore, Ahroon, Baffuto, Friedman, Puleo and Fink (1977). These last investigators extended their work into vision, demonstrating categorical perception of flicker, with a sharp boundary at the critical flicker-fusion threshold. They also induced clearly categorical perception of a sine-wave intensity series by providing listeners with a constant-reference tone, or "pedestal," at the center of the series. Pastore *et al.* (1977) conclude that a continuum may be categorically divided either by a sensory threshold (as in flicker-fusion) or by a stimulus-internal reference (as in the intensity series). Presumably, the portion of the signal with the earlier onset serves as a reference in a VOT series, while in a place-of-articulation series, cued by direction and extent of formant transitions, a reference is provided by the fixed vowel. If this last point is correct, we perceive a place series categorically precisely because the consonants are judged relationally rather than absolutely – an interpretation not compatible with the notion of invariant property detectors.

Just how a stimulus-internal reference suppresses discrimination within categories is not clear, but the results of Carney, Widin and Viemeister (1977) suggest that it may simply serve to divert the listener's attention from detailed variations within the series. To Carney *et al.* (1977) (see also Pisoni and Lazarus, 1974; Samuel, 1977), we owe the demonstration that a VOT continuum *need* not be perceived categorically. Each of their subjects displayed good within-category discrimination after moderate training on a bilabial VOT continuum. Indeed discrimination was so good that subjects were able to shift category boundaries on request and assign consistent labels to arbitrary subsets of the stimuli. The outcome suggests that "... utilization of acoustic differences between speech stimuli may be determined primarily by attentional factors, ... distinct from the perceptual *capacities* of the organism" (Carney *et al.*, p. 969).

This is precisely what is suggested by the numerous instances in which speakers of different languages perceive an acoustic continuum in different ways. (For a review, see Strange and Jenkins, 1978). For example, while American English speakers perceive an [r] to [l] continuum categorically, Japanese speakers do not (Miyawaki, Strange, Verbrugge, Liberman, Jenkins and Fujimura, 1975). For another example, not only do Spanish and American English speakers place their category boundaries at different points along the VOT continuum (Abramson and Lisker, 1973; Williams, 1977), but also Spanish-English bilinguals can be induced to shift their boundaries by a shift in language set within a single test (Elman, Diehl and Buchwald, 1977). Not unrelated, perhaps, is the recent demonstration by Ganong (1978) that listeners have a bias for words over nonwords: offered a continuum of which one end is a word (e.g., *gift*) and the other not (e.g., *kift*), they shift their normal boundary away from the word, thus increasing the number of words they hear.

Presumably there are limits to this sort of thing. With adequate synthesis, the range of uncertainty must be limited and we may still use synthetic continua to assess "the auditory tolerance of phonological categories" (Brady and Darwin, 1978, p. 1556) – precisely the use for which they were first designed over 25 years ago.

Feature or Property Detectors

The demonstration that listeners can be trained to hear a supposedly categorical continuum noncategorically undercuts the original evidence for acoustic feature, or property, detectors in speech perception, namely, categorical perception itself. Moreover, it throws into doubt the interpretation of a substantial body of work on selective adaptation of speech sounds that has appeared in the past five years.

The series began with a paper by Eimas and Corbit (1973). They asked listeners to categorize members of a synthetic voice onset time (VOT) continuum (Lisker and Abramson, 1964) and demonstrated that the perceptual boundary between voiced and voiceless categories along that continuum was shifted by repeated exposure to (that is, adaptation with) either of the endpoint stimuli: there was a decrease in the frequency with which stimuli close to the original boundary were assigned to the adapted category and a consequent shift of the boundary toward the adapting stimulus. Since the effect could be obtained on a labial VOT continuum after adaptation with a syllable drawn from an alveolar VOT continuum, and vice versa, adaptation was clearly neither of the syllable as a whole nor of the unanalyzed phoneme, but of a feature within the syllable. Eimas and Corbit therefore termed the adaptation "selective" and attributed their result to the fatigue of specialized detectors and to the relative "sensitization" of opponent detectors. Subsequent studies replicated the results for VOT and extended them to other feature oppositions, such as place and manner of articulation. These studies have been reviewed by Cooper (1975), Ades (1976), and Eimas and Miller (1978).

Unfortunately, there are many grounds for doubting the opponent detector model. First, as already remarked, is the demonstration that listeners can be trained to discriminate at least some speech continua within categories. Second, the model lacks behavioral or neurological motivation. For, while the facts of additive color mixture make an opponent detector account of after-effects entirely plasuible, the facts of laryngeal timing or spectral scatter at stop consonant onset certainly do not. Third, the hypothesis is rendered implausible by dozens of reports of contextual effects: adaptation of consonantal features is apparently specific to following vowel, to syllable position, to syllable structure (Hall and Blumstein, 1978) and even to fundamental frequency (Ades, 1977). As Simon and Studdert-Kennedy (1978) remark, "... the theoretical utility of selectively tuned feature detectors goes down as the number of contexts to which they must be tuned goes up." Moreover, the degree of adaptation varies quite generally with the acoustic distance between adaptor and test syllables, an effect typical of psychophysical contrast studies. In fact, Simon and Studdert-Kennedy (1978), drawing on their own work and that of Sawusch (1977), marshall evidence to argue that selective adaptation along speech continua reflects a combination of peripheral auditory fatigue and central auditory contrast. They do not deny that selective adaptation has possible fruitful use in isolating functional channels of analysis. But if their argument is correct, we now have no evidence at all for specialized speech detector mechanisms tuned to the acoustic correlates of abstract linguistic features.

Scaling Studies and Feature Interactions

This conclusion sits nicely with the results of many studies in which phoneme confusions or similarity judgments have been used to characterize the psychological representation of speech sounds. Although results vary widely with experimental method (van den Broecke, 1976), these studies typically find that vowels (e.g., Terbeek, 1977) and consonants (e.g., Singh, Woods and Becker, 1972) fall readily into low-confusion/high-similarity groups isomorphic with some standard phonological feature set. However, as Goldstein (1977) has pointed out, relations within these feature groups are usually not random. Rather, the psychological space is structured in such a way as to suggest a continuous auditory representation within feature groups. Presumably, since the continuous auditory representation derives from an acoustic structure shaped by articulation, we could describe an analogous articulatory space by scaling articulatory errors. It was Goldstein's (1977) insight to hypothesize that the variance common to the auditory and articulatory spaces would then prove to be categorical. His study – too complicated for summary here – largely supported that hypothesis. We may fairly conclude that our models of perception should allow for continuous auditory and articulatory representations from which categories can only be derived by some abstract metric common to both.

The idea that speech sounds (perhaps unsegmented syllables) may be internally represented in a continuous auditory space (at some point before classification) is compatible with the repeated finding of interaction between features during perceptual processing (e.g., Sawusch and Pisoni, 1974; Miller, 1977). There is, in fact, no good reason to refer to these auditory processes as "featural" at all (Parker, 1977). Repp

(1977) and Oden and Massaro (1978), for example, have already proposed specific models of integration based on a continuous spatial representation.

STEPS TOWARD AN AUDITORY-ARTICULATORY SPACE

The view of speech perception that seems to be emerging from the studies we have reviewed is of an active, continuous process. We turn now to several studies of perceptual integration across the syllable which seem to call for just such an interpretation.

Perhaps the most familiar example is provided by voicing cues for stops in initial position. The concept of voice onset time (VOT) originally offered an *articulatory* account of how a range of disparate and incommensurable acoustic cues (including, as it happens, the interval between release burst and the onset of voicing) comes to signal the voiced-voiceless distinction. In fact, as Abramson (1977) has recently reminded us, VOT is itself simply a special case of the laryngeal timing mechanisms by which voicing distinctions are, in general, implemented.

To illustrate the underlying articulatory rationale, consider the suggestion by Stevens and Klatt (1974) that the duration of the first formant voiced transition might be a more potent cue than VOT itself. The motivation for the proposal seems to have been to coordinate the voicing cue with Stevens' hypothesized cues to place of articulation (rapid spectral scatter), and perhaps to avoid saddling the infant with a delicate timing mechanism. As it happens, Simon and Fourcin (1978) have shown that English speaking children do not learn to use the F1 cue until they are five years old, while French-speaking children never use it at all. In any event, careful analysis by Lisker (1975) and by Summerfield and Haggard (1977) has shown that the principal first formant cue is not transition duration, but frequency at onset: the higher the frequency, the less likely is a sound to be judged voiced. Listeners apparently take a high first formant onset as a cue that the mouth was relatively wide open (and release therefore well past) when voicing began.

A less familiar set of cues to another distinction has recently been studied by Repp, Liberman, Eccardt and Pesetsky (1978). They recorded the utterance: "Did anybody see the *gray ship?*" Then, by varying the durations of fricative noise at the onset of *ship* and of the silent interval between *gray* and *ship,* they explored the conditions under which the utterance was heard as ending with "gray chip," "great ship" or "great chip." Among their results was the finding that whether or not a syllable final stop was heard (*gray* v. *great*) depended not only on the duration of the silence, but also on the duration of the noise following the silence. Just such an equivalence between a spectral property and silence emerges from an analysis of the trading relation between silence and formant transition in the cues for the medial [p] of [splɪt] (Liberman and Pisoni, 1977). How are we to rationalize such an equivalence? Repp *et al.* (1978) point out that neither a single feature detector nor a set of feature detectors, integrated by some higher level decision mechanism (as proposed by Massaro and Cohen, 1976), nor, it would seem, any purely auditory principle can explain why such phenomenologically diverse cues can be traded off and integrated into a unitary percept.

As a final example, consider a positively daedalian series of experiments by Bailey and Summerfield (1978). They explored the conditions under which a particular voiceless stop ([p], [t] or [k]) is perceived if a silence is introduced between [s] and a following vowel. Whether a stop is heard at all depends, of course, on the duration of the silence, but the effect of that duration itself depends on the onset frequency of F1, while the perceived place of articulation depends on the duration of the closure, on spectral properties at the offset of [s] and on the relation between those properties and the following vowel (cf. Dorman *et al.*, 1977). Bailey and Summerfield (1978, p. 55) suggest that, "... given sufficiently precise stimulus control, perceptual sensitivity could be demonstrated to every difference between two articulations" (cf. Haggard, 1977). Again, the problem is to understand the principles by which such heterogeneous collections of spectral and temporal cues are combined into a percept. What rationalizes their integration?

The answer, explicitly proposed by the authors of these several studies, is that the cues are held together by their origin in the integral, articulatory gesture. We should be absolutely clear that this is *not* a form of motor theory. Rather, it is a description of what the perceptual system appears to do. The system follows the moment-to-moment acoustic flow, apprehending an auditory "motion picture," as it were, of the articulation, in a manner totally analogous to that by which the visual system might follow the optic flow to apprehend the articulation by reflected light rather than by radiated sound (cf. Fowler, 1979; Studdert-Kennedy, 1977).

Reading Lips and Reading Spectrograms

The argument is clarified, and developed, in a recent study of lip reading by Summerfield (1979). Subjects were asked to write down a series of sentences spoken over an audio system, but simultaneously masked by the talker's own voice reading another text. There were three conditions of interest to the present discussion: (1) audio alone; (2) audio with full video of the speaker's face; (3) audio with a display of the speaker's lips. Without any training, naïve subjects scored 23%, 65% and 54% correct, respectively. In a second experiment, Summerfield analyzed errors made against deliberately conflicting video. He found, as did McGurk and McDonald (1976), that subjects frequently made judgments reflecting a compound between the auditory and visual information. Summerfield (as also Haggard, 1977) points out that such instantaneous interplay between modalities seems to require a common metric by which the two streams of information can be combined (cf. Campbell and Dodd, in press). (The problem, incidentally, is quite general and may apply to any sound-producing visual event.)

It is instructive to compare the ease with which naïve subjects used the visual display of face or lips with the obvious difficulty experienced by even the most skilled spectrogram reader. Cole, Rudnicky, Reddy and Zue (1978) report a systematic study of subject VZ who has been studying acoustic phonetics for more than seven years and has logged some 2000–2500 hours reading spectrograms – perhaps as many hours as a child of two years has spent listening to speech. Despite the fact that VZ is free to use the ample context of vision (rather than the narrow window of audition) and that he reports conscious, acoustic-phonetic interpretation of visual context at least 18% of the time; despite the fact that he came to the spectrograms knowing that their visual segments were not isomorphic with phonetic segments (a crucial piece of knowledge that cannot be derived from the spectrograms themselves); despite the fact that, in the hours devoted to spectrograms, he could probably have learned to read several foreign languages with fair proficiency, VZ now transcribes spectrograms at a rate some 20 to 40 times real time (Cole, personal communication).

One is not surprised. There are, after all, biological constraints on learning (see Hinde and Stevenson-Hinde, 1973): pigeons learn more readily to peck plastic keys for grain and to jump to avoid shock than vice versa. The visual display of talking lips and face is natural and its code is known to every sighted speaker of a natural language, as the code of a spectrographic display is not. Watching its mother's face and listening to her speak, the infant learns to perceive articulation directly, whether by light or by sound.

EXTRACTING INFORMATION FROM THE SYLLABLE

The primary unit of perception is evidently the unsegmented syllable (the rhythmic unit of nursery rhymes), and there is ample evidence for perceptual interaction among its components (see Studdert-Kennedy, 1976, for a review). For a recent example, Hasegawa and Daniloff (1976) synthesized two fricative continua, /s/ – /ʃ/, after two different vowels, /i/ and /u/, and found a significant shift in the phoneme boundary as a function of preceding vowel. Kunisaki and Fujisaki (1977) developed the finding by showing that contextual dependency in perception corrects for a mirror-image contextual dependency in production: just as the frequencies of fricative poles and zeros are lower before /u/ than before /ɑ/, so, in perception, the frequencies of the poles and zeros at the synthetic boundary between /s/ and /ʃ/ are higher before /ɑ/ than before /u/. These results mesh neatly with our earlier conclusion that consonantal onset is judged as part of a dynamic, temporal pattern.

Just such a process has recently been shown to play an important role also in vowel perception. Strange, Jenkins and Edman (1978) recorded tokens of /b/–vowel–/b/ syllables with ten different medial vowels, spoken by several speakers. They edited out the steady-state syllable nuclei (50% to 65% of the entire syllable, depending on the vowel) and presented various fragments of the syllables for identification. The results varied with both speaker and vowel, but overall, for three speakers of the same dialect as the listeners, error rates on the original syllables, on the syllables without their centers ("silent centers") and on the isolated centers were 4%, 10% and 18% respectively. The error rates for either the initial or the final transitions alone were approximately 60%. Evidently, the dynamic sweep of the spectral information and its temporal distribution across the syllable was the principal source of listener information in identifying these vowels, even when that portion usually said to characterize a vowel (namely, its steady state) was completely missing.

Results such as these return us to the segmentation issue. Clearly, there was little basis for peripheral segmentation in these syllables. In fact, one is tempted to suppose that listeners recognized syllables (Massaro, 1975) or perhaps phone-pairs (Klatt, 1978) rather than phonemes. Mermelstein (1978) reports a subtle experiment that speaks to this issue. He varied the duration and first formant frequency of the steady-state nucleus of synthetic syllables to yield /bɛd/, /bæd/, /bɛt/, /bæt/. Notice that exactly the same acoustic information (namely, duration of the steady-state nucleus) controls both vowel and final consonant decision. Accordingly, if subjects are asked to determine duration boundaries for both consonant voicing and vowel quality as a function of F1 frequency, and if the boundaries prove to be correlated, then we can conclude that listeners made a single – presumably syllabic – decision. However, if the boundary values prove independent, we can conclude that listeners recognized phonemes rather than syllables, and that they made two phonetic decisions on the basis of a single piece of acoustic information. This was, in fact, the outcome. If this is the normal mode of speech perception, it would seem that, even if syllabic segmentation is peripheral (cf. Myers *et al.*, 1975), phonemic segmentation may be a central process consequent upon classification. Usually, this process is facilitated by auditory contrast within the syllable (cf. Bondarko, 1969).

CONTINUOUS SPEECH

We now come full circle to continuous speech with its prosody, syntax and "real world" constraints. Here, the main question is whether the perceptual processes we have been discussing up to this point have any bearing at all. Is it possible, for example, that, given the contextual aids of prosody, syntax, semantics, the listener needs no more than the "auditory contour" of a word (Nooteboom *et al.*, 1976; cf. Morton and Long, 1976) or perhaps a few "invariant features" (Cole and Jakimik, 1978) to gain access to his lexicon?

I have no space for a full discussion of this issue (see Liberman and Studdert-Kennedy, 1978). But a good place to start is with a paper by Shockey and Reddy (1975) who studied speech recognition in the absence of phonological and all other higher order constraints. They recorded some 50 short utterances, spoken by native speakers of 11 different languages, and presented them to four phoneticians for transcription. The transcriptions were then compared with a "target" description, determined from native speakers and spectral analysis. The average "correct" score for the four transcribers was 56% and their average agreement, 50%. Comparable scores for transcription of a familiar language, without contextual or syntactic constraints, would be roughly 90% – the level reached by the three transcribers of Cole *et al.* (1978), in their spectrogram-reading study, cited above, and, moreover, a level close to that of VZ himself when reading spectrograms. The difference of roughly 40% is evidently due to the transcribers' knowledge of the phonology of the language being transcribed.

The point of this example is that the main difference between listening to continuous speech in a familiar language and to isolated words in a foreign one may not be in the syntax, semantics or real world constraints so much as in the phonology. This is a simplification, since phonology and syntax are not independent. But it serves to emphasize that phonology makes linguistic communication possible by setting limits on how a speaker is permitted to articulate and what a listener can expect to hear (Liberman and Studdert-Kennedy, 1978). The problem of how the listener extracts and combines information from the signal to arrive at a unitary percept is, of course, exactly the same for continuous speech as for isolated words.

The function of the other higher order constraints – syntax, context, semantics – is facilitative. They serve to delimit the sampling space from which the listener's percepts may be drawn. This is well illustrated by several experiments of Cole and Jakimik (1978), using the ingenious "listening for mispronunciations" technique, devised by Cole (1973). Subjects are asked to listen to a recorded story into which mispronunciations have been systematically introduced. Their accuracy and speed of detection is then measured as a function of different variables. Mispronunciations prove to be more rapidly reported for high than for low transitional probability words (cf. Morton and Long, 1976), for words appropriate to a theme than for words inappropriate, for words implied by previous statements than for words not implied, and so on. Presumably the more rapid reports reflect the varied ways in which thresholds for words are lowered by contextual factors. Of course, the fact that listeners recover the words at all means that they can do so without a full phonetic analysis. But this does not mean that they regularly do so without any phonetic analysis at all.

By far the fullest and most careful account of the interactive processes of word recognition in continuous speech is offered by Marslen-Wilson (1975; Marslen-Wilson and Welsh, 1978). His experimental procedure also involves mispronunciations, but the subjects' task is to shadow the text as rapidly as possible. Marslen-Wilson examines the effects of context on the frequency of fluent restorations. These restorations are often so fast that the shadower begins to say the correct word (e.g., "company") before the second syllable of the mispronounced word (e.g., "compsiny") has begun (cf. Kozhevnikov and Chistovich, 1965). Since such restorations only occur when the disrupted word is syntactically and semantically apt, it is evident that these higher order factors have facilitated recovery of the correct word. However, they cannot do so for an entire utterance in the absence of all phonetic information. It is reassuring to read as the conclusion of a lengthy and subtle discussion of these matters: ". . . word-recognition in continuous speech is fundamentally data-driven, in the specific sense that the original selection of word-candidates is based on the acoustic-phonetic properties of the initial segment of the incoming word" (Marslen-Wilson and Welsh, 1978, p. 60; cf. Nakatani and Dukes, 1977). Perhaps all these years of studying CV syllables have not been wasted after all.

Cerebral Specialization

Nonetheless, opposition between the two modes of lexical access – holistic, from "auditory contour," analytic, from phonetic segments – should not be too sharply drawn. The work of Zaidel (1978a,b) with "split-brain" patients has demonstrated that holistic access is certainly possible. The cerebral hemispheres of such patients have been surgically separated by section of the connecting pathways (corpus callosum) for relief of epileptic seizure. The separation permits an investigator to assess the linguistic capacities of each hemisphere independently. Zaidel (1978a,b) has shown that the isolated right hemisphere of such a patient, though totally mute, can recognize a sizeable auditory lexicon and has a rudimentary syntax sufficient for understanding phrases of up to three or four words in length. However, it is incapable of identifying nonsense syllables or of performing tasks that call for phonetic analysis, such as recognizing rhyme (cf. Levy, 1974). This phonetic deficit evidently precludes short-term verbal store, thus limiting the right hemisphere's capacity for syntactic analysis of lengthy utterances, and forces organization of language around meaning. Whether we assume a similar, subsidiary organization in the left hemisphere or some process of interhemispheric collaboration, it is clear that normal language comprehension could, at least in principle, draw on both holistic and analytic mechanisms.

At the same time, Zaidel's work provides striking support for the hypothesis, originally derived from dichotic studies, that the distinctive linguistic capacity of the left hemisphere is for phonological analysis of auditory pattern (Studdert-Kennedy and Shankweiler, 1970). Further support has come from electroencephalography (Wood, 1975) and, quite recently, from studies of the effects of electrical stimulation during craniotomy (Ojemann and Mateer, 1979). The latter work isolated, in four patients, left frontal, temporal and parietal sites, surrounding the final cortical motor pathway for speech, in which stimulation blocked both sequencing of oro-facial movements and phoneme identification.

This fascinating discovery meshes neatly with a growing body of data and theory that has sought, in recent years, to explain the well-known link between lateralizations for hand control and speech. Semmes (1968) offered a first account of the association by arguing, from a lengthy series of gunshot lesions, that the left hemisphere is focally organized for fine motor control, the right hemisphere diffusely organized for broader control. Subsequently, Kimura and her associates reported that skilled manual movements (Kimura and Archibald, 1974) and non-verbal oral movements (Mateer and Kimura, 1977) tend to be impaired in cases of non-fluent aphasia. These impairments are specifically for the sequencing of fine motor movements and are consistent with other behavioral evidence that motor control of the hands and of the speech apparatus is vested in related neural centers (Kinsbourne and Hicks, 1979). In fact, Kimura (1976, p. 154) has proposed that ". . the left hemisphere is particularly well adapted, not for symbolic function *per se*, but for the execution of some categories of motor activity which happened to lend themselves readily to communication." Among these categories we must, incidentally, include those that support the complex "phonological" and morphological processes of manual sign languages, now being discovered by the research of Klima, Bellugi and their colleagues (Klima and Bellugi, 1979).

The drift of all this work is toward a view of the left cerebral hemisphere as the locus of interrelated sensorimotor centers, essential to the development of language, whether spoken or signed. To understanding of the speech sensorimotor system, perceptual studies of dichotic listening will doubtless contribute. Indeed, important dichotic studies have recently found evidence for the double dissociation of left and right hemisphere, speech and music, in infants as young as two or three months (Entus, 1977; Glanville, Best and Levenson, 1977). However, dichotic work has not fulfilled its early promise, largely because it has proved extraordinarily difficult to partial out the complex of factors, behavioral and neurological, that determine the degree of observed ear advantage (cf. Shankweiler and Studdert-Kennedy, 1975). For the future, we may increasingly rely on instrumental techniques for monitoring brain activity, such as the blood-flow studies of Lassen and his colleagues (Lassen, Ingvar and Skinhøj, 1978), induced reversible lesions by focal cooling (Zaidel, 1978b), improved methods of electroencephalographic analysis, auditory evoked potentials (Molfese, Freeman and Palermo, 1975) and, perhaps infrequently, direct brain stimulation.

REFERENCES

ABERCROMBIE, D. (1967). *Elements of General Phonetics* (Chicago).

ABRAMSON, A.S. (1977). Laryngeal timing in consonant distinctions. *Phonetica*, **34**, 295-303.

ABRAMSON, A.S. and LISKER, L. (1973). Voice timing perception in Spanish word-initial stops. *Journal of Phonetics*, **1**, 1-8.

ADES, A.E. (1976). Adapting the property detectors for speech perception. In R.J. Wales and E. Walker (eds.), *New Approaches to Language Mechanisms* (Amsterdam).

ADES, A.E. (1977). Source assignment and feature extraction in speech. *Journal of Experimental Psychology: Human Perception and Performance*, **3**, 673-685.

AINSWORTH, W.A. (1972). Duration as a cue in the recognition of synthetic vowels. *Journal of the Acoustical Society of America*, **15**, 72-100.

ALLEN, G. (1972). The location of rhythmic stress beats in English: An experimental study. *Language and Speech*, **15**, 72-100.

ASLIN, R.N. and PISONI, D.B. (1978). Some developmental processes in speech perception. Paper presented at NICHD Conference on Child Phonology: Perception, Production and Deviation, Bethesda, Maryland, May 28-31, 1978.

BAILEY, P.J. and SUMMERFIELD, Q. (1978). Some observations on the perception of [s] + stop clusters. *Haskins Laboratories Status Report on Speech Research*, SR-53, 25-60.

BLUMSTEIN, S.E. and STEVENS, K.N. (1979). Acoustic invariance in speech production. *Journal of the Acoustical Society of America*, **66**, 1001-1017.

BONDARKO, L.V. (1969). The syllable structure of speech and distinctive features of phonemes. *Phonetica*, **20**, 1-40.

BRADY, S.A. and DARWIN, C.J. (1978). Range effect in the perception of voicing. *Journal of the Acoustical Society of America*, **63**, 1556-1558.

CAMPBELL, R. and DODD, B. (in press). Hearing by eye. *Quarterly Journal of Experimental Psychology*.

CARNEY, A.E., WIDIN, G.P. and VIEMEISTER, N.F. (1977). Noncategorical perception of stop consonants differing in VOT. *Journal of the Acoustical Society of America*, **62**, 961-970.

CHISTOVICH, L.A., FYODOROVA, N.A., LISSENKO, D.M. and ZHUKOVA, M.G. (1975). Auditory segmentation of acoustic flow and its possible role in speech processing. In G. Fant and M.A.A. Tathem (eds.), *Auditory Analysis and the Perception of Speech* (New York), pp. 221-232.

COHEN, A. and NOOTEBOOM, S.G. (eds.) (1975). *Structure and Process in Speech Perception* (New York).

COLE, R.A. (1973). Listening for mispronunciations: A measure of what we hear during speech. *Perception and Psychophysics*, **1**, 153-156.

COLE, R.A. and JAKIMIK, J. (1978). Understanding speech: How words are heard. In G. Underwood (ed.), *Strategies of Information Processing* (New York).

COLE, R.A. and SCOTT, B. (1973). Perception of temporal order in speech: The role of vowel transitions. *Canadian Journal of Psychology*, **27**, 441-449.

COLE, R.A., RUDNICKY, A., REDDY, R. and ZUE, V.W. (1978). Speech as patterns on paper. In R.A. Cole (ed.), *Perception and Production of Fluent Speech* (Hillsdale, New Jersey).

COLLIER, R. and 't HART, J. (1975). The role of intonation in speech perception. In A. Cohen and S.G. Nooteboom (eds.), *Structure and Process in Speech Perception* (New York), pp. 107-123.

COOPER, W.E. (1975). Selective adaptation of speech. In F. Restle, R.M. Shiffrin, N.J. Castellan, H. Lindman and D.B. Pisoni (eds.), *Cognitive Theory* (Hillsdale, New Jersey).

COOPER, W.E. (1976). Syntactic control of timing in speech production: A study of complement clauses. *Journal of Phonetics*, **4**, 151-171.

COOPER, F.S., DELATTRE, P.C., LIBERMAN, A.M. BORST, J.M. and GERSTMAN, L.J. (1952). Some experiments on the perception of synthetic speech sounds. *Journal of the Acoustical Society of America*, **24**, 597-606.

CUTLER, A. (1976). Phoneme-monitoring reaction time as a function of preceding intonation contour. *Perception and Psychophysics*, **20**, 55-60.

CUTLER, A. and FOSS, D.J. (1977). On the role of sentence stress in sentence processing. *Language and Speech*, **20**, 1-10.

CUTTING, J.E. and ROSNER, B.S. (1974). Categories and boundaries in speech and music. *Perception and Psychophysics*, **16**, 564-570.

DARWIN, C.J. (1975). On the dynamic use of prosody in speech perception. In A. Cohen and S.G. Nooteboom (eds.), *Structure and Process in Speech Perception* (New York).

DARWIN, C.J. and BETHELL-FOX, C.E. (1977). Pitch continuity and speech source attribution. *Journal of Experimental Psychology: Human Perception and Performance*, **3**, 665-672.

DORMAN, M.F., CUTTING, J.E. and RAPHAEL, L. (1975). Perception of temporal order in vowel sequences with and without formant transitions. *Journal of Experimental Psychology: Human Perception and Performance*, **1**, 121-129.

DORMAN, M.F., STUDDERT-KENNEDY, M. and RAPHAEL, L.J. (1977). Stop consonant recognition: Release bursts and formant transitions as functionally equivalent, context-dependent cues. *Perception and Psychophysics*, **22**, 109-122.

EIMAS, P.D. and CORBIT, J.D. (1973). Selective adaptation of linguistic feature detectors. *Cognitive Psychology*, **4**, 99-109.

EIMAS, P.D. and MILLER, J.L. (1978). Effects of selective adaptation on the perception of speech and visual patterns: Evidence for feature detectors. In R.D. Walk and H.L. Pick, Jr. (eds.), *Perception and Experience* (New York).

ELMAN, J.L., DIEHL, R.L. and BUCHWALD, S.E. (1977). Perceptual switching in bilinguals. *Journal of the Acoustical Society of America*, **62**, 971-974.

ENTUS, A.K. (1977). Hemispheric asymmetry in processing dichotically presented speech and nonspeech stimuli by infants. In S.J. Segalowitz and F.A. Gruber (eds.), *Language Development and Neurological Theory* (New York), pp. 64-73.

FISCHER-JØRGENSEN, E. (1972). Tape cutting experiments with Danish stop consonants in initial position. *Annual Report of the Institute of Phonetics, University of Copenhagen*, 7.

FOWLER, C.A. (1979). Perceptual centers in speech production and perception. *Perception and Psychophysics*, **25**, 375-388.

FUJISAKI, H., NAKAMURA, K. and IMOTO, T. (1975). Auditory perception of duration of speech and non-speech stimuli. In G. Fant and M.A.A. Tatham (eds.), *Auditory Analysis and Perception of Speech* (New York), pp. 197-220.

GANONG, F. (1978). A word advantage in phoneme boundary experiments. *Journal of the Acoustical Society of America*, **63**, 520(A).

GLANVILLE, B.B., BEST, C.T. and LEVENSON, R. (1977). A cardiac measure of asymmetries in infant auditory perception. *Developmental Psychology*, **13**, 54-59.

GOLDSTEIN, L. (1977). Categorical features in speech perception and production. To appear in V. Fromkin (ed.), *Proceedings of the Workshop on Slips of the Tongue and Ear* (Vienna). (Also *UCLA Working Papers in Phonetics*, 39).

HAGGARD, M.P. (1977). Do we want a theory of speech perception? Paper presented to the Research Conference on Speech-Processing Aids for the Deaf, Gallaudet College, Washington, D.C., May 23-26.

HALL, L.L. and BLUMSTEIN, S.E. (1978). The effect of syllabic stress and syllabic organization on the identification of speech sounds. *Perception and Psychophysics*, **24**, 137-144.

't HART, J. and COHEN, A. (1973). Intonation by rule: A perceptual quest. *Journal of Phonetics*, **1**, 309-327.

HASEGAWA, A. and DANILOFF, R.G. (1976). Effects of vowel context upon labeling the /s/ – /ʃ/ continuum. *Journal of the Acoustical Society of America*, **59**, 525(A).

HINDE, R.A. and STEVENSON-HINDE, J. (1973). *Constraints on Learning* (New York).

KIMURA, D. (1976). The neural basis of language *qua* gesture. In H. Whitaker and H.A. Whitaker (eds.), *Studies in Neurolinguistics*, vol. III (New York).

KIMURA, D. and ARCHIBALD, Y. (1974). Motor functions of the left hemisphere. *Brain*, **97**, 337-350.

KINSBOURNE, M. and HICKS, R.E. (1979). Mapping cerebral functional space: Competition and collaboration in human performance. In M. Kinsbourne (ed.), *Asymmetrical Function of the Brain* (New York), pp. 267-273.

KLATT, D.H. (1978). Speech perception: A model of acoustic phonetic analysis and lexical access. In R.A. Cole (ed.), *Perception and Production of Fluent Speech* (Hillsdale, New Jersey).

KLATT, D.H. (1977). Review of the ARPA Speech Understanding Project. *Journal of the Acoustical Society of America*, **62**, 1345-1366.

KLATT, D.H. (1976). Linguistic uses of segmental duration in English: Acoustic and perceptual evidence. *Journal of the Acoustical Society of America*, **59**, 1208-1221.

KLATT, D.H. and COOPER, W.E. (1975). Perception of segment duration in sentence context. In A. Cohen and S.G. Nooteboom (eds.), *Structure and Process in Speech Perception* (New York), pp. 69-89.

KLIMA, E.S. and BELLUGI, U. (1979). *The Signs of Language* (Cambridge, Mass.).

KOZHEVNIKOV, V.A. and CHISTOVICH, L.A. (1965). Rech 'artikuliatsiia i vospriiatiie. Transl. as *Speech Articulation and Perception*. Clearinghouse for Federal Scientific and Technical Information, JPRS 30-543 (Washington, D.C.).

KUHN, G.M. (1975). On the front cavity resonance and its possible role in speech perception. *Journal of the Acoustical Society of America*, **58**, 428-433.

KUNISAKI, O. and FUJISAKI, H. (1977). On the influence of context upon perception of voiceless fricative consonants. *Annual Bulletin of the Research Institute of Logopedics and Phoniatrics, University of Tokyo*, **11**, 85-91.

LASSEN, N.A., INGVAR, D.H. and SKINHØJ, E. (1978). Brain function and blood flow. *Scientific American*, **239**, 62-71.

LEHISTE, I. (1977). Isochrony reconsidered. *Journal of Phonetics*, **5**, 253-264.

LEVY, J. (1974). Psychobiological implications of bilateral asymmetry. In S.J. Dimond and J.G. Beaumont (eds.), *Hemisphere Function in the Human Brain* (London).

LIBERMAN, A.M. and PISONI, D.B. (1977). Evidence for a special speech-perceiving subsystem in the human. In T.H. Bullock (ed.), *Recognition of Complex Acoustic Signals* (Berlin), pp. 59-76.

LIBERMAN, A.M. and STUDDERT-KENNEDY, M. (1978). Phonetic perception. In R. Held, H. Leibowitz and H.L. Teuber, *Handbook of Sensory Physiology*. Vol. VIII (Heidelberg).

LIBERMAN, A.M., HARRIS, K.S., HOFFMAN, H.S. and GRIFFITH, B.C. (1957). The discrimination of speech sounds within and across phoneme boundaries. *Journal of Experimental Psychology*, **53**, 358-368.

LINDBLOM, B. and RAPP, K. (1973). Some temporal regularities of spoken Swedish. *Papers from the Institute of Linguistics, University of Stockholm*, **21**, 1-59.

LISKER, L. (1975). Is it VOT or a first-formant transition detector? *Journal of the Acoustical Society of America*, **57**, 1547-1551.

LISKER, L. and ABRAMSON, A. (1964). A cross-language study of voicing in initial stops: Acoustical measurements. *Word*, **20**, 384-422.

MARSLEN-WILSON, W.D. (1975). Sentence perception as an interactive parallel process. *Science*, **189**, 226-228.

MARSLEN-WILSON, W.D. and WELSH, A. (1978). Processing interactions and lexical access during word recognition in continuous speech. *Cognitive Psychology*, **10**, 29-63.

MARTIN, J.G. (1972). Rhythmic (hierarchical) versus serial structure in speech and other behavior. *Psychological Review*, **79**, 487-509.

MASSARO, D.W. (1975). Preperceptual images, processing time, and perceptual units in speech perception. In D.W. Massaro (ed.), *Understanding Language* (New York).

MASSARO, D.W. and COHEN, N.M. (1976). The contribution of fundamental frequency and voice onset time to the /zi/–/si/ distinction. *Journal of the Acoustical Society of America*, **60**, 704-717.

MATEER, C. and KIMURA, D. (1977). Impairment of non-verbal oral movements in aphasia. *Brain and Language*, **4**, 262-276.

MCGURK, H. and MCDONALD, J. (1976). Hearing lips and seeing voices. *Nature*, **264**, 746-748.

MERMELSTEIN, P. (1975). A phonetic context-controlled strategy for segmentation and phonetic labeling of speech. *IEEE Transactions on Acoustics, Speech and Signal Processing*, ASSP-23, 79-82.

MERMELSTEIN, P. (1978). On the relationship between vowel and consonant identification when cued by the same acoustic information. *Perception and Psychophysics*, **23**, 331-336.

MILLER, J.L. (1977). Nonindependence of feature processing in initial consonants. *Journal of Speech and Hearing Research*, **20**, 519-528.

MILLER, J.D., WIER, C.C., PASTORE, R., KELLY, W.J. and DOOLING, D.J. (1976). Discrimination and labeling of noise-buzz sequences with varying noise-lead times: An example of categorical perception. *Journal of the Acoustical Society of America*, **60**, 410-417.

MIYAWAKI, K., STRANGE, W., VERBRUGGE, R., LIBERMAN, A.M., JENKINS, J.J. and FUJIMURA, O. (1975). An effect of linguistic experience: The discrimination of [r] and [l] by native speakers of Japanese and English. *Perception and Psychophysics*, **18**, 331-340.

MOLFESE, D.L., FREEMAN, R.B. and PALERMO, D.S. (1975). The ontogeny of brain lateralization for speech and nonspeech stimuli. *Brain and Language*, **2**, 356-368.

MORTON, J. and LONG, J. (1976). Effect of word transition probability on phoneme identification. *Journal of Verbal Learning and Verbal Behavior*, **15**, 43-51.

MORTON, J., MARCUS, S. and FRANKISH, C. (1976). Perceptual centers (P-centers). *Psychological Review*, **83**, 405-408.

MYERS, T.F., ZHUKOVA, M.G., CHISTOVICH, L.A. and MUSHNIKOV, V.N. (1975). Auditory segmentation and the method of dichotic stimulation. In G. Fant and M.A.A. Tatham (eds.), *Auditory Analysis and the Perception of Speech* (New York), pp. 243-274.

NAKATANI, L.H. and DUKES, K.D. (1977). Locus of segmental cues for word juncture. *Journal of the Acoustical Society of America*, **62**, 714-719.

NOOTEBOOM, S.G., BROKX, J.P.L. and DE ROOIJ, J.J. (1976). Contributions of prosody to speech perception. Institute for Perception Research Preprint, Eindhoven, Netherlands. (To be published in W.J.M. Levelt and G.B. Flores d'Arcais, *Studies in Language Perception* (New York).

ODEN, G.C. and MASSARO, D.W. (1978). Integration of featural information in speech perception. *Psychological Review*, **85**, 172-191.

ÖHMAN, S.E.G. (1975). What is it that we perceive when we perceive speech? In A. Cohen and S.G. Nooteboom (eds.), *Structure and Process in Speech Perception* (New York), pp. 36-48.

OJEMANN, G. and MATEER, C. (1979). Human language cortex: Localization of memory, syntax and sequential motor-phoneme identification systems. *Science*, **205**, 1401-1403.

PARKER, F. (1977). Distinctive features and acoustic cues. *Journal of the Acoustical Society of America*, **62**, 1051-1054.

PASTORE, R.E., AHROON, W.A., BAFFUTO, K.J., FRIEDMAN, C., PULEO, J.S. and FINK, E.A. (1977). Common factor model of categorical perception. *Journal of Experimental Psychology: Human Perception and Performance*, **3**, 686-696.

PISONI, D.B. (1977). Identification and discrimination of the relative onset of two component tones: Implications for the perception of voicing in stops. *Journal of the Acoustical Society of America*, **61**, 1352-1361.

PISONI, D.B. and LAZARUS, J.H. (1974). Categorical and non-categorical modes of speech perception along the voicing continuum. *Journal of the Acoustical Society of America*, **55**, 328-333.

REPP, B.H. (1977). Dichotic competition of speech sounds: The role of acoustic stimulus structure. *Journal of Experimental Psychology: Human Perception and Performance*, **3**, 37-50.

REPP, B.H., LIBERMAN, A.M., ECCARDT, T. and PESETSKY, D. (1978). Perceptual integration of cues for stop, fricative and affricate manner. *Haskins Laboratories Status Report on Speech Research*, SR-53(2), 61-83.

SAMUEL, A.G. (1977). The effect of discrimination training on speech perception: Noncategorical perception. *Perception and Psychophysics*, **22**, 321-330.

SAWUSCH, J.R. (1977). Peripheral and central processing in speech perception. *Journal of the Acoustical Society of America*, **62**, 738-750.

SAWUSCH, J.R. and PISONI, D.B. (1974). On the identification of place and voicing features in synthetic stop consonants. *Journal of Phonetics*, **2**, 181-194.

SEARLE, C.L., JACOBSON, J.Z. and RAYMENT, S.G. (1979). Stop consonant discrimination based on human audition. *Journal of the Acoustical Society of America*, **65**, 799-809.

SEMMES, J. (1968). Hemispheric specialization: A possible clue to mechanism. *Neuropsychologia*, **6**, 11-26.

SHANKWEILER, D.P. and STUDDERT-KENNEDY, M. (1975). A continuum of lateralization for speech perception? *Brain and Language*, **2**, 212-225.

SHIELDS, J.L., MCHUGH, A. and MARTIN, J.G. (1974). Reaction time to phoneme targets as a function of rhythmic cues in continuous speech. *Journal of Experimental Psychology*, **102**, 250-255.

SHOCKEY, L. and REDDY, R. (1975). Quantitative analysis of speech perception. In G. Fant (ed.), *Proceedings of the Speech Communication Seminar (Stockholm, Sweden)* (New York).

SIMON, C. and FOURCIN, A.J. (1978). Cross-language study of speech-pattern learning. *Journal of the Acoustical Society of America*, **63**, 925-935.

SIMON, H.J. and STUDDERT-KENNEDY, M. (1978). Selective anchoring and adaptation of phonetic and nonphonetic continua. *Journal of the Acoustical Society of America*, **64**, 1338-1368.

SINGH, S. (1978). Distinctive features: A measurement of consonant perception. In S. Singh (ed.), *Measurement Procedures in Speech, Hearing and Language* (Baltimore), pp. 93-155.

SINGH, S., WOODS, D.R. and BECKER, G.M. (1972). Perceptual structure of 22 prevocalic English consonants. *Journal of the Acoustical Society of America*, **52**, 1698-1713.

STEVENS, K.N. (1975). The potential role of property detectors in the perception of consonants. In G. Fant and M.A.A. Tatham (eds.), *Auditory Analysis and Perception of Speech* (New York), pp. 303-330.

STEVENS, K.N. and BLUMSTEIN, S. (1978). Invariant cues for place of articulation. *Journal of the Acoustical Society of America*, **64**, 1358-1368.

STEVENS, K.N. and KLATT, D.H. (1974). Role of formant transitions in the voiced-voiceless distinction for stops. *Journal of the Acoustical Society of America*, **55**, 653-659.

STRANGE, W. and JENKINS, J.J. (1978). The role of linguistic experience in the perception of speech. In H.L. Pick, Jr. and R.D. Walk (eds.), *Perception and Experience* (New York).

STRANGE, W., JENKINS, J.J. and EDMAN, T.R. (1978). Dynamic information specifies vowel identity. *Journal of the Acoustical Society of America*, **63**, S5 (A).

STRANGE, W., VERBRUGGE, R., SHANKWEILER, D.P. and EDMAN, T.R. (1976). Consonant environment specifies vowel identity. *Journal of the Acoustical Society of America*, **60**, 213-224.

STUDDERT-KENNEDY, M. (1976). Speech perception. In N.J. Lass (ed.), *Contemporary Issues in Experimental Phonetics* (New York).

STUDDERT-KENNEDY, M. (1977). Universals in phonetic structure and their role in linguistic communication. In T.H. Bullock (ed.), *Recognition of Complex Acoustic Signals* (Berlin), pp. 37-48.

STUDDERT-KENNEDY, M. and SHANKWEILER, D.P. (1970). Hemispheric specialization for speech perception. *Journal of the Acoustical Society of America*, **48**, 579-594.

STUDDERT-KENNEDY, M., LIBERMAN, A.M., HARRIS, K.S. and COOPER, F.S. (1970). Motor theory of speech perception: A reply to Lane's critical review. *Psychological Review*, **77**, 234-249.

SUMMERFIELD, Q. (1975). How a full account of segmental perception depends on prosody and vice versa. In A. Cohen and S.G. Nooteboom (eds.), *Structure and Process in Speech Perception* (New York), pp. 51-68.

SUMMERFIELD, Q. (1979). Use of visual information for phonetic perception. *Phonetica*, **36**, 314-331.

SUMMERFIELD, Q. and HAGGARD, M. (1977). On the dissociation of spectral and temporal cues to the voicing distinction in initial stop consonants. *Journal of the Acoustical Society of America*, **62**, 436-448.

SVENSSON, S.G. (1974). *Prosody and Grammar in Speech Perception*. University of Stockholm, MILUS 2.

TERBEEK, D. (1977). Across-language multi-dimensional scaling of study of vowel perception. *UCLA Working Papers in Phonetics*, 37.

VAN DEN BROECKE, M.P.R. (1976). *Hierarchies and Rank Orders in Distinctive Features* (Utrecht).

WILLIAMS, L. (1977). The perception of stop consonant voicing by Spanish-English bilinguals. *Perception and Psychophysics*, **21**, 289-297.

WOOD, C.C. (1975). Auditory and phonetic levels of processing in speech perception: Neurophysiological and information-processing analysis. *Journal of Experimental Psychology: Human Perception and Performance*, **104**, 3-20.

ZAIDEL, E. (1978a). Lexical organization in the right hemisphere. In P.A. Buser and A. Rougeul-Buser (eds.), *Cerebral Correlates of Conscious Experience*, (Amsterdam), pp. 177-197.

ZAIDEL, E. (1978b). Concepts of cerebral dominance in the split-brain. In P.A. Buser and A. Rougeul-Buser (eds.), *Cerebral Correlates of Conscious Experience* (Amsterdam), pp. 263-285.

ZHUKOV, S.Ya., ZHUKOVA, M.G. and CHISTOVICH, L.A. (1974). Some new concepts in the auditory analysis of acoustic flow. *Soviet Physics and Acoustics*, **20**, 237-240 [*Akust. Ah.*, **20**, 386-392].

ZUE, V.W. (1976). Acoustic characteristics of stop consonants: A controlled study. *Lincoln Laboratory Technical Report*, *523*.

[8]

The Hearsay-II Speech-Understanding System: Integrating Knowledge to Resolve Uncertainty

LEE D. ERMAN

USC/Information Sciences Institute, Marina del Rey, California 90291

FREDERICK HAYES-ROTH

The Rand Corporation, Santa Monica, California 90406

VICTOR R. LESSER

University of Massachusetts, Amherst, Massachusetts 01003

D. RAJ REDDY

Carnegie-Mellon University, Pittsburgh, Pennsylvania 15213

The Hearsay-II system, developed during the DARPA-sponsored five-year speech-understanding research program, represents both a specific solution to the speech-understanding problem and a general framework for coordinating independent processes to achieve cooperative problem-solving behavior. As a computational problem, speech understanding reflects a large number of intrinsically interesting issues. Spoken sounds are achieved by a long chain of successive transformations, from intentions, through semantic and syntactic structuring, to the eventually resulting audible acoustic waves. As a consequence, interpreting speech means effectively inverting these transformations to recover the speaker's intention from the sound. At each step in the interpretive process, ambiguity and uncertainty arise.

The Hearsay-II problem-solving framework reconstructs an intention from hypothetical interpretations formulated at various levels of abstraction. In addition, it allocates limited processing resources first to the most promising incremental actions. The final configuration of the Hearsay-II system comprises problem-solving components to generate and evaluate speech hypotheses, and a focus-of-control mechanism to identify potential actions of greatest value. Many of these specific procedures reveal novel approaches to speech problems. Most important, the system successfully integrates and coordinates all of these independent activities to resolve uncertainty and control combinatorics. Several adaptations of the Hearsay-II framework have already been undertaken in other problem domains, and it is anticipated that this trend will continue; many future systems necessarily will integrate diverse sources of knowledge to solve complex problems cooperatively.

Discussed in this paper are the characteristics of the speech problem in particular, the special kinds of problem-solving uncertainty in that domain, the structure of the Hearsay-II system developed to cope with that uncertainty, and the relationship between Hearsay-II's structure and those of other speech-understanding systems. The paper is intended for the general computer science audience and presupposes no speech or artificial intelligence background.

Keywords and Phrases: artificial intelligence, blackboard, focus of control, knowledge-based system, multiple diverse knowledge sources, multiple levels of abstraction, problem-solving system, speech-understanding systems, uncertainty resolving

CR Categories: 3.6, 3.64

CONTENTS

INTRODUCTION

The Hearsay-II speech-understanding system (SUS) developed at Carnegie-Mellon University recognizes connected speech in a 1000-word vocabulary with correct interpretations for 90 percent of test sentences. Its basic methodology involves the application of symbolic reasoning as an aid to signal processing. A marriage of general artificial intelligence techniques with specific acoustic and linguistic knowledge was needed to accomplish satisfactory speech-understanding performance. Because the various techniques and heuristics employed were embedded within a general problem-solving framework, the Hearsay-II system embodies several design characteristics that are adaptable to other domains as well. Its structure has been applied to such tasks as multisensor interpretation [NII78], protein-crystallographic analysis [ENGE77], image understanding [HANS76], a model of human reading [RUME76], and dialogue comprehension [MANN79]. This paper discusses the characteristics of the speech problem in particular, the special kinds of problem-solving uncertainty in that domain, the structure of the Hearsay-II system developed to cope with that uncertainty, and the relationship between Hearsay-II's structure and the structures of other SUSs.

Uncertainty arises in a problem-solving system if the system's knowledge is inadequate to produce a solution directly. The fundamental method for handling uncertainty is to create a space of candidate solutions and search that space for a solution. "Almost all the basic methods used by intelligent systems can be seen as some variation of search, responsive to the particular knowledge available" [NEWE77, p. 13]. In a difficult problem, i.e., one with a large search space, a problem solver can be effective only if it can search efficiently. To do so, it must apply knowledge to guide the search so that relatively few points in the space need be examined before a solution is found. A key way of accomplishing this is by augmenting the space of candidate solutions with candidate *partial* solutions and then constructing a complete solution by extending and combining partial candidates. A candidate partial solution represents all complete candidates that contain it. By considering partial solution candidates, we can often eliminate whole subspaces from further consideration and simultaneously focus the search on more promising subspaces.

To solve a problem as difficult as speech understanding, a problem solver requires several kinds of capabilities in order to search effectively: It must collect and analyze data, set goals to guide the inferential search processes, produce and retain appro-

This research was supported chiefly by Defense Advanced Research Projects Agency contract F44620-73-C-0074 to Carnegie-Mellon University. In addition, support for the preparation of this paper was provided by USC/ISI, Rand, and the University of Massachusetts. We gratefully acknowledge their support. Views and conclusions contained in this document are those of the authors and should not be interpreted as representing the official opinion or policy of DARPA, the U.S. government, or any other person or agency connected with them.

priate inferences, and decide when to stop working for a possibly better solution. Years ago, when AI problem solvers first emerged, they attempted to provide these capabilities through quite general domain-independent methods, the so-called *weak* methods [NEWE69]. A prime example of such a problem solver is GPS [ERNS69]. More recently, several major problem-solving accomplishments, such as Dendral [FEIG71] and Mycin [SHOR76], have reflected a different philosophy: Powerful problem solvers depend on extensive amounts of knowledge about both the problem domain and the problem-solving strategies effective in that domain [FEIG77]. Much of what we view as expertise consists of these two types of knowledge; without capturing and implementing this knowledge, we could not create effective computer problem solvers. Because knowledge plays a crucial role in these kinds of tasks, many people call the corresponding problem solvers *knowledge-based systems* [BARN77]. The design of Hearsay-II is responsive to both concerns. While formulated as a general system-building framework that would structure and control problem-solving behavior involving multiple, diverse, and error-full sources of knowledge, the current Hearsay-II system consists of a particular collection of programs embedding speech knowledge that are capable of solving the understanding problem.[1]

The difficulty of the speech-understanding problem, and hence the need for powerful problem-solving methods, derives from two inherent sources of uncertainty or error. The first includes ordinary variability and noise in the speech waveform, and the second includes the ambiguous and inaccurate judgments arising from an application of incomplete and imprecise theories of speech. Because we cannot resolve these uncertainties directly, we structure the speech-understanding problem as a space in which our problem solver searches for a solution. The space is the set of (partial and complete) *interpretations* of the input acoustic signal, i.e., the (partial and complete) mappings from the signal to the possible messages. The goal of our problem-solving system is to find a complete interpretation (i.e., a message and mapping) which maximizes some evaluation function based on knowledge about such things as acoustic-phonetics, vocabulary, grammar, semantics, and discourse. This resolution of the combined sources of uncertainty requires the generation, evaluation, and integration of numerous partial interpretations. The need to consider many alternative interpretations without spawning an explosive combinatorial search thus becomes a principal design objective. Each of these issues is discussed in more detail in the following section.

Dimensions of the Problem: Uncertainty and Hypothetical Interpretations

The first source of difficulty in the speech problem arises from the speaking process itself. In the translation from intention to sound, a speaker transforms concepts into speech through processes that introduce variability and noise (see Figure 1). If, for example, we consider the semantic, syntactic, lexical, and phonemic stages, the types of variance introduced from one level to the next would correspond to errors or peculiarities of conceptualization, grammar, word choice, and articulation. In addition

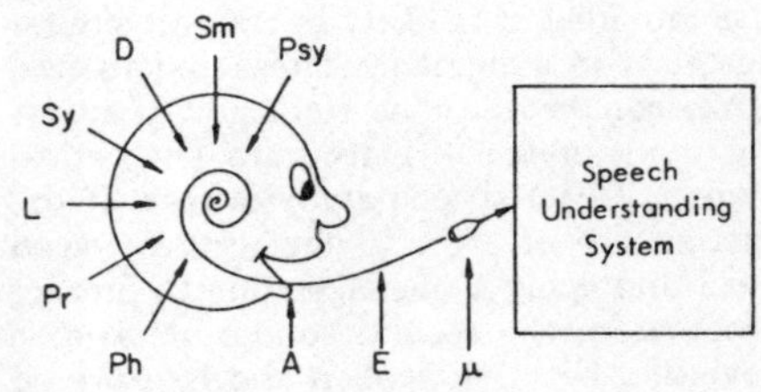

FIGURE 1. Some of the mechanisms that affect the message: psychology of the speaker, semantics, rules of discourse, syntax, lexicon, prosodic system, phonemic system, speaker's articulatory apparatus, ambient environmental noise, and microphone and system. [After NEWE75.]

[1] The problem of speech understanding has been actively pursued recently [REDD75, REDD76, CMU77, BERN76, WALK78, WOOD76, KLAT77, MEDR78, LEA80]. With the exception of HARPY [LOWE80], however, none of the other efforts has been presented as a structure for problem solving in other domains.

to these sources of variability, speech is often affected by pauses, extraneous sounds, or unnecessary phrase repetitions. The effect of these factors upon the physical sound signal is to distort it significantly from the *ideal message*, that is, from the message that would be produced if the production mechanisms did not introduce variability and noise. Accordingly, we speak of the disparity between the ideal and actual signals as *error*, and of the variety of factors that contribute to such distortion as *sources of error*. Thus the first source of error is inherent in the speaker and his environment.

The second source of error in the understanding process is intrinsic to the listener. Just as the speaker must transform his intention through successive intermediate levels of representation, so we presume the listener must accomplish the inverse of those transformations; from the physical signal the listener must detect acoustic-phonetic elements, syllables, words, and syntactic and conceptual structures corresponding to the speaker's intentions. At each step in this reconstruction the listener may introduce new errors corresponding to incorrect perceptual or interpretive judgments.[2] Because a machine speech-understanding system must also develop interpretations of what was spoken and what was intended, it is likely to commit similar mistakes in judgment. These judgmental errors can be viewed as the result of applying inadequate or inaccurate theoretical models to the speech-analysis task. If the first source of error is deviation between ideal and spoken messages due to inexact production, the second source of error is deviation between spoken and interpreted messages due to imprecise rules of comprehension.

To comprehend an utterance in the context of such errors, a speech-understanding system must formulate and evaluate numerous candidate interpretations of speech fragments. Understanding a message requires us to isolate and recognize its individual words and parse their syntactic and conceptual relationships. Each intermediate state of this process can be viewed as either the generation or evaluation of symbolic interpretations for portions of the spoken utterance. We use the term *hypothesis* to refer to a partial interpretation actually constructed. During the process of speech interpretation, hypotheses may vary from highly confident identification of particular words to great confusion concerning particular portions of the utterance. Between these two extremes, the listener may entertain simultaneously several competing hypotheses for what was said. Competing alternatives might occur at any of several levels of abstraction. For example, at the word level the listener may struggle to distinguish whether "till" or "tell" was spoken in one portion of the utterance while simultaneously attempting to differentiate the words "brings" and "rings" in another interval. These uncertainties derive from comparable uncertainties at lower levels of interpretation, such as syllabic and acoustic, where multiple competing hypotheses can also exist simultaneously. Similarly, uncertainty among word hypotheses at the lexical level engenders uncertainty at higher levels of interpretation. Thus the previously discussed inability to distinguish between alternative words may be the underlying cause of an inability to distinguish between the four hypothetical phrase interpretations:

till Bob rings
tell Bob rings
till Bob brings
tell Bob brings

Just as this example suggests, higher level interpretations incorporate lower level ones. A phrase-level hypothesis consists of a selection of word hypotheses from each interval of time spanned by the higher level hypothesis. Only one lower level hypothesis in any time interval can be incorporated into the higher level interpretation. Thus a phrase consists of a sequence of words, a word consists of a sequence of syllables, a syllable consists of a sequence of acoustic-phonetic segments, and so on. An overall interpretation of an entire utterance would consist of a syntactic or semantic analysis that recursively incorporated one hypoth-

[2] Though the levels of representation appear to be linearly ordered, the encoding and decoding processes do not necessarily operate sequentially through this ordering.

esis from each level of interpretation for each temporal interval of the utterance.

A fundamental assumption underlying the understanding problem is that a correct interpretation of an utterance should minimize the difference between those properties of the speech that the hypothetical interpretation would predict and those that are observed. This gives rise to the notion of the *consistency* between an interpretation and its supporting data. Thus certain parameter values derived from an acoustic waveform are more or less consistent with various phonetic classifications, particular sequences of phones are more or less consistent with various monosyllabic categorizations, and various syllable sequences are more or less consistent with particular lexical and phrase interpretations. The concept of consistency between two adjacent levels of interpretation can be generalized to permit consideration of the consistency between hypotheses at any two levels and, in particular, the consistency between an overall interpretation of the utterance and its supporting hypotheses at the lowest, acoustic-parametric level. A central assumption is that the greater the consistency between the overall interpretation and the acoustic data, the more likely the interpretation is to be correct.

We refer to the likelihood that some hypothesis is correct as its *credibility*. As the preceding suggests, the credibility of each hypothesis is a measure of consistency between the data generating the hypothesis and the expectations it engenders. A credibility calculation involves a judgment about the knowledge used in creating the hypothesis and therefore is itself subject to uncertainty.

To assess the credibility of a hypothesis, we need basically to evaluate two things: all plausible alternatives to this hypothesis and the degree of support each receives from data. Consider, for example, the evaluation of word hypotheses. Initially, nearly all words in the language are plausible candidates for occurring within any time interval. As a consequence, our uncertainty at the outset, as approximated by the number of equally plausible alternatives, is maximal. Over time we accrue evidence to eliminate some of these alternatives. Moreover, by eliminating one particular hypothesis, we may logically exclude others that are in temporally adjacent regions and that depend directly on that hypothesis. For example, if we have ruled out all possible adjectives and nouns in a particular location, we can also rule out adjectives in the preceding interval. Conversely, if we can identify a particular word as an adjective, we can increase our belief that the following word will be an adjective or noun. In general, each individual hypothesis is strengthened by its apparent combinability with others. Thus we say uncertainty is reduced by detecting mutually supporting hypotheses that are consistent with the acoustic data. Equivalently, the credibility of hypotheses increases as a function of their involvement in such mutually supportive clusters.

This technique for reducing uncertainty leads to the following incremental problem-solving method: The goal of the problem solver is to construct the most credible overall interpretation. The fundamental operations in the construction are hypothesis generation, hypothesis combination, and hypothesis evaluation. At each step in the construction, sources of knowledge use these operations to build larger partial interpretations, adding their constraints to the interpretation. The accrual of constraints reduces the uncertainty inherent in the data and in the knowledge sources themselves.

Three requirements must be met for such a problem solver to be effective:

(1) At least one possible sequence of knowledge-based operations must lead to a correct overall interpretation.
(2) The evaluation procedure should assess the correct overall interpretation as maximally credible among all overall interpretations generated.
(3) The cost of problem solving must satisfy some externally specified limit. Usually this limit restricts the time or space available for computing. As a consequence, it leads to restrictions on the number of alternative partial interpretations that can be considered. Alternative partial solutions must be considered in order to ensure that a correct

one is included. The greater the uncertainty in the knowledge used to generate and evaluate hypotheses, the greater the number of alternatives that must be considered, leading to a possible combinatorial explosion.

As we have seen, the speech-understanding problem is characterized by the need for highly diverse kinds of knowledge for its solution and by large amounts of uncertainty and variability in input data and knowledge. The diversity of knowledge leads to a search space of multilevel partial solutions. The uncertainty and variability mean that the operators used for searching the space are themselves error-prone; therefore many competing alternative hypotheses must be generated. To avoid a combinatorial explosion, a powerful control scheme is needed to exploit selectively the most promising combinations of alternatives. As systems tackle more such difficult real-world problems, such multilevel representations and powerful control schemes will become increasingly important [HAYE78a]. The next section discusses how the Hearsay-II system copes with these representation and control problems.

Hearsay-II Problem-Solving Model

The key functions of generating, combining, and evaluating hypothetical interpretations are performed by diverse and independent programs called *knowledge sources* (KSs). The necessity for diverse KSs derives from the diversity of transformations used by the speaker in creating the acoustic signal and the corresponding inverse transformations needed by the listener for interpreting it. Each KS can be schematized as a condition–action pair. The condition component prescribes the situations in which the KS may contribute to the problem-solving activity, and the action component specifies what that contribution is and how to integrate it into the current situation.[3] According to the original conception of the diverse stages and processes involved in speech understanding, KSs have been developed to perform a variety of functions. These include extracting acoustic parameters, classifying acoustic segments into phonetic classes, recognizing words, parsing phrases, and generating and evaluating predictions for undetected words or syllables. Figure 2 presents a schematic view of the KSs in the September 1976 configuration of the Hearsay-II speech-understanding system. Figure 3 gives a brief functional description of these KSs.

Because each KS is an *independent* condition–action module, KSs communicate through a global database called the *blackboard*. The blackboard records the hypotheses generated by KSs. Any KS can generate a hypothesis (record it on the blackboard) or modify an existing one. These actions in turn may produce structures that satisfy the applicability conditions of other KSs. In this framework the blackboard serves in two roles: It represents intermediate states of problem-solving activity, and it communicates messages (hypotheses) from one KS that activate other KSs.

The blackboard is subdivided into a set of information levels corresponding to the intermediate representation levels of the decoding processes (phrase, word, syllable, etc.). Each hypothesis resides on the blackboard at one of the levels and bears a defining label chosen from a set appropriate to that level (e.g., the word FLYING, the syllable ING, or the phone NG). The hypothesis contains additional information, including its time coordinates within the spoken utterance and a credibility rating. The sequence of levels on the blackboard forms a loose hierarchical structure: hypotheses at each level aggregate or abstract elements at the adjacent lower level. The possible hypotheses at a level form a search space for KSs operating at that level. A partial

[3] The condition and action components of a KS are realized as arbitrary programs. To minimize reevaluating the condition programs continuously, each condition program declares to the system the primitive kinds of situations in which it is interested. The condition program is triggered only when there occur changes that create such situations (and is then given pointers to all of them). This changes a polling action into an interrupt-driven one and is more efficient, especially for a large number of KSs. When executed, the condition program can search among the set of existing hypothetical interpretations for arbitrarily complex configurations of interest to its KS.

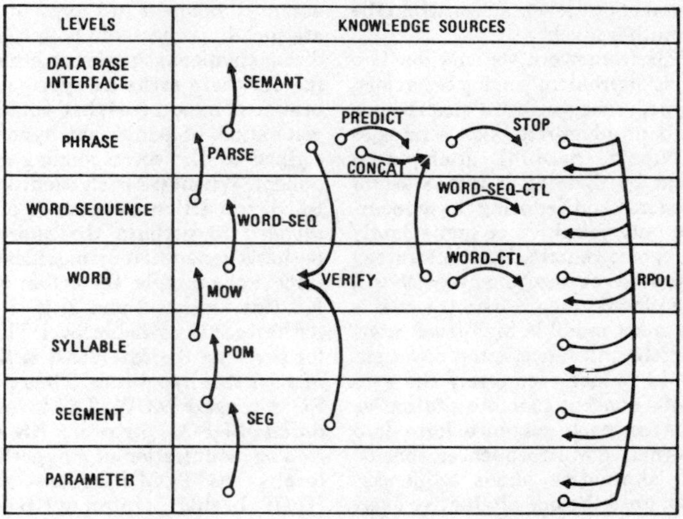

FIGURE 2. The levels and knowledge sources of September 1976. KSs are indicated by vertical arcs with the circled ends indicating the input level and the pointed ends indicating output level.

FIGURE 3. Functional description of the speech-understanding KSs.

Signal Acquisition, Parameter Extraction, Segmentation, and Labeling:
- SEG: Digitizes the signal, measures parameters, and produces a labeled segmentation.

Word Spotting:
- POM: Creates syllable-class hypotheses from segments.
- MOW: Creates word hypotheses from syllable classes.
- WORD-CTL: Controls the number of word hypotheses that MOW creates.

Phrase-Island Generation:
- WORD-SEQ: Creates word-sequence hypotheses that represent potential phrases from word hypotheses and weak grammatical knowledge.
- WORD-SEQ-CTL: Controls the number of hypotheses that WORD-SEQ creates.
- PARSE: Attempts to parse a word sequence and, if successful, creates a phrase hypothesis from it.

Phrase Extending:
- PREDICT: Predicts all possible words that might syntactically precede or follow a given phrase.
- VERIFY: Rates the consistency between segment hypotheses and a contiguous word–phrase pair.
- CONCAT: Creates a phrase hypothesis from a verified contiguous word–phrase pair.

Rating, Halting, and Interpretation:
- RPOL: Rates the credibility of each new or modified hypothesis, using information placed on the hypothesis by other KSs.
- STOP: Decides to halt processing (detects a complete sentence with a sufficiently high rating, or notes the system has exhausted its available resources) and selects the best phrase hypothesis or set of complementary phrase hypotheses as the output.
- SEMANT: Generates an unambiguous interpretation for the information-retrieval system which the user has queried.

interpretation at one level can constrain the search at another level.

Within this framework we consider two general types of problem-solving behaviors. The first type, associated with means–ends analysis and problem-reduction strategies [ERNS69, NILS71, SACE74], attempts to reach a goal by dividing it into a set of simpler subgoals and reducing these recursively until only primitive or immediately solvable subgoals remain. Such a strategy is called *top-down* or *analysis-by-synthesis*. In speech understanding, where the goal is to find the most credible high-level interpretation of the utterance, a top-down approach would reduce recursively the general sentential concept goal into alternative sentence forms, each sentence form into specific alternative word sequences, specific words into alternative phone sequences, and so forth, until the one alternative overall interpretation most consistent with the observed acoustic parameters is identified. The second, or *bottom-up*, method attempts to synthesize interpretations directly from characteristics of the data provided. One type of bottom-up method would employ procedures to classify acoustic segments within phonetic categories by comparing their observed parameters with the ideal parameter values of each phonetic category. Other bottom-up procedures might generate syllable or word hypotheses directly from sequences of phone hypotheses, or might combine temporally adjacent word hypotheses into syntactic or conceptual units. For a hypothesis generated in either the top-down or bottom-up mode, we would like to represent explicitly its relationship to the preexisting hypotheses that suggested it. *Links* are constructed between hypotheses for this purpose.

Both types of problem-solving behaviors can be accommodated simultaneously by the condition–action schema of a Hearsay-II KS. Top-down behaviors represent the reduction of the higher level goal as the condition to be satisfied and the generation of appropriate subgoals as the associated action. Bottom-up behaviors employ the condition component to represent the lower level hypothesis configurations justifying higher level interpretations, and employ the action component to represent and generate such hypotheses. In both cases the condition component performs a test to determine if there exists an appropriate configuration of hypotheses that would justify the generation of additional hypotheses prescribed by the corresponding action component. Whenever such conditions are satisfied, the action component of the KS is *invoked* to perform the appropriate hypothesis generation or modification operations. For example, the action of the POM KS (see Figures 2 and 3) is to create hypotheses at the syllable level. The condition for invoking the MOW KS is the creation of a syllable hypothesis. Thus the action of POM triggers MOW. The invocation condition of RPOL, the rating KS, is the creation or modification of a hypothesis at any level; thus POM's actions also trigger RPOL. In short, control of KS activation is determined by the blackboard actions of other KSs, rather than explicit calls from other KSs or some central sequencing mechanism. This *data-directed* control regime permits a more flexible scheduling of KS actions in response to changing conditions on the blackboard. We refer to such an ability of a system to exploit its best data and most promising methods as *opportunistic* problem solving [NII78, HAYE79a].

While it is true that each condition–action knowledge source is logically independent of the others, effective problem-solving activity depends ultimately on the capability of the individual KS actions to construct cooperatively an overall interpretation of the utterance. This high-level hypothesis and its recursive supports represent the *solution* to the understanding *problem*. Since each KS action simply generates or modifies hypotheses and links based on related information, a large number of individual KS invocations may be needed to construct an overall interpretation.

Any hypothesis that is included in the solution is *cooperative* with the others. Conversely, any hypothesis that is unincorporated into the solution is *competitive*. In a similar way, KS invocations can be considered cooperative or competitive depending on whether their potential actions would or would not contribute to the same solution. Because of the inherent uncertainty in the speech-understanding task, there are inevitably large numbers of plausible alternative actions in each time interval of the utterance. Before the correct interpretation has been found, we cannot evaluate with certainty the prospective value of any potential action. Actions appear cooperative to the extent to which they contribute to the formation and support of increasingly comprehensive interpretations. Conversely, any hypothesis occupying the same time interval as another hypothesis but not part of its support set must be considered competitive. That is, two hypotheses compete if they represent incompatible interpretations of the same portion of the utterance. As a result, KS invocations can be viewed as competitive if their likely actions would generate inconsistent hypotheses, and they can be viewed as cooperative if their actions would combine to form more comprehensive or more strongly supported hypotheses.

The major impediment to discovery of the best overall interpretation in this scheme is the combinatorial explosion of KS invocations that can occur. From the outset, numerous alternative actions are warranted. A purely top-down approach would generate a vast number of possible actions, if unrestrained. Because certainty of recognition is practically never possible and substantial numbers of competing hypotheses must be entertained at each time interval of analysis, any bottom-up approach generates a similarly huge number of competing possible actions. Thus additional constraints on the problem-solving activity must be enforced. This is accomplished by selecting for execution only a limited subset of the invoked KSs.

The objective of *selective attention* is to allocate limited computing resources (processing cycles) to the most important and most promising actions. This selectivity involves three components. First, the probable effects of a potential KS action must be estimated before it is performed. Second, the global significance of an isolated action must be deduced from analysis of its cooperative and competitive relationships with existing hypotheses; *globally significant actions* are those that contribute to the detection, formation, or extension of combinations of redundant hypotheses. Third, the desirability of an action must be assessed in comparison with other potential actions. While the inherent uncertainty of the speech task precludes error-free performance of these component tasks, there have been devised some approximate methods that effectively control the combinatorics and make the speech-understanding problem tractable.

Selective attention is accomplished in the Hearsay-II system by a heuristic scheduler which calculates a priority for each action and executes, at each time, the waiting action with the highest priority [HAYE77a]. The priority calculation attempts to estimate the usefulness of the action in fulfilling the overall system goal of recognizing the utterance. The calculation is based on information provided when the condition part of a KS is satisfied. This information includes the *stimulus frame*, which is the set of hypotheses that satisfied the condition, and the *response frame*, a stylized description of the blackboard modifications that the KS action is likely to perform. For example, consider a syllable-based word hypothesizer KS (such as MOW); its stimulus frame would include the specific syllable hypothesis which matched its condition, and its response frame would specify the expected action of generating word hypotheses in a time interval spanning that of the stimulus frame. In addition to this action-specific information, the scheduler uses global state information in its calculations and considers especially the credibility and duration of the best hypotheses in each level and time region and the amount of processing required from the time the current best hypotheses were generated. The latter information allows the system to reappraise its confidence in its current best hypotheses if they are not quickly incorporated into more comprehensive hypotheses.

Hearsay-II Architecture

Figure 4 illustrates the primary architectural features of the Hearsay-II system. At the start of each cycle, the scheduler, in

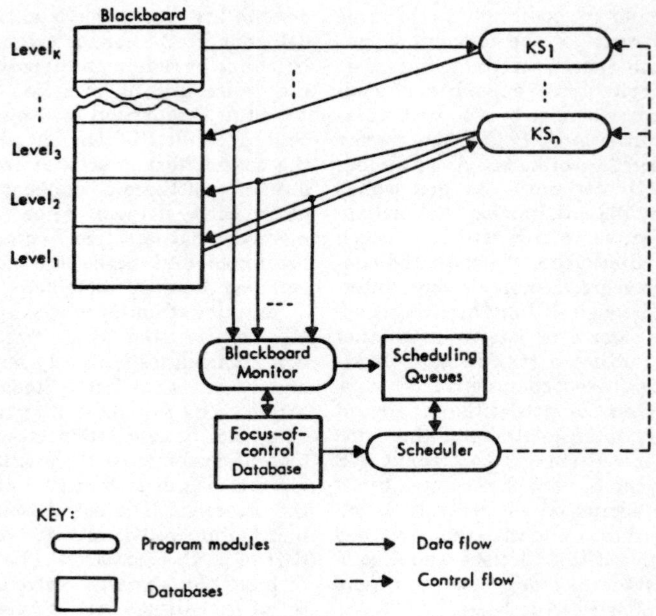

FIGURE 4. Schematic of the Hearsay-II architecture.

accordance with the global state information, calculates a priority for each activity (KS condition program or action program) in the scheduling queues. The highest priority activity is removed from the queues and executed. If the activity is a KS condition program, it may insert new instances of KS action programs into the scheduling queues. If the activity is a KS action program, the blackboard monitor notices the blackboard changes it makes. Whenever a change occurs that would be of interest to a KS condition program, the monitor creates an activity in the scheduling queues for that program. The monitor also updates the global state information to reflect the blackboard modifications.

1. AN EXAMPLE OF RECOGNITION

In this section we present a detailed description of the Hearsay-II speech system understanding one utterance. The task for the system is to answer questions about and retrieve documents from a collection of computer science abstracts (in the area of artificial intelligence). Example sentences:

"Which abstracts refer to theory of computation?"

"List those articles."

"What has McCarthy written since nineteen seventy-four?"

The vocabulary contains 1011 words (in which each extended form of a root, e.g., the plural of a noun, is counted separately if it appears). The grammar defining the legal sentences is context-free and includes recursion. The style of the grammar is such that there are many more nonterminals than in conventional syntactic grammars; the information contained in the greater number of nodes imbeds semantic and pragmatic constraint directly within the grammatical structure. For example, in place of 'Noun' in a conventional grammar, this grammar includes such nonterminals as 'Topic', 'Author', 'Year', and 'Publisher'. Because of its emphasis on semantic categories, this type of grammar is called a *semantic template grammar* or simply a *semantic grammar* [HAYE75, BURT76,

HAYE80]. The grammar allows each word to be followed, on the average, by 17 other words of the vocabulary.[4] The standard deviation of this measure is very high (about 51), since some words (e.g., "about" or "on") can be followed by many others (up to 300 in several cases).

1.1 Introduction to the Example

We will describe how Hearsay-II understood the utterance "ARE ANY BY FEIGENBAUM AND FELDMAN?"[5] Each major *step* of the processing is shown; a step usually corresponds to the action of a knowledge source. Executions of the condition programs of the KSs are not shown explicitly, nor do we list those potential knowledge-source actions which are never chosen by the scheduler for execution. Executions of RPOL are also omitted; in order to calculate credibility ratings for hypotheses, RPOL runs in high priority immediately after any KS action that creates or modifies a hypothesis.

The waveform of the spoken utterance is shown in Figure 5a. The "correct" word boundaries (determined by human inspection) are shown in Figure 5b for reference. The remaining sections of Figure 5 contain all the hypotheses created by the KSs. Each hypothesis is represented by a box; the box's horizontal position indicates the location of the hypothesis within the utterance. The hypotheses are grouped by level: segment, syllable, word, word sequence, and phrase. Links between hypotheses are not shown. The processing will be described in terms of a sequence of *time steps*, where each step corresponds approximately to KS execution governed by one scheduling decision. Within each hypothesis, the number preceding the colon indicates the time step during which the hypothesis was created. The symbol following the colon names the hypothesis. At the word level and above, an asterisk (*) following the symbol indicates that the hypothesis is correct. The trailing number within each hypothesis is the credibility rating on an arbitrary scale ranging from 0 to 100.

In the step-by-step description, the name of the KS executed at each step follows the step number. An asterisk following the KS name indicates that the hypotheses in the stimulus frame of this KS instantiation are all correct. Single numbers in parentheses after hypotheses are their credibility ratings. All times given are in centisecond units; thus the duration of the whole utterance, which was 2.25 seconds, is marked as 225. When begin- and end-times of hypotheses are given, they appear as two numbers separated by a colon (e.g., 52:82). As in the figure, correct hypotheses are marked with an asterisk.

1.2 The Example

The utterance is recorded by a medium-quality Electro-Voice RE-51 close-speaking headset microphone in a moderately noisy environment (>65 dB). The audio signal is low-pass filtered and 9-bit sampled at 10 kHz. All subsequent processing, including the control of the A/D converter, is performed digitally on a time-shared PDP-10 computer. Four acoustic parameters (called ZAPDASH) are derived by simple algorithms operating directly on the sampled signal [GOLD77]. These parameters are extracted in real time and are used initially to detect the beginning and end of the utterance.

Step 1. KS: SEG.
Stimulus: Creation of ZAPDASH parameters for the utterance.
Action: Create segment hypotheses.

The ZAPDASH parameters are used by the *SEG* knowledge source as the basis for an acoustic segmentation and classification of the utterance [GILL78]. This segmentation is accomplished by an iterative refinement technique: First, silence is separated from nonsilence; then the nonsilence is broken down into the sonorant and nonsonorant regions, and so on. Eventually five

[4] Actually, a family of grammars, varying in the number of words (terminals) and in the number and complexity of sentences allowed, was generated. The grammar described here and used in most of the testing is called X05.

[5] To improve clarity, the description differs from the actual computer execution of Hearsay-II in a few minor details.

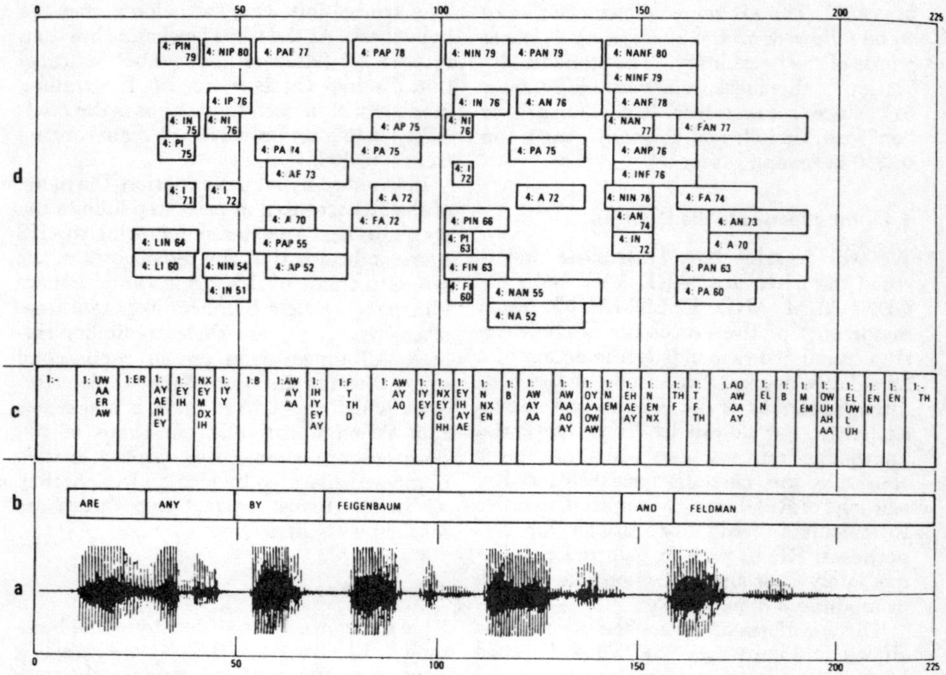

FIGURE 5. The example utterance: (a) the waveform of "Are any by Feigenbaum and Feldman?"; (b) the correct words (for reference); (c) segments; (d) syllable classes; (e) words (created by MOW); (f) words (created by VERIFY); (g) word sequences; (h) phrases. (See facing page for Figure 5e–h.)

classes of segments are produced: silence, sonorant peak, sonorant nonpeak, fricative, and flap. Associated with each classified segment is its duration, absolute amplitude, and amplitude relative to its neighboring segments (i.e., local peak, local valley, or plateau). The segments are contiguous and nonoverlapping, with one class designation for each.

SEG also does a finer labeling of each segment, using a repertory of 98 phonelike labels. Each of the labels is characterized by a vector of autocorrelation coefficients [ITAK75]. These template vectors were generalized from manually labeled speaker-specific training data. The labeling process matches the central portion of each segment against each of the templates using the Itakura metric and produces a vector of 98 numbers. The ith number is an estimate of the (negative log) probability that the segment represents an occurrence of the ith allophone in the label set. For each segment, SEG creates a hypothesis at the segment level and associates with it the vector of estimated allophone probabilities. The several highest rated labels of each segment are shown in Figure 5c.

Step 2. KS: WORD-CTL.
Stimulus: Start of processing.
Action: Create goal hypotheses at the word level. These will control the amount of hypothesization that MOW will do. (The goal hypotheses are not shown in Figure 5.)

Step 3. KS: WORD-SEQ-CTL.
Stimulus: Start of processing.
Action: Create goal hypotheses at the word-sequence level. These will control the amount of hypothesization that WORD-SEQ will do.

Step 4. KS: POM.
Stimulus: New segment hypotheses.
Action: Create syllable-class hypotheses.

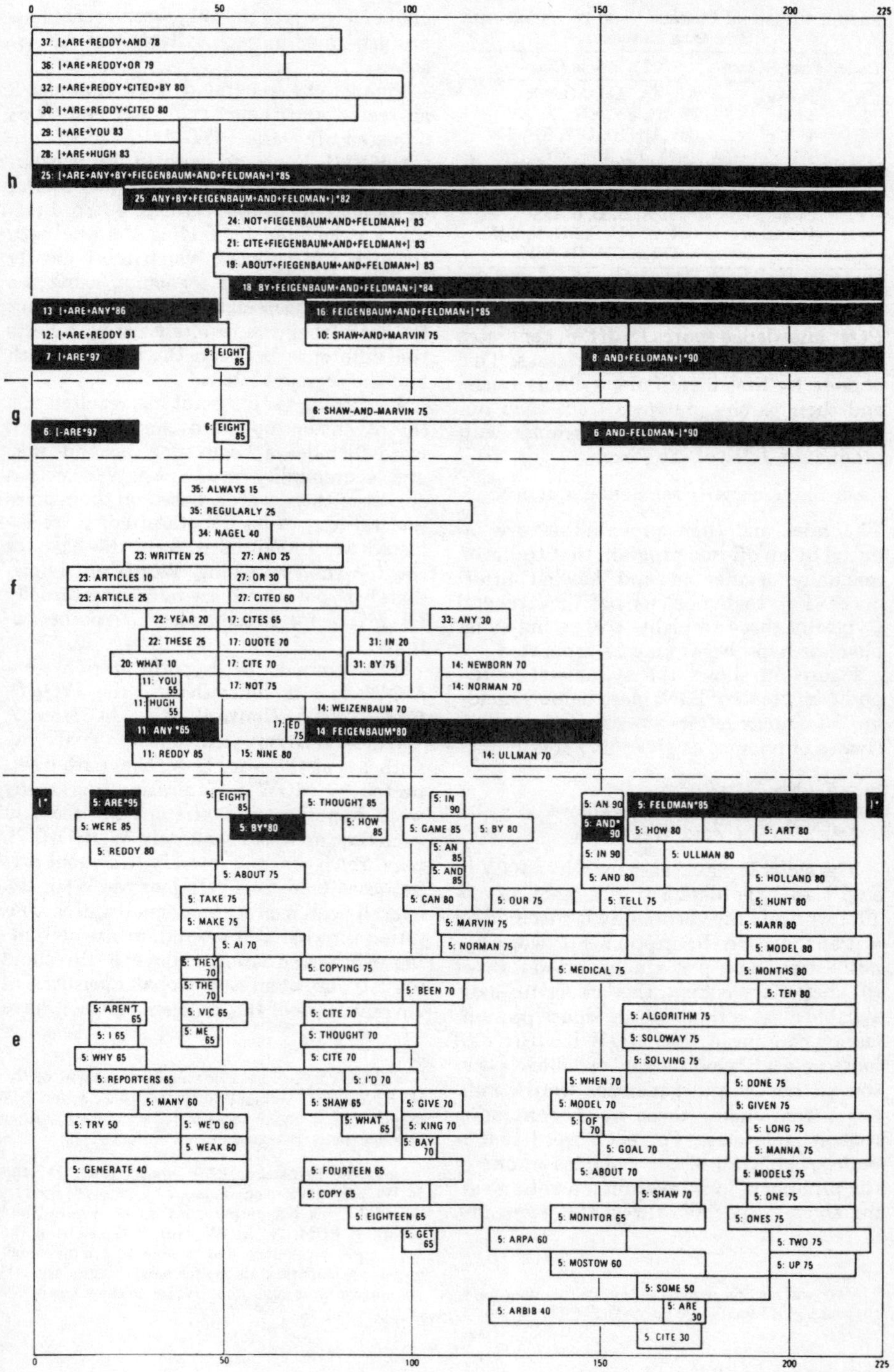
0
50
100
150
200
225
37: [+ARE+REDDY+AND 78
36: [+ARE+REDDY+OR 79
32: [+ARE+REDDY+CITED+BY 80
30: [+ARE+REDDY+CITED 80
29: [+ARE+YOU 83
28: [+ARE+HUGH 83
h
25: [+ARE+ANY+BY+FIEGENBAUM+AND+FELDMAN+]*85
25: ANY+BY+FEIGENBAUM+AND+FELDMAN+]*82
24: NOT+FIEGENBAUM+AND+FELDMAN+] 83
21: CITE+FIEGENBAUM+AND+FELDMAN+] 83
19: ABOUT+FIEGENBAUM+AND+FELDMAN+] 83
18: BY+FEIGENBAUM+AND+FELDMAN+]*84
13: [+ARE+ANY*86
16: FEIGENBAUM+AND+FELDMAN+]*85
12: [+ARE+REDDY 91
10: SHAW+AND+MARVIN 75
7: [+ARE*97
9: EIGHT 85
8: AND+FELDMAN+]*90
g
6: SHAW-AND-MARVIN 75
6: [-ARE*97
6: EIGHT 85
6: AND-FELDMAN-]*90
f
35: ALWAYS 15
35: REGULARLY 25
34: NAGEL 40
23: WRITTEN 25
27: AND 25
23: ARTICLES 10
27: OR 30
23: ARTICLE 25
27: CITED 60
22: YEAR 20
17: CITES 65
33: ANY 30
22: THESE 25
17: QUOTE 70
31: IN 20
20: WHAT 10
17: CITE 70
31: BY 75
14: NEWBORN 70
11: YOU 55
17: NOT 75
14: NORMAN 70
11: HUGH 55
14: WEIZENBAUM 70
11: ANY *65
17: ED 75
14: FEIGENBAUM*80
11: REDDY 85
15: NINE 80
14: ULLMAN 70
e
]*
5: ARE*95
5: EIGHT 85
5: THOUGHT 85
5: IN 90
5: AN 90
5: FELDMAN*85
]*
5: WERE 85
5: BY*80
5: HOW 85
5: GAME 85
5: BY 80
5: AND* 90
5: HOW 80
5: ART 80
5: REDDY 80
5: AN 85
5: IN 90
5: ULLMAN 80
5: ABOUT 75
5: AND 85
5: AND 80
5: HOLLAND 80
5: TAKE 75
5: CAN 80
5: OUR 75
5: TELL 75
5: HUNT 80
5: MAKE 75
5: MARVIN 75
5: MARR 80
5: AI 70
5: NORMAN 75
5: MODEL 80
5: THEY 70
5: COPYING 75
5: MEDICAL 75
5: MONTHS 80
5: THE 70
5: BEEN 70
5: TEN 80
5: AREN'T 65
5: VIC 65
5: CITE 70
5: ALGORITHM 75
5: I 65
5: ME 65
5: THOUGHT 70
5: SOLOWAY 75
5: WHY 65
5: CITE 70
5: SOLVING 75
5: REPORTERS 65
5: I'D 70
5: WHEN 70
5: DONE 75
5: MANY 60
5: SHAW 65
5: GIVE 70
5: MODEL 70
5: GIVEN 75
5: TRY 50
5: WE'D 60
5: WHAT 65
5: KING 70
5: OF 70
5: LONG 75
5: WEAK 60
5: BAY 70
5: GOAL 70
5: MANNA 75
5: GENERATE 40
5: FOURTEEN 65
5: ABOUT 70
5: MODELS 75
5: COPY 65
5: SHAW 70
5: ONE 75
5: EIGHTEEN 65
5: MONITOR 65
5: ONES 75
5: GET 65
5: ARPA 60
5: TWO 75
5: MOSTOW 60
5: UP 75
5: SOME 50
5: ARE 30
5: ARBIB 40
5: CITE 30

TABLE 1. PHONE CLASSES USED TO DEFINE THE SYLLABLE CLASSES

Code	Phone Class	Phones in Class
A	A-like	AE, AA, AH, AO, AX
I	I-like	IY, IH, EY, EH, IX, AY
U	U-like	OW, UH, U, UW, ER, AW, OY, EL, EM, EN
L	Liquid	Y, W, R, L
N	Nasal	M, N, NX
P	Stop	P, T, K, B, D, G, DX
F	Fricative	HH, F, TH, S, SH, V, DH, Z, ZH, CH, JH, WH

Using the labeled segments as input, the *POM* knowledge source [SMIT76] generates hypotheses for likely syllable classes. This is done by first identifying syllable nuclei and then parsing outward from each nucleus, using a probabilistic grammar with production rules of the form:

syllable-class → segment-sequence.

The rules and their probabilities are induced by an off-line program that trains on manually segmented and labeled utterances. For each nucleus position, several (typically three to eight) competing syllable-class hypotheses may be generated.

Figure 5d shows the syllable-class hypotheses created. Each class name is made up of single-letter codes representing classes of phones, as given in Table 1.

Step 5. KS: MOW.
Stimulus: New syllable hypotheses.[6]
Action: Create word hypotheses.

The syllable classes are used by MOW in step 5 to hypothesize words. Each of the 1011 words in the vocabulary is specified by a pronunciation description. For word hypothesization purposes, an inverted form of the dictionary is kept; this associates each syllable class with all words whose pronunciation contains it. The *MOW* KS [SMIT76] looks up each hypothesized syllable class in the dictionary and generates word candidates from among those words containing that syllable class. For each word that is multisyllabic, all of the syllables in one of the pronunciations must match with a rating above a specified threshold. Typically, about 50 words of the 1011-word vocabulary are generated at each syllable nucleus position.

Finally, the generated word candidates are rated and their begin- and end-times adjusted by the *WIZARD* procedure [McKE77]. For each word in the vocabulary, WIZARD has a network description of its possible pronunciations. A word rating is calculated by finding the one path through the network which most closely matches the labeled segments, using the probabilities associated with the segment for each label; the resultant rating reflects the difference between this optimal path and the segment labels.[7]

Processing to this point has resulted in a set of bottom-up word candidates. Each word includes a begin-time, an end-time, and a credibility rating. MOW selects a subset of these words, based on their times and ratings, to be hypothesized; these selected word hypotheses form the base for the top-end processing. Words not immediately hypothesized are retained internally by MOW for possible later hypothesization.[8]

The amount of hypothesization that MOW does is controlled by the *WORD-CTL* (Word Control) KS. At step 2, WORD-CTL created initial goal hypotheses at the word level; these are interpreted by MOW as indicating how many word hypotheses to attempt to create in each time area. Subsequently, WORD-CTL may retrigger and modify the goal hypotheses (and thus retrigger MOW) if the overall search process stagnates; this condition is recognized when there are no waiting KS instantiations above a threshold priority or when the global measures of current state of the problem solution have

[6] MOW will also be reinvoked upon a modification to the word goal hypotheses by WORD-CTL.

[7] WIZARD is, in effect, a miniature version of the HARPY speech-recognition system (see Section 2.3), except that it has a network for each word, rather than one network containing all sentences.

[8] Since the September 1976 version, the POM and MOW KSs have been replaced by *NOAH* [SMIT77, SMIT81]. This KS outperforms, in both speed and accuracy, POM and MOW (with WIZARD) on the 1011-word vocabulary and is able to handle much larger vocabularies; its performance degradation is only logarithmic in vocabulary size, in the range of 500 to 19,000 words.

not improved in the last several KS executions.

WORD-CTL (and WORD-SEQ-CTL) are examples of KSs not directly involved in the hypothesizing and testing of partial solutions. Instead, these KSs control the search by influencing the activations of other KSs. These *policy* KSs impose global search strategies on the basic priority scheduling mechanism. For example, MOW is a generator of word hypotheses (from the candidates it creates internally) and WORD-CTL controls the number to be hypothesized. This clear separation of policy from mechanism has facilitated experimentation with various control schemes. A trivial change to WORD-CTL such that goal hypotheses are generated only at the start of the utterance (left-hand end) results in MOW creating word hypotheses only at the start, thus forcing all top-end processing to be left-to-right (see Section 3.2).

In this example four words (ARE, BY, AND, and FELDMAN) of the six in the utterance were correctly hypothesized; 86 incorrect hypotheses were generated (see Figure 5e). The 90 words that were hypothesized represent approximately 1.5 percent of the 1011-word vocabulary for each one of the six words in the utterance.

In addition, two unique word-level hypotheses are generated before the first and after the last segment of the utterance to denote the start and end of utterance, respectively. They are denoted by [and].

Step 6. KS: WORD-SEQ.
Stimulus: New words created bottom-up.
Action: Create four word-sequence hypotheses:
[–ARE*(97, 0:28),
AND–FELDMAN–]*(90, 145:225),
EIGHT(85, 48:57).
SHAW–AND–MARVIN(75, 72:157),

The *WORD-SEQ* knowledge source [LESS77a] has the task of generating, from the bottom-up word hypotheses, a small set (about three to ten) of word-sequence hypotheses. Each of these sequences, or *islands*, can be used as the basis for expansion into larger islands, which it is hoped will culminate in a hypothesis spanning the entire utterance. Multiword islands are used rather than single-word islands because of the relatively poor reliability of ratings of single words. With multiword islands, syntactic and coarticulation constraints can be used to increase the reliability of the ratings.

WORD-SEQ uses three kinds of knowledge to generate multiword islands efficiently:

(1) A table derived from the grammar indicates for every ordered pair of words in the vocabulary (1011 × 1011) whether that pair can occur in sequence within some sentence of the defined language. This binary table, whose density of ones for the X05 grammar is 1.7 percent, defines a *language-adjacent* relation.
(2) Acoustic-phonetic knowledge, embodied in the *JUNCT* (juncture) procedure [CRON77], is applied to pairs of word hypotheses and is used to decide if that pair might be considered to be *time-adjacent* in the utterance. JUNCT uses the dictionary pronunciations, and examines the segments at their juncture (gap or overlap) in making its decision.
(3) Statistical knowledge is used to assess the credibility of generated alternative word sequences and to terminate the search for additional candidates when the chance of finding improved hypotheses drops. The statistics are generated from previously observed behavior of WORD-SEQ and are based on the number of hypotheses generable from the given bottom-up word hypotheses and their ratings.

WORD-SEQ takes the highest rated single words and generates multiword sequences by expanding them with other hypothesized words that are both time- and language-adjacent. This expansion is guided by credibility ratings generated by using the statistical knowledge. The best of these word sequences (which occasionally includes single words) are hypothesized.

The *WORD-SEQ-CTL* (Word-Sequence-Control) KS controls the amount of hypothesization that WORD-SEQ does by creating "goal" hypotheses that WORD-SEQ interprets as indicating how many hypotheses to create. This provides the same kind of separation of policy and mechanism achieved in the MOW/WORD-CTL pair of KSs. WORD-SEQ-CTL fired at the start of processing, at step 3, in order to create the goal hypotheses. Subsequently, WORD-SEQ-CTL may trigger if stagnation is recognized; it then modifies the word-sequence goal hypotheses, thus stimulating WORD-SEQ to generate new islands from which the search may prove more fruitful. WORD-SEQ may generate the additional hypotheses by decomposing word sequences already on the blackboard or by generating islands previously discarded because their ratings seemed too low.

Step 6 results in the generation of four multiword sequences (see Figure 5g). These are used as initial, alternative anchor points for additional searching. Note that two of these islands are correct, each representing an alternative search path that potentially can lead to a correct interpretation of the utterance. This ability to derive the correct interpretation in multiple ways makes the system more robust. For example, there have been cases in which a complete interpretation could not be constructed from one correct island because of KS errors but was derived from another island.

High-level processing on the multiword sequences is accomplished by the following KSs: PARSE, PREDICT, VERIFY, CONCAT, STOP, and WORD-SEQ-CTL. Since an execution of the VERIFY KS will often immediately follow the execution of the PREDICT KS (each on the same hypothesis), we have combined the descriptions of the two KS executions into one step for ease of understanding.

Because the syntactic constraint used in the generation of the word sequences is only pairwise, a sequence longer than two words might not be syntactically acceptable. The *PARSE* knowledge source [HAYE77b] can parse a word sequence of arbitrary length, using the full grammatical constraints. This parsing does not require that the word sequence form a complete nonterminal in the grammar or that the sequence be sentence-initial or sentence-final; the words need only occur contiguously in some sentence of the language. If a sequence hypothesis does not parse, it is marked as "rejected." Otherwise a phrase hypothesis is created. Associated with the phrase hypothesis is the word sequence that supports it, as well as information about the parse(s).

Steps 7 through 10 show the PARSE KS processing each of the multiword sequences. In this example all four multiword sequences were verified as valid language fragments. However, if a multiword sequence had been rejected, the WORD-SEQ KS might have been reinvoked to generate additional multiword sequences in the time area of the rejected one. WORD-SEQ would generate the additional hypotheses by decomposing (shortening) word-sequence islands already on the blackboard or by regenerating islands which may not have been hypothesized initially owing to low ratings. Additional word-sequence hypotheses might also be generated in response to the modification of "goal" hypotheses at the word-sequence level by the WORD-SEQ-CTL. Such a structuring of a KS as a *generator* is a primary mechanism in Hearsay-II for limiting the number of hypotheses created on the blackboard and thereby reducing the danger of a combinatorial explosion of KS activity in reaction to those hypotheses.

The scheduling strategy is parameterized to delay phrase-level processing until an adequate number of highly rated phrase hypothesis islands is generated. This strategy is not built directly into the scheduler, but rather is accomplished by (1) appropriately setting external scheduling parameters (i.e., the high setting of the priorities of WORD-SEQ and PARSE KS actions in contrast to those of PREDICT, VERIFY, and CONCAT),[9] and (2) taking into account the current state of hypotheses on the phrase level of the blackboard in evaluating the usefulness of potential KS actions as described by their response frames.

Step 7. KS: PARSE*.
Stimulus: [−ARE* (word sequence).
Action: Create phrase: [+ARE* (97, 0:28).

Step 8. KS: PARSE*.
Stimulus: AND−FELDMAN−]* (word sequence).

[9] These settings are determined empirically by observing a number of training runs. They are not adjusted during test runs of the system.

Action: Create phrase:
AND+FELDMAN+]* (90, 145:225).

Step 9. KS: PARSE.
Stimulus: EIGHT (word sequence).
Action: Create phrase: EIGHT (85, 48:57).

Step 10. KS: PARSE.
Stimulus: SHAW–AND–MARVIN (word sequence).
Action: Create phrase: SHAW+AND+MARVIN (75, 72:157).

Each of the four executions of the PARSE KS (steps 7–10) results in the creation of a phrase hypothesis; these are shown in Figure 5h. Each of these hypotheses causes an invocation of the PREDICT KS.

The *PREDICT* knowledge source [HAYE 77b] can, for any phrase hypothesis, generate predictions of all words which can immediately precede and all which can immediately follow that phrase in the language. In generating these predictions this KS uses the parsing information attached to the phrase hypothesis by the parsing component. The action of PREDICT is to attach a "word-predictor" attribute to the hypothesis which specifies the predicted words. Not all of these PREDICT KS instantiations are necessarily executed (and thus indicated as a step in the execution history). For instance, further processing on the phrases [+ARE and AND+FELDMAN+] is sufficiently positive that the scheduler never executes the instantiation of PREDICT for the phrase SHAW+AND+MARVIN (created in step 10).

The *VERIFY* KS can attempt to verify the existence of or reject each such predicted word in the context of its predicting phrase. If verified, a confidence rating for the word is also generated. The verification proceeds as follows: First, if the word has been hypothesized previously and passes the test for time-adjacency (by the JUNCT procedure), it is marked as verified and the word hypothesis is associated with the prediction. (Note that some word hypotheses may thus become associated with several different phrases.) Second, a search is made of the internal store created by MOW to see if the prediction can be matched by a previously generated word candidate which had not yet been hypothesized. Again, JUNCT makes a judgment about the plausibility of the time-adjacency relationship between the predicting phrase and the predicted word. Finally, WIZARD compares its word-pronunciation network with the segments in an attempt to verify the prediction.

For each of these different kinds of verification, the approximate begin-time (end-time if verifying an antecedent prediction) of the word being predicted following (preceding) the phrase is taken to be the end-time (begin-time) of the phrase. The end-time (begin-time) of the predicted word is not known, and in fact one function of the verification step is to generate an approximate end-time (begin-time) for the verified word. In general, it is possible to generate several different "versions" of the word which differ primarily in their end-times (begin-times); since no context following (preceding) the predicted word is given, several different estimates of the end (beginning) of the word may be plausible solely on the basis of the segmental information. These alternatives give rise to the creation of competing hypotheses.

VERIFY is invoked when a KS (PREDICT) places a "word-predictor" attribute on a phrase hypothesis. For each word on the attribute list that it verifies, it creates a word hypothesis (if one does not already exist), and the word is placed on a "word-verification" attribute of the phrase hypothesis. Word hypotheses created by VERIFY are shown in Figure 5f.

The *CONCAT* KS [HAYE77b] is invoked on phrase hypotheses which have word-verification attributes attached. For each verified word, the phrase and new word are parsed together and a new, extended phrase hypothesis is created (as shown in Figure 5h). The new hypothesis receives a rating based on the ratings of its component words. If all word predictions preceding or following the phrase had been rejected, the phrase hypothesis itself would be marked as "rejected." Similarly, the underlying word-sequence hypothesis is rejected if all the phrase hypotheses it supports are rejected. (This action would retrigger WORD-SEQ to generate more word sequences.)

Step 11. KS: PREDICT & VERIFY*.
Stimulus: [+ARE* (phrase).

Action: Predict (from the grammar) 292 words following. Reject (using the acoustic information) 277 of them. The four highest rated of the 15 verified words are
REDDY(85, 26:52),
ANY*(65, 24:49),
HUGH(55, 30:39), and
YOU(55, 28:39).

Step 12. KS: CONCAT.
Stimulus: [+ARE* (phrase), REDDY (word).
Action: Create phrase: [+ARE+REDDY (91, 0:52).

Step 13. KS: CONCAT*.
Stimulus: [+ARE* (phrase), ANY* (word).
Action: Create phrase: [+ARE+ANY* (86, 0:49).

In steps 11 through 13 the highly rated phrase [+ARE is extended and results in the generation of the additional phrases [+ARE+REDDY and [+ARE+ANY. These phrases, however, are not immediately extended because the predicted words REDDY and ANY are not rated sufficiently high. Instead, the scheduler, pursuing a strategy more conservative than strict best-first, investigates phrases that look almost as good as the best one. This scheduling strategy results in the execution of the PREDICT and VERIFY KSs on two of the other initial phrase islands: AND+FELDMAN+] and EIGHT.

Step 14. KS: PREDICT & VERIFY*.
Stimulus: AND+FELDMAN+]* (phrase).
Action: Predict 100 words preceding. Reject 76 of them. The best of the verified 24 (in descending rating order) are
FEIGENBAUM*(80, 72:150),
WEIZENBAUM(70, 72:150),
ULLMAN(70, 116:150),
NORMAN(70, 108:150), and
NEWBORN(70, 108:150).

Step 15. KS: PREDICT & VERIFY.
Stimulus: EIGHT (phrase).
Action: Predict the word NINE following and verify it (80, 52:82). Predict SEVEN preceding, but reject this because of mismatch with the acoustic segments.

The attempt to extend the phrase EIGHT at step 15 is not successful; none of the grammatically predicted words are acoustically verified, even using a lenient threshold. Thus this phrase is marked rejected and is dropped from further consideration.

Step 16. KS: CONCAT*.
Stimulus: FEIGENBAUM* (word), AND+FELDMAN+]* (phrase).
Action: Create phrase: FEIGENBAUM+AND+FELDMAN+]* (85, 72:225).

Beginning with step 16, extending the phrase AND+FELDMAN+] with the highly rated word FEIGENBAUM looks sufficiently promising for processing to continue now in a more depth-first manner along the path FEIGENBAUM+AND+FELDMAN+] through step 25.[10] Processing on the path [+ARE+REDDY does not resume until step 26.

Step 17. KS: PREDICT & VERIFY*.
Stimulus: FEIGENBAUM+AND+FELDMAN+]* (phrase).
Action: Predict eight preceding words. Reject one (DISCUSS). Find two already on the blackboard:
BY*(80, 52:72) and
ABOUT(75, 48:72).
Verify five others:
NOT(75, 49:82),
ED(75, 67:72),
CITE(70, 49:82),
QUOTE(70, 49:82),
CITES(65, 49:82).

In steps 18 through 24, alternative word extensions of FEIGENBAUM+AND+FELDMAN+] are explored. As a result of this exploration the phrase BY+FEIGENBAUM+AND+FELDMAN+] is considered the most credible.

Step 18. KS CONCAT*.
Stimulus: BY* (word), FEIGENBAUM+AND+FELDMAN+]* (phrase).
Action: Create phrase: BY+FEIGENBAUM+AND+FELDMAN+]* (84, 52:225).

Step 19. KS: CONCAT.
Stimulus: ABOUT (word), FEIGENBAUM+AND+FELDMAN+]* (phrase).
Action: Create phrase: ABOUT+FEIGENBAUM+AND+FELDMAN+] (83, 48:225).

Step 20. KS: PREDICT & VERIFY.
Stimulus:
ABOUT+FEIGENBAUM+AND+ FELDMAN+] (phrase).

[10] The rating on a hypothesis is only one parameter used by the scheduler to assign priorities to waiting KS instantiations. In particular, the length of a hypothesis is also important. Thus, FEIGENBAUM with a rating of 80 looks better than REDDY with a rating of 85 because it is much longer.

Action: Predict one preceding word: WHAT. Verify it (10, 20:49).

Step 21. KS: CONCAT
Stimulus: CITE (word), FEIGENBAUM+AND+FELDMAN+] (phrase).
Action: Create phrase: CITE+FEIGENBAUM+AND+FELDMAN+] (83, 49:225).

Step 22. KS: PREDICT & VERIFY.
Stimulus: CITE+FEIGENBAUM+AND+FELDMAN+] (phrase).
Action: Predict four preceding words. Reject two of them: BOOKS, PAPERS. Verify
THESE (25, 28:49),
YEAR (20, 30:49).

Step 23. KS: PREDICT & VERIFY*.
Stimulus: BY+FEIGENBAUM+AND+FELDMAN+]* (phrase).
Action: Predict ten preceding words. Reject five: ABSTRACTS, ARE, BOOKS, PAPERS, REFERENCED. Find two already on the blackboard:
ANY* (65, 24:49),
THESE (25, 28:49).
Verify three more:
ARTICLE (25, 9:52),
WRITTEN (25, 24:52),
ARTICLES (10, 9:52).

Step 24. KS: CONCAT.
Stimulus: NOT (word), FEIGENBAUM+AND+FELDMAN+]*.
Action: Create phrase: NOT+FEIGENBAUM+AND+FELDMAN+] (83, 49:225).

Step 25. KS: CONCAT*.
Stimulus: ANY* (word), BY+FEIGENBAUM+AND+FELDMAN+]* (phrase).
Action: Create phrase: ANY+BY+FEIGENBAUM+AND+FELDMAN+]* (82, 24:225).
[+ARE+ANY+BY+FEIGENBAUM+AND+FELDMAN+]* (85, 0:225) is also created, from [+ARE+ANY and BY+FEIGENBAUM+ AND+FELDMAN+].

In step 25 the word ANY is concatenated onto the phrase BY+FEIGENBAUM+AND+FELDMAN+]. However, instead of only creating this new combined phrase, the CONCAT KS also notices that the word ANY is the last word of the phrase [+ARE+ANY; this leads the CONCAT KS to merge the two adjacent phrases [+ARE+ANY and BY+FEIGENBAUM+AND+FELDMAN+] into a single phrase, after first ascertaining that the resulting phrase is grammatical. This merging bypasses the several single-word PREDICT, VERIFY, and CONCAT actions needed to generate the enlarged hypothesis from either of the two original hypotheses in an incremental fashion. Thus the recognition process is sped up in two ways: (1) several single-word actions are eliminated, and (2) the scheduler postpones KS actions on competing (incorrect) hypotheses since these potential actions seem less attractive than actions on the new, enlarged hypothesis. Such mergings occur in approximately half of the runs on the 1011-word grammar with the small branching factor (X05); in grammars with higher branching factors, the merging of phrase hypotheses occurs with even higher frequency.

It has been our experience that just as a multiword island is more credible than the individual words that compose it, so a merged phrase hypothesis is more credible than its two constituent phrases. For example, about 80 percent of the mergings in X05 runs produce correct hypotheses. In more complex grammars this statistic drops to about 35 percent, but correspondingly more phrase mergings occur.

The newly created merged phrase also happens to be a complete sentence; i.e., it has begin- and end-of-utterance markers as its extreme constituents. Thus it is a candidate for the interpretation of the utterance.

Step 26. KS: STOP.
Stimulus: [+ARE+ANY+BY+FEIGENBAUM+AND+FELDMAN+]* (complete sentence).
Action: Deactivation of several dozen competing hypotheses.

These start- and end-of-utterance "words" (denoted by [and]) appear in the syntactic specification of the language as the first and last terminals of every complete sentence. Thus any verified phrase that includes these as its extreme constituents is a complete sentence and spans the entire utterance. Such a sentence becomes a candidate for selection as the system's recognition result.

In general, the control and rating strategies do not guarantee that the first such complete spanning hypothesis found will have the highest rating of all possible spanning sentence hypotheses that might be found if the search were allowed to continue, so the system does not just stop with the first one generated.[11] However the characteristics of such a hypothesis are used by the *STOP* knowledge source to prune from further consideration other partial hypotheses which, because of their low ratings, are unlikely to be extendible into spanning hypotheses with ratings higher than the best already-discovered spanning sentence. This heuristic pruning procedure is based on the form of the ratings function (i.e., how the rating of the phrase is derived from its constituent words). The pruning procedure considers each partial phrase and uses the ratings of other word hypotheses in the time areas not covered by the phrase to determine if the phrase might be extendible to a phrase rated higher than the spanning hypothesis; if not, the partial phrase is pruned. This pruning process and the rating and halting policies are discussed in MOST77.

Of the hypotheses not discarded, extensions to the phrase [+ARE now appear as the most likely candidates to produce new and better complete phrases. This search for better complete phrases results, in steps 27 through 36, in the examination of numerous alternative extensions, each of which is promising.

Step 27. KS: PREDICT & VERIFY.
Stimulus: [+ARE+REDDY.
Action: Predict three following words. Verify CITED (60, 52:86), OR (30, 52:67), AND (25, 52:82).

Step 28. KS: CONCAT.
Stimulus: [+ARE* (phrase), HUGH (word).
Action: Create phrase: [+ARE+HUGH (83, 0:39).

Step 29. KS: CONCAT.
Stimulus: [+ARE (phrase), YOU (word).
Action: Create phrase: [+ARE+YOU (83, 0:39).

Step 30. KS: CONCAT.
Stimulus: [+ARE+REDDY (phrase), CITED (word).
Action: Create phrase: [+ARE+REDDY+CITED (80, 0:86).

Step 31. KS: PREDICT & VERIFY.
Stimulus: [+ARE+REDDY+CITED (phrase).
Action: Predict two following words. Verify BY(75, 83:98), IN(20, 86:114).

Step 32. KS: CONCAT.
Stimulus: [+ARE+REDDY+CITED (phrase), BY (word).
Action: Create phrase: [+ARE+REDDY+CITED+BY (80, 0:98).

Step 33. KS: PREDICT & VERIFY.
Stimulus: [+ARE+REDDY+CITED+BY (phrase).
Action: Predict one following word. Verify ANY (30, 105:126).

Step 34. KS: PREDICT & VERIFY.
Stimulus: [+ARE+HUGH (phrase).
Action: Predict one following word. Verify NAGEL (40, 42:63).

Step 35. KS: PREDICT & VERIFY.
Stimulus: [+ARE+YOU (phrase).
Action: Predict three following words. Reject USUALLY. Verify
REGULARLY (25, 39:116),
ALWAYS (15, 39:72).

Step 36. KS: CONCAT.
Stimulus: [+ARE+REDDY (phrase), OR (word).
Action: Create phrase: [+ARE+REDDY+OR (79, 0:67).

Step 37. KS: CONCAT.
Stimulus: [+ARE+REDDY (phrase), AND (word).
Action: Create phrase: [+ARE+REDDY+AND (78, 0:82).

Step 38. KS: STOP.
Stimulus: Stagnation
Action: Stop search and accept [+ARE+ANY+BY+FEIGENBAUM+AND+FELDMAN+]*.

The recognition processing finally halts in one of two ways: First, there may be no more partial hypotheses left to consider for prediction and extension. Because of the combinatorics of the grammar and the likelihood of finding some prediction rated at least above the absolute rejection threshold, this termination happens when the heuristic pruning procedure used by STOP and RPOL has eliminated all competitors. Such a halt occurs here as STOP decides to terminate the search process and accept the phrase [+ARE+ANY+BY+FEIGEN-

[11] An alternative control strategy based on the Shortfall Density scoring function and assumptions of the completeness of processing at the word level can guarantee that the first complete interpretation generated is the best possible (see Section 2.1). This results in a more conservative, breadth-first search, in which more alternatives are explored.

BAUM+AND+FELDMAN+] as the correct interpretation. In general there might be more than one complete sentence hypothesis at this point; STOP would select the one with the highest rating.

A second kind of halt occurs if the system expends its total allowed computing resources (time or space). (The actual thresholds used are set according to the past performance of the system on similar sentences, i.e., of the given length and over the same vocabulary and grammar.) In that case a selection of several of the highest rated phrase hypotheses is the result of the recognition process, with the selection biased toward the longest phrases which overlap (in time) the least.

Step 39. KS: SEMANT*.
Stimulus: Recognized utterance: [+ARE+ANY+BY+FEIGENBAUM+AND+FELDMAN+]*.
Action: Generate an interpretation for the database retrieval system.

The *SEMANT* knowledge source [Fox77] takes the word sequence(s) result of the recognition process and constructs an interpretation in an unambiguous format for interaction with the database that the speaker is querying. The interpretation is constructed by actions associated with "semantically interesting" nonterminals (which have been prespecified for the grammar) in the parse tree(s) of the recognized sequence(s). In our example the following structure is produced:

```
F:[U:([ARE ANY BY FEIGENBAUM AND
      FELDMAN])
  N:($PRUNE!LIST
     S:($PRUNE!LIST!AUTHOR   K:(A:
       ((FEIGENBAUM * FELDMAN)))))]
```

F denotes the total message. U contains the utterance itself. N indicates the main type of the utterance (e.g., PRUNE a previously specified list of citations, REQUEST, HELP), S the subtype (e.g., PRUNE a list according to its author). K denotes the different attributes associated with the utterance (e.g., A is the author, T is the topic).

If recognition produces more than one partial sequence, SEMANT constructs a maximally consistent interpretation based on all of the partial sentences, taking into account the rating, temporal position, and semantic consistency of the partial sentences.

The *DISCO* (discourse) knowledge source [HAYE77c] accepts the formatted interpretation of SEMANT and produces a response to the speaker. This response is often the display of a selected portion of the queried database. In order to retain a coherent interpretation across sentences, DISCO maintains a finite-state model of the ongoing discourse.

2. COMPARISON WITH OTHER SPEECH-UNDERSTANDING SYSTEMS

In addition to Hearsay-II, several other speech-understanding systems were also developed as part of the Defense Advanced Research Projects Agency (DARPA) research program in speech understanding from 1971 to 1976 [MEDR78]. As a way of concretely orienting the research, a common set of system performance goals, shown in Figure 6, was established by the study committee that launched the project [NEWE73]. All of the systems are based on the idea of diverse, cooperating KSs to handle the uncertainty in the signal and processing. They differ in the types of knowledge, interactions of knowledge, representation of search space, and control of the

FIGURE 6. DARPA speech-understanding-system performance goals set in 1971. [After NEWE73 and MEDR78.]

The system should
- Accept connected speech
- from many
- cooperative speakers of the General American Dialect
- in a quiet room
- using a good-quality microphone
- with slight tuning per speaker
- requiring only natural adaptation by the user
- permitting a slightly selected vocabulary of 1000 words
- with a highly artificial syntax and highly constrained task
- providing graceful interaction
- tolerating less than 10 percent semantic error
- in a few times real time on a 100-million-instructions-per-second machine
- and be demonstrable in 1976 with a moderate chance of success.

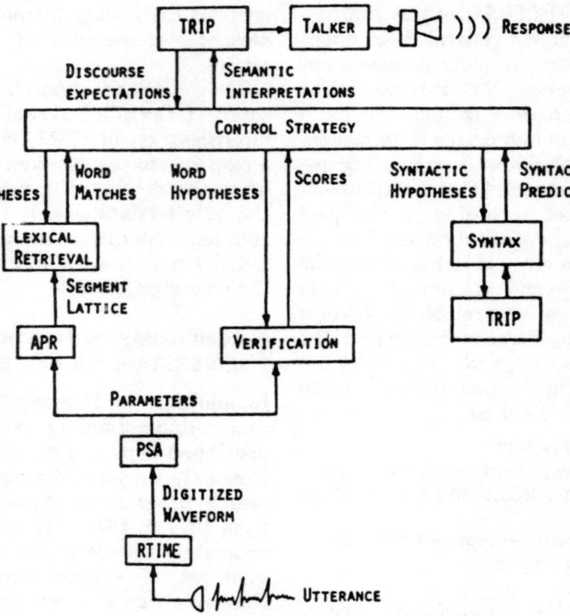

FIGURE 7. Structure of HWIM. [From WOLF80.]

search. (They also differ in the tasks and languages handled, but we do not address those here.) In this section we describe three of these systems, Bolt Beranek and Newman's (BBN's) HWIM, Stanford Research Institute's (SRI's) system, and Carnegie-Mellon University's (CMU's) HARPY, and compare them with Hearsay-II along those dimensions. For consistency we will use the terminology developed in this paper in so far as possible, even though it is often not identical to that used by the designers of each of the other systems.[12]

Although the performance specifications had the strong effect of pointing the various efforts in the same directions, the backgrounds and motivations of each group led to different emphases. For example, BBN's expertise in natural-language processing and acoustic-phonetics led to an emphasis on those KSs; SRI's interest in semantics and discourse strongly influenced its system design; and CMU's predilection for system organization placed that group in the central position (and led to the Hearsay-II and HARPY structures).

[12] IBM has been funding work with a somewhat different objective [BAHL76]. Its stated goals mandate little reliance on the strong syntactic/semantic/task constraints exploited by the DARPA projects. This orientation is usually dubbed *speech recognition* as distinguished from *speech understanding*.

2.1 BBN's HWIM System

Figure 7 shows the structure of BBN's *HWIM* (Hear What I Mean) system [WOOD76, WOLF80]. In overall form, HWIM's general processing structure is strikingly similar to that of Hearsay-II. Processing of a sentence is bottom-up through audio signal digitization, parameter extraction, segmentation and labeling, and a scan for word hypotheses; this phase is roughly similar to Hearsay-II's initial bottom-up processing up through the MOW KS.

Following this initial phase, the Control Strategy module takes charge, calling the Syntax and Lexical Retrieval KSs as subroutines:

- The grammar is represented as an augmented transition network [WOOD70], and, as in Hearsay-II, includes semantic and pragmatic knowledge of the domain (in this case, "travel planning"). The Syntax KS combines the functions of Hearsay-II's PREDICT and CONCAT KSs. Like them, it handles contiguous sequences of words in the language, independently of the phrase structure nonterminal boundaries, as well as the merging of phrase hypotheses (i.e., island collision).

- The Lexical Retriever functions in this phase much like Hearsay-II's VERIFY KS, rating the acoustic match of a predicted word at one end of a phrase. Some configurations of HWIM also have a KS which does an independent, highly reliable, and very expensive word verification; that KS is also called directly by the Control Strategy.
- The Control Strategy module schedules the Syntax and Lexical Retrieval KSs opportunistically. To this end it keeps a task agenda that prioritizes the actions on the most promising phrase hypotheses. The task agenda is initialized with single-word phrase hypotheses constructed from the best word hypotheses generated in the bottom-up phase.

Given these similarities between HWIM and Hearsay-II, what besides the content of the KSs (which we do not address) are the differences? The most significant differences involve the mechanisms for instantiating KSs, scheduling KSs (i.e., selective attention for controlling the search), and representing, accessing, and combining KS results. These differences stem primarily from differing design philosophies:

- The Hearsay-II design was based on the assumption that a very general and flexible model for KS interaction patterns was required because the type, number, and interaction patterns of KSs would change substantially over the lifetime of the system [LESS75, LESS77b]. Thus we rejected an explicit subroutine-like architecture for KS interaction because it reduces modularity. Rather, the implicit data-directed approach was taken, in which KSs interact uniformly and anonymously via the blackboard.
- The HWIM design evolved out of an *incremental simulation* methodology [WOOD73]. In this methodology the overall system is implemented initially with some combination of computer programs and human simulators, with the latter filling the role of components (i.e., KSs and scheduling) not fully conceptualized. As experience is gained, the human simulators are replaced by computer programs. Thus by the time the system has evolved into a fully operational computer program, the type of KSs and their interaction patterns are expected to be stable. Modifications after this point aim to improve the performance of individual KSs and their scheduling, with only minor changes expected in KS interaction patterns. From this perspective, developing specific explicit structures for explicit KS interactions is reasonable.

Thus HWIM has an explicit control strategy, in which KSs directly call each other, and in which the scheduler has built-in knowledge about the specific KSs in the system. The Hearsay-II scheduler has no such built-in knowledge but rather is given an abstract description of each KS instantiation by its creator condition program.

Similarly, one KS communicates with another in HWIM via ad hoc KS-specific data structures. The introduction of a new KS is expected to occur very rarely and requires either that it adopt some other KS's existing data representation or that its new formats be integrated into those KSs that will interact with it. Hearsay-II's blackboard, on the other hand, provides a uniform representation which facilitates experimentation with new or highly modified KSs.

When one KS in a hierarchical structure like that in HWIM calls another, it provides the called KS with those data it deems relevant. The called KS also uses whatever data it has retained internally plus what it might acquire by calling other KSs. Hearsay-II's blackboard, on the other hand, provides a place for all data known to all the KSs; one KS can use data created by a previous KS execution without the creator of the data having to know which KS will use the data and without the user KS having to know which KS might be able to create the data.

The ability to embed into the HWIM system a detailed model of the KSs and their interaction patterns has had its most profound effect on the techniques developed for scheduling. Several alternative scheduling policies were implemented in the Control Strategy module. The most interesting of these, the "shortfall density scoring strategy" [WOOD77], can be shown formally to guarantee that the first complete sentence hypothesis constructed by the system is the best possible (i.e., highest rated) such hypothesis that it will ever be able to construct. Heuristic search strategies with this property are called *admissible* [NILS71]. This contrasts with the *approximate* Hearsay-II scheduling strategy, in which there is no guarantee at any point that a better interpretation cannot be found by continued search. Thus Hearsay-II requires a heuristic stopping decision, as described in Section 1.2. In HWIM an admissible strategy is possible because the scheduler can make some strong assumptions about the nature of KS processing: in particular, the algorithms used by the Lexical Retriever KS are such that it does not subsequently generate a higher rating for a predicted word than that of the highest rated word predicted in that utterance location by the initial, bottom-up processing.

An admissible strategy eliminates errors which an approximate strategy may make by stopping too soon. However, even when an admissible strategy can be constructed, it may not be preferable if it generates excessive additional search in order to guarantee its admissibility. More discussion of this issue in speech understanding can be found in WOLF80, WOOD77, MOST77, and HAYE80. Discussions of it in more general cases can be found in POHL70, HARR74, and POHL77.

Given that hypotheses are rated by KSs, combining on a single hypothesis several ratings generated by different KSs is a problem. A similar problem also occurs within a KS when constructing a hypothesis from several lower level hypotheses; the rating of the new one should reflect the combination of ratings of its components. Hearsay-II uses ad hoc schemes for such rating combinations [HAYE77d]. HWIM takes a formal approach, using an application of Bayes' theorem. To implement this, each KS's ratings are calibrated by using performance statistics gathered on test data. This uniform scheme for calibration and combination of ratings facilitates adding and modifying KSs. The issue of evaluating the combination of evidence from multiple sources is a recurrent problem in knowledge-based systems [SHOR75, DUDA78].

2.2 SRI's System

The SRI system [WALK78, WALK80], though never fully operational on a large vocabulary task, presents another interesting variant on structuring a speech-understanding system. Like the HWIM system, it uses an explicit control strategy with, however, much more control being centralized in the Control Strategy module. The designers of the system felt there was "a large potential for mutual guidance that would not be realized if all knowledge source communication was indirect" [WALK78, p. 84]. Part of this explicit control is embedded within the rules that define the phrases of the task grammar; each rule, in addition to defining the possible constituent structure for phrases in an extended form of BNF, contains procedures for calculating attributes of phrases and factors used in rating phrases. These procedures may, in turn, call as subroutines any of the knowledge sources in the system. The attributes include acoustic attributes related to the input signal, syntactic attributes (e.g., mood and number), semantic attributes such as the representation of the meaning of the phrase, and discourse attributes for anaphora and ellipsis. Thus the phrase itself is the basic unit for integrating and controlling knowledge-source execution.

The interpreter of these rules (i.e., the Syntax module) is integrated with the scheduling components to define a high-level Control Strategy module. Like Hearsay-II and HWIM, this control module opportunistically executes the syntax rules to predict new phrases and words from a given phrase hypothesis and executes the word verifier to verify predicted words. This module maintains a data structure, the "parse-net," containing all the word and phrase hypotheses constructed, and the at-

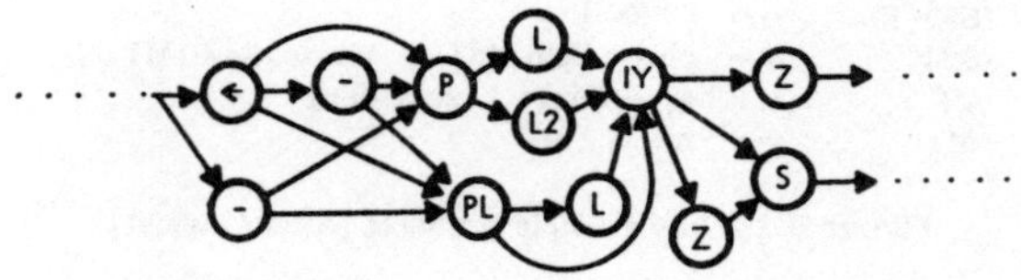

FIGURE 8. HARPY pronunciation network for the word "Please." [After LOWE80.]

tributes and factors associated with each hypothesis. This data structure is similar to a Hearsay-II blackboard restricted to the word and phrase levels. Like the blackboard, it serves to avoid redundant computation and facilitates the detection of possible island collisions.

As with Hearsay-II and HWIM, the SRI Control Strategy module is parameterized to permit a number of different strategies, such as top-down, bottom-up, island-driving, and left-to-right. Using a simulated word recognizer, SRI ran a series of experiments with several different strategies. One of the results, also substantiated by BBN experiments with HWIM, is that island-driving is inferior to some forms of left-to-right search. This appears to be in conflict with the Hearsay-II experimental results, which show island-driving clearly superior [LESS77a]. We believe the difference to be caused by the reliability of ratings of the initial islands: Both the HWIM and SRI experiments used single-word islands, but Hearsay-II uses multiword islands, which produce much higher reliability. (See the discussion at step 6 in Section 1.2 and in HAYE78b.) Single-word island-driving proved inferior in Hearsay-II as well.

2.3 CMU's HARPY System

In the systems described so far, knowledge sources are discernible as active components during the understanding process. However, if one looks at Hearsay-II, HWIM, and the SRI system in that order, there is clearly a progression of increasing integration of the KSs with the control structure. The HARPY system [LOWE76, LOWE80] developed at Carnegie-Mellon University is located at the far extreme of that dimension: Most of the knowledge is precompiled into a unified structure representing all possible utterances; a relatively simple interpreter then compares the spoken utterance against this structure to find the utterance that matches best. The motivation for this approach is to speed up the search so that a larger portion of the space may be examined explicitly. In particular, the hope is to avoid errors made when portions of the search space are eliminated on the basis of characteristics of small partial solutions; to this end, pruning decisions are delayed until larger partial solutions are constructed.

To describe HARPY, we describe the knowledge sources, their compilation, and the match (search) process. The *parameterization* and *segmentation* KSs are identical to those of Hearsay-II [GOLD77, GILL78]; these are not compiled into the network but, as in the other systems, applied to each utterance as it is spoken. As in Hearsay-II, the *syntax* is specified as a set of context-free production rules; HARPY uses the same task and grammar definitions. *Lexical* knowledge is specified as a directed pronunciation graph for each word; for example, Figure 8 shows the graph for the word "please." The nodes in the graph are names of the phonelike labels also generated by the labeler KS. A graph is intended to represent all possible pronunciations of the word. Knowledge about phonetic phenomena at *word junctures* is contained in a set of rewriting rules for the pronunciation graphs.

For a given task language, syntax and lexical and juncture knowledge are combined by a *knowledge compiler* program to form a single large network. First, the grammar is converted into a directed graph, the "word network," containing only terminal symbols (i.e., words); because of heuristics

⟨SENT⟩	::==	[⟨SS⟩]
⟨SS⟩	::==	please help ⟨M⟩ | please show ⟨M⟩ ⟨Q⟩
⟨Q⟩	::==	everything | something
⟨M⟩	::==	me | us

FIGURE 9. A tiny example grammar. [After LOWE80.]

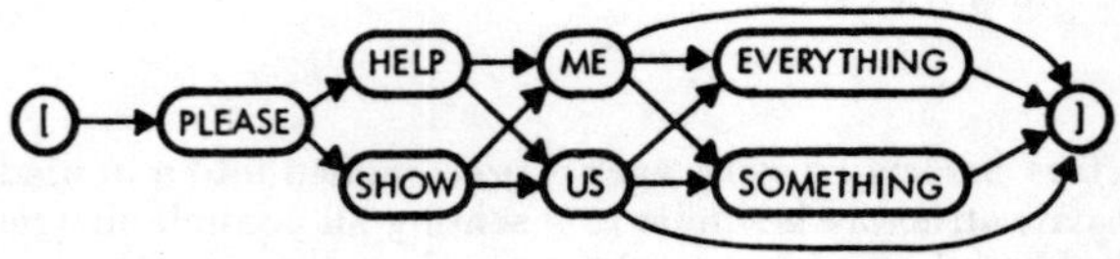

FIGURE 10. Word network for example language. [After LOWE80.]

used to compact this network, some of the constraint of the original grammar may be lost. Figure 9 shows a toy grammar, and Figure 10 the resulting word network. Next, the compiler replaces each word by a copy of its pronunciation graph, applying the word-juncture rules at all the word boundaries. Figure 11 shows part of the network for the toy example. The resulting network has the name of a segment label at each node. For the same 1011-word X05 language used by Hearsay-II, the network has 15,000 nodes and took 13 hours of DEC PDP-10(KL10) processing time to compile.

In the network each distinct path from the distinguished start node to the distinguished end node represents a sequence of segments making up a "legal" utterance. The purpose of the search is to find the sequence which most closely matches the segment sequence of the input spoken utterance. For any given labeled segment and any given node in the network, a primitive match algorithm can calculate a score for matching the node to the segment. The score for matching a sequence of nodes with a sequence of segments is just the sum of the corresponding primitive matches.

The search technique used, called *beam search*, is a heuristic form of dynamic programming, with the input segments processed one at a time from left to right and matched against the network. At the beginning of the *i*th step, the first $i - 1$ segments have been processed. Some number of nodes in the network are *active*; associated with each active node is a path to it from the start node and the total score of the match between that path and the first $i - 1$ segments of the utterance. All nodes

FIGURE 11. Partial final network for example language. [After LOWE80.]

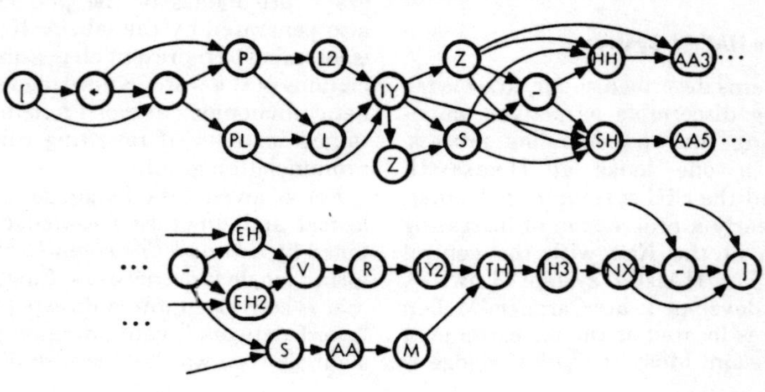

in the network that are successors of the active nodes are matched against the *i*th segment and become the new active nodes. The score for a new active node is the best path score that reaches the node at the *i*th segment, i.e., the sum of the primitive match at the segment plus the best path score to any of its predecessor nodes.

The best path score among all the new active nodes is taken as the target, and any new active nodes with path scores more than some threshold amount from the target are pruned away. This pruning rule is the heuristic heart of the search algorithm. It reduces the number of active nodes at each step and thus reduces the amount of processing time (and storage) needed in the search; typically only about 3 percent of the nodes in the net need to be matched. Note that the heuristic does not fix the number of active nodes retained at each step but allows it to vary with the density of competitors with scores near the best path. Thus in highly uncertain regions, many nodes are retained, and the search slows down; in places where one path is significantly better than most others, few competitors are kept, and the processing is rapid. The search strategy, therefore, is automatically cautious or decisive in response to the partial results. The threshold, i.e., the "beam width," is tuned ad hoc from test runs.

There are two major concerns about the extensibility of HARPY. First, the compilation process requires all knowledge to be represented in a highly stylized form; adding new kinds of knowledge strains the developer's ingenuity. So far, however, several kinds of knowledge have been added within the basic framework of expanding a node by replacing it with a graph. For example, as mentioned previously, phonetic phenomena at word junctures are handled. Also, the expected length of each segment is stored at each node and influences the match score. The second concern is with the size and compilation cost of the compiled network; both grow very large as the task language becomes more complex. There have been proposals that the word network not be expanded explicitly, but rather that the word pronunciation graphs be interpreted dynamically, as needed. An alternative response to this concern is that computer memory and processing costs continue to decline, so that using larger networks becomes increasingly feasible.

HARPY's novel structure is also interesting in its own right and is beginning to have effects beyond speech-understanding systems. Newell has done a speculative but thorough analysis of HARPY as a model for human speech understanding, using the production system formalism [NEWE80]; Rubin has successfully applied the HARPY structure to an image-understanding task [RUBI78].

3. SYSTEM PERFORMANCE AND ANALYSIS

3.1 Overall Performance of Hearsay-II

Overall performance of the Hearsay-II speech-understanding system at the end of 1976 is summarized in Table 2 in a form paralleling the goals given in Figure 6.

TABLE 2. HEARSAY-II PERFORMANCE

Number of speakers	One
Environment	Computer terminal room (>65 dB)
Microphone	Medium-quality, close-talking
System speaker-tuning	20–30 training utterances
Speaker adaptation	None required
Task	Document retrieval
Vocabulary	1011 words, with no selection for phonetic discriminability
Language constraints	Context-free semantic grammar, based on protocol analysis, with static branching factor of 10
Test data	23 utterances, brand-new to the system and run "blind." 7 words/utterance average, 2.6 seconds/utterance average, average fanout[a] of 40 (maximum 292)
Accuracy	9 percent sentence semantic error,[b] 19 percent sentence error (i.e., not word-for-word correct)
Computing resources	60 MIPSS (million instructions per second of speech) on a 36-bit PDP-10

[a] The *static branching factor* is the average number of words that can follow any initial sequence as defined by the grammar. The *fanout* is the number of words that can follow any initial sequence in the test sentences.

[b] An interpretation is *semantically correct* if the query generated for it by the SEMANT KS is identical to that generated for a sentence which is *word-for-word* correct.

Active development of the Hearsay-II speech system ceased at the end of 1976 with the conclusion of the speech-understanding program sponsored by DARPA [MEDR78, KLAT77]. Even though the configuration of KSs at that point was young, having been assembled in August 1976, the performance described in Table 2 comes close to meeting the ambitious goals, shown in Figure 6, established for the DARPA program in 1971 [NEWE73]. This overall performance supports our assertion that the Hearsay-II architecture can be used to integrate knowledge for resolving uncertainty. In the following sections we relate some detailed analyses of the Hearsay-II performance to the resolution of uncertainty. We finish with some comparison with the performances of the other systems described in Section 2.

3.2 Opportunistic Scheduling

In earlier KS configurations of the system, low-level processing (i.e., at the segment, syllable, and word levels) was not done in the serial, lock-step manner of steps 1, 4, and 5 of the example, that is, level-to-level, where each level is completely processed before work on the next higher level is begun. Rather, processing was opportunistic and data-directed as in the higher levels; as interesting hypotheses were generated at one level, they were immediately propagated to and processed by KSs operating at higher and lower levels. We found, however, that opportunistic processing at the lower levels was ineffective and harmful because the credibility ratings of hypotheses were insufficiently accurate to form hypothesis islands capable of focusing the search effectively. For example, even at the relatively high word level, the bottom-up hypotheses created by MOW include only about 75 percent of the words actually spoken; and the KS-assigned ratings rank each correct hypothesis on the average about 4.5 as compared with the 20 or so incorrect hypotheses that compete with it (i.e., which overlap it in time significantly). It is only with the word-sequence hypotheses that the reliability of the ratings is high enough to allow selective search.

Several experiments have shown the effectiveness of the opportunistic search. In one [HAYE77a] the opportunistic scheduling was contrasted with a strategy using no ordering of KS activations. Here, all KS precondition procedures were executed, followed by all KS activations they created; this cycle was repeated. For the utterances tested, the opportunistic strategy had a 29 percent error rate (word for word), compared with a 48 percent rate for the nonopportunistic. Also, the opportunistic strategy took less than half as much processing time.[13]

In another experiment [LESS77a] the island-driving strategy, which is opportunistic across the whole utterance, was compared with a left-to-right strategy, in which the high-level search was initiated from single-word islands in utterance-initial position. For the utterances tested, the opportunistic strategy had a 33 percent error rate as compared with 53 percent for the left-to-right; for those utterances correctly recognized by both strategies, the opportunistic one used only 70 percent as much processing time.

3.3 Use of Approximate Knowledge

In several places the Hearsay-II system uses approximate knowledge, as opposed to its more complete form also included in the system. The central notion is that even though the approximation increases the likelihood of particular decisions being incorrect, other knowledge can correct those errors, and the amount of computational resources saved by first using the approximation exceeds that required for subsequent corrections.

The organization of the POM and MOW KSs is an example. The bottom-up syllable and word-candidate generation scheme approximates WIZARD matching all words in the vocabulary at all places in the utterance, but in a fraction of the time. The errors show up as poor ratings of the can-

[13] The performance results given here and in the following sections reflect various configurations of vocabularies, grammars, test data, halting criteria, and states of development of the KSs and underlying system. Thus the absolute performance results of each experiment are not directly comparable to the performance reported in Section 3.1 or to the results of the other experiments.

grammar:	05	15	F
vocabulary			
S 250 words	err = 5.9 percent comp = 1.0 fanout = 10	err = 20.6 percent comp = 2.7 fanout = 17	err = 20.6 percent comp = 3.4 fanout = 27
M 500 words	err = 5.9 percent comp = 1.1 fanout = 18		
X 1011 words	err = 11.8 percent comp = 2.0 fanout = 36		

err = = semantic error rate
comp = = average ratio of execution time to that of S05 case, for correct utterances
fanout = = fanout of the test sentences (see note a of Table 2, Section 3.1)
N = 34 utterances

FIGURE 12. Hearsay-II performance under varying vocabularies and grammars.

didate words and as missing correct words among the candidates. The POM-MOW errors are corrected by applying WIZARD to the candidates to create good ratings and by having the PREDICT KS generate additional candidates.

Another example is the WORD-SEQ KS. Recall that it applies syntactic and acoustic-phonetic knowledge to locate sequences of words within the lattice of bottom-up words and statistical knowledge to select a few most credible sequences. The syntactic knowledge only approximates the full grammar, but takes less than 1 percent as much processing time to apply. The errors WORD-SEQ makes because of the approximation (i.e., generating some nongrammatical sequences) are corrected by applying the full grammatical knowledge of the PARSE KS, but only on the few, highly credible sequences WORD-SEQ identifies.

3.4 Adaptability of the Opportunistic Strategy

The opportunistic search strategy adapts automatically to changing conditions of uncertainty in the problem-solving process by changing the breadth of search. The basic mechanism for this is the interaction between the KS-assigned credibility ratings on hypotheses and scheduler-assigned priorities of pending KS activations. When hypotheses have been rated approximately equal, KS activations for their extension are usually scheduled together. Thus where there is ambiguity among competing hypotheses, the scheduler automatically searches with more breadth. This delays the choice among competing hypotheses until further information is brought to bear.

This adaptiveness works for changing conditions of uncertainty, whether it arises from the data or from the knowledge. The data-caused changes are evidenced by large variations in the numbers of competing hypotheses considered at various locations in an utterance, and by the large variance in the processing time needed for recognizing utterances. The results of changing conditions of knowledge constraint can be seen in Figure 12, which shows the results of one experiment varying vocabulary sizes and grammatical constraints.[14]

3.5 Performance Comparisons

It is extremely difficult to compare the reported performances of existing speech-understanding systems. Most have operated in different task environments and hence can apply different amounts of constraint

[14] Note that Figure 12 shows imperfect correlation between fanout and performance; compare, for example, X05 and SF. Fanout is an approximate measure of language complexity that reflects the average uncertainty between adjacent words. While X05 has a large fanout, it may be a simpler language to interpret than SF because most of the fanout is restricted to a few loci in the language, as opposed to the lower but more uniform uncertainty of SF.

GOAL: ACCEPT CONTINUOUS SPEECH FROM MANY COOPERATIVE SPEAKERS,

HARPY:	tested with	184	sentences from	3 Male, 2 Female	Speakers
Hearsay-II:		22		1 Male	
HWIM:		124		3 Male	
SDC:		54		1 Male	

GOAL: IN A QUIET ROOM, WITH A GOOD MIC, AND SLIGHT TUNING/SPEAKER,

HARPY:	in a computer terminal room, with a close-talking mic,	and	20	training sentences per speaker,
Hearsay-II:			20	
HWIM:			NO	
SDC:	in a quiet room, with a good mic,	and	NO	

GOAL: ACCEPTING 1000 WORDS, USING AN ARTIFICIAL SYNTAX & CONSTRAINING TASK,

HARPY:	1011 words, context-free grammar,	BF = 33	for document retrieval,
Hearsay-II:		BF = 33, 46	
HWIM:	1097 words, restricted ATN grammar,	BF = 196,	for travel planning,
SDC:	1000 words, context-free grammar,	BF = 105,	for data retrieval,

GOAL: YIELDING <10% SEMANTIC ERROR, IN A FEW TIMES REAL-TIME (=300 MIPSS)

HARPY:	yielding	5%	semantic error, using	28	million instructions per second of speech (MIPSS)
Hearsay-II:		9%,26%		85	
HWIM:		56%		500	
SDC:		76%		92	

FIGURE 13. Goals and performance for final (1976) DARPA systems. [After Lea79.]

from the task language to help the problem solving. Although some progress has been made [GOOD76, SOND78, BAHL78], there is no agreed-upon method for calibrating these differences. Also, the various systems use different speakers and recording conditions. And finally, none of the systems has reached full maturity; the amount that might be gained by further debugging and tuning is unknown, but often clearly substantial.

LEA79 contains an extensive description of the systems developed in the DARPA speech-understanding project and includes the best existing performance comparisons and evaluations. Figures 13 and 14, reproduced here from that report, show some comparison of the performances of Hearsay-II, HARPY, HWIM, and the SDC system [BERN76].[15]

The Hearsay-II and HARPY results are directly comparable, the two systems having been tested on the same tasks using the same test data. HARPY's performance here dominates Hearsay-II's in both accuracy and computation speed. And, in fact, HARPY was the only system clearly to meet and exceed the DARPA specifications (see Figure 6). It is difficult to determine the exact reasons for HARPY's higher accuracy, but we feel it is caused primarily by a combination of three factors:

(1) Because of its highly compiled efficiency, HARPY can afford to search a relatively large part of the search space. In particular, it can continue pursuing partial solutions even if they contain several low-rated segments (and its pruning threshold is explicitly set to ensure this). Thus HARPY is less prone to catastrophic errors, that is, pruning away the correct path. Hearsay-II, on the other hand, cannot afford to delay pruning decisions as long and thus is more likely to make such errors.

(2) Some knowledge sources are weaker in Hearsay-II than in HARPY. In particular, Hearsay-II's JUNCT KS has only a weak model of word juncture phenomena as compared with the more

[15] Performance of the SRI system is not included because that system was run only with a simulated bottom-end. Also, there are slight differences between the Hearsay-II performance shown in Figure 13 and that of Section 3.1; the former shows results from the official end of the DARPA project in September 1976, while the latter reflects some slight improvements made in the subsequent three months.

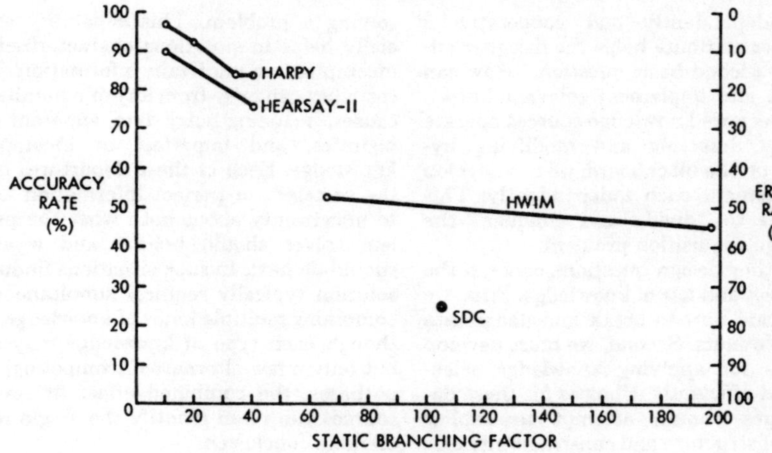

FIGURE 14 Effects of static branching factor on recognition error rate. [After LEA79.]

comprehensive and sophisticated juncture rules in HARPY. This disparity is an accident of the systems' development histories; there is no major conceptual reason why HARPY's juncture rules could not be employed by Hearsay-II.

(3) HARPY was debugged and tuned much more extensively than Hearsay-II (or any of the other DARPA SUSs, for that matter).This was facilitated by the lower processing costs for running tests. It was also helped by fixing the HARPY structure at an earlier point; Hearsay-II's KS configuration underwent a massive modification very late in the DARPA effort, as did HWIM's.

It seems clear that for a performance system in a task with a highly constrained and simply structured language, the HARPY structure is an excellent one. However, as we move to tasks that require more complex kinds of knowledge to constrain the search, we expect conceptual difficulties incorporating those kinds of knowledge into HARPY's simple integrated network representation.

4. CONCLUSIONS

Hearsay-II represents a new approach to problem solving that will prove useful in many domains other than speech. Thus far, however, we have focused on the virtues, and limitations, of Hearsay-II as a solution to the speech-understanding problem per se. In this section we consider what Hearsay-II suggests about problem-solving systems in general. To do so, we identify aspects of the Hearsay-II organization that facilitate development of "expert systems." Before concluding, we point out some apparent deficiencies of the current system that suggest avenues of further research. A more detailed discussion of these issues can be found in LESS77b.

4.1 Problem-Solving Systems

The designer of a knowledge-based problem-solving system faces several typical questions, many of which motivate the design principles evolved by Hearsay-II. The designer must first represent and structure the problem in a way that permits decomposition. A general heuristic for solving complex problems is to "divide and conquer" them. This requires methods to factor subproblems and to combine their eventual solutions. Hearsay-II, for example, divides the understanding problem in two ways: It breaks the total interpretation into separable hypotheses, and it modularizes different types of knowledge that can operate independently and cooperatively. This latter attribute helps the designer address the second basic question, "How can I acquire and implement relevant knowledge?" Because knowledge sources operate solely by detecting and modifying hypotheses on the blackboard, we can develop and implement each independently. This allows us to "divide and conquer" the knowledge acquisition problem.

Two other design questions concern the *description* and *use* of knowledge. First, we must decide how to break knowledge into executable units. Second, we must develop strategies for applying knowledge selectively and efficiently. Choices for these design issues should attempt to exploit sources of structure and constraint intrinsic to the problem domain and knowledge available about it. In the current context this means that a speech-understanding system should exploit many alternative types of speech knowledge to reduce uncertainty inherent in the signal. Moreover, the different types of knowledge should apply, ideally, in a best-first manner. That is, the most credible hypotheses should stimulate searches for the most likely adjoining hypotheses first. To this end, the Hearsay-II focusing scheduler considers the quality of hypotheses and potential predictions in each temporal interval and then selectively executes only the most marginally productive KS actions. Accomplishing this type of control required several new sorts of mechanisms. These included explicit interlinked hypothesis representations, declarative descriptions of KS stimulus and response frames, a dynamic problem state description, and a prioritized schedule of pending KS instantiations.

4.2 Specific Advantages of Hearsay-II as a Problem-Solving System

This paper has covered an extensive set of issues and details. From these we believe the reader should have gained an appreciation of Hearsay-II's principal benefits, summarized briefly as follows.

Multiple Sources of Knowledge

Hearsay-II provides a framework for diverse types of knowledge to cooperate in solving a problem. This capability especially helps in situations characterized by incomplete or uncertain information. Uncertainty can arise from any of a number of causes, including noisy data, apparent ambiguities, and imperfect or incomplete knowledge. Each of these departures from the certainty of perfect information leads to uncertainty about both what the problem solver should believe and what it should do next. In such situations finding a solution typically requires simultaneously combining multiple kinds of knowledge. Although each type of knowledge may rule out only a few alternative (competing) hypotheses, the combined effect of several sources can often identify the single most credible conclusion.

Multiple Levels of Abstraction

Solving problems in an intelligent manner often requires using descriptions at different levels of abstraction. After first finding an approximate or gross solution, a problem solver may work quickly toward a refined, detailed solution consistent with the rough solution. In its use of multiple levels of abstraction, Hearsay-II provides rudimentary facilities for such variable-granularity reasoning. In the speech task particularly, the different levels correspond to separable domains of reasoning. Hypotheses about word sequences must satisfy the constraints of higher level syntactic phrase-structure rules. Once these are satisfied, testing more detailed or finely tuned word juncture relations would be justified. Of course the multiple levels of abstraction also support staged decision making that proceeds from lower level hypotheses up to higher levels. Levels in such bottom-up processing support a different type of function, namely, the sharing of intermediate results, discussed separately in the following paragraph.

Shared Partial Solutions

The blackboard and hypothesis structures allow the knowledge sources to represent and share partial results. This proves especially desirable for complex problems where no a priori knowledge can reliably foretell the best sequence of necessary decisions. Different attempts to solve the same problem may require solving identical subproblems. In the speech domain these problems correspond to comparable hypotheses (same level, type, time). Hearsay-II provides capabilities for the KSs to recognize a hypothesis of interest and to incorporate it into alternative competing hypotheses at higher levels. Subsequent changes to the partial result then propagate to all of the higher level constructs that contain it.

Independent Knowledge Sources Limited to Data-Directed Interactions

Separating the diverse sources of knowledge into independent program modules provides several benefits. Different people can create, test, and modify KSs independently. In addition to the ordinary benefits of modularity in programming, this independence allows human specialists (e.g., phoneticians, linguists) to operationalize their diverse types of knowledge without concern for the conceptual framework and detailed behavior of other possible modules. Although the programming style and epistemological nature of several KSs may vary widely, Hearsay-II provides for all of them a single uniform programming environment. This environment constrains the KSs to operate in a data-directed manner—reading hypotheses from the blackboard when situations of interest occur, processing them to draw inferences, and recording new or modified hypotheses on the blackboard for others to process further. This paradigm facilitates problem-oriented interactions while minimizing complicated and costly design interactions.

Incremental Formation of Solutions

Problem solving in Hearsay-II proceeds incrementally through the accretion and integration of partial solutions. KSs generate hypotheses based on current data and knowledge. By integrating adjacent and consistent hypotheses into larger composites, the system develops increasingly credible and comprehensive partial solutions. These in turn stimulate focused efforts that drive the overall system toward the final goal, one most credible interpretation spanning the entire interval of speech. By allowing information to accumulate in this piecemeal fashion, Hearsay-II provides a convenient framework for heuristic problem solving. Diverse heuristic methods can contribute various types of assistance in the effort to eliminate uncertainty, to recognize portions of the sequence, and to model the speaker's intentions. Because these diverse methods exist in the form of independent, cooperating KSs, each addition to the current problem solution consists simply of an update to the blackboard.

Opportunistic Problem-Solving Behavior

Whenever good algorithms do not exist for solving a problem, we must apply heuristic methods or "rules-of-thumb" to search for a solution. In problems where a large number of data exist to which a large number of alternative heuristics potentially apply, we need to choose each successive action carefully. We refer to a system's ability to exploit selectively its best data and most promising methods as "opportunistic" problem solving [NII78, HAYE79b]. Hearsay-II developed several mechanisms to support such opportunistic behavior. In particular, its focus policies and prioritized scheduling allocate computation resources first to those KSs that exploit the most credible hypotheses, promise the most significant increments to the solution, and use the most reliable and inexpensive methods. Similar needs to focus intelligently will arise in many comparably rich and complex problem domains.

Experimentation in System Development

Whenever we attempt to solve a previously unsolved problem, the need for experimentation arises. In the speech-understanding task, for example, we generated several different types of KSs and experimentally tested a variety of alternative system configurations (specific sets of KSs) [LESS77b]. A solution to the overall problem depended on both developing powerful individual KSs and organizing multiple KSs to cooperate effectively to reduce uncertainty. These requirements necessitated a trial-and-error evaluation of alternative system designs. Throughout these explorations, the basic Hearsay-II structure proved robust and sufficient. Alternative configurations were constructed with relative ease by inserting or removing specific KSs. Moreover, we could test radically different high-level control concepts (e.g., depth-first versus breadth-first versus left-to-right searches) simply by changing the focus policy KS. The need for this kind of flexibility will probably arise in many future state-of-the-art problem-solving tasks. To support this flexibility, systems must be able to apply the same KSs in different orders and to schedule them according to varying selection criteria. These requirements directly motivate KS data-directed independence, as well as autonomous scheduling KSs that can evaluate the probable effects of potential KS actions. Because it supports these needs, Hearsay-II provides an excellent environment for experimental research and development in speech and other complex tasks.

4.3 Disadvantages of the Hearsay-II Approach

We can identify two different but related weaknesses of the Hearsay-II approach to problem solving. One weakness derives from the system's generality, and the other concerns its computational efficiency. Each of these is considered briefly in turn.

Generality Impedes Specialization and Limits Power

The Hearsay-II approach suggests a very general problem-solving paradigm. Every inference process reads data from the blackboard and places a new hypothesis also on the blackboard. Thus blackboard accesses mediate each decision step. While this proved desirable for structuring communications between different KSs, it proved undesirable for most intermediate decision tasks arising within a single KS. Most KSs employed private, stylized internal data structures different from the single uniform blackboard links and hypotheses. For example, the word recognizer used specialized sequential networks, whereas the word sequence recognizer exploited a large bit-matrix of word adjacencies. Each KS also stored intermediate results, useful for its own internal searches, in appropriately distinctive data structures. Attempts to coerce these specialized activities into the general blackboard-mediated style of Hearsay-II either failed completely or caused intolerable performance degradation [LESS77b].

Interpretive Versus Compiled Knowledge

Hearsay-II uses knowledge interpretively. That is, it actively evaluates alternative actions, chooses the best for the current situation, and then applies the procedure associated with the most promising KS instantiation. Such deliberation takes time and requires many fairly sophisticated mechanisms; its expense can be justified whenever an adequate, explicit algorithm does not exist for the same task. Whenever such an algorithm emerges, equal or greater performance and efficiency may be obtained by compiling the algorithm and executing it directly. For example, recognizing restricted vocabulary and grammatical spoken sentences from limited syntax can now be accomplished faster by techniques other than those in Hearsay-II. As described in Section 2.3, by compiling all possible interlevel substitutions (sentence to phrase to word to phone to segment) into one enormous finite-state Markov network, the HARPY system uses a modified dynamic programming search to find the one network path that most closely approximates the segmented speech signal. This type of systematic, compiled, and broad search becomes increasingly desirable as problem-solving knowledge improves. Put another way, once a satisfactory specific method for solving any problem is found, the related procedure can be "algorithmetized," compiled, and applied repetitively. In such a case the flexibility of a system like Hearsay-II may no longer be needed.

4.4 Other Applications of the Hearsay-II Framework

Both the advantages and disadvantages of Hearsay-II have stimulated additional research. Several researchers have applied the general framework to problems outside the speech domain, and others have begun to develop successors to the Hearsay-II system. We will briefly discuss one of these new applications and then mention the other types of activities underway.

Although the Hearsay-II framework developed around an understanding task, B. and F. Hayes-Roth et al. have extended many of its principal features to develop a model of planning [HAYE79b, HAYE79c]. While understanding tasks require "interpretive" or "analytic" processes, planning belongs to a complementary set of "generative" or "synthetic" activities. The principal features of the Hearsay-II system which make it attractive as a problem-solving model for speech understanding also suggest it as a model of planning.

The planning application shares all the principal features of the Hearsay-II system summarized in Section 4.2, but, as Figure 15 suggests, the planning model differs from the Hearsay-II framework in several ways. In particular, the designers found it convenient to distinguish five separate blackboard "planes," reflecting five qualitatively different sorts of decisions. The Plan plane corresponds most closely to Hearsay-II's single blackboard, holding the decisions that combine to form a solution to the planning problem, i.e., what low-level operations can be aggregated to achieve the high-level outcomes of the plan. These kinds of decisions in generative tasks can be thought of as the dual of the successively higher level, more aggregated hypotheses constituting the blackboard for interpretation tasks. In the speech task, corresponding hypotheses ex-

FIGURE 15 The planning blackboard and the actions of illustrative knowledge sources. [From HAYE79b.]

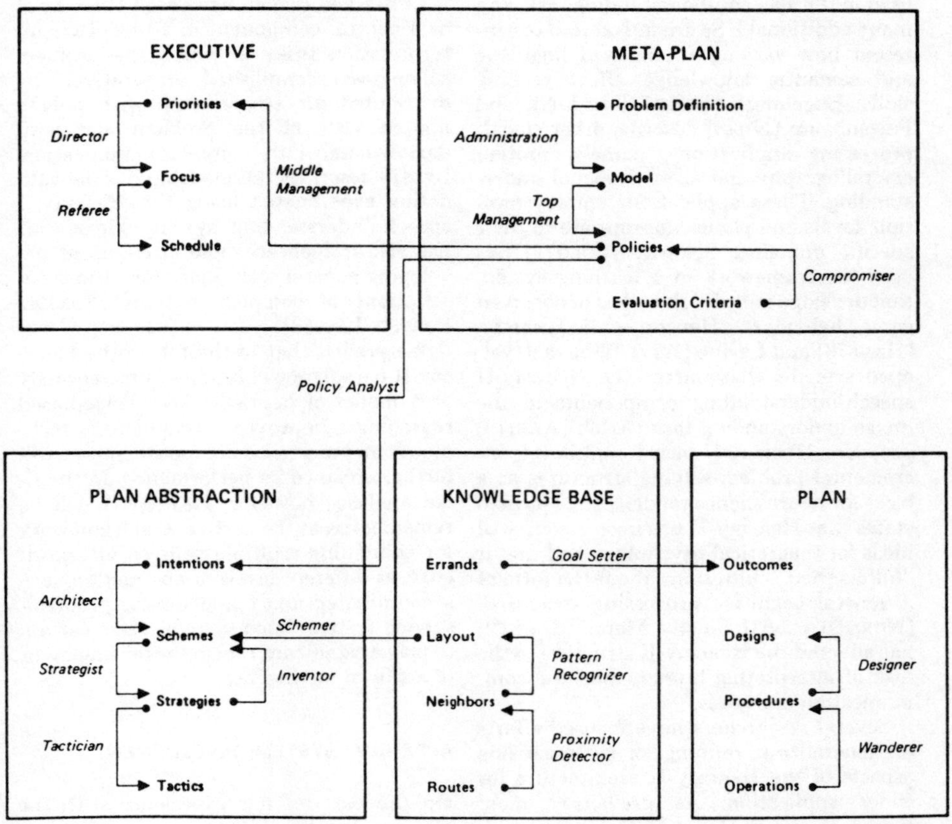

press how low-level segments and phones can be aggregated to form the high-level phrases and sentences intended by the speaker. The other four planes of the planning blackboard hold intermediate decisions that enter into the planning process in various ways. For example, based on the Hearsay-II experience with selective attention strategies, resource allocation strategies were formalized and associated explicitly with an Executive plane.

Although the planning model is the only current application of the Hearsay-II framework to generative tasks, several interesting applications that transfer the approach to other interpretation problems have been made. Rumelhart [RUME76] has proposed to apply the Hearsay-II framework to model human reading behavior. In this application only one blackboard plane is used, the levels closely approximate those used in the speech-understanding task, and many additional KSs are introduced to represent how varying amounts of linguistic and semantic knowledge affect reading skills. Engelmore [ENGE77] and Nii and Feigenbaum [NII78] describe other signal-processing applications, namely, protein crystallography and acoustic signal understanding. These applications employ multiple levels and planes appropriate to their specific domains. Soloway [SOLO77] has used the framework in a learning system that develops multilevel models of observed game behaviors. Hanson and Riseman [HANS78] and Levine [LEVI78] have developed systems that mirror the Hearsay-II speech-understanding components in the image-understanding task. Arbib [ARBI79] proposes Hearsay-II-based multilevel, incremental problem-solving structures as a basis for neuroscience models, and Norman states that Hearsay-II has been a source of ideas for theoretical psychology and that it "fulfills [his]. . .intuitions about the form of a general cognitive processing structure" [NORM80, p. 383]. Finally, Mann [MANN79] has adapted the Hearsay-II structure to the task of interpreting human–machine communication dialogues.

Several researchers have focused efforts on generalizing, refining, or systematizing aspects of the Hearsay-II architecture for wider application. As previously mentioned, B. and F. Hayes-Roth have formalized some aspects of meta-planning and executive control and have treated this type of problem solving within one uniform framework. Nii [NII79] has developed a system that assists a programmer in developing a new special-purpose variant of a Hearsay-II system suitable for some particular new task. Balzer and others [BALZ80] have implemented a more formalized, domain-independent version of Hearsay-II and are applying it to an automatic-programming-like task. This system uses one blackboard for interpretation and another for scheduling decisions, in a manner akin to that proposed for the Executive decisions in the Hayes-Roth planning system. In a similar way, Stefik uses three distinct planes to record the plan, meta-plan, and executive decisions arising in a system that incrementally plans genetic experiments [STEF80].

Lesser and Erman have used Hearsay-II as a central component in a model for interpretation tasks in which the problem solving is accomplished cooperatively by distributed processors, each with only a limited view of the problem and with narrow-bandwidth intercommunication; LESS79 describes the model and some validating experiments using the Hearsay-II speech-understanding system. Hearsay-II has also influenced some attempts at developing general techniques for formal descriptions of complex systems [FOX79a, FOX79b, LESS80].

We predict that in the future the Hearsay-II paradigm will be chosen increasingly as a model of heuristic, knowledge-based reasoning. Improved compilation techniques and increased computing power will further enhance its performance. In the final analysis, however, Hearsay-II will be remembered as the first general framework for combining multiple sources of knowledge, at different levels of abstraction, into a coordinated and opportunistic problem-solving system. Such systems seem certain to play a significant role in the development of artificial intelligence.

APPENDIX. SYSTEM DEVELOPMENT

On the basis of our experience with the Hearsay-I system [REDD73a, REDD73b], at the beginning of the Hearsay-II effort in 1973 we expected to require and evolve types of knowledge and interaction patterns whose details could not be anticipated. Because of this, the development of the system was marked by much experimentation and redesign. This uncertainty characterizes the development of knowledge-based systems. Instead of designing a specific speech-understanding system, we considered Hearsay-II as a model for a class of systems and a framework within which specific configurations of that general model could be constructed and studied [LESS75, ERMA75].

On the basis of this approach a high-level programming system was designed to provide an environment for programming knowledge sources, configuring groups of them into systems, and executing them. Because KSs interact via the blackboard (triggering on patterns, accessing hypotheses, and making modifications) and the blackboard is uniformly structured, KS interactions are also uniform. Thus one set of facilities can serve all KSs. Facilities are provided for

- defining levels on the blackboard,
- configuring groups of KSs into executable systems,
- accessing and modifying hypotheses on the blackboard,
- activating and scheduling KSs,
- debugging and analyzing the performance of KSs.

These facilities collectively form the Hearsay-II "kernel." One can think of the Hearsay-II kernel as a high-level system for programming speech-understanding systems of a type conforming to the underlying Hearsay-II model.

Hearsay-II is implemented in the SAIL programming system [REIS76], an Algol-60 dialect with a sophisticated compile-time macro facility as well as a large number of data structures (including lists and sets) and control modes which are implemented fairly efficiently. The Hearsay-II kernel provides a high-level environment for KSs at compile-time by extending SAIL's data types and syntax through declarations of procedure calls, global variables, and macros. This extended SAIL provides an explicit structure for specifying a KS and its interaction with other KSs (through the blackboard). The high-level environment also provides mechanisms for KSs to specify (usually in nonprocedural ways) information used by the kernel when configuring a system, scheduling KS activity, and controlling researcher interaction with the system.

The knowledge in a KS is represented using SAIL data structures and code, in whatever stylized form the KS developer chooses. The kernel environment provides the facilities for structuring the interface between this knowledge and other KSs, via the blackboard. For example, the syntax KS contains a grammar for the specialized task language to be recognized; this grammar is coded in a compact network form. The KS also contains procedures for searching this network, for example, to parse a sequence of words. The kernel provides facilities (1) for triggering this KS when new word hypotheses appear on the blackboard, (2) for the KS to read those word hypotheses (in order to find the sequence of words to parse), and (3) for the KS to create new hypotheses on the blackboard, indicating the structure of the parse.

Active development of Hearsay-II extended for three years. About 40 KSs were developed, each a one- or two-person effort lasting from two months to three years. The KSs range from about 5 to 100 pages of source code (with 30 pages typical), and each KS has up to about 50 kbytes of information in its local database.

The kernel is about 300 pages of code, roughly one-third of which is the declarations and macros that create the extended environment for KSs. The remainder of the code implements the architecture: primarily activation and scheduling of KSs, maintenance of the blackboard, and a variety of other standard utilities. During the three years of active development, an average of about two full-time-equivalent research programmers were responsible for the implementation, modification, and maintenance of the kernel. Included during this period were a half-dozen major reimplementations and scores of minor ones; these changes usually were specializations or selective optimizations, designed as experience with the system led to a better understanding of the usage of the various constructs. During this same period about eight full-time-equivalent researchers were using the system to develop KSs.

Implementation of the first version of the kernel began in the autumn of 1973, and was completed by two people in four months. The first major KS configuration, though incomplete, was running in early 1975. The first complete configuration, "C1," ran in January 1976. This configuration had very poor performance, with more than 90 percent sentence errors over a 250-word vocabulary. Experience with this configuration led to a substantially different KS configuration, "C2," completed in September 1976. C2 is the configuration described in this paper.

Implementing a general framework has a potential disadvantage: the start-up cost is relatively high. However, if the framework is suitable, it can be used to explore different configurations within the model more easily than if each configuration were built in an ad hoc manner. Additionally, a natural result of the continued use of any high-level system is its improvement in terms of enhanced facilities, increased stability, reliability, and efficiency, and greater familiarity on the part of the researchers using it.

Hearsay-II has been successful in this respect; we believe that the total cost of creating the high-level system and using it to develop KS configurations C1 and C2 (and intermediate configurations) was less than it would have been to generate them in an ad hoc manner. It should be stressed that the construction of even one configuration is itself an experimental and evolving process. The high-level programming system provides a framework, both conceptual and physical, for developing a configuration in an incremental fashion. The speed with which C2 was developed is some indication of the advantage of this system-design approach. A more detailed description of the development philosophy and tools can be found in ERMA78, and a discussion of the relationships between the C1 and C2 configurations can be found in LESS77b.

ACKNOWLEDGMENTS

The success of the Hearsay-II project depended on many persons, especially the following members of the Carnegie-Mellon University Computer Science Department "Speech Group": Christina Adam, Mark Birnbaum, Robert Cronk, Richard Fennell, Mark Fox, Gregory Gill, Henry Goldberg, Gary Goodman, Bruce Lowerre, Paul Masulis, David McKeown, Jack Mostow, Linda Shockey, Richard Smith, and Richard Suslick. Daniel Corkill, David Taylor, and the reviewers made helpful comments on early drafts of this paper.

Figure 1 is adapted from A. Newell, "A tutorial on speech understanding systems," in *Speech recognition: Invited papers of the IEEE symposium*, D. R. Reddy, Ed., Academic Press, New York, 1975. Figure 6 is adapted from M. F. Medress et al., "Speech understanding systems, Report of a steering committee," *Artif. Intell.* **9** (1978). Figure 7 is reprinted from J. J. Wolf and W. A. Woods, "The HWIM speech understanding system," in *Trends in speech recognition*, W. A. Lea, Ed., © 1980, by permission of Prentice-Hall, Inc., Englewood Cliffs, N.J. Figures 8–11 are reprinted from B. T. Lowerre and R. Reddy, "The HARPY speech understanding system," in *Trends in speech recognition*, W. A. Lea, Ed., © 1980, by permission of Prentice-Hall, Inc., Englewood Cliffs, N.J. Figure 15 originally appeared in B. Hayes-Roth and F. Hayes-Roth, "A cognitive model of planning," *Cognitive science*, 1979, **3** 275–310. Ablex Publishing Corporation, Norwood, N.J.

REFERENCES

ARBI79 ARBIB, M. A., AND CAPLAN, D. "Neurolinguistics must be computational," *Behav. Brain Sci.* **2**, 3 (1979).

BAHL76 BAHL, L. R., BAKER, J. K., COHEN, P. S., DIXON, N. R., JELINEK, F., MERCER, R. L., AND SILVERMAN, H. F. "Preliminary results on the performance of a system for the automatic recognition of continuous speech," in *1976 IEEE Int. Conf. Acoustics, Speech, and Signal Processing*, Philadelphia, Apr. 1976, pp. 425–433.

BAHL78 BAHL, L. R., BAKER, J. K., COHEN, P. S., COLE, A. G., JELINEK, F., LEWIS, B. L., AND MERCER, R. L. "Automatic recognition of continuously spoken sentences from a finite state grammar," in *Proc. IEEE Int. Conf. Acoustics, Speech, and Signal Processing*, Tulsa, Okla., Apr. 1978, pp. 418–421.

BALZ80 BALZER, R., ERMAN, L. D., AND WILLIAMS, C. *Hearsay-III: A domain-independent base for knowledge-based problem-solving*, Tech. Rep., USC/Information Sciences Institute, Marina del Rey, Calif., 1980. To appear.

BARN77 BARNETT, J. A., AND BERNSTEIN, M. I. *Knowledge-based systems: A tutorial*, Tech. Rep. TM-(L)-5903/000/00 (NTIS: AD/A-044-883), System Development Corp., Santa Monica, Calif., June 1977.

BERN76 BERNSTEIN, M. I. *Interactive systems research: Final report to the Director, Advanced Research Projects Agency*, Tech. Rep. TM-5243/006/00, System Development Corp., Santa Monica, Calif., Sept. 1976.

BURT76 BURTON, R. R. *Semantic grammar: An engineering technique for constructing natural language understanding systems*, Tech. Rep. BBN Rep. No. 3453, Bolt Beranek and Newman, Cambridge, Mass., 1976.

CMU77 CMU COMPUTER SCIENCE SPEECH GROUP. *Summary of the CMU five-year ARPA effort in speech understanding research*, Tech. Rep., Computer Science Dep., Carnegie-Mellon Univ., Pittsburgh, Pa., 1977.

CRON77 CRONK, R. "Word pair adjacency acceptance procedure in Hearsay-II," in CMU77, pp. 15–16.

DUDA78 DUDA, R. O., HART, P. E., NILSSON, N. J., AND SOUTHERLAND, G. L. "Semantic network representation in rule-based inference systems," in *Pattern-directed inference systems*, D. A. Waterman and F. Hayes-Roth, Eds., Academic Press, New York, 1978, pp. 203–222.

ENGE77 ENGELMORE, R. S., AND NII, H. P. *A knowledge-based system for the interpretation of protein X-ray crystallographic data*, Tech. Rep. Stan-CS-77-589, Computer Science Dep., Stanford Univ., Stanford, Calif., 1977.

ERMA75 ERMAN L. D., AND LESSER, V. R. "A multi-level organization for problem solving using many diverse cooperating sources of knowledge," in *Proc. 4th Int. Jt. Conf. Artificial Intelligence*, Tbilisi, USSR, 1975, pp. 483–490.

ERMA78 ERMAN, L. D., AND LESSER, V. R. "System engineering techniques for artificial intelligence systems," in *Computer vision systems*, A. Hanson and E. Riseman, Eds., Academic Press, New York, 1978, pp. 37–45.

ERNS69 ERNST, G., AND NEWELL, A. *GPS: A case study in generality and problem solving*, Academic Press, New York, 1969.

FEIG71 FEIGENBAUM, E. A., BUCHANAN, B. G., AND LEDERBERG, J. "On generality and problem solving: A case study using the DENDRAL program," in *Machine intelligence 6*, D. Michie, Ed., Edinburgh Univ. Press, Edinburgh, Scotland, 1971.

FEIG77 FEIGENBAUM, E. A. "The art of artificial intelligence: Themes and case studies of knowledge engineering," in *Proc. 5th Int. Jt. Conf. Artificial Intelligence*, Cambridge, Mass., 1977, pp. 1014–1029.

FOX77 FOX, M. S., AND MOSTOW, D. J. "Maximal consistent interpretations of errorful data in hierarchically modelled domains," in *Proc. 5th Int. Jt. Conf. Artificial Intelligence*, Cambridge, Mass., 1977, pp. 65–171.

FOX79a FOX, M. S. "An organizational view of distributed systems," in *Proc. Int. Conf. Systems and Cybernetics*, Denver, Colo., Oct. 1979.

FOX79b FOX, M. S. *Organization structuring: Designing large, complex software*, Tech. Rep. CMU-CS-79-115, Computer Science Dep., Carnegie-Mellon Univ., Pittsburgh, Pa., 1979.

GILL78 GILL, G., GOLDBERG, H., REDDY, R., AND YEGNANARAYANA, B. *A recursive segmentation procedure for continuous speech*, Tech. Rep. CMU-CS-78-134, Computer Science Dep., Carnegie-Mellon Univ., Pittsburgh, Pa., May 1978.

GOLD77 GOLDBERG, H., REDDY, R., AND GILL, G. "The ZAPDASH parameters, feature extraction, segmentation, and labeling for speech understanding systems," in CMU77, pp. 10–11.

GOOD76 GOODMAN, G. *Analysis of languages for man-machine voice communication*, Tech. Rep., Computer Science Dep., Carnegie-Mellon Univ., Pittsburgh, Pa., May 1976.

HANS78 HANSON, A. R., AND RISEMAN, E. M. "VISIONS: A computer system for interpreting scenes," in *Computer vision systems*, A. Hanson and E. Riseman, Eds., Academic Press, New York, 1978, pp. 303–333.

HARR74 HARRIS, L. R. "The heuristic search under conditions of error," *Artif. Intell.* **5**, 3 (1974), 217–234.

HAYE75 HAYES-ROTH, F., AND MOSTOW, D. J. "An automatically compilable recognition network for structured patterns," in *Proc. 4th Int. Jt. Conf. Artificial Intelligence*, Tbilisi, USSR, 1975, pp. 246–252.

HAYE77a HAYES-ROTH, F., AND LESSER, V. R. "Focus of attention in the Hearsay-II system," in *Proc. 5th Int. Jt. Conf. Artificial Intelligence*, Cambridge, Mass., 1977, pp. 27–35.

HAYE77b HAYES-ROTH, F., ERMAN, L. D., FOX, M., AND MOSTOW, D. J. "Syntactic processing in Hearsay-II," in CMU77, pp. 16–18.

HAYE77c HAYES-ROTH, F., GILL, G., AND MOSTOW, D. J. "Discourse analysis and task performance in the Hearsay-II speech understanding system," in CMU77, pp. 24–28.

HAYE77d HAYES-ROTH, F., LESSER, V. R., MOSTOW, D. J., AND ERMAN, L. D. "Policies for rating hypotheses, halting, and selecting a solution in Hearsay-II," in CMU77, pp. 19–24.

HAYE78a HAYES-ROTH, F., WATERMAN, D. A., AND LENAT, D. B. "Principles of pattern-directed inference systems," in *Pattern-directed inference systems*, D. A. Waterman and F. Hayes-Roth, Eds., Academic Press, New York, 1978.

HAYE78b HAYES-ROTH, F. "The role of partial and best matches in knowledge systems," in *Pattern-directed inference systems*, D. A. Waterman and F. Hayes-Roth, Eds., Academic Press, New York, 1978.

HAYE79a HAYES-ROTH, B., AND HAYES-ROTH, F. *Cognitive processes in planning*, Tech. Rep. R-2366-ONR, The RAND Corp., Santa Monica, Calif., 1979.

HAYE79b HAYES-ROTH, B., AND HAYES-ROTH, F. "A cognitive model of planning," *Cognitive Sci.* **3** (1979), 275–310.

HAYE79c HAYES-ROTH, B., HAYES-ROTH, F., ROSENSCHEIN, S., AND CAMMARATA, S. "Modeling planning as an incremental opportunistic process," in *Proc. 6th Int. Jt.*

Conf. Artificial Intelligence, Tokyo, 1979, pp. 375–383.

HAYE80 HAYES-ROTH, F. "Syntax, semantics, and pragmatics in speech understanding," in *Trends in speech recognition*, W. A. Lea, Ed., Prentice-Hall, Englewood Cliffs, N.J., 1980.

ITAK75 ITAKURA, F. "Minimum prediction residual principle applied to speech recognition," *IEEE Trans. Acoust., Speech, Signal Proc.* **23** (1975), 67–72.

KLAT77 KLATT, D. H. "Review of the ARPA speech understanding project," *J. Acoust. Soc. Am.* **62** (Dec. 1977), 1345–1366.

LEA79 LEA, W. A., AND SHOUP, J. E. *Review of the ARPA SUR Project and survey of current technology in speech understanding*, Final Rep., Office of Naval Research Contract No. N00014-77-C-0570, Speech Communications Research Lab., Los Angeles, Calif., Jan. 1979.

LEA80 LEA, W. A., ED. *Trends in speech recognition*, Prentice-Hall, Englewood Cliffs, N.J., 1980.

LESS75 LESSER, V. R., FENNELL, R. D., ERMAN, L. D., AND REDDY, D. R. "Organization of the Hearsay-II speech understanding system," *IEEE Trans. Acous., Speech, Signal Proc.* **23** (1975), 11–23.

LESS77a LESSER, V. R., HAYES-ROTH, F., BIRNBAUM, M., AND CRONK, R. "Selection of word islands in the Hearsay-II speech understanding system," in *Proc. IEEE Int. Conf. Acoustics, Speech, and Signal Processing*, Hartford, Conn., 1977, pp. 791–794.

LESS77b LESSER, V. R., AND ERMAN, L. D. "A retrospective view of the Hearsay-II architecture," in *Proc. 5th Int. Joint Conf. Artificial Intelligence*, Cambridge, Mass., 1977, pp. 790–800.

LESS79 LESSER, V. R., AND ERMAN, L. D. "An experiment in distributed interpretation," in *1st Int. Conf. Distributed Computing Systems*, IEEE Computer Society, Huntsville, Ala., Oct. 1979, pp. 553–571.

LESS80 LESSER, V. R., PAVLIN, J., AND REED, S. *Quantifying and simulating the behavior of knowledge-based systems*, Tech. Rep., Dep. Computer and Information Sciences, Univ. Massachusetts, Amherst, Mass., 1980.

LEVI78 LEVINE, M. D. "A knowledge-based computer vision system," in *Computer vision systems*, A. Hanson and E. Riseman, Eds., Academic Press, New York, 1978, pp. 335–352.

LOWE76 LOWERRE, B. T. *The HARPY speech recognition system*, Ph.D. thesis, Computer Science Dep., Carnegie-Mellon Univ., Pittsburgh, Pa., 1976.

LOWE80 LOWERRE, B. T., AND REDDY, R. "The HARPY speech understanding system," in *Trends in speech recognition*, W. A. Lea, Ed., Prentice-Hall, Englewood Cliffs, N.J., 1980, Chap. 15.

LOWR80 LOWRANCE, J. *Dependence-graph models of evidential support*, Ph.D. thesis, Dep. Computer and Information Sciences, Univ. Massachusetts, 1980 (forthcoming).

MANN79 MANN, W. C. "Design for dialogue comprehension," in *17th Ann. Meeting Assoc. Computational Linguistics*, La Jolla, Calif., Aug. 1979.

MCKE77 MCKEOWN, D. M. "Word verification in the Hearsay-II speech understanding system," in *Proc. IEEE Int. Conf. Acoustics, Speech, and Signal Processing*, Hartford, Conn., 1977, pp. 795–798.

MEDR78 MEDRESS, M. F., COOPER, F. S., FORGIE, J. W., GREEN, C. C., KLATT, D. H., O'MALLEY, M. H., NEUBURG, E. P., NEWELL, A., REDDY, D. R., RITEA, B., SHOUP-HUMMEL, J. E., WALKER, D. E., AND WOODS, W. A. "Speech understanding systems: Report of a steering committee," *Artif. Intell.* **9** (1978), 307–316.

MOST77 MOSTOW, D. J. "A halting condition and related pruning heuristic for combinatorial search," in CMU77, pp. 158–166.

NEWE69 NEWELL, A. "Heuristic programming: Ill-structured problems," in *Progress in operations research 3*, J. Aronofsky, Ed., Wiley, New York, 1969, pp. 360–414.

NEWE73 NEWELL, A., BARNETT, J., FORGIE, J., GREEN, C., KLATT, D., LICKLIDER, J. C. R., MUNSON, J., REDDY, R., AND WOODS, W. *Speech understanding systems: Final report of a study group*, North-Holland, Amsterdam, 1973.

NEWE75 NEWELL, A., "A tutorial on speech understanding systems," in *Speech recognition: Invited papers of the IEEE symposium*, D. R. Reddy, Ed., Academic Press, New York, 1975, pp. 3–54.

NEWE77 NEWELL, A., MCDERMOTT, J., AND FORGIE, C. *Artificial intelligence: A self-paced introductory course*, Computer Science Dep., Carnegie-Mellon Univ., Pittsburgh, Pa., 1977.

NEWE80 NEWELL, A. "HARPY, production systems and human cognition," in *Perception and production of fluent speech*, R. Cole, Ed., L. Erlbaum, Hillsdale, N.J., 1980, Chap. 11.

NII78 NII, H. P., AND FEIGENBAUM, E. A. "Rule-based understanding of signals," in *Pattern-directed inference systems*, D. A. Waterman and F. Hayes-Roth, Eds., Academic Press, New York, 1978.

NII79 NII, H. P., AND AIELLO, N. "AGE (Attempt to Generalize): A knowledge-based program for building knowledge-based programs," in *Proc. 6th Int. Jt. Conf. Artificial Intelligence*, Tokyo, Feb. 1979, pp. 645–655.

NILS71 NILSSON, N. *Problem-solving methods in artificial intelligence*, McGraw-Hill, New York, 1971.

NORM80 NORMAN, D. A. "Copycat science or does the mind really work by table look-up?," in *Perception and production of fluent speech*, R. Cole, Ed., L. Erlbaum, Hillsdale, N.J., 1980, Chap. 12.

POHL70 POHL, I. "First results on the effects of error in heuristic search," in *Machine in-*

telligence 5, B. Meltzer and D. Michie, Eds., Edinburgh Univ. Press, Edinburgh, Scotland, 1970.

POHL77 POHL, I. "Practical and theoretical considerations in heuristic search algorithms," in *Machine intelligence 8*, E. Elcock and D. Michie, Eds., Ellis Horwood, Chichester, England, 1977.

REDD73a REDDY, D. R., ERMAN, L. D., AND NEELY, R. B. "A model and a system for machine recognition of speech," *IEEE Trans. Audio and Electroacoustics* **AU-21** (1973), 229–238.

REDD73b REDDY, D. R., ERMAN, L. D., FENNELL, R. D., AND NEELY, R. B. "The Hearsay speech understanding system: An example of the recognition process," in *Proc. 3rd Int. Jt. Conf. Artificial Intelligence*, Stanford, Calif., 1973, pp. 185–193.

REDD75 REDDY, D. R., ED. *Speech recognition: Invited papers presented at the 1974 IEEE Symposium*, Academic Press, New York, 1975.

REDD76 REDDY, D. R. "Speech recognition by machine: A review," *Proc. IEEE* **64** (Apr. 1976), 501–531.

REIS76 REISER, J. F. *SAIL*, Tech. Rep. AIM-289, AI Lab., Stanford Univ., Stanford, Calif., 1976.

RUBI78 RUBIN, S. *The ARGOS image understanding system*, Ph.D. thesis, Computer Science Dep., Carnegie-Mellon Univ., Pittsburgh, Pa., 1978.

RUME76 RUMELHART, D. E. *Toward an interactive model of reading*, Tech. Rep. 56, Center for Human Information Processing, Univ. California, San Diego, 1976.

SACE74 SACERDOTI, E. E. "Planning in a hierarchy of abstraction spaces," *Artif. Intell.* **5** (1974), 115–135.

SHOR75 SHORTLIFFE, E. H., AND BUCHANAN, B. G. "A model of inexact reasoning in medicine," *Math. Bio. Sci.* **23** (1975).

SHOR76 SHORTLIFFE, E. *Computer-based medical consultation: MYCIN*, Elsevier, New York, 1976.

SMIT76 SMITH, A. R. "Word hypothesization in the Hearsay-II speech system," in *Proc. IEEE Int. Conf. Acoustics, Speech, and Signal Processing*, Philadelphia, Pa., 1976, pp. 549–552.

SMIT77 SMITH, A. R. *Word hypothesization for large-vocabulary speech understanding systems*, Ph.D. thesis, Computer Science Dep., Carnegie-Mellon Univ., Pittsburgh, Pa., 1977.

SMIT81 SMITH, A. R., AND ERMAN, L. D. "NOAH: A bottom-up word hypothesizer for large-vocabulary speech-understanding systems," *IEEE Trans. Pattern Anal. Mach. Intell.* (1981), to be published.

SOLO77 SOLOWAY, E. M., AND RISEMAN, E. M. "Levels of pattern description in learning," in *Proc. 5th Int. J. Conf. Artificial Intelligence*, Cambridge, Mass., 1977, pp. 801–811.

SOND78 SONDHI, M. M., AND LEVINSON, S. E. "Computing relative redundancy to measure grammatical constraint in speech recognition tasks," in *Proc. IEEE Int. Conf. Acoustics, Speech, and Signal Processing*, Tulsa, Okla., Apr. 1978.

STEF80 STEFIK, M. *Planning with constraints*, Ph.D. thesis, Computer Science Dep., Stanford Univ., Stanford, Calif., Jan. 1980.

WALK78 WALKER, D. E., ED. *Understanding spoken language*, Elsevier North-Holland, New York, 1978.

WALK80 WALKER, D. E. "SRI research on speech understanding," in *Trends in speech recognition*, W. A. Lea, Ed., Prentice-Hall, Englewood Cliffs, N.J., 1980, Chap. 13.

WOLF80 WOLF, J. J., AND WOODS, W. A. "The HWIM speech understanding system," in *Trends in speech recognition*, W. A. Lea, Ed., Prentice-Hall, Englewood Cliffs, N.J., 1980, Chap. 14.

WOOD70 WOODS, W. A. "Transition network grammars for natural language analysis," *Commun. ACM* **13**, 10 (Oct. 1970), 591–606.

WOOD73 WOODS, W. A., AND MAKHOUL, J. "Mechanical inference problems in continuous speech understanding," in *Proc. 3rd Int. Jt. Conf. Artificial Intelligence*, Stanford, Calif., 1973, pp. 73–91, also *Artif. Intell.* **5**, 1 (Spring 1974), 73–91.

WOOD76 WOODS, W., BATES, M., BROWN, G., BRUCE, B., COOK, C., KLOVSTAD, J., MAKHOUL, J., NASH-WEBBER, B., SCHWARTZ, R., WOLF, J., AND ZUE, V. *Speech understanding systems: Final technical progress report*, Tech. Rep. 3438, Bolt Beranek and Newman, Cambridge, Mass., Dec. 1976 (in five volumes).

WOOD77 WOODS, W. A. "Shortfall and density scoring strategies for speech understanding control," in *Proc. 5th Int. Jt. Conf. Artificial Intelligence*, Cambridge, Mass., 1977, pp. 13–26.

RECEIVED AUGUST 1979; FINAL REVISION ACCEPTED FEBRUARY 1980

Part II
Production

[9]

THE ANALYSIS OF SENTENCE PRODUCTION[1]

M. F. Garrett

MASSACHUSETTS INSTITUTE OF TECHNOLOGY, CAMBRIDGE, MASSACHUSETTS

I. Introduction

When one is speaking, one is thinking;[2] the problem of language production has often been thought of as determining what necessary relations hold between the former and the latter. However, no attempt will be made here to wrestle with such hoary issues as whether the structure of language determines general cognitive structure. Rather, it is taken as a point of departure that there exists a nonlinguistic representational system in terms of which significant cognitive functions can be performed (for recent discussions of this and related issues, see Fodor, 1975; Fodor, Bever, & Garrett, 1974, Chapter 7).

Given this assumption, we may raise the more tractable problem of *sentence* production. That is, we distinguish between the general problem of language production, which must include message formu-

[1] The research reported here was supported in part by NIMH grant HD 05/68-02,03,04. Some of the work was done in collaboration with S. R. Shattuck and her contribution is gratefully acknowledged. This paper was written while the author was a Senior Research Fellow of the Australian-American Educational Foundation (Fulbright-Hayes program) at Monash University, Melbourne.

[2] Barring appeals to certain chimerical pathologies and rote recitation or reading.

lation, and the specific problem of sentence production, which may be viewed as *translation* process (Fig. 1).

We assume that somewhere in the recesses of our central nervous system an interaction takes place among our current motor and perceptual experiences, our stored information, our motivational systems, and, doubtless, sundry other variables. That interaction gives rise to what we might call a "communicative intention"—hereafter *message*. In some manner, that message must be translated into a set of instructions sufficient to guide our articulatory apparatus.

Beyond knowing that some translation exists—there is nothing so clear as that we do not think with our lips or velum—what can be said of the translation process(es)? If we first focus on the form of messages when they are uttered, we can begin to explore some aspects of the translation process.

Constraints on the form of utterances which are to count as sentences are determined by formal linguistic analyses. How does this bear on the translation process we wish to analyze? We know that whatever models of the translation process we may postulate, one thing they must do is to distinguish among all the meaningfully distinct messages. In mapping from any of the indefinitely many possible messages $M_1, M_2, M_3, \ldots M_n$ onto an utterance, the translation process must preserve the bases for computing distinctions among $M_1, M_2, M_3, \ldots M_n$ if communication is to take place. Thus, the translation process must preserve whatever aspects of the manifest form of sentences can be shown to contribute to their semantic interpretation. That is tantamount to the claim that the structural descriptions of sentences must be preserved by the translation process, for virtually all the structural features encompassed by the surface phonological description and the underlying and surface syntactic descriptions of sentences[3] can be shown to contribute in one way or another to their semantic analysis. Thus, the structural analyses assigned to sentences by an adequate grammar provide one powerful constraint on the formulation of models of the translation process from messages to utterances.

[3] Excluded from this claim are "intermediate" syntactic trees and underlying phonological descriptions. This is not because these aspects of structural descriptions are taken to be less well motivated linguistically, but only because they cannot be given a prima facie defense on grounds of maintaining distinct representation of semantically distinct utterances. The relevance of such levels of representation to the processes of producing or comprehending sentences must appeal to other empirical support, some of which is mentioned or discussed in subsequent portions of this paper.

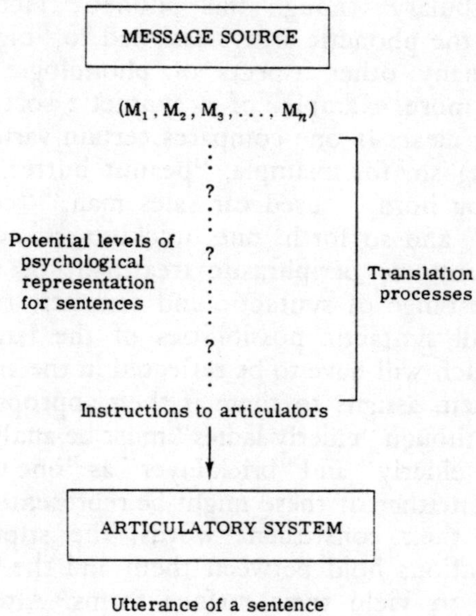

Fig. 1

But even though this constraint on the character of possible production models is a very powerful one, it tells us less than we might wish about the computational procedures which will effect the translation. For example, we know that the structural descriptions for sentences express regularities in the distribution of a variety of structural types—sounds, syllables, words, phrases and phrase types, sentences and sentence types. Such structural distinctions must be recoverable from the output of the translation process (if they are semantically relevant), but the way in which they are reflected in the information flow which gives rise to that output is an open question. We know for example that the structural analysis of sentences requires that "words" be decomposed into more elementary meaning bearing elements (morphemes); but the computational system may take account of this in a variety of ways. In particular, there may or may not be an independent, phonological representation of bound morphs in the computational vocabulary of the production system. Thus although the analysis of the word "bigger" must mark it as the comparative form of the adjective "big," that analysis may not be reflected in the existence of an element "comparative:—er" in the processing vocabulary (though this phonetic fact would still be inferrable from the phonetic arrays assigned to "big" and "bigger"). Similarly for many other aspects of phonological and syntactic structure. One more example, of a syntactic sort, may suffice to make the point clear. If one compares certain varieties of complex nominals in English, for example, "peanut butter," "brick layer," "cast iron," "fog horn," "used car sales man," "college students," "elderly ladies," and so forth, one quickly sees, even with a rough attempt at a uniform periphrastic treatment of them, that they embody a wide range of syntactic and semantic relations, (indeed virtually the full syntactic possibilities of the language; see Lees, 1960) all of which will have to be reflected in the analysis which the production system assigns to them if their appropriate use is to be explained. But, though "elderly ladies" must be analyzed, roughly, as "ladies who are elderly" and "brick layer" as "one who lays bricks," both, either, or neither of these might be represented in the processing system by their constituent words, the stipulation of what grammatical relations hold between them and the derivational processes required to yield their surface forms. Alternatively, both, either, or neither might be entered in their derived form as elements of the computational vocabulary in the same way as words like "dog," "symbol," or "apricot." The alternatives sketched do not begin to exhaust the possibilities, of course, but for our purposes it is enough that the compatibility of structural descriptions with a variety of computational procedures be appreciated.

The final point that should be made on this aspect of the analysis of sentence production is that, even were we to have good evidence about the vocabulary of the computational system that mediates the translation from messages into their realization as instructions to the articulatory system, we would still be unsatisfied. For one would want to be able to characterize the information flow in the system in terms of interactions between the various structural types represented in the computational vocabulary. It would, for example, be important to know whether there were "stages" or levels in the processing which correspond to the distinct linguistic types characteristic of semantic, syntactic, and phonetic descriptions of sentences. It is entirely consonant with our acceptance of structural descriptions as constraints on translation models that there should be *no* stage in the processing at which the "decisions" are primarily syntactic or semantic. That is an empirical question about information flow in the system.

How does one go about assessing the behavior of speakers in order to provide answers to the sorts of questions raised above? How do we discover what sorts of structure are computed on-line by the speaker, and in what sequence? Such questions are not easily amenable to experimental attack (although there have been a few useful efforts, some of which I will take up in the discussion section). On the one hand, we are handicapped by our inability to control or manipulate the input to the sentence production system (messages) and, on the other, by the arduousness of analyzing the experimental results (large amounts of "spontaneous" speech). The recourse for people interested in language production processes has, largely, been to "observational" techniques, and, in particular, to the study of various sorts of departures from ideal speech (e.g., nonfluencies and speech errors of a variety of types); and that is the approach adopted for the work described in this paper. This is not because I doubt the possibility or need for specifically experimental enquiry into production processes. It is, rather, because I feel both that experimental enquiry can be better undertaken against a background of empirically supported working hypotheses and that the study of speech errors is a good way to generate and support such hypotheses. For, as has often been observed, speech "errors" have a powerful claim to face validity as indicants of the on-line processing that underlies speech. Beyond this, the study of speech errors is a larger enterprise than might be imagined on initial consideration. Thus, despite the interest of several investigators past and present, much remains to be determined about the nature and explanation of the several ways in which "natural speech" departs from the idealization of prose or the precision of formal address.

II. Speech Errors: The Corpus

There are a variety of departures from normally fluent speech which could count as "errors"; for example, hesitations, changes of mind, repetition of sounds or words, and so forth. The data I will be discussing, however, concerns less common, but by no means rare, error types. Consider the entries in Table I. Each of the varieties of error listed has a single example given, and that example is, perforce, one which involves only one of the several types of linguistic elements that appear in speech errors. But each of these error patterns (e.g. addition, exchange, etc.) does have exemplars for nearly every linguistic type (e.g., sounds, morphemes, words, etc.). It is evident that the sorts of errors represented in Table I are *not* part of the

TABLE I

SOME PATTERNS OF SPEECH ERRORS

a. Addition:
"I don't see any many paddocks around here."
(intended either *any* or *many* but not both)

b. Deletion:
"I'll just get up and mutter _intelligibly."
(intended *unintelligibly*)

c. Substitution:
"At low speeds it's too light."
(intended *heavy*)

d. Complex addition:
"The one exPosner experiment that. . . ."
(intended *Posner*)

e. Complex deletion:
"That would be _having like Harry."
(intended *behaving*)

f. Shift:
"That's so she'll be ready in case she decide_ to hits it."
(intended *decides to hit it*)

g. Exchange:
"Fancy getting your model renosed."
(intended *nose remodeled*)

h. Fusion:
"At the end of todays lection. . . ."
(intended *lecture* or *lesson*)

i. Double whammy:
"He's a laving runiac."
(intended *raving lunatic* or *maniac*)

same class of events as hesitations, changes of mind, or "sloppy articulation" (as that characteristic of inebriated, hurried, or very casual speech). These sorts of errors occur (although not exclusively) in fluent speech, and although they may well be affected by changes of communicative intention on the part of speakers, they are not readily interpretable as the simple consequence of having abandoned one form of expression in mid-utterance in order to shape another nearer to intent's desire. These errors represent cases in which either all and only the intended elements of the speaker's utterance appear, but in the wrong order, or some intended element is missing, or some element not intended intrudes in one way or another.

These error types are illustrative of those in an error corpus of some 3400 errors I gathered, together with my colleague S. R.

Shattuck, principally over the past 3½ years. The errors are with minor exceptions, all from spontaneous speech, and all are (a) those observed by myself or S. R. S., or (b) those reported by our friends, colleagues, and students.[4] For most of the errors, as much was recorded (written down) of the utterance surrounding the error locus as was consonant with the preservation of friendships or the exigencies of the recording situation. In the sections that follow I will report some aspects of the analysis of that corpus (hereafter called "MIT corpus"), and an interpretation of its apparent regularities. Relatively little discussion is offered of errors of sound structure (see however Fromkin, 1971; Shattuck, 1974); the primary focus is on syntactic processes and their relation to semantic and phonological variables.

There are a number of methodological caveats to be entered before discussing the structure of the error types. First of all, there will rarely be any mention in what follows of the statistical significance of such differences as I may call attention to. There are a number of good reasons for this. For the most part, the behavior of speech elements that I have tried to interpret has virtually binary conditions—either a certain class of speech errors is conditioned by a given structural variable, or it is not sensitive to that variable at all. This is not to say that one cannot find "statistical error patterns," nor that such may not be of importance. It is only to say that I have avoided their interpretation for the most part (only occasionally succumbing to the temptation of the implications of "trends"). This fastidiousness is prompted by an appreciation of the hazards of "naturalistic" data collection and the paucity of reliable information on the frequency of occurrence of elements of the language at levels higher than sound elements or "word forms" (and even there the information is primarily orthographically based). Even if one had standards against which to assess error frequencies, one could not be assured that the corpus one has amassed provides a reasonable basis for inferring the incidence of different error types. The "sampling" must be decidedly nonrandom given the reasonable assumption that certain sorts of errors are perceptually more salient than others.[5]

[4] Errors were accepted from contributors other than myself or S.R.S. but the "source" of all errors was recorded. Subsequent analysis shows no apparent basis for distinguishing among the errors contributed by others and those observed by myself or S.R.S. so far as the generalizations discussed in this paper are concerned.

[5] For example, there is a clear difference in the sensitivity of myself and S.R.S. to vowel errors. I rarely hear them unless they are metathesized. In general, it seems likely that vowel changes will be less readily noted than consonantal changes for most listeners.

Similarly there are very likely contributions of situational and individual variability in speakers. Since a large proportion of the errors are those committed by myself and my most frequent colleagues, considerable caution is indicated in interpreting differences in the incidence of different error types. I believe that one is on somewhat firmer ground, however, in discussing the apparent interactions between structural features of sentences and a particular error type, given a fairly large number of errors and very pronounced differences within the class in its sensitivity to given structural constraints.

III. Speech Errors: The Analysis

As the preceding remarks may suggest, my approach to the analysis of speech errors departs in some measure from that of most previous investigators. This is true in two primary respects: (1) a focus on syntactic variables rather than sound structure, semantics, or motivational processes, and (2) a focus on constraints on particular error types rather than on the existence of errors involving particular linguistic types. Roughly, in what follows I will be interested in generalizations concerning limitations on the range of movement of various linguistic types, the nature of immediate environmental determinants of errors, and the relevant description of elements which are (apparently) interacting, and I will be exploring the implications of such generalizations for a characterization of sentence production as a succession of quasi-independent "levels" or stages of processing activity. We will begin by loosely characterizing some of the properties of speech errors that emerge from past research and then proceed, through emendations and exceptions based on the current data, to some working hypotheses about sentence production.

One of the most striking regularities encountered on first reading through a set of speech errors, especially those involving sound elements, is the degree to which interacting elements occur in similar environments, both positionally and, to some extent, phonetically. Errors such as the following

(1) *So while you do the cooking, Bill snovels show, does he?* (shovels snow)
(2) *The little burst of beadan is. . . .* (beast of burden)
(3) *You're not a poojin pitter-downer . . . (are you)?* (pigeon putter-downer)

(4) *That's an easer-eagy sentence.* (eager-easy)
(5) *Children interfere with your nife lite. . . .* (night life)
(6) *I think I met someone who had a spet pider.* (pet spider)
(7) *There's a lot of flee floating anxiety. . . .* (free floating)
(8) *The straight lawn drawn through. . . .* (line drawn)
(9) *We will go down to the sound roof proo. . . .* (sound proof room)

illustrate the tendency of word initial segments to exchange with, or intrude into, other word initial positions, and correspondingly for medial and final segments;[6] and they are more likely to do so when the elements preceding or following those positions are phonetically similar (Fromkin, 1971; Mckay, 1970). These examples also illustrate another powerful condition on elements that interact: namely, that they be themselves similar in certain aspects. In particular, for sound errors, identity as consonant or vowel seems crucial; consonants exchange with other consonants but not with vowels, and conversely. There are virtually no plausible exceptions to this generalization in the MIT corpus and none that I am aware of in published reports of other error corpora. So far as these varieties of sound errors are concerned, these two generalizations seem secure: interacting elements are similar to each other, and their environments are also similar. Their application to other types of errors involving other linguistic elements is, as we shall see later, sometimes straightforward, sometimes obscure.

A third general aspect of these sorts of errors is what might be termed the preservation of phonological well-formedness. Nearly every investigation of speech errors has remarked upon the rarity of error outputs which violate the constraints on sequencing of sounds in whatever language is being spoken. Thus, in English, one simply does not find movement errors which yield, for example, illegal initial clusters like /sd/ or /sg/;[7] this in spite of the fact that such clusters are perfectly pronounceable and, indeed, regularly occur in casual speech (e.g.,/sdən/for *it's done* or /sgʌnə/ for *it's going to,*

[6] This might be couched (and has been) as a syllable-position constraint. The data, however, do not really seem to distinguish between a syllable structure constraint, on the one hand, and the joint effects of constraints on word (or morpheme) position and the vowel/consonant identity of exchanged elements, on the other (see later in text). See Shattuck (1974) for some discussion of this point.

[7] Though no such errors occur in the MIT corpus, this is apparently because they are rare, not that they are "impossible"; for example Kermit Shaefer's recorded collection of TV and radio errors contains the following: *. . . forks and spoons . . . sforks and sfoons . . .*

etc.). There seems to be either an aspect of the articulatory program which prevents such errors from occurring or an editing function which selectively picks up errors that violate sequencing constraints before they are emitted, thus barring them from the error corpus. On the current evidence, there is not any very satisfactory way to decide which of these is in fact the correct account (although I will argue below that there are some grounds for adopting a modified editing view).

The fourth, and final, general remark I wish to make in this preliminary characterization of speech errors concerns the role of prosodic features of sentences. There is a clear relation between the involvement of a speech segment in an error and both word and phrasal stress. Boomer and Laver (1968) reported that the majority of the sound errors they analyzed[8] involved an intrusion from the tonic word of a major phrase group ("phonemic clause"). And further, they observed that the two syllables involved in an error interaction were metrically similar for at least two degrees of stress; for example, the two syllables were either both weak or both salient, usually the latter, but rarely was one weak and the other salient. McKay (1970) has observed similar sorts of effects in his analysis of the corpus of German speech errors collected by Merringer (1908). Fromkin's (1971) report of her own corpus of errors and my analysis of the MIT corpus are also consonant with Boomer and Laver's generalizations. One of the things we will explore here is the degree to which these relations between stress features and errors may be accountable in terms of planning systems at the syntactic and morphological levels.

In short, where there is an error interaction involving two elements of the intended utterance, as those in examples (1)–(9), such elements

(10) a. occur in similar environments, both with respect to word (or syllable) positions and preceding or following phonetic elements;
b. are similar to each other both phonetically and prosodically;

[8] Boomer and Laver's corpus consisted of 200 errors; 100 were tape-recorded and 100 from written records. Their errors, judging from the examples they provided, were not just exchanges, but included what I have called shifts and complex additions (e.g., their *frunds have been frozen,* would be a shift in my terminology if the second underlined element had been omitted; as it stands, it is a complex addition).

c. will very likely involve the tonic word of a major phrase group;
d. will yield from their interaction a phonologically permissable sequence.

It will be clear from a comparison of the errors in (1)–(9) with the categories given in Table I that these generalizations do not cover the full range of error phenomena, nor, it should be pointed out, do they exhaust the claims that can be made of such "interactive errors" themselves. They are, however, well-supported generalizations and they provide an initial illustration of the sorts of descriptive and environmental constraints on error interactions that we may expect to find at levels other than that of sound structure.

IV. Sound Exchange Errors

The first errors from the MIT corpus that we will examine in some detail are those like examples (1)–(5). These are *full exchanges* (i.e., enough of the utterance was completed by the speaker to allow unambiguous determination of the complementary elements and their positions) of separate segments of the intended utterance. Exchange errors of every sort (i.e., those involving morphemes, words, or phrases as well as those involving sounds) are of paramount importance in our analysis for a number of reasons. They first of all provide us with an unambiguous instance of the simultaneous existence of temporally discrete elements of the intended utterance; and second, to the extent that the error interacts with the intervening structures and elements, we may also argue for their simultaneous existence. Thus, such errors enable us to make inferences both about the size of the units over which production processes are normally integrated, and about the structural properties of those units. It is, therefore, of considerable importance to establish the character of such errors.

We remarked earlier that our results were generally compatible with those of previous analyses both with respect to the stress of interacting elements and the prominent role of the tonic word of major phrase groups; this is very strikingly the case for the pure sound exchange errors. Of the total of 137 such errors in the MIT corpus, almost all (92%) were interactions between two salient sylla-

TABLE II

SOUND EXCHANGE ERRORS

A. Stress Relations

	Exchanges within words	Exchanges between words	Total
Word stress			
Salient/salient	8	118	126
Weak/weak	0	1	1
Weak/strong	7	0	7
Indeterminate	3	0	3
Phrase stress			
Tonic word	12	94	106
Other	4	15	19
Indeterminate	2	10	12

B. Syntactic Relations

	Within clause	Between surface clauses
Form class preserved	21	2
Form class changed	89	7
Totals:	110	9

bles, and the large majority involved the tonic word (80%) of a major phrase (see Table II).[9]

At this level of description our results reinforce those of Boomer and Laver and others. There is, however, another powerful constraint on sound exchanges which may be related to the results for stress relations, and which may lead us to a reinterpretation of their significance. That constraint, as it is reported in Table II, is a

[9] The within-word exchanges are not included in the analysis of the syntactic relations. Note that the exceptions to the word stress regularity are all within-word exchanges. This may indicate that such errors are the consequence of different processing mechanisms than are other exchanges. Note further that the exceptions to the tonic word involvement are usually interactions between two modifiers of the tonic word.

syntactic one: the interacting elements of an exchange, with very rare exception,[10] are both members of the same surface clause.

This fact relates most obviously to another conclusion reached by Boomer and Laver. They analyzed their error corpus in terms of tone groups (or "phonemic clauses") which are units defined by a single tonic center and certain junctural phenomena characteristic of the boundaries of major phrase groups. Boomer and Laver found not only that the tonic word was usually involved in the errors they analyzed, but that it usually interacted with a word from its tone group, and hence, the results in Table II would be seen in their terms as an effect of tone groups. Thus, the findings of stress involvement at word and phrase levels coupled with the limiting effects of tone group boundaries conspire to suggest a prosodically defined encoding unit, and at the same time, an account of these sound errors in terms of neural correlates of the physical differences representing stress variations. Boomer and Laver's remarks are certainly in this spirit, [11] "but McKay (1969, 1970) has been most explicit on this point. He attributes the intrusion of error elements to the level of activation of speech motor units, with " . . . stressed elements taking a higher level of subthreshold activation than unstressed ones [McKay, 1970, p. 40] ." If this sort of account is correct, it is important, for it suggests both a hypothesis about the planning in speech and it provides a mechanism which accounts for a significant error type. There is, however, an alternative account of these same facts, as the classification in Table II suggests; namely, that they are a consequence of the syntactic and morphological structure which underlies prosodic features. What can be done by way of discovering which of these levels of description is the preferred one for accounts of error interactions, and by hypothesis, for the existence of particular varieties of planning in the sentence production system?

Two preliminary points might be made. First, the account of errors in terms of stress-correlated levels of activation in the motor system would lead one to expect only errors of anticipation to occur.

[10] The sound exchange exceptions to the clausal constraint are usually clause adjuncts (e.g., *Did you get yoursev socks, Clease*?") or idioms; e.g., *With this wing, I thee red*. (Surprisingly, Fromkin reports an error in a variant of the same idiom, *With this wing, I do red*.) In these respects the clause exceptions for sound errors differ from those for word exchanges.

[11] My interpretation of Boomer and Laver's remarks may be too strong. They speak of "neurophysiological prominence" of stressed elements, and of the "strength" of representations. These remarks do not necessarily commit them to a motor level of representation.

Though we have yet to discuss them, there are numerous nonexchange sound errors that appear to be perseveratory, for instance, example (9) and (f) in Table I), and a different account would be necessary for them. Similarly, one would expect that in exchange errors the higher stressed element should follow the lower stressed one in the intended output. Though this tends to be true by the report of most investigators, and it is true of the MIT corpus (62% of the cases where stress of the exchanged elements differs are anticipatory), it is by no means so impressive a regularity as the other constraints we have been discussing.

Second, if physical level of stress is the significant variable in precipitating speech errors, one might expect to find a greater incidence of errors in sentences with emphatic or contrastive stress than in sentences with normal stress, for normal stress variations are much less impressively reflected in actual acoustic energy differences than are emphatic and contrastive stress. It is difficult properly to evaluate this expectation since one has no good idea of the relative incidence of the relevant sentence types in normal conversation. In any event, there is no striking support for this prediction in the MIT corpus.

Though these sorts of observations might give one pause, they are, of course, by no means sufficient to render untenable the interpretation that motor correlates of physical stress differences are a major causal factor in these sorts of speech errors. There is, however, another set of observations which does seem to conclusively rule out that sort of explanation. We need to recall another of the strong generalizations about errors which was noted above; namely, that the result of an error interaction is phonologically well formed. This constraint is, in fact, a good deal more general than the examples used to present it might suggest. For not only are the sequencing constraints honored, but so too are various stress regularities and phonetic accommodations of elements to their error-induced environments.[12] To illustrate briefly:

(11) a. Vowels are restored or reduced when the deletion or addition of an intended element requires; e.g.,

easily	*easy*
/isɨli/ *enough*	→ /izi/ *enoughly*
	easuh
	↛ /izɨ) *enoughly*

[12] Examples of one or another kind of accommodation are reported by most observers. In the MIT corpus there are about 20 cases where an accommodation is clearly demanded by an error and in every case it occurred.

b. The phonetic form of an indefinite article accommodates to its environment; e.g.,
an aunt's money → *a money's aunt*
↛ *an money's aunt*
(see also Fromkin, 1971)

c. The phonetic forms of tense and number morphemes are appropriate to the stems to which errors attach them; e.g.,
. . . *add/z/ up to* → *add up/s/ to*
↛ *add up/z/ to*
(see also Fromkin, 1971)

d. Word stress is appropriate to the error forms resulting from syllable movements, additions or deletions; e.g.,
treméndously → *trémenly*
specifícity → *specífity*
compúted → *cómputated*
marsúpials → *musárpials*

e. Phrasal stress is preserved; e.g.,
. . . *stop beating your heád against a bríck wall*
→ . . . *bríck against a head wall*
. . . *avoid the trée pruning* → *trúe preening*
(see also Boomer and Laver, 1968).

These examples and those given earlier all indicate a late condition on output which insures that the actual uttered form will be phonologically acceptable (although it may depart markedly from "sensibleness" or syntactic well-formedness). This conditioning of the output must occur *after* the level at which the errors occur, and must be either prior to or identical with the level at which particular phonetic forms are being translated into motor commands.[13] But that fact presents an insupportable problem for the view that it is interactions among the motor elements underlying utterances which account for errors, and, in particular, for the prominence of stress factors in error regularities.

It is undeniable that there is a level of representation of speech events in the motor system at which there is a reflection of physical differences of output energy in levels of neural activity; subglottal air pressure does rise prior to the emission of a stressed syllable, and thoracic and laryngeal muscle groups will reflect this. But, it is

[13] The suggestion of an ordering of error types by ordered levels of processing can be taken as an "editing" explanation of accommodations, as opposed to a view that they are somehow antecedently prevented.

equally certain to be fatuous to look for such a correspondence at much higher levels; my intention to speak softly surely does not have a notably lesser level of neural excitation than does my intention to speak loudly. One may legitimately wonder at what point the isomorphisism breaks down between these two extremes. One thing we can be confident of, however, is that for speech, that point must fall somewhere before the level at which accommodations like those of (11) (*a*)–(*e*) take place, for all those are cases in which errors have demonstrably occurred prior to the realization of the particular phonetic form of the elements involved in the error.

Put another way, if the neural correlates of physical differences between stressed and unstressed speech elements were the primary causal factor in movement errors, one would expect those differences to appear as features of the emitted error elements. But that is precisely what we do not find. Rather, it appears to be the position which an element occupies within an independently specified structure which endows it with the detail of its physical form.

We may regard the foregoing discussion as the first step in a contrast of the notion of "forcible intrusion" with that of "descriptive error" as ways of accounting for speech errors. The former notion stresses the physical concommitants of structural features and exploits their differences of "strength" in explanations of the etiology of errors. The latter notion relates structural differences to planning differences and the consequent opportunities for interaction among similarly described elements of the ultimate speech event. The evidence that we have so far considered indicates that the regularities of stress involvement in errors should not be accounted for by an appeal to a notion like forcible intrusion, and that we should consider them further aspects of the importance of similarities in the description of interacting elements, that is, as intimately related to the generalizations in (10) (*a*) and (*b*). Two things suggest that we should look to syntactic descriptions for the relevant analyses. One is the fact that prosodic features are determined by surface syntactic descriptions, and the other is the powerful constraint exerted on sound exchange errors by surface clausal structure. What one needs is additional independent evidence both for the operation of the clausal constraint and for the existence of a specifically syntactic processing level which could give rise to it.

V. Word and Morpheme Exchanges

The most obvious place to begin looking for such evidence is with exchanges involving elements that have themselves syntactic features:

TABLE III

WORD AND COMBINED FORM EXCHANGES

	Within clause	Between clause
Form class preserved		
Independent words	67	17
Combined forms	13	3
Form class changed		
Independent words	13	0
Combined forms	29	1
Totals:	122	21

morphemes, words, and phrases. Very little can be made of phrase exchanges, for they are too rare; exchanges of words and morphemes, however, are relatively frequent. In the MIT corpus there are 143 such errors that are unambiguously exchanges between words or parts of words that are morphemes. There are an additional 72 errors in which the words are in incorrect serial order, but because the misordered elements are adjacent, one cannot determine whether an exchange has taken place, or simply a shift in the location of a single word. These three classes of word movement errors (exchanges of independent forms, exchanges of combined forms, and "exchanges" of adjacent forms) turn out to have quite different properties, and these differences provide us with a basis for evaluating the role of syntactic factors in sentence production.

Table III presents the distribution of the first two of these three error types with respect to clausal structure and form class; examples (12)–(19) are typical of these error categories.

(12) *Slips and kids—I've got* **both** *of* **enough**.
(intended: *enough of both*)

(13) *I broke a* **dinghy** *in the* **stay** *yesterday*.
(intended: *stay in the dinghy*)

(14) *Although* **murder** *is a form of* **suicide**, . . .
(intended: *suicide is a form of murder*)

(15) *I've got to go home and give my* **bath** *a hot* **back**.
(intended: *back a hot bath*)

(16) *McGovern favors* **push***ing* **bust***ers*.
(intended: *busting pushers*)

(17) *I hate working on two-***word letters**.
(intended: *two-letter words*)
(18) *It just* **sound***ed to* **start**.
(intended: *started to sound*)
(19) *Oh, that's just a* **back truck***ing out*.
(intended: *truck backing out*)

If one simply looks at the total frequencies of between- and within-clause errors, it is apparent that the clausal constraint we observed for sound exchanges is also present for word forms. There are, however, differences in the behavior of sound errors and word errors with respect to that constraint.

We note first of all that even though the word exchanges are predominantly between members of the same surface clause, this restriction is not quite so sharply drawn as for the sound errors (7% sound exceptions versus 15% for word exceptions). But what is more interesting is that these exceptions to the clausal constraint appear to be disciplined. In particular, such exceptions are almost invariably interactions between words of the same form class. If one compares Table II with Table III, one finds no apparent form class effect for sound exchanges, but a strong one for word exchanges that span clauses (and perhaps for those within clauses as well). This immediately suggests two things: (a) though both sound and word exchanges show sensitivity to clausal structure, they may do so for somewhat different reasons, and (b) the description which governs interactions between words in different clauses is one that includes at least their form class designation, while that for sound interactions does not. There are a number of aspects of these two possibilities which deserve exploration.

The most obvious first question is whether the form class constraint so evident for the between-clause errors is, in fact, restricted to such errors and, hence, whether we are entitled to infer that form class designations are only relevant to a level of planning at which elements of more than a single clause are being manipulated. A look at Table III suggests that this is very likely not true, for there is indication that form class is preserved both between and within clauses for exchanges between independent words. Of the 80 such within-clause exchanges, 84% preserve form class. This contrasts sharply with the pattern for those exchanges in Table III which involve combined word forms, as well as with the pattern for sound errors. For the combined form exchanges, within-clause errors show the reverse tendency; 70% of these errors violate form class (this is similar to the pattern for sound exchanges—see Table II).

In a limited way, these results provide what we were looking for, namely, an indication that elements (words) with specifically syntactic properties (e.g., form class) show a clausal constraint, and thereby justify the attribution of both the clausal limitation on the excursion of sound exchanges and their prosodic involvement to a syntactic level of sentence production processes. This resolution of the apparent facts is, however, unsatisfactory in a number of respects. Most glaringly, it does not countenance the existence of a class of word exchanges (the combined forms) which do not observe the form class restriction. To this must be added the difference in the behavior of sound errors with respect to clause boundaries and form class. When sound errors do transcend the clausal limitation, there is no indication of a change in their respect for form class (seven of nine such errors change form class). This distinguishes them from the word exchanges, where the form class restriction on interclause errors is virtually absolute (extending even to the combined forms, although there are really too few cases to make a firm judgment).[14] Thus, simply to attribute limitations on sound exchanges to the manifest existence of a syntactic level of processing would be to ignore an inconsistency in the behavior of the putatively syntactically organized elements with respect to our diagnostic variable, form class.

The inference that one is tempted to make from all this is that the very strong form class constraint observed for between-clause errors is, in fact, not solely the consequence of the distinct clausal membership of the elements involved, but is instead a characteristic of the level of processing at which most word exchanges take place. But this suggests that the combined forms are exchanged at a different level of processing than are the independently occurring word forms, and *that* is surprising, for, as the examples in (12)–(19) indicate, these errors are, barring the differences which define their initial classifica-

[14] It should be obvious that the error classifications that are being explored are classifications in terms of superficial properties (e.g., the presence of a bound morph) that may be reasonably supposed to be *diagnostic* of syntactic processes. The absence of such a diagnostic variable does not preclude the occurrence of a given error at a different level than the typical one. Hence, the frequencies of the error types are not "pure," but reflect some potential overlap in levels.

TABLE IV

FREQUENCY OF EXCHANGE ERRORS BY CHANGE IN FORM CLASS AND DEGREE OF SEPARATION IN WORDS

Error type		Degree of separation 0	1	2	3 or more	Σ
Sound exchanges	Form class same	6	15	2	0	23
	Form class diff.	45	41	7	2	96
Combined-form Exchanges	Form class same	9	5	1	1	16
	Form class diff.	20	8	2	0	30
Independent word exchanges	Form class same	–	37	30	17	84
	Form class diff.	–	12	1	0	13

tions, seemingly very similar. Indeed, the combined forms are almost always free forms, that is, the permuted elements are not bound morphemes but forms that can occur grammatically as independent words.

The theoretical seat becomes, at this point, uncomfortably warm. We seem well on the way toward claiming not one, but two syntactically organized levels of sentence production processes. The empirical thread from which these speculations depend is admittedly a slender one, but there is one immediately available property of these errors which suggests that we ought not resist this particular proliferation of theoretical entities.

It appears that, though all are clause limited, sound exchanges, and combined and independent word exchanges nevertheless "span" different intervals. So far, our discussion has been in terms of the interactions between errors and certain structural variables and in terms of the type of interacting elements. We now add to those variables the notion of degree of separation in the intended utterance. This might be expressed temporally (a measure for which we have no data at present), or in terms of the number of intervening elements–phones, syllables, morphemes, or words; since we will be comparing the three error types, discussion here is in terms of word separation.[15] Table IV provides the frequencies of errors for each of

[15] A distance measure in terms of syllables would not change the comparison except to magnify the distance of the word exchanges. This is because the words that intervene between the elements of a combined form or sound exchange are closed class (and almost always single syllables) while those intervening between independent word exchanges are often open class (although not often multisyllable).

the error types at varying degrees of separation. The comparison is vexed by the fact that the number of independent word exchanges is indeterminate for 0 degrees of separation. I will argue below that, in fact, Table IV represents the frequencies of these types fairly well as it stands (i.e., that there are very few 0-degree exchanges of independent words). But for the moment, we may simply confine our attention to those errors of 1 degree or more. It is very clear that the likelihood of an error spanning two or three elements is greater for the word errors than for the sound or combined form exchanges ($p < .05$, χ^2 for both the word versus sound and the word versus combined forms comparisons). What seems equally clear is that these longer error spans are confined to elements of a common form class for word exchanges.[16]

I have ventured onto dangerous ground in this comparison of the spans of different error types—and been properly punished by the necessity of resorting to a statistical defense of the distinction. Nowhere is one more at the mercy of the undoubted vegaries of sampling in this research than in comparisons of the incidence of different error types.[17] Nonetheless, in this case, the risk may prove justified by the conclusions I will argue for subsequently. I believe the clear implication of the differences among the error types we have been discussing, with respect to form class and error span, is the existence of two processing levels for word forms, one under "real

[16] The effect of distance on form class might be taken to explain the failure of combined form and sound exchanges to honor form class. That would, I think, be mistaking a symptom for the disease. Consider: (a) even at 1 degree of separation the word exchanges are predominantly form class preserving, and (b) clause-spanning errors of 0 and 1 degree are form class preserving. What appears likely is that the form class constraint and the clausal constraint are diagnostic of processing levels, and so too are the types of error elements involved (sounds, combined forms, and words); but none of these are definitionally related to a processing level. Thus, for example, we may reasonably expect some errors which involve combined forms (especially those which cross clause boundaries) to occur at the level characteristic of word exchanges and vice versa.

[17] This is particularly true for the present argument since I am using differences in error spans as an indication of differences in processing. For example, one might reasonably expect an "editing" function to catch long span errors more often than short span errors, thus excluding most of the latter from our analysis (which is based on completed errors). It is certain that there is such an editor, for incomplete exchanges/anticipations (one cannot tell which it might have been) are much more frequent than are complete ones. But though one might readily grant that an editor would be "length sensitive," it would still be left open why such an editor would more often pass long word exchanges than long combined-form exchanges. Similarly, why should both the long and the short-word exchanges show the same properties? Most important, however, is the fact that an analysis of the *in*complete exchange/anticipations shows the same patterns as the analysis of the complete ones (making the best guess one can about the likely nature of the completed error).

time" constraints and, hence, affected by adjacency of elements in the intended output, and one under "functional" constraints, and, hence, primarily sensitive to the structural relations upon which form class distinctions among the words of the intended output are attendant. These points may be more elaborately stated as follows:

(20) a. Exchanged words that are (relatively) widely separated in the intended output or that are members of distinct surface clauses will serve similar roles in the sentence structures underlying the intended utterance and, in particular, will be of the same form class. These exchange errors represent interactions of elements at a level of processing for which functional relations are the determinant of "computational simultaneity"; similarity of the descriptions that govern the selection among elements at this level is assumed to determine error interactions.
b. Exchanged elements that are (relatively) near to each other and which violate form class represent interactions at a level of processing for which the serial order of the elements of an intended utterance is the determinant of computational simultaneity. Similarity of left and right adjacent elements both phonetically and syntactically, as well as similarity of the elements themselves, is assumed to determine the likelihood of error interactions.

The facts about production errors which we have discussed to this point are not by themselves sufficient to sustain the claims of (20) (*a*) and (*b*). There are, however, a variety of other regularities of production errors which indicate the truth of something like these generalizations. We turn now to a systematic discussion of such additional properties of errors as bear on (20) (a) and (b) and on related claims.

VI. Stranded Morphemes, Shifts, and Exchanges

A somewhat unexpected consequence of the preceding analysis has been the postulation of two levels of nonphonological structural processing. That postulation rests primarily on the failure of words that are combined with bound morphs to show the same error behavior as independently occurring words, both with respect to form class and span of error movement. In (20) (a) and (b), these facts have been interpreted as the reflections of a planning level in which structural relations among words are being manipulated, as opposed to one in which the serial deployment of lexical and grammatical formatives is being worked out. (We leave open for the moment the question of how sound errors fit into this system.) The data we wish to consider now strengthen the basis for claiming the existence of these two processing levels and provide some further indication of their character.

We will begin by a closer examination of the structures in which the various error types occur. We note first of all that a striking regularity of the sound errors converges with a "definitional" property of the combined forms: they are both restricted to open-class elements (primarily to nouns, verbs, and adjectives). Since, barring some pronouns and elements of the auxiliary expansion, closed-class words are not inflected or combined, the combined form errors will necessarily involve only open-class words. It is significant that sound exchanges are similarly restricted, but in one particular respect, word exchanges are not. Only two of the 137 sound exchanges in the MIT corpus involve closed-class items (a qualifier, "few," and a subordinate conjunction, "until"); twenty of the 97 word exchanges involve closed classes. The most interesting for present purposes are the ten that exchange prepositions (the remaining ten such exchanges are distributed: six between pronouns, two between determiners, one between the elements of a complementizer, and one between a negative and a temporal qualifier). Their interest lies not simply in the closed versus open contrast, although that has its significance, but in the fact that these closed-class elements of word exchanges, like the open-class ones, come from corresponding positions within their respective structures. But the open-class items of sound and combined-form exchanges do *not* show any such parallelism of structure; they are typically interactions between the head of a construction and one of its modifiers. (For this contrast, see the earlier examples (1)–(9), (12)–(19), and (21)–(28), following; the elements that have been permuted are boldfaced.

(21) *I have to fill up the* **gas** *with* **car**.
(22) *Prior to the operation they had to shave all the* **head** *off my* **hair**.
(23) *She donated a* **library** *to the* **book**.
(24) *Older men* **choose** *to* **tend** *younger wives*.
(25) *. . . which was parallel* **to** *a certain sense,* **in** *an experience. . . .*
(26) *Everytime I put one of these buttons* **off**, *another one comes* **on**.
(27) *she* **sings** *everything she* **writes**.
(28) *. . . read the newspapers,* **watch** *the radio, and* **listen to** *T.V.*

The parallelism of structure is most strikingly evident for the word exchanges that cross clause boundaries, but even the within-clause exchanges show strong correspondence, usually involving two similarly placed words from distinct phrases. These phrases are quite often, for example, the noun phrases (NPs) of direct and indirect objects, or the NPs from a direct object and an adverbial phrase, or from successive adverbial phrases. This contrasts with the dominant pattern for sound or combined-form errors, in which the interacting elements are members of the same phrase. These differences in the structural features of the error types are summarized in Table V.

There are two aspects of this contrast which bear on the characterization given in (20) (a). One, the similarity of structural roles played by exchanged words, speaks for itself. The other turns on the fact that both verbs and prepositions are (relatively) frequently involved in the word exchanges. The prepositions are of particular interest since they do not enter into sound exchanges at all (nor, obviously into combined form exchanges). Both prepositions and verbs are important determinants of the relations between phrase types. It seems fairly clear that the sorts of structures typically involved in a sound or combined-form exchange are different from those of word exchanges, and that the nature of this difference is consonant with a characterization of the processing level at which word exchanges take place as one in which the functional relations between the words within constructions, and of the relations between the constructions themselves, are paramount.

It should be recognized that the facts about the error types we have just been discussing are not independent of the earlier observations about form class and error span. Though it is perfectly possible for adjacent element exchanges, or exchanges of one degree separation, to involve different phrases, it is clear that an increase in the separation of error elements will increase the likelihood of distinct phrases being involved. And, similarly, the likelihood that an error will involve elements of the same form class is decreased if both are members of the same phrase. Thus, one should see the error span and form class properties of exchanges on the one hand, and the structural properties in Table V on the other, as different, but closely related reflections of the same causes.

In the discussion thus far, we have focussed most attention on the level of processing described in (20) (a). There remain two principal issues to be dealt with before attempting to relate the predominantly syntactic processes now under examination to the more general framework of sentence production. These are (a) the character of the

TABLE V

STRUCTURAL SUMMARY FOR EXCHANGES[a]

Types of structures in which exchanges occur	Error class		
	(n=119) Sound	(n=46) Comb. form	(n=97) Word
Within NPs (or Pred. Phrase)	85	19	19
Names (e.g., *Sen Diago, Jahn Honcock,* Treato Seaty)	13	4	–
N and N (e.g., *cheeps and twirts*)[b]	9	2	4
Nominal Compounds (e.g., *skay pale; pre-truning*)	25	6	1
Adj (Adj) N (e.g., *stilthy finking drunk; sollow hound*)	25	5	–
N of *N* (e.g., *bate of dirth*)[b]	10	2	15
Pred. Phrase: (Adv Adj)	3	–	–
Between NPs	6	7	45
NP V NP	2	4	8
V NP (prep) NP	4	3	37
Between V and Pred. (or Subj. Phrases)	19	16	4
V NP, or V Adj Phrase or Adv Phrase (e.g., *he was slowing shides; that would pick in steoples mind*)	14	9	3
NP V (e.g., *you should have your brekes chācked*)	5	7	1
Between closed classes *not* in NPs	–	–	12
V *Prep* NP *Prep* NP	–	–	10
V (VP) *Adv Adv*, or V *comp.* NP*comp.* VP	–	–	2
Between clauses	9	4	17
Relatives	2	1	2
Complements	3	3	4
Conjunctions	–	–	11
Clause adjuncts (e.g. *Helf, Helf, the wolp is after me*)	4	–	–

[a]The errors used as examples consist mostly of sound errors simply because the other error types have ample numbers of illustrations in the text.

[b]May be better analyzed as *NP* and *NP* or *NP* of *NP* in some cases.

level of processing outlined in (20) (b) and (b) the relation of the levels in (20) (a) and (b) to sound exchanges. We first take up the evidence relevant to (20) (b).

The errors we have been referring to as "combined form" exchanges are errors of a rather remarkable sort. They might, as a matter of fact, have been more aptly designated as "morpheme stranding" errors, for not only are the permuted elements nearly always free forms, but the elements that are left behind are as often bound morphemes, as examples: (16)–(19) and (29)–(32) show (permuted elements boldfaced).[18]

(29) . . . *but the* **clean***'s* **two***er*.
(30) . . . *I'm not in the* **read** *for* **mood***ing*.
(31) . . . *he made a lot of money in***telephon***ing* **stall***s*.
(32) *She's already* **trunk***ed two* **pack***s*.

Moreover, as the examples also attest, not just any bound morph gets stranded; it is, rather, those we might term "syntactically active" affixes; for example, tense and number morphemes, possessive morpheme, comparative morphemes, the *-er* of noun formation, and so forth. There are also a very few nonderivational morphemes and a nonmorphemic maverick or two which are intermixed in a fashion discussed below. Table VI provides a summary of these errors in terms of the sort of elements stranded by the exchange. Of the 46 errors classified as combined form exchanges, 33 involve only the bound morphs I have called "syntactically active," that is, the morphemes that are directly involved in syntactic rules (as, auxiliary elements, or possessives, or comparatives), or that affect the form class of the stem to which they are attached (as the *-er* of noun formation). Of the remaining thirteen errors, a further six also involve syntactically active bound morphs, but include as well a nonmorphemic segment (e.g., the *er* in *shoulder*), or what might be called "moribund" affixes, that is, those of dubious productivity and very likely not semantically analyzed (as, the prefixes in *along, install,* and *intend*).

[18] It will be apparent that the basis of classification of errors as "combined form errors" is crucial to any evaluation of the nature of stranded elements. If only errors which result in the movement of a morpheme are included, then, of course, only morphemes can be stranded. Such a classification does not require, however, that the moved elements be potentially free forms, nor that the stranded elements be bound morphs of a particular subvariety. In fact, the classification used was more conservative: an error was counted as a combined-form error just in case the permuted elements had the *form* of morphemes (i.e., could have been morphemes in a different environment), and the shorter of the two forms created by the error was the one which maintained its serial position. Thus the actual description of the elements involved in these errors is not dictated by the basis of the classification. Errors like, *errples of examors* (examples of errors), or *chainse playges* (place changes) *would* have been included had they been in the corpus. It simply turns out that no such errors were observed.

TABLE VI

SUMMARY OF STRANDED ELEMENTS IN COMBINED FORM ERRORS

No. of such errors	Type of stranded elements	Examples
9	Number only	*Make it so the tree has less apples* → . . . *apple has less trees*; also (17) above
8	Tense only	*It just started to sound* → . . . *sounded to start*; also (18) and (19) above
2	Tense and number	. . . *Windows rolled up* → *rolls windowed up*; also (32) above
6	Tense and/or number, and a derivational suffix	*All the starters scored in double figures* → . . . *scorers started*. . . .
4	Gerundive	*O.J. is thirst quenching* → . . . *quench thirsting*. . . .
3	Comparative and/or possessive	*the Cognitive Center's study of* . . . → . . . *Cognitive study's Center* . . . ; also (29) above
1	Tense and a free form	. . . *the flashlight he smashed* → . . . *smash light he flashed*
	A syntactically active bound morph and	
4	(a) a "moribund" morph	*I had intended staying* → . . . *instayed tending*; also, (g) in Table I and (31) above
2	(b) a nonmorph	*My shoulders are frozen* → . . . *frozers are shoulden*
2	Two free forms	*Paperback Booksmith* → *Paperbook backsmith*
1	A nonmorph and a free form	*get ready for bedtime* → . . . *bedy for red time*
3	Nonmorphs only	*Cambridge Fenway* → *Fenwidge Cāmway*, *Seato Treaty* → *Treato Seaty*, *Lackner and Goldstein* → *Goldner and Lackstein*
1	Moribund morph only	*He didn't get along so well* → . . . *awell so long*

It is a striking fact that involvement of a moribund morph or a nonmorph in these errors is usually accompanied by a syntactically introduced morpheme. A possibly related fact is that even among the affixes I have dubbed syntactically active, the derivational affixes also do not appear alone in stranding errors, but nearly always in conjunction with a morpheme introduced by a syntactic process; the syntactically introduced morphs, on the other hand, often appear singly in errors. It almost appears that the bound morphs in the domain of syntactic rules "catalyze" the involvement of other affixal elements or of elements that have the phonetic shape of affixes (e.g., the *er* in *shoulder*, or the *y* in *epitome*). At any event, it is true that 85% of the stranding errors involve syntactically introduced bound morphs (and the exceptions are of limited sort: three could be sound errors and four of them are proper names).

It would be unwise to conclude from the preceding that strictly derivational affixes *cannot* appear alone in stranding errors. The sample of stranding errors that we have is, after all, rather small. None the less, it is surprising that when morphemic decomposition of a word occurs in an exchange, it should be so nearly restricted to cases in which that decomposition implicates syntactic processes. This suggests two things. First, the level of representation at which stranding errors take place—by hypothesis, the level described in (20)(b)—is one for which the computational vocabulary is of morphologically complete types. Thus, at this level words like *compassion, inept, unwieldy, preamble,* and so forth, that are undoubtedly represented in the lexicon as sequences of morphemes, are single vocabulary units, coming apart at their morphemic seams only when an error in the integration of such an element with its syntactic frame occurs. Second, as the just preceding remark in fact assumes, the nature of the stranded elements in the combined-form errors indicates that we are, in fact, dealing with a processing level for which the syntactic organization of the sentence is at issue. This is important, for without this evidence the basis for imputing specifically syntactic significance to level (20)(b) is weak. We have only the fact that *word* forms are involved in the errors to go on, coupled with the clausal restriction on the error span. On the other hand, there is a clear failure of these errors to honor the sort of structural features we have taken as diagnostic of the syntactic processes for level (20)(a).

Why should the presence of a syntactically active bound morpheme be associated with an error at the level described by (20)(b)? Precisely because the attachment of a syntactic morpheme to a particular lexical item reflects a mapping from the "functional" level to the "positional" level of sentence planning. Why should stranding errors fail to honor form class? Because that is exactly the nature of the error being made; a given pair of free forms is assigned to the wrong position within the syntactic "frame" which determines the order of their surface appearance and their form class.

Finally, why should the elements we have called syntactically active bound morphemes be stranded? Not simply because their morphemic analysis is represented at (20)(b), for after all, so too is some sound structure, and the structure of some other morphemes. A stronger claim seems required, and it is suggested by the earlier observation that the stranding errors are restricted to open class items. That might be taken not as a definitional consequence of the error classification, but as a substantive claim about processing at this level; namely, that the open class elements are being "inserted" into a grammatical framework defined by closed-class elements, including syntactically active bound morphs. Such an assumption makes a good deal of sense in light of the sort of observations we made in discussing the phenomena of accommodation, particularly the preservation of phrasal stress. For as Boomer and Laver pointed out (1968, p. 12), the appearance of appropriate phrasal stress on misplaced elements requires that the sentence processing system provide for an independent (of lexical items) specification of that stress. A syntactic frame which specifies the serial order of content words would provide a vehicle for such an independent specification. A "list" of such frames might be thought of as a part of a heuristic system for mapping from an underlying functional representation of sentences onto a representation constrained by the real time occurrence of the words.

This separation of "form" and "content" items into distinct aspects of the processing system is attractive, for it provides not only for the facts discussed above, but also accounts in a natural way for the very strong restriction on *all* exchanges that has not yet been mentioned. That restriction precludes the interchange of open- and closed-class items (even when form class is violated, this restriction is not). Further, if we assume that the syntactic morphs are parts of such structural frames, it accounts for another feature of their behavior—they do not themselves exchange. It would, of course, be perfectly possible for stranded morphs to have the characteristics we have so far described and still undergo exchange; for example, it is logically possible to have an error like *the boys shouting disturbed us* → *the boying shouts disturbed us*, but no such errors appear in the

MIT corpus. The rarity of exchanges between the final portions of words makes the firm evaluation of this "error lacuna" difficult, but, the frequency with which these elements occur as the terminal portions of words, coupled with their behavior in stranding errors and their involvement in "shift" errors (mentioned later), makes their complete absence from the corpus of such final position exchanges noteworthy. Rather than prepare a table, I have simply listed all the cases of final position exchanges in the MIT corpus (permuted sequences boldfaced).[19]

(33) a. *Jill and Mike* → *J***ike** *and M***ill**
b. *Howard and Claire* → *H***aire** *and Cl***oward**
c. *Structure and function* → *stru***nction** *and fu***cture**
d. *expect and persuade* → *exs***pade** *and persw***ect**

(34) a. *I'm about to spill beer* → . . . *s***peer** *b***ill**
b. *I won't hit the ball hard* → . . . *b***ard** *h***all**
c. *you should have your sholders forward* → . . . *shol***wards** *for***ders**
d. *in the next ten minutes* → . . . *n***en** *t***ext**
e. *listen to me give* → . . . *l***ive** *me g***issen**
f. *night life* → . . . *n***ife** *l***ight**

(35) a. *the girl who called* → . . . *g***all** *who c***url***ed*
b. *a monkey's uncle* → . . . *monk***le***'s unk***ey**

(36) a. *the single biggest problem* → . . . *sing***est** *bigg***le**
b. *Singer Sewing Machine* → *Sing***ing** *Sew***er** *Machine*
c. *passive usage* → *pass***age** *us***ive**

Note first of all that most of these segments do not have a morphemic analysis at all, and of the cases, (36), where a morpheme is involved, it exchanges with a nonmorph, suggesting that the morphemic status of the element is irrelevant and that these are in fact sound errors. The errors in (35) (a) and (b) are especially instructive, for these are cases where the final portion of two words *does* permute, but one of them is inflected; the inflection stays behind.

[19] The four errors in (33) are suspect cases since either output order of the nouns is grammatically acceptable, and there is therefore no basis for classifying these as word final exchanges, rather than word initial, other than the speaker's report of his intention.

Note further that (36) (*c*) *might* have been classed as a morpheme-stranding error since both *-age* and *-ive* are morphemes (although it seems unlikely that *passive* should be so analyzed—i.e., it is a moribund morph in this case). If (36) (*c*) were so classified, it would then constitute the single case of a morpheme-stranding error which moved the final rather than the initial or medial portions of a word, and which moved bound rather than free forms.

This suggests not only that syntactically active bound morphs do not permute even when a sound exchange involving their word position occurs, but also that perhaps sound-exchange errors may be conditioned by the syntactic frames we have been discussing. (I will return to this point momentarily).

The final observation about syntactically active bound morphs concerns their involvement in what I have called "shifts"; for example, Table I, entry (f) and examples (6) and (9). Shifts in the location of a sound seem to fall into roughly two classes: those that involve nonfinal sounds and usually result in the formation or destruction of a cluster, and those that involve final sounds and are almost exclusively confined to syntactically active bound morphs. That is, one does *not* find errors like: *final sound* → *fina soundl,* or *tax restriction* → *tak restrictions,* or *end run* → *en rund*. One *does* find errors like (37)–(40), however.

(37) *they get wierder every day* → . . . *wierd everi*er . . .
(38) *he gets it done* → . . . *get it*s *done*
(39) *he goes back to* → *go back*s *to*
(40) *pull its gene*s *down* → . . . *gene down*s

This difference could be a "sampling error," and claims about the nonoccurrence of a given error type have a dismaying way of being disconfirmed by the next sleepy speaker one encounters. At any event, of the 25 cases of a word-final sound shift, 25 are syntactically active bound morphs (this includes both derivational and syntactic morphs, but principally the latter). It is clear that one cannot explain this by an appeal to phonologic constraints or to a prohibition against producing nonwords. There *are* final clusters in English (if shifts were restricted to cluster formation), and sound errors yield nonwords as often as they yield word forms. There are also numerous end-ryme errors in the MIT corpus (e.g., *Wenatchie is the app***ital** *capital of the world*; . . . *and so*l *did Tolman*; *half a cu*ff *of coffee*; etc.), so it cannot be a "constraint" against distortions of the ends of words that restricts such shifts to bound morphs. The obvious inference is that these are not sound errors at all—their status as bound morphs is related to their appearance in this error pattern.

If this is the case, it would make sense in light of the notion of a syntactic frame, of which these morphs are features. One would not expect exchanges of such morphs at the same level as exchanges of the elements (words) being fitted into the structural framework of which the bound morphs are a part. Hence, the stranding errors. But, one might very well expect to get errors of placement of these syntactic features when they are phonetically realized and attached to particular lexical items. Further, these morpheme shifts, unlike the stranding exchanges, frequently involve closed-class words—e.g., (37)–(40); 16 of the 25 shifts are between an open-class and a closed-class word, or between two closed-class words. This provides a convincing independent argument against considering these shifts as exchanges between a "zero" morph (phonetically unrealized) and a phone; for, of course, the words to which the shifts attach the bound morphs are often not capable of inflection. This involvement of closed class words in the bound-morpheme shifts is also typical of word shifts. I will presently argue (in discussing word shifts) that the shifts of bound morphemes and of free forms arise from similar processes, and that both are distinct from exchanges.

What we have discussed of the behavior of stranding exchanges and of morpheme shifts to this point seems to support the characterization offered in (20) (b) of a processing "level" constrained by the serial deployment of the lexical items in a sentence, and indicates that this is accomplished through the integration of two computationally distinct types of information—that connected with the open-class 'content" items on the one hand, and that connected with the closed-class structural elements which "define" the surface form of sentences on the other.

In the preceding discussion, I have used the behavior of sound exchanges as a touchstone of contrast with word exchanges, but have not explicitly addressed the question of how sound planning in general, and sound exchanges in particular, relate to the processing levels of (20)(a) and (b). Such comment as can be made is limited, but I will take up that issue now and at the same time summarize some of the contrasts discussed so far.

It will have been obvious that, by the diagnostic variables we have been using, sound exchanges are more akin to the combined-form exchanges than to the word exchanges. It should be emphasized that this is an empirical consequence of the error data and *not* a necessary condition on sound errors. There is no a priori reason why sound errors should not occur at any level of processing for which there exists some specification of phonetic information about lexical items. On the current evidence, one cannot rule out the possibility of sound errors at the level of word exchanges, as in (20)(a); but the assignment of morpheme-stranding exchanges and some sound errors (exchanges) to a common level is not only indicated by the facts we have, it makes a certain sort of sense as well. Consider the "facts" first, and then the sense:

	Sound	Combined form	Word
Phonetic similarity:			
of exchanged elements	Yes	(Possibly)	No
of environment	Yes	(Possibly)	No
Structural similarity:			
of exchanged elements (form class)	No	No	Yes
of environment (structural role)	No	No	Yes
Closed-class words involved	No	No	Yes
Error spans open-class words (error span frequently exceeds 1 word)	No	No	Yes
Clausally constrained	Yes	Yes	Yes

I have not commented previously on the phonetic relation between exchanged words, although the phonetic similarity of exchanged sounds and their environments has been mentioned. It is clear, however, that in this respect as in the others word exchanges contrast with sound exchanges. (There is also an indication that the combined-form exchanges are more likely to be phonetically similar than are exchanged words, especially at their final segments, i.e., the point of their attachment to stranded morphs. It is not a powerful constraint, however, if it does exist at all.) In every respect, save that of the clausal constraint, the sound exchanges and combined form exchanges are similar to each other and contrast with the word exchanges.

To this list of similarities in the behavior of sound and combined form exchanges we might also recall the pair of examples in (35). We were examining the behavior of syntactically active bound morphs, and we noted that (35)(a) and (b) were cases of word-final sound exchanges which stranded morphemes. This strongly suggests that at the point where sound exchanges occur, the bound morphs are still marked for their syntactic status. In terms of our earlier discussion, that would entail that they are still features of the "nodes" of the syntactic frame to which lexical items are attached. One other aspect of the sound exchanges which deserves stress in this connection is their apparent restriction to open-class words, especially since most other sound errors *do* include closed-class words as sources. This is somewhat surprising, barring some reason for "insulating" closed-class and open-class sound elements from interaction in exchanges. But the assumption that sound exchanges take place at the point of inserting lexical items into their "surface" structural frames, just as we have argued combined-form exchanges do, provides such a reason.

Finally, if we assume that the phenomena of phonetic accommodation cannot take place at the same level as sound exchanges and stranding errors, we need to postulate two levels at which sound-structure errors can occur. Under the current suggestion the first such level is coincident with the planning for the serial deployment of lexical items—that is, (20)(b)—and would be the point at which the phonetic consequences of syntactic variables are accommodated. Detailed phonetic regularities that are the consequence of the phonetic environment of a sound segment would be worked out at a subsequent level of phonological processing, and that level would be the locus of sound errors other than exchanges (barring the possibility of a fourth category arising out of motor interactions—like "tongue twisters" and some common word-internal errors of similar ilk; e.g., *tachistoscope* → *taskistoscope*).

This reconstruction of the role of sound exchanges also accommodates the clausal constraint displayed by such errors. It should be noted, however, that it is not necessary to assign these errors to a specifically syntactic level of processing in order to rationalize their sensitivity to a syntactic variable. For example, both their error span (an open-class word never intervenes between the source words) and the character of the constructions they occur in (see Table V) suggest that sound exchanges may very well take place in a system with units smaller than clause size. One might suppose, on these grounds, that sound-exchange interactions take place when the sound structure of the lexical items to be inserted into the hypothesized syntactic frames of level (20)(b) is specified in a limited capacity store and that store has a length of two content items. The *input* to this store might be clausally constrained such that two elements from different clauses are never simultaneously present in it. This would preclude the occurrence of interclause errors even though the size of the sound planning system was itself smaller than a clause and, indeed, even quite insensitive to syntactic processes in its internal operation. Given that combined form errors have so much in common with sound exchanges, in fact, one might even suppose that the stranding errors arise because of a misassignment of the output of the words from such a store to their positions in the syntactic frames that specify their surface order. This is no more than speculation, of course, for we have no grounds in the present error data to decide matters of such detail. It is sufficient to emphasize the point that the decision to assign sound exchanges to the level (20)(b) is compatible with the clausal constraint but not demanded by it.

There remains one more class of syntactic errors that bears in an interesting way on the issues we have been considering. Those are the "word shifts" we referred to earlier in our discussion of word exchanges. Errors such as those in (41)–(44) are evidently ambiguous as to their status—exchanges of adjacent elements, or shifts in the position of a single element.

(Elements that would be involved if an exchange is assumed are bracketed; the element presumably moved if a shift is assumed is boldfaced)

(41) *Who* (**else**) (*would*) *like one?* → *Who would else . . .*

(42) *There's (something)* (**very**) *peculiar about this* → *There's very something . . .*

(43) *They (might)* (**not**) *be too closely glued together* → *They not might be . . .*

(44) *I'll bet you (that)* (**what**) *he said was . . .* → *I'll bet you what that he . . .*

More than a terminological quibble is at issue here, for I wish to argue that word exchanges are typical of a level of processing at which the functional relations between vocabulary elements is at issue, while shifts are a consequence of the operations which map from a level that captures such functional relations to one that represents the serial constraints on vocabulary items, that is, that word shifts are more related to the processes of (20)(b) than those of (20)(a).

Though we cannot tell about cases like (41)–(44), we can, I believe, safely consider (45)–(50) as shifts for several reasons. Eighteen of the 72 shifts errors are like (45)–(50) in that an analysis of them as exchanges would require an exchange between a single word and a word sequence. An inspection of the properties of these errors (which I will refer to as "word/string" shifts) may help decide the unclear cases.

(45) *When the number of letters (in common) (gets) to be large → . . . letters gets in common to . . .*
(46) *I've got something (to tell you)* **(all)** → *. . . something all to tell you*
(47) **(Little)** *(beads of) blood will pop out on my brow,* → *beads of little blood will . . .*
(48) *If you can't figure* **(out)** *(what that) is,* → *. . . figure what that out is*
(49) *Who* **(else)** *(did I think) had left?* → *Who did I think else had left*
(50) *(This place is)* **(hardly)** *well run* → *Hardly this place is well run*

Consider what we would be required to claim if these were taken as exchanges. We would first of all have to accept that words may exchange with phrases, and presumably, therefore, that at some level they have similar descriptions. In principle, there is no bar to this move, and if it could be sustained for a putatively syntactic level of sentence processing it would be of considerable theoretical interest independent of our current concerns.[20] There are a number of grounds for rejecting this sort of account, however. For example, if words are exchanged with phrases in the same fashion as words are exchanged with other words (if they are, in other words, typically at the same level of description for the sentence processor), one might expect the phrase types that are exchanged with words to fulfill structurally similar roles, just as do the words in word exchanges. And, indeed, for those very few cases of clear exchanges where a single word is exchanged with a construction, this is true (e.g., *Did you ever go to the F & T with Bob → . . . to Bob with the F & T?*). Moreover, in the MIT corpus, all such cases proper names or idioms. Neither of these properties is true of the putatively exchanged elements in (45)–(50) or of the other similar cases. Even more telling, in (47)–(50) the word string that must be assumed to exchange is not even properly analyzed as a phrase. Indeed, much of the earlier discussion of bound morphemes can be seen as bearing on

[20] There are independent grounds for doubting that lexical and syntactic structures should be interchangeable within a given language (see Fodor, Fodor, & Garrett, 1975).

the relevant generalization here: exchanges do not take place between elements at significantly different levels of linguistic description, words do not exchange with bound morphs, sounds do not exchange with words, words do not exchange with phrases, We will thus consider the sort of errors in (45)–(50) to be shifts in the position of a single word or phrase, usually the former.

It is immediately apparent that the word/string shifts have properties (other than those just discussed) which distinguish them from exchanges and, further, that these are properties which they have in common with most of the 54 cases, like (41)–(44), that misplace single adjacent words. Consider the prominence of closed-class words. These errors more than any others seem to involve closed-class items: fourteen of the eighteen word versus word/string shifts involve apparent movement of closed-class words, and 42 of the 54 single-word cases. By contrast, only the word exchanges showed any closed-class involvement at all, and it was in a substantially smaller proportion of the errors, confined to elements in structurally similar roles. Thus, in both proportion (24% of word exchanges as against 77% of shifts) and structural role, the closed-class items of shifts contrast with those of word exchange. But it is in the contribution of adverbs to shift errors that the sharpest difference emerges. Adverbs are conspicuously absent (both open and closed class) from all exchanges, but in the shifts, 40 of the 72 errors involve misplaced adverbs or abverbials (nine of the 18 word versus string cases, and 31 of 54 single word cases). As all the preceding would suggest, form class is not preserved between the putatively exchanged adjacent words, and neither are they parts of the sort of structures typical of exchanges (e.g., noun phrases of various sorts). On every count save that of the clausal constraint (discussed later) they contrast with word exchanges.

At this point, it would seem that a prima facie case has been made for assigning all the 54 ambiguous movement cases to the shift category established by the 18 clear cases. But one may do a bit better than that if one takes seriously the arguments from diagnostic variables that we have used in the preceding several sections. We can "purge" the ambiguous 54 of just those cases which could reasonably be exchanges—of either level (20)(a) or (20)(b)—on grounds of form class and structural role. There are, in fact, twelve such cases; (51)–(54) give four that are typical.

(51) *It's just* **plain heavy** *stop-and-go traffic down there* →
. . . just heavy plain stop . . .

(52) *Though the long* **Labor-day** *weekend . . .* → *. . . long day-Labor week . . .*
(53) *I'd like to make a credit-***card call** → *. . . a credit-call card*
(54) *You're not allowed to* **mix meat** *with milk* → *. . . to meat mix with . . .*

The contrast of these with the errors of (41)–(50), and their similarity to the errors of word and, especially, combined-form exchanges is obvious.[21] Whether one confines one's attention to the presumably pristine remainder of 60 shifts or contemplates the entire set of 72, the account of this error pattern is transparent. They are predominantly instances in which a word in the surface positional string must intervene between two words that are in immediate construction at the functional level. Another way of reflecting this is to note the frequency with which movement transformations are involved in the linguistic derivation of the structures in which shift errors take place. As the frequency of adverbial involvement indicates, adverb placement will figure in the majority of the shift cases. Note, too, that these are almost entirely closed-class adverbs (qualifiers and quantifiers like *so, very, really, quite, all, else,* etc.). Moreover, the adverb-placement transformations very often occur in conjunction with other movement transforms (e.g., "*wh-* fronting," "particle movement," "question," "extraposition of relatives," etc.).

There are several very important aspects and implications of these errors which should be stressed. Note first of all the parallel between the word shifts and the bound-morpheme shifts discussed earlier. Both types of shifts seem to be strongly restricted to *syntactic* morphemes. The interpretation of the bound morphemes as features of the syntactic frames which organize the serial order of the open-class lexical formatives might thus quite reasonably be extended to the closed-class syntactic elements of word shifts as well.[22]

[21] If one wished to carry this bootstrap operation further, of course, the obvious move would be to sort out those combined form errors that arise at the functional level and those word exchanges that arise at the positional level. Though one is filled with an ineffable sense of tidiness when this operation is performed on the data, it does not shed any new light on the mechanisms underlying the error classes.

[22] The similarities of the bound morpheme shifts and the word shifts are perhaps stronger than I have suggested in the text. Jackendoff (1969) has argued that the bound morphs I have referred to as "syntactic" are, in fact, the only vehicles for *transformationally* altering word form. Thus, *all* the morpheme shifts, both bound and free forms, would implicate transformational placements by this argument.

Second, note that these sorts of word shifts reflect an aspect of the relations between the functional and the positional levels of representation that was hinted at in the discussion of differences among the error types in terms of error span. Elements which are near to each other in the surface form of a sentence need not be adjacent at underlying levels and conversely. If adjacancy is an important determinant of the occurrence of sound exchange and combined form exchanges, one might expect to find grammatical formatives and "movable" open-class elements as the only types of elements which intervene between the source words of an exchange between sounds or combined forms. That appears, in fact, to be true. Error spans of one or two words are found for sound exchanges, but the intervening elements are closed class, grammatical formatives, or movable elements like adverbs. This is not true of independent word exchanges, for which open-class words frequently intervene between the two error positions. When this difference is taken account of, the differences in the error spans of word exchanges and sound and combined forms are even more striking.

Closely related to the preceding observation is the obvious implication that the surface position of words does not predict their "position" at the functional level of representation. Consider the error in example (41) where the actual utterance was, *Who did I think else would come?* The position occupied by the word *else* corresponds to the location it would have in an underlying description of the sentence prior to *wh-* movement in the question form. One might almost think of this error and a number of very similar ones as "failures" of movement transformations (but see later). In this same vein, we might reconsider the matter of structural parallels between the two elements of a word exchange. When the main verb and the verb of an infinitival complement change places, the parallelism of structure at the surface level is not readily apparent; but when one considers an underlying, functional level of description, the parallelism is obvious. For example, (24) would have, very roughly, the representation: *Older men tend* (for *to older men choose younger wives*).

Finally we should explicitly note that the shift errors do not violate the clausal constraint: 64 of the 72 cases of shifts are intraclause errors. Thus, it is clear that the involvement of movement operations in the processing system is very likely defined over clause-length units, just as are the operations of inserting lexical elements into their structural frames.

One should not assume that because transformations figure in the linguistic derivation of the structures in which shift errors occur that, therefore, the linguistic transforms are part of the information processing that underlies these sentences. Having noted that, however, the evidence of these errors strongly indicate that the "movement" of lexical items is involved in the mapping from the "functional" level of representation that is the result of processing referred to in (20)(a), to the "positional" level of representation that results from the processes of (20)(b).

We should at this point take explicit notice of a systematic and heretofore harmless ambiguity in the way I have used the term "level." Clearly, it may refer to both a structural representation and to the processing activity presumed to give rise to it. The shift errors simply provide the most obvious demonstration of this distinction. If we restated the implications of the preceding several sections in terms of that distinction, we would characterize the error processes of (20)(a) as attendant upon the mapping from the message level of representation to a functional level of representation; the error processes of (20)(b) would be assumed to arise from the mapping of that level onto the surface positional level given by the "syntactic frames" we discussed earlier. I will argue in the next section that there are other errors which should be assigned to the message-to-functional level mapping.

VII. Fusions and Substitutions: Semantic Errors

This discussion of semantic factors in production will be limited; it is presented primarily for purposes of contrast with the preceding discussion of errors I have deemed syntactic. The first thing of import to note is the absence of any indication that specifically semantic factors influence the exchange errors of various sorts.

One may look for semantic relations between the two words involved in a sound, combined form, or word exchange just as one looks for phonetic and syntactic relations.[23] But while the evidences

[23] The phonetic and semantic "indices" of similarity are my own judgments. They are conservative, and will err on the side of ignoring any "subtle" involvements of sound or meaning. Unless there was an obvious phonetic overlap (e.g., *Did you know you have a* **sole** *in the* **hole** *of your shoe?*), the case was counted as phonetically unrelated (e.g., *I left the* **briefcase** *in my* **cigar.**) Similarly for semantic cases; unless there was near synonomy or antonymy (e.g., *lion/tiger* or *car/truck*; *start/stop, on/off*) cases were counted as semantically unrelated (e.g. *car/gas, barns/cats, idea/guy* etc.)

for involvement of the latter factors are plentiful, no suggestion of a semantic relation (that is not exhausted by correlated syntactic features) is to be found. The many examples of the various error types given in the preceding discussions are quite typical in this respect. The only place where there is a hint of semantic regularity appears in the cross-clausal word exchanges. There are, however, too few such cases to make an effective assessment, and even within the small set available there are exchanges that are hard to rationalize semantically. It thus seems very likely, not only on the positive grounds of syntactic and phonological involvement in exchanges and related error types, but also on the negative grounds of no apparent semantic effects, that we are dealing with syntactic processes in exchanges.

One has only to turn to errors like those of (53)–(62) to see what a "semantic involvement" yields (see also Table I). These errors are of two sorts: *fusions,* in which two expressions destined for the same functional role are in competition, and portions of both are output, and *substitutions,* in which roughly the same circumstance yields a different word than that intended by the speaker.

(53) *I just* **snabbed** *it* (intended *nabbed it* or *snapped it up*)
(54) *Nobody gets very* **upcited** *about that . . .* (intended: *excited* or *upset*)
(55) *I don't want to* **intervere** (intended: *intervene* or *interfere*)
(56) *I don't know what the* **outshot** *of that is* (intended: *outcome* or *upshot*)
(57) *Do you want to try just a* **tab** *. . . uh, a* **dad,** *uh, Jesus! Do you want a dab or a tad of this stuff?* (speaker intended only one of the two expressions initially)
(58) *I'm chronically on the* **fringe** *of . . . on the verge of making a break.*
(59) *Now that's what I call a full* **cup** *of tea.* (speaker was commenting on an overflowing **pot** of tea)
(60) *What I've done here is torn* **together** *three . . . uh, torn apart three issues that . . .*
(61) *I have to leave in at* **least** *. . . in at most an hour.*
(62) *I would like to see it now that I've* **written** *the book—uh, read the book.*

The fact there are semantic relations between the conflated words is very clear. A discussion of the precise nature of those relations would take us beyond the scope of this paper. Two or three comments are in order, however.

First of all, recall the discussion of shift errors, in which I argued against the assumption of equivalence between words and phrases at the syntactic level of processing. In the fusion errors, however, we find fairly frequent cases like the one in (53). Thus, one might wish to argue that where semantic (message level) selection among expressions is concerned, there is a word–phrase equivalence. The indications from the errors in the MIT corpus are, however, that this is of a very limited sort. Nearly all the cases of such a conflation involve idioms or very highly practiced "formula phrases."

Second, we note that a very large percentage of the substitutions involve outright antonyms or "pragmatic" opposites (e.g., *write/read, speak/listen, see/hear* etc.),[24] while the majority of the fusions turn on a rough sort of synonomy (e.g., *athlete/player, grip/fist, bet/guess, bottles/jars, shout/yell*). And related to this, most of the cases of word/phrase conflation arise as fusions. This, plus one further point raised later, suggests that though the fusions and substitutions ought both be considered a consequence of mapping operations from messages onto the first syntactic level, they may nevertheless arise from rather different aspects of that process—roughly, that the substitutions arise from outright errors of lexical selection, while the fusions may arise from failures of an evaluation or checking procedure which determines the appropriateness of the mapping from messages to the functional level.

It is worthwhile commenting on the fact that it is a loose sort of synonomy that connects the two expressions involved in a fusion. But that approximateness is a bit deceptive in one respect: *within the context of the intended utterance,* the two words almost invariably have equal currency. That is, they will equally well convey the communicative intent of the speaker. This would seem to rule out

[24] The tendency for substitutions to be between opposites is true for cases (most of those in the MIT corpus) for which there is no apparent "motivational" account of the error or its direction. The so-called "Freudian" interpretations of word substitutions (e.g., *Who said that liars . . . uh that lawyers make a living by shoveling smoke?*) and sound errors (e.g., *We can make passionate love while the bed breaks, . . . I mean, while the* **bread** *bakes!*) should perhaps be tempered by such facts. For, if there is a general tendency to substitute opposites, it dictates caution in interpreting cases of such substitution where one has, antecedently, an expectation of a motivational account of some particular person's error. Similarly, for sound errors, there really does not seem to be any particular evidence that sound errors which yield words (let alone situationally apposite words) are more likely than sound errors which yield nonwords. There undoubtedly are "Freudian slips," but they may be a good deal rarer than might be supposed.

the possibility that the representation of messages consists, in part, of the vocabulary items of the language. For it is hard to see how one could accommodate the sort of situationally apposite confusions so evident in fusion errors. If messages are framed in words, all the fusion errors ought, as most of the substitutions are, to be very closely related semantically, and in particular, that relation ought to be context free.

The last point I wish to raise is one that further indicates that these errors are not message-level interactions, but are a part of the process of moving from that level to the representation level of (20)(a). Both fusions and substitutions are strongly constrained by their ultimate form class role: nouns substitute or fuse with other nouns, verbs with other verbs, and so forth. This constraint on the word substitutions seems to require an interaction at the level where syntactic relations have been at least partially formulated.

The comments I have made on the semantic errors are meant to be little more than suggestive. Their proper evaluation requires, I believe, both a clearer understanding of the syntactic processes of production and more information concerning the organization of our stored lexical information. A good deal of experimental interest in this latter area has been manifest in recent years and the evaluation of errors like those of (53)–(62) within the context of, for example, proposals for the organization of the lexicon which arise out of lexical access studies (e.g., Conrad, 1972; Rubenstein, Garfield, & Millikan, 1970) would be worthwhile.

VIII. Summary

We can conveniently draw together the several threads of the sections on the various error types in terms of Fig. 1. We might now replace some of those question marks as in Fig. 2. In the figure the various error types that have been used to motivate the postulated levels of representation are indicated.

At several points in the preceding pages I have gone beyond the data to speculate on the possible character of the processing systems under discussion. Those matters aside, it seems clear that a satisfactory reconstruction of the error data requires the postulation of two distinct levels of syntactic analysis, and that these levels will differ in that one is sensitive to the functional grammatical relations among words and phrases, while the other is primarily responsive to the integration of grammatical formatives with the serial ordering of

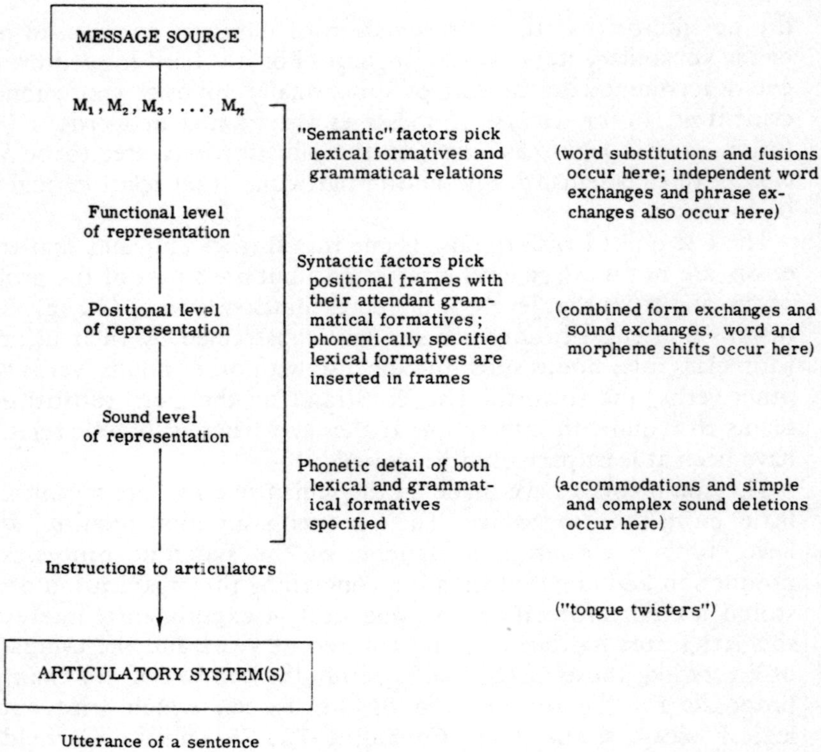

Fig. 2

"content words." Further, there is strong indication that this second level involves certain aspects of sound structure. Finally, it is clear that the clausal restriction on sound errors first suggested by Boomer and Laver is, in fact, a very much more general constraint on errors of a variety of types, and that it arises out of specifically syntactic processes. Such clausally constrained syntactic processes seem most likely to be connected with the mapping from the functional level of sentence planning to the positional one.

These and related observations about the role of motor integrations in error patterns will, I hope, prove amenable to experimental investigation. Ultimately it is only upon the results of such experimental work that detailed answers to the nature of the sentence production process can be worked out.

REFERENCES

Boomer, D. S., & Laver, J. D. M. Slips of the tongue. *British Journal of Disorders of Communication,* 1968, 3, 2–12.

Conrad, C. Cognitive economy in semantic memory. *Journal of Experimental Psychology,* 1972, 92, 149–154.

Fodor, J. A. *The Language of thought.* New York: Thos. Y. Crowell, 1975.

Fodor, J. A., Bever, T. G., & Garrett, M. F. *The psychology of language.* New York: McGraw-Hill, 1974.

Fodor, J. D., Fodor, J. A., & Garrett, M. F. The psychological unreality of semantic representations. *Lingusitic Enquiry,* 1975, in press.

Fromkin, V. A. The non-anomalous nature of anomalous utterances. *Language,* 1971, 47, 27–52.

Jackendoff, R. S. Some rules for semantic interpretation for English. Unpublished doctoral dissertation, Massachusetts Institute of Technology, 1969.

Lees, R. B. *The grammar of English nominalizations.* Publ. No. 12. Bloomington: Indiana University Research Center in Anthropology, Folklore, and Linguistics, 1960.

McKay, D. G. Forward and backward masking in motor systems. *Kybernetik,* 1969, 2, 57–64.

McKay, D. G. Spoonerisms: the structure of errors in the serial order of speech. *Neuropsychologia,* 1970, 8, 323–350.

Merringer, S. M. *Aus dem Leben der Sprache.* Berlin: Behr's, 1908.

Nooteboom, S. G. *Some regularities in phonemic speech errors.* Annual Progress Report No. 2. Eindhoven: Institut voor Perceptie Onderzock, 1967.

Rubenstein, H., Garfield, L., & Millikan, J. A. Homographic entries in the internal lexicon. *Journal of Verbal Learning and Verbal Behavior,* 1970, 9, 487–494.

Shattuck, S. R. Speech errors: An analysis. Unpublished doctoral dissertation, Massachusetts Institute of Technology, 1974.

[10]

COGNITIVE SCIENCE 3, 177–212 (1979)

Elements of a Plan-Based Theory of Speech Acts*

PHILIP R. COHEN
Bolt Beranek and Newman Inc.
AND
C. RAYMOND PERRAULT
University of Toronto

This paper explores the truism that people think about what they say. It proposes that, to satisfy their own goals, people often *plan* their speech acts to affect their listeners' beliefs, goals, and emotional states. Such language use can be modelled by viewing speech acts as operators in a planning system, thus allowing both physical and speech acts to be integrated into plans.

Methodological issues of how speech acts should be defined in a plan-based theory are illustrated by defining operators for requesting and informing. Plans containing those operators are presented and comparisons are drawn with Searle's formulation. The operators are shown to be inadequate since they cannot be composed to form questions (requests to inform) and multiparty requests (requests to request). By refining the operator definitions and by identifying some of the side effects of requesting, compositional adequacy is achieved. The solution leads to a metatheoretical principle for modelling speech acts as planning operators.

1. INTRODUCTION

The Sphinx once challenged a particularly tasty-looking student of language to solve the riddle: "How is saying 'My toe is turning blue,' as a request to get off my toe, similar to slamming a door in someone's face?" The poor student stammered that in both cases, when the agents are trying to communicate something, they have analogous intentions. "Yes indeed" countered the Sphinx, "but what are those intentions?" Hearing no reply, the monster promptly devoured the poor student and sat back smugly to wait for the next oral exam.

*The research described herein was supported primarily by the National Research Council of Canada, and also by the National Institute of Education under Contract US-NIE-C-400-76-0116, the Department of Computer Science of the University of Toronto, and by a summer graduate student associateship (1975) to Cohen from the International Business Machines Corporation.

Contemporary philosophers have been girding up for the next trek to Giza. According to Grice (1957)[1], the slamming of a door communicates the slammer's anger only when the intended observer of that act realizes that the slammer wanted both to slam the door in his face and for the observer to believe that to be his intention. That is, the slammer intended the observer to recognize his intentions. Slamming caused by an accidental shove or by natural means is not a communicative act. Similarly, saying "My toe is turning blue" only communicates that the hearer is to get off the speaker's toe when the hearer has understood the speaker's intention to use that utterance to produce that effect.

Austin (1962) has claimed that speakers do not simply produce sentences that are true or false, but rather perform speech actions such as requests, assertions, suggestions, warnings, etc. Searle (1969) has adapted Grice's (1957) recognition of intention analysis to his effort to specify the necessary and sufficient conditions on the successful performance of speech acts. Though Searle's landmark work has led to a resurgence of interest in the study of the pragmatics of language, the intentional basis of communicative acts requires further elaboration and formalization; one must state for any communicative act, precisely which intentions are involved and on what basis a speaker expects and intends those intentions to be recognized.

The Sphinx demands a competence theory of speech act communication—a theory that formally models the possible intentions underlying speech acts. This paper presents the beginnings of such a theory by treating intentions as plans and by showing how plans can link speech acts with nonlinguistic behavior. In addition, an adequacy test for plan-based speech act theories is proposed and applied.

1.1 A Plan-based Theory of Speech Acts

Problem solving involves pursuing a goal state by performing a sequence of actions from an initial state. A human problem-solver can be regarded as "executing" a *plan* that prespecifies the sequence of actions to be taken. People can construct, execute, simulate, and debug plans, and in addition, can sometimes infer the plans of other agents from their behavior. Such plans often involve the communication of beliefs, desires and emotional states for the purpose of influencing the mental states and actions of others. Furthermore, when trying to communicate, people expect and want others to recognize their plans and may attempt to facilitate that recognition.

Formal descriptions of plans typically treat actions as *operators*, which are defined in terms of applicability conditions, called *preconditions*, *effects* that will be obtained when the corresponding actions are executed, and *bodies* that describe the means by which the effects are achieved. Since operators are repre-

[1]See also (Strawson, 1964; Schiffer, 1972)

sentations, their preconditions, effects, and bodies are evaluated relative to the problem-solver's model of the world. We hypothesize that people maintain, as part of their models of the world, symbolic descriptions of the world models of other people. Our plan-based approach will regard speech acts as operators whose effects are primarily on the models that speakers and hearers maintain of each other.[2]

Any account of speech acts should answer questions such as:

—Under what circumstances can an observer believe that a speaker has sincerely and successfully performed a particular speech act in producing an utterance for a hearer? (The observer could also be the hearer or speaker.)

—What changes does the successful performance of a speech act make to the speaker's model of the hearer, and to the hearer's model of the speaker?

—How is the meaning (sense/reference) of an utterance x related to the acts that can be performed in uttering x?

To achieve these ends, a theory of speech acts based on plans should specify at least the following:

—A planning system: a formal language for describing states of the world, a language for describing operators, a set of plan construction inferences, a specification of legal plan structures. Semantics for the formal languages should also be given.

—Definitions of speech acts as operators in the planning system. What are their effects? When are they applicable? How can they be realized in words?

As an illustration of this approach, this paper presents a simple planning system, defines the speech acts of requesting and informing as operators within that system, and develops plans containing direct requests, informs and questions (which are requests to inform). We do not, however, discuss how those speech acts can be realized in words.

We argue that a plan-based theory, unlike other proposed theories of speech acts, provides formal adequacy criteria for speech act definitions: given an initial set of beliefs and goals, the speech act operator definitions and plan construction inferences should lead to the generation of plans for those speech acts that a person could issue appropriately under the same circumstances.[3] This adequacy criterion should be used in judging whether speech act definitions pass certain tests, in particular, the test of compositionality. For instance, since a speaker can request that a hearer do some arbitrary action, the operator definitions should show how a speaker can request a hearer to perform a speech act. Similarly, since one can inform a hearer that an action was done, the definitions should capture a speaker's informing a hearer that a speech act was performed. We show how a number of previous formulations of requesting and informing are

[2]This approach was inspired by Bruce and Schmidt (1974) and Bruce (1975). This paper can be viewed as supplying methodological foundations for the analyses of speech acts and their patterned use that they present.

[3]Though this could perhaps be an empirical criterion, it will be used intuitively here.

compositionally inadequate, and then develop definitions of informing that can be composed into questions.

Another goal of this research is to develop metatheoretical principles that state how to formulate speech act definitions to pass these adequacy tests. This paper proposes such a principle and shows how its application leads to compositionally adequate definitions for multiparty requests (as in "Ask Tom to open the door").

To simplify our problems in the early stages of theory construction, several restrictions on the communication situation that we are trying to model have been imposed:

—Any agent's model of another will be defined in terms of "facts" that the first believes the second believes, and goals that the first believes the second is attempting to achieve. We are not attempting to model obligations, feelings, etc.

—The only speech acts we try to model are requests, informs, and questions since they appear to be definable solely in terms of beliefs and goals. Requesting and informing are prototypical members of Searle's (1976) "directive" and "representative" classes, respectively, and are interesting since they have a wide range of syntactic realizations, and account for a large proportion of everyday utterances.

—We have limited ourselves to studying "instrumental dialogues"—conversations in which it is reasonable to assume that the utterances are planned and that the topic of discourse remains fixed. Typically, such dialogues arise in situations in which the conversants are cooperating to achieve some task-related goal (Deutsch, 1974), for example, the purchasing of some item. The value of studying such conversations relative to the structure of a task is that the conversants' plans can be more easily formalized.

1.2 A Competence Theory of Speech Acts

At least two interdependent aspects of a plan-based theory should be examined—the plans themselves, and the methods by which a person could construct or recognize those plans. This paper will be concerned with theories of the first aspect, which we shall term *competence* theories, analogous to competence theories of grammar (Chomsky, 1965). A plan-based competence theory of speech acts describes the *set of possible plans* underlying the use of particular kinds of speech acts, and thus states the conditions under which speech acts of those types are appropriate. Such descriptions are presented here in the form of a set of operator definitions (akin to grammatical "productions") and a specification of the ways in which plans are created from those operators.

The study of the second aspect aims for a *process* theory, which concerns *how* an ideal speaker/hearer chooses one (or perhaps more than one) plan out of the set of possible plans. Such a theory would characterize how a speaker decides what speech act to perform and how a hearer identifies what speech act was performed by recognizing the plan(s) in which that utterance was to play a part.

By separating out these two kinds of theoretical endeavors we are not claiming that one can study speech act competence totally divorced from issues of processing. On the contrary, we believe that for a (careful) speaker to issue a particular speech act appropriately, she must determine that the hearer's speech act recognition process(es) will correctly classify her utterance. Thus, a competence theory would state the conditions under which a speaker can make that determination—conditions that involve the speaker's beliefs about the hearer's beliefs, goals, and inferential processes.

Our initial competence theory has been embodied in a computer program (Cohen, 1978) that can construct most of the plans presented here. Programs often point out weaknesses, inconsistencies, and incorrect assumptions in the statement of the competence theory, and can provide an operational base from which to propose process theories. However, we make no claims that computational models of plan construction and recognition are cognitive process theories; such claims would require empirical validation. Moreover, it is unclear whether there could be just one process theory of intentional behavior since each individual might use a different method. A more reasonable goal, then, is to construct computational models of speech act use for which one could argue that a person could employ such methods and converse successfully.

1.3 Outline of the Paper

The thread of the paper is the successive refinement of speech act definitions to meet the adequacy criteria. First, we introduce in sections 2 and 3 the tools needed to construct plans: the formal language for describing beliefs and goals, the form of operator definitions, and a set of plan construction inferences.

As background material, section 4 summarizes Austin's and Searle's accounts of speech acts. Then, Searle's definitions of the speech acts of requesting and informing are reformulated as planning operators in section 5 and plans linking those speech acts to beliefs and goals are given. These initial operator definitions are shown to be compositionally inadequate and hence are recast in section 6 to allow for the planning of questions. Section 7 shows how the definitions are again inadequate for modelling plans for composed requests. After both revising the preconditions of requests and identifying their side effects, compositional adequacy for multiparty requests is achieved. The solution leads to a metatheoretical "point of view" principle for use in formulating future speech act definitions within this planning system. Finally, section 8 discusses the limitations of the formalism and ways in which the approach might be extended to handle indirect speech acts.

2. ON MODELS OF OTHERS

In this section, we present criteria that an account of one agent's (AGT1) model of another's (AGT2's) beliefs and goals ought to satisfy.[4] A theory of speech acts need not be concerned with what is actually true in the real world; it should

[4]The representations used by Meehan (1976), and Schank and Abelson (1977) do not, in a principled way, maintain the distinctions mentioned here for belief or want.

describe language use in terms of a person's beliefs about the world. Accordingly, AGT1's model of AGT2 should be based on "believe" as described, for example, in Hintikka (1962; 1969). Various versions of the concept "know" can then be defined to be agreements between one person's beliefs and another's.

2.1 Belief

Apart from simply distinguishing AGT1's beliefs from his beliefs about AGT2's beliefs, AGT1's belief representation ought to allow him to represent the fact that AGT2 knows *whether* some proposition P is true, without AGT1's having to know which of P or ~ P it is that AGT2 believes. A belief representation should also distinguish between situations like the following:

1. AGT2 believes that the train leaves from gate 8.
2. AGT2 believes that the train has a departure gate.
3. AGT2 knows what the departure gate is for the train.

Thus, case 3 allows AGT1 to believe *that* AGT2 knows what the departure gate is without AGT1's actually knowing which gate AGT2 thinks that is. This distinction will be useful for the planning of questions and will be discussed further in section 6.

Following Hintikka (1969), belief is interpreted as a model operator A BELIEVE(P), where A is the believing agent, and P the believed proposition.[5] This allows for an elegant, albeit too strong, axiomatization and semantics for BELIEVE. We shall point out uses of various formal properties of BELIEVE as the need arises.

A natural question to ask is how many levels of belief embedding are needed by an agent capable of participating in a dialogue? Obviously, to be able to deal with a disagreement, AGT1 needs two levels (AGT1 BELIEVE and

[5] The following axiom schemata will be assumed:

B.1 aBELIEVE(all axioms of the predicate calculus)
B.2 aBELIEVE(P) => aBELIEVE(aBELIEVE(P))
B.3 aBELIEVE(P) OR aBELIEVE (Q) => aBELIEVE(P OR Q)
B.4 aBELIEVE(P&Q) <=>aBELIEVE(P) & aBELIEVE(Q)
B.5 aBELIEVE(P) => ~ aBELIEVE(~ P)
B.6 aBELIEVE(P => Q) => (aBELIEVE(P) => aBELIEVE(Q))
B.7 ∃x aBELIEVE(P(x)) => aBELIEVE(∃x P(x))
B.8 all agents believe that all agents believe B.1 to B.7

These axioms unfortunately characterize an idealized "believer" who can make all possible deductions from his beliefs, and doesn't maintain contradictory beliefs. Clearly, the logic should be weakened. However, we shall assume the usual possible worlds semantics of BELIEVE in which the axioms are satisfied in a model consisting of a *universe* U, a subset A of U of *agents*, a set of *possible worlds* W, and *initial world* WO in W, a *relation* R on the cross-product A× W × W, and for each world w and predicate P, a subset Pw of U called the *extension* of P in w. The truth functional connectives *and*, *or*, *not*, and => have their usual interpretations in all possible worlds. aBELIEVE(P) is true in world w if P is true in all worlds w1 such that R(a', w,w1), where a' is the interpretation of a in w. ∃x P(x) is true in world w if there is some individual i in U such that P(x) is true in w when all free occurrences of x in P are interpreted as i.

AGT1 BELIEVE AGT2 BELIEVE). If AGT1 successfully lied to AGT2, he would have to be able to believe some proposition P, while believing that AGT2 believes that AGT1 believes P is false (i.e., AGT1 BELIEVE AGT2 BELIEVE AGT1 BELIEVE (~ P)). Hence, AGT1 would need at least three levels. However, there does not seem to be any bound on the possible embeddings of BELIEVE. If AGT2 believes AGT1 has lied, he would need four levels. Furthermore, Lewis (1969) and Schiffer (1972) have shown the ubiquity of *mutual belief* in communication and face-to-face situations—a concept that requires an infinite conjunction of beliefs.[6] Cohen (1978) shows how a computer program that plans speech acts can represent beliefs about mutual beliefs finitely.

2.2 Want

Any representation of AGT2's goals (wants) must distinguish such information from: AGT2's beliefs, AGT1's beliefs and goals, and (recursively) from AGT2's model of someone else's beliefs and goals. The representation for WANT must also allow for different scopes of quantifiers. For example, it should distinguish between the readings of "AGT2 wants to take a train" as "There is a specific train that AGT2 wants to take" or as "AGT2 wants to take any train." Finally, it should allow arbitrary embeddings with BELIEVE. Wants of beliefs (as in "AGT1 WANTS AGT2 BELIEVE P") become the reasons for AGT1's telling P to AGT2, while beliefs of wants (i.e., AGT1 BELIEVES AGT1 WANTS P) will be the way to represent AGT1's goals P.[7] In modelling planning behavior, we are not concerned with goals that the agent does not think he has, nor are we concerned with the subtleties of "wish," "hope," "desire," and "intend" as these words are used in English. The formal semantics of WANT, however, are problematic.

3. MODELS OF PLANS

In most models of planning (e.g., Fikes & Nilsson, 1971; Newell & Simon, 1963), real world actions are represented by *operators* that are organized into plans.[8] To execute a plan, one performs the actions corresponding to the

[6]Lewis (1969) and Schiffer (1972) talk only about mutual or common knowledge, but the extension to mutual belief is obvious.

[7]This also allows a third place to vary quantifier scope, namely:

$$\exists x \text{ aBELIEVE aWANT } P(x)$$
$$\text{aBELIEVE } \exists x \text{ aWANT } P(x),$$
$$\text{aBELIEVE aWANT } \exists x P(x)$$

[8]One usually generalizes operators to *operator schemata* in correspondence with *types* of actions; operator instances are then formed by giving values to the parameters of an operator schema. Since only operator instances are contained in plans we will not distinguish between the operator schema and its instances unless necessary. The same schema/instance, type/token distinction applies as well to speech acts modelled as planning operators.

operators in that plan. An operator will be regarded as transforming the planner's model of the world, the *propositions* that the planner believes, in correspondence with the changes to the real world made by the operator's associated action.[9] An operator is *applicable* to a model of the world in which that operator's *preconditions* hold. Operators can be defined in terms of others, as stated in their *bodies* (Sacerdoti, 1975). The changes that an operator makes to the world model in which it is evaluated to produce a new world model are called that operator's *effects*.

We shall view plans for an arbitrary agent S to be constructed using (at least) the following heuristic principles of purposeful behavior:

At the time of S's planning:

1. S should not introduce in the plan actions whose effects S believes are (or will be) true at the time the action is initiated.
2. If E is a goal, an operator A that achieves E can be inserted into the plan.
3. If an operator is not applicable in the planner's belief model, all the preconditions of that operator that are not already true can be added to the plan.

   The previous two inferences reflect an agent's reasoning "in order to do this I must achieve that."
4. If the planner needs to know the truth-value of some proposition, and does not, the planner can create a goal that it know whether that proposition is true or false.
5. If the planner needs to know the value of some description before planning can continue, the planner can create a goal that it find out what the value is.

   The previous two inferences imply that the planner does not have to create an entire plan before executing part of it.
6. Everyone expects everyone else to act this way.

   Since agents can sometimes recognize the plans and goals of others, and can adopt others' goals (or their negations) as their own, those agents can plan to facilitate or block someone else's plans. Bruce and Newman (1978) and Carbonell (1978) discuss these issues at length.

The process of planning to achieve a goal is essentially a search through this space of inferences to find a temporal sequence of operators such that the first operator in the sequence is applicable in the planner's current world model and the last produces a world model in which the goal is true. A new world model is obtained by the execution of each operator.

3.1 The Form of Operators

Early approaches to problem-solving based on first order logic (Green, 1969; McCarthy & Hayes, 1969) have emphasized the construction of provably correct

[9]We are bypassing the fact that people need to observe the success or failure of their actions before being able to accurately update their beliefs. The formalism thus only deals with operators and models of the world rather than actions and the real world. Operators names will be capitalized while their corresponding actions will be referred to in lower case.

plans. Such approaches formalize the changes an action makes to the state of the world model by treating an operator as a predicate of one whose arguments is a *state variable*, which ranges over states of the world model. Unfortunately, to be able to reason about what is true in the world after an action is executed, one must give axiom schemata that describe which aspects of the state of the world are *not* changed by each operator. For instance, calling someone on the telephone does not change the height of the Eiffel Tower. This thorny "frame problem" (McCarthy & Hayes, 1969) occurs because individual states of the world are not related to one another *a priori*.

To overcome this problem, Fikes and Nilsson (1971) in their STRIPS planning system assume that all aspects of the world stay constant except as described by the operator's effects and logical entailments of those effects. Such an assumption is not formalized in the reasoning system, making it difficult to prove the correctness of the resulting plans. Nevertheless, it has become the standard assumption upon which to build problem-solvers. We too will make it and thus shall describe an operator's effects by the propositions that are to be added to the model of the world.[10]

All operator schemata will have two kinds of preconditions—"cando" and "want" preconditions. The former, referred to as CANDO.PRs, indicate proposition schemata that, when instantiated with the parameter values of an operator instance, yield propositions that must be true in the world model for that operator instance to be applicable. We do not discuss how they can be proven true. The "want" precondition, henceforth WANT.PR, formalizes a principle of intentional behavior—the agent of an action has to want to do that action.

The following example serves to illustrate the form of such definitions.

<u>MOVE(AGT,SOURCE,DESTINATION)</u>

CANDO.PR:	LOC(AGT,SOURCE)
WANT.PR:	AGT BELIEVE AGT WANT move-instance
EFFECT:	LOC(AGT, DESTINATION)

The parameters of an operator scheme are stated in the first line of the definitions and it is assumed that values of these parameters satisfy the appropriate selectional restrictions, (here, a person, and two locations, respectively). The WANT.PR uses a parameter "move-instance" that will be filled by any instance of the MOVE operator schema that is currently being planned, executed, or recognized. The CANDO.PR states that before an agent can move from the SOURCE location, he must be located there. The EFFECT of the MOVE indicates that the agent's new location is the DESTINATION.

S's plan to achieve goal G is pictured schematically in Figure 1 (P and Q are arbitrary agents, A1 and A2 are arbitrary actions). Instead of indicating the entire state of the planner's beliefs after each operator, those propositions that are effects of an operator and are preconditions of some other operator in the plan are presented.

[10]Those propositions that need to be deleted (or somehow made "invisible" in the *current* worlmodel) will not be discussed here.

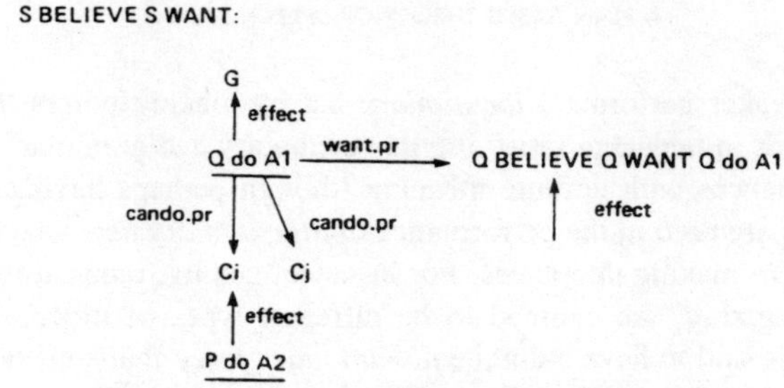

Figure 1. A schematic of S's plan to achieve G.

This diagram illustrates the building block of plans—given goal G, S applies an inference of type 2 and selects operator A1, whose agent is Q as a producer of that effect. That operator is applicable when preconditions Ci and Cj hold and when agent Q wants to perform A1. Type 3 inferences allow each of the preconditions to be achieved by other actions (e.g., A2), which may be performed by another agent (e.g., P). This chaining of operators continues until all preconditions are satisfied. Plan diagrams are thus read from "top" to "bottom".

To indicate that this schematic is part of agent S's plan, the plan components are "embedded" in what S BELIEVE S WANTs. The truth or falsity of preconditions is evaluated with respect to S's beliefs. For example, verifying the WANT.PR of operator A1 (i.e., Q BELIEVE Q WANT Q do A1) would involve establishing that S BELIEVE Q BELIEVE Q WANT Q do A1. If Q is the same person as S (i.e., S is planning her own action A1) then this condition is trivially true since A1 is already part of S's plan, and since for all agents R, we assume that if R BELIEVE (P) then R BELIEVE R BELIEVE (P). However, if Q is not the same as S, the WANT.PR also needs to be achieved, leading, as we shall see, to S's planning a speech act.

4. SPEECH ACTS

4.1 Austin's Performatives

Austin (1962) notes a peculiar class of declarative utterances, which he termed *performatives*, that do not state facts but rather constitute the performance of an action. For instance saying, "I hereby suggest you leave" is an act of suggesting. Unlike the usual declaratives, such sentences are not true or false, but rather are subject to the same kinds of failures ("infelicities") as nonlinguistic actions—such as being applied in the wrong circumstances or being performed insincerely.

Generalizing further, Austin claims that in uttering any sentence, one performs three types of speech acts: the *locutionary*, *illocutionary*, and *perlocution-*

ary acts. A speaker performs a *locutionary* act by making noises that are the uttering of words in language satisfying its vocabulary and grammar, and by the uttering of sentences with definite meaning (though perhaps having more than one). Such acts are used in the performance of *illocutionary acts* which are those acts performed *in* making utterances. For instance, stating, requesting, warning, ordering, apologizing, are claimed to be different types of illocutionary acts, each of which is said to have a unique *illocutionary force* that somehow characterizes the nature of the act. Each illocutionary act contains *propositional content* that specifies what is being requested, warned about, ordered, etc.

New distinctions, however, bring new problems. Frequently, when performative verbs are not used, the utterance's illocutionary force is not directly interpretable from its content. For example, to understand the force of the utterance "The door," the hearer may need to use his beliefs that the door is currently closed, that the speaker has two arm-loads of groceries, and that he wants to be on the other side of the door in determining that the speaker has requested that the door be opened. Furthermore, a speaker may appear to be performing one illocutionary act, and actually may be trying to use it to do something else. Thus, "We have to get up early tomorrow" may simply be an assertion but when said at a party, may be intended as an excuse to the host for leaving, *and* may be intended as a request that the hearer leave. Such *indirect speech acts* (Gordon & Lakoff, 1971; Searle, 1975) are the touchstone of any theory of speech acts.

The last major kind of act identified by Austin is the *perlocutionary* act—the act performed *by* making an utterance. For instance, with the illocutionary act of asserting something, I may *convince* my audience of the truth of the corresponding proposition (or *insult* or *frighten* them). Perlocutionary acts produce *perlocutionary effects*: convincing produces belief and frightening produces fear. While a speaker often has performed illocutionary acts with the goal of achieving certain perlocutionary effects, the actual securing of those effects is beyond his control. Thus, it is entirely possible for a speaker to make an assertion, and for the audience to recognize the force of the utterance as an assertion and yet not be convinced.

4.2 Speech Acts à la Searle

Searle (1969) presents a formulation of the structure of illocutionary acts (henceforth referred to simply as speech acts) by suggesting a number of necessary and sufficient conditions on their successful performance. He goes on to state rules corresponding to these conditions, for a speaker's using any "indicator of illocutionary force" to perform a particular speech act.

As an example, let us consider Searle's conditions for a speaker S, in uttering T, to request that some hearer H do action A. The conditions are grouped as follows:

Normal Input/Output Conditions. These include such conditions as: H is not deaf and S is not mute, joking, or acting.

Propositional Content Conditions. Literal speech acts only use propositions of certain forms. The restrictions on these forms are stated in the *propositional content conditions*. For a request, the proposition must predicate a future act of H.

Preparatory Condition. A preparatory condition states what must be true in the world for a speaker to felicitously issue the speech act. For a request, the preparatory conditions include:

—H is able to do A.
—S believes H is able to do A.
—It is not obvious to S and H that H will do A in the normal course of events (the "non-obviousness" condition).

Searle claims the non-obviousness condition is not peculiar to illocutionary acts. This paper will support his claim by showing how the condition can be applied more generally to rational, intentional behavior.

Sincerity Condition. A *sincerity condition* distinguishes a sincere performance of the speech act from an insincere one. In the case of a request, S must want H to do A; for a promise, S must intend to do the promised action; for an assertion, S must believe what he is asserting.

Essential Condition. An *essential condition* specifies what S was trying to do. For a request, the act is an attempt to get H to do A.

Force Condition (our terminology). The purpose of the *force condition* is to require that the speaker utter a speech act only if he intends to communicate that he is performing that act. "Intending to communicate" involves having certain intentions regarding how the hearer will recognize the force of the utterance. The basic idea is that it is intended that the hearer recognize that the speaker is trying to bring about the satisfaction of the essential condition. For a request this amounts to the speaker's wanting the hearer to realize the speaker intends for him to do A.

5. A FIRST REFORMULATION OF SEARLE'S CONDITIONS

Searle (1969) unfortunately does not supply justifications for the adequacy of his definitions for various kinds of speech acts. A primary goal of this paper is to show how a plan-based theory provides the basis for such adequacy criteria by allowing one to see clearly how changes in speech act definitions affect the plans that can be generated.

A second, more specific point of this formulation exercise is to show which of Searle's conditions are better regarded as pertaining to more general aspects of intentional behavior than to particular speech acts. In this spirit, we show how the sincerity condition, which we shall argue is a misnomer, and the propositional content and "non-obviousness" conditions arise during the course of planning. Concerning the remaining conditions, we assume the "normal input/output conditions," but have chosen not to deal with the force condition until we have a better understanding of the plans for speech acts and how they can be recognized. The remaining conditions, the preparatory and essential conditions, will be mapped into the formalism as the preconditions and effects of speech act operators.

5.1 First Definition of REQUEST

Searle claims the preparatory conditions are required for the "happy" performance of the speech act—where "happy" is taken to be synonymous with Austin's use of "felicitous." Austin was careful to distinguish among infelicities, in particular, misapplications (performing the act in the wrong circumstances), and flaws (incorrectly performing the act). We take Searle's preparatory conditions as conditions guaranteeing applicability rather than successful performance, allowing them to be formalized as preconditions. Thus if an operator's preconditions are not satisfied when it is performed, then the operator was "misapplied." Before expressing preconditions in a formalism, a systematic "point of view" must be adopted. Since the applicability conditions affect the planning of that speech act, the preconditions are stated as conditions on the speaker's beliefs and goals. Correspondingly, the effects describe changes to the hearer's mental state.[11] We establish a *point-of-view principle*, that is intended to be a guideline for constructing speech act definitions in *this* planning system—namely: preconditions begin with "speaker believe" and effects with "hearer believe."

Let us consider Searle's preparatory conditions for a request: H is able to do ACT, and S believes H is able to do ACT. From our discussion of "belief," it should be clear what H can *in fact* do, i.e., what the real world is like is not essential to the success of a request. What may be relevant is that S and/or H thinks H can do ACT. To formalize "is able to do A," we propose a predicate CANDO (Q,ACT) that is true if the CANDO.PR's of ACT are true (with person Q bound to the agent role of ACT).[12]

The essential condition, which is modeled as the EFFECT of a REQUEST,

[11]This does not violate our modelling just one person's view since a speaker, after having issued a speech act, will update his beliefs to include the effects of that speech act, which are defined in terms of the hearer's beliefs.

[12]This should be weakened to " . . . are true or are easily achievable"—i.e. if Q can plan to make them true.

is based on a separation of the illocutionary act from its perlocutionary effect. Speakers, we claim, cannot influence their hearers' beliefs and goals directly. The EFFECTs of REQUEST are modeled so that the hearer's actually wanting to do ACT is not essential to the successful completion of the speech act. Thus, the EFFECT is stated as the hearer's believing the speaker wants him to do the act. For important reasons, to be discussed in section 5.7, this formulation of the essential condition will prove to be a major stumbling block.

The operator REQUEST from SPEAKER to HEARER to do action ACT, which represents a literal request, can now be defined as:

```
REQUEST(SPEAKER,HEARER,ACT)
CANDO.PR:     SPEAKER BELIEVE HEARER CANDO ACT
                  AND
              SPEAKER BELIEVE
                  HEARER BELIEVE HEARER CANDO ACT
WANT.PR:      SPEAKER BELIEVE SPEAKER WANT request-instance
EFFECT:       HEARER BELIEVE
                  SPEAKER BELIEVE SPEAKER WANT ACT
```

5.2 Mediating Acts and Perlocutionary Effects

To bridge the gap between REQUESTs and the perlocutionary effect for which they are planned, a mediating step named CAUSE-TO-WANT is posited, that models what it takes to get someone to want to do something. Our current analysis of this "act" trivializes the process it is intended to model by proposing that to get someone to want to do something, one need only get that person to know that you want them to do it.

The definition of an agent's (AGT1) causing another agent (AGT) to want to do ACT is:

```
CAUSE-TO-WANT (AGT1,AGT,ACT)
CANDO.PR:     AGT BELIEVE
                  AGT1 BELIEVE AGT 1 WANT ACT
EFFECT:       AGT BELIEVE AGT WANT ACT
```

The plan for a REQUEST is now straightforward. REQUEST supplies the necessary precondition for CAUSE-TO-WANT (as will other act combinations). When the WANT.PR of some action that the speaker is planning for someone else to perform, is not believed to be true, the speaker plans a REQUEST. For example, assume a situation in which there are two agents, SYSTEM[13](S) and JOHN, who are located inside a room (i.e., they are at location INROOM). Schematically, to get JOHN to leave the room by moving himself to location

[13]The agent who creates plans will often be referred to as "SYSTEM," which should be read as "planning system."

OUTROOM, the plan would be as in Figure 2. Notice that the WANT.PR of the REQUEST itself, namely

S BELIEVE
S WANT
REQUEST(S,JOHN,MOVE(JOHN,INROOM,OUTROOM))

is trivially true since that particular REQUEST is already part of S's plan. The CANDO.PR's of the REQUEST are true if S believes JOHN is located INROOM and if it believes JOHN thinks so too. Thus, once the planner chooses someone else, say H, to do some action that it believes H does not yet want to do, a directive act (REQUEST) may be planned.

5.3 Comparison with Searle's Conditions for a REQUEST

Searle's "non-obviousness" condition for the successful performance of a request stated that it should not be obvious to the speaker that the hearer is about to

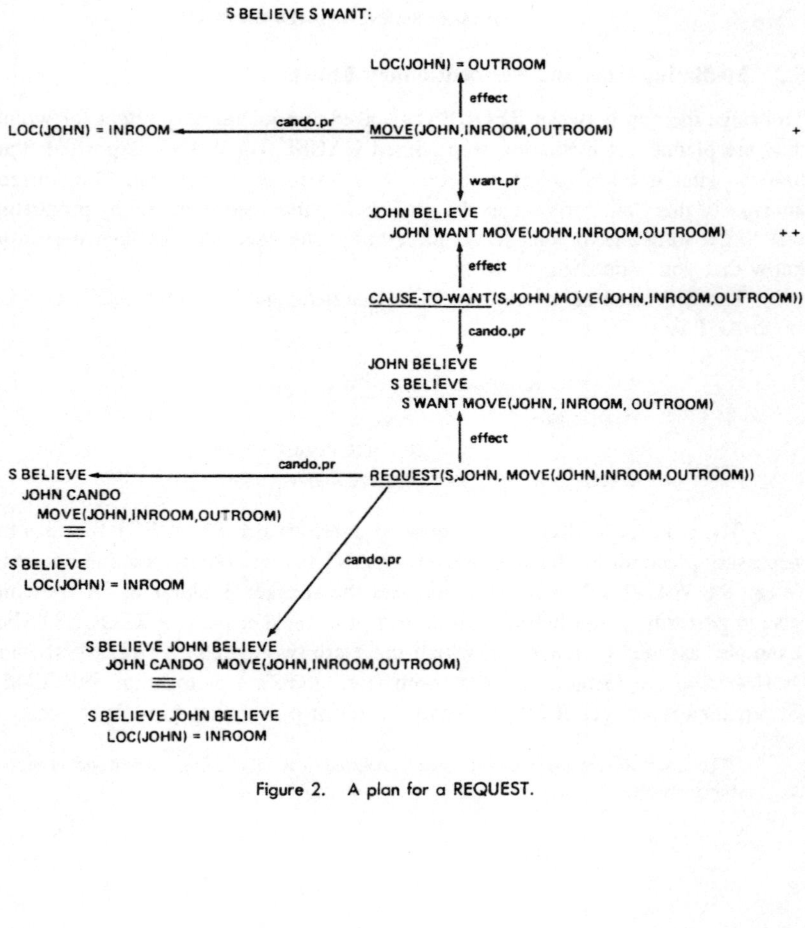

Figure 2. A plan for a REQUEST.

do the action being requested, independently of the request. If that were obvious to the speaker, the request would be pointless. However, as Searle noted, the non-obviousness condition applies more generally to rational, intentional behavior than to speech acts alone. In our formalism, it is the WANT.PR of the act being requested (goal "++" in Figure 2). If the planning system believed the WANT.PR were already true, i.e., if it believed that John already wanted to leave the room, then the plan would proceed no further; no REQUEST would take place.

Searle's "sincerity" condition, stated that the speaker had to want the requested act to be performed. The sincerity condition in the plan of Figure 2 is the goal labeled "+." The speaker's wanting the hearer to move is the reason for planning a REQUEST.

Notice also that the propositional content of the REQUEST, a future act to be performed by the hearer, is determined by prior planning—i.e., by a combination of that act's WANT.PR, the mediating act CAUSE-TO-WANT, and by the EFFECT of a REQUEST. Searle's propositional content condition thus seems to be a function of the essential condition (which is approximated by the EFFECTs of the speech act operator), as Searle claimed. So far, we have factored out those aspects of a request that Searle suggested were eliminable. Future revisions will depart more significantly.

5.4 Definition of INFORM

The speech act of informing is represented by the operator INFORM, which is defined as a speaker's stating a proposition to a hearer for the purpose of getting the hearer to believe that the speaker believes that proposition to be true. Such acts will usually be planned on the basis of wanting the hearer to believe that proposition. For a SPEAKER to INFORM a HEARER that proposition PROP is true, we have:

<u>INFORM(SPEAKER, HEARER, PROP)</u>

CANDO.PR:	SPEAKER BELIEVE PROP
WANT.PR:	SPEAKER BELIEVE
	SPEAKER WANT inform-instance
EFFECT:	HEARER BELIEVE
	SPEAKER BELIEVE PROP

The CANDO.PR simply states that the only applicability condition to INFORMing someone that proposition PROP is true is that the speaker believes PROP.[14] The EFFECT of an INFORM is to communicate what the speaker believes. This allows for the hearer to refuse to believe the proposition without

[14]Other preconditions to the INFORM act could be added—for instance, to talk to someone one must have a communication link (Schank & Abelson, 1977); which may require telephoning or going to that person's location, etc. However, such preconditions would apply to *any* speech act, and hence probably belong on the locutionary act of making noises to someone.

invalidating the speaker's action as an INFORM. Therefore, an intermediate "act," termed CONVINCE, is necessary to get the hearer to believe the proposition.

For a person AGT 1 to CONVINCE another person AGT that proposition PROP is true, we define:

CONVINCE(AGT1, AGT, PROP)
CANDO.PR: AGT BELIEVE
AGT1 BELIEVE PROP
EFFECT: AGT BELIEVE PROP

This operator says that for AGT 1 to convince AGT of the truth of PROP AGT need only believe that AGT1 thinks PROP is true. Though this may be a necessary prerequisite to getting someone to believe something, it is clearly not sufficient. For a more sophisticated precondition of CONVINCE, one might state that before AGT will be convinced, she needs to know the justifications for AGT1's belief, which may require that AGT believe (or be CONVINCE of) the justifications for believing those justifications, etc. Such a chain of reasons for believing might be terminated by mutual beliefs that people are expected to have or by a belief AGT believes AGT1 already has. Ideally, a good model of CONVINCE would allow one to plan persuasive arguments.[15]

5.5 Planning INFORM Speech Acts

The planning of INFORM speech acts now becomes a simple matter. For any proposition PROP, S's plan to achieve the goal H BELIEVE PROP would be that of Figure 3. Notice that it is unnecessary to state as a precondition to inform, that the hearer H does not already believe PROP. Again, this non-obviousness condition that can be eliminated by viewing speech acts in a planning context.

What would be Searle's sincerity condition for the INFORM above (S BELIEVE PROP) turns out to be a precondition for the speech act rather than a reason for planning the act as we had for REQUEST's sincerity condition, (i.e., SPEAKER BELIEVE SPEAKER WANT HEARER do ACT). If we were to use REQUEST as a model, the sincerity condition for an INFORM would be SPEAKER BELIEVE SPEAKER WANT HEARER BELIEVE PROP. One may then question whether Searle's sincerity condition is a consistent naming of distinctive features of various kinds of speech acts. Insincerity is a matter of falsely claiming to be in a psychological state, which for this model is either belief or want. By this definition, both conditions, SPEAKER BELIEVE PROP

[15] Without a specification of the justifications for a belief, this operator allows one to become convinced of the truth of one's own lie. That is, after speaker S lies to hearer H that P is true, and receives H's acknowledgment indicating H has been convinced, S can decide to believe P because he thinks H thinks so. Further research needs to be done on CONVINCE and BELIEVE to eliminate such bizarre behavior.

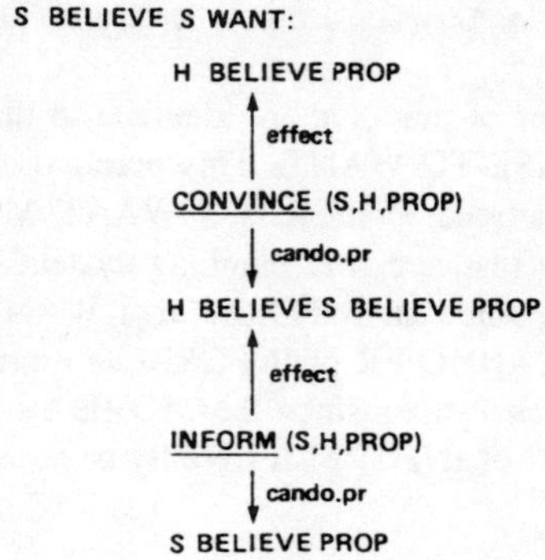

Figure 3. A plan for an INFORM.

and SPEAKER BELIEVE SPEAKER WANT HEARER BELIEVE PROP, are subject to insincerity.

5.6 Planning an INFORM of a WANT

As stated earlier, there are other ways to satisfy the precondition to CAUSE-TO-WANT. Since REQUEST was taken as a prototypical directive act, all members of that class share the same EFFECT (Searle's (1976) "illocutionary point"). However, issuing an INFORM of a WANT, as in "I want you to do X," also achieves it. Another plan to get John to move appears in Figure 4.

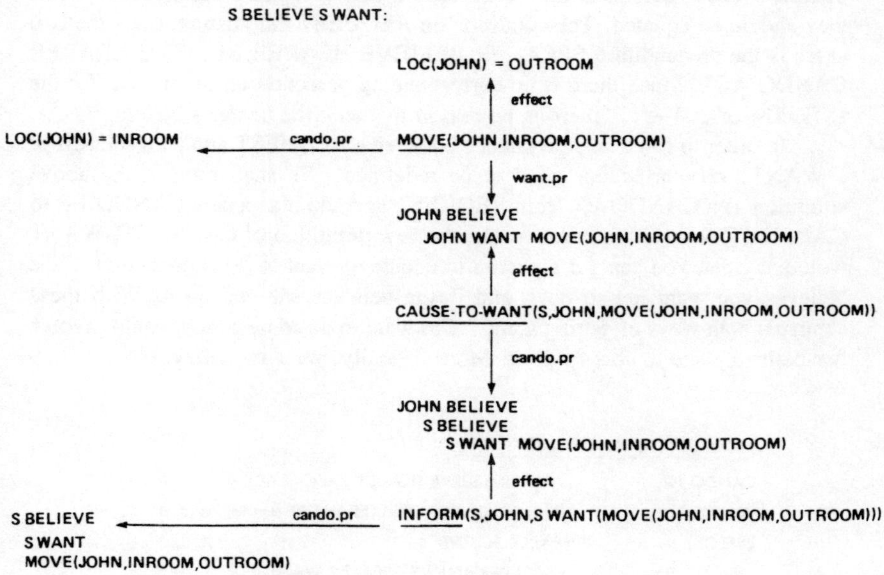

Figure 4. A plan for an INFORM of a WANT.

The initial stages of this plan are identical to that of Figure 2 through the CANDO.PR of CAUSE-TO-WANT. This precondition is achieved by an INFORM whose propositional content is S WANT MOVE (JOHN, INROOM, OUTROOM). In this instance, the planning system does not need to proceed through CONVINCE since an INFORM of a WANT produces the necessary effects. Testing the CANDO.PR of INFORM determines if the system believes this proposition, which it does since the MOVE by John is already one of its goals. The WANT.PR of INFORM is trivially true, as before, and thus the plan is complete.

5.7 REQUEST vs. INFORM of WANT

Searle claimed that the conditions he provided were necessary and jointly sufficient for the successful and nondefective performance of various illocutionary acts. Any behavior satisfying such a set of conditions was then said to be a particular illocutionary act. Thus, if two utterances have the same illocutionary force, they should be equivalent in terms of the conditions on their use. We believe that the two utterances "please open the door" and "I want you to open the door (please)" *can* have the same force as directives, differing only in their politeness. That is, they both *can be* planned for the same reasons. However, our treatment does not equate the literal speech acts that could realize them when they should be equated. The condition on REQUEST that distinguishes the two cases is the precondition SPEAKER BELIEVE HEARER BELIEVE HEARER CANDO ACT. Since there is no corresponding precondition in the plan for the INFORM of a WANT, there is no reason to check the hearer's beliefs.

In order to force an equivalence between a REQUEST and an INFORM of a WANT, various actions need to be redefined. We shall remove the above condition as a CANDO.PR from REQUEST and add it as a new CANDO.PR to CAUSE-TO-WANT. In other words, the new definition of CAUSE-TO-WANT would say that you can get a person to decide to want to do some action if she believes you want her to do it and if she believes she can do it. With these changes, both ways of getting someone to want to do some action would involve her believing she is able to do it. More formally, we now define:

<u>REQUEST (SPEAKER, HEARER, ACT)</u>

CANDO.PR: SPEAKER BELIEVE HEARER CANDO ACT
WANT.PR: SPEAKER BELIEVE SPEAKER WANT request-instance
EFFECT: HEARER BELIEVE
SPEAKER BELIEVE SPEAKER WANT ACT

and

CAUSE-TO-WANT (AGT1, AGT, ACT)

CANDO.PR: AGT BELIEVE
AGT1 BELIEVE AGT1 WANT ACT
AND
AGT BELIEVE AGT CANDO ACT

EFFECT: AGT BELIEVE AGT WANT ACT

Though REQUEST and INFORM of a WANT can achieve the same effect, they are not interchangeable. A speaker (S), having previously said to a hearer (H) "I want you to do X," can deny having the intention to get H to want to do X by saying "I simply told you what I wanted, that's all." It appears to be much more difficult, however, after having requested H to do X, to deny the intention of H's wanting to do X by saying "I simply requested you to do X, that's all." S usually plans a request for the purpose of getting H to want to do some act X by means of getting H to believe that S wants H to do it. While maintaining the distinction between illocutionary acts and perlocutionary effects, thus allowing for the possibility that H could refuse to do X, we need to capture this distinction between REQUEST and INFORM of WANT. The solution (Allen, 1979; Perrault & Allen, forthcoming) lies in formulating speech act bodies as plans achieving the perlocutionary effect—plans that a hearer is intended to recognize.

In the next two sections, we investigate the compositional adequacy of these operator definitions via the planning of REQUESTs that a hearer perform REQUEST or INFORM speech acts.

6. COMPOSITIONAL ADEQUACY: QUESTIONS

We are in agreement with many others, in proposing that questions be treated as requests for information. In terms of speech act operators, the questioner is performing a REQUEST that the hearer perform an INFORM. That is, the REQUEST leads to the satisfaction of INFORM's "want precondition." However, for a wh-question, the INFORM operator as defined earlier cannot be used since the questioner does not know the full proposition of which he is to be informed. If he did know what the proposition was there would be no need to ask; he need only decide to believe it.

Intuitively, one plans a wh-question to find out the value of some expression and a yes/no question to find out whether some proposition is true. Such questions are planned, respectively, on the basis of believing *that* the hearer knows what the value of that expression is or *that* the hearer knows whether the proposition is true, without the speaker's having to know what the hearer believes.

Earlier we stated that a person's (AGT1) belief representation should represent cases like the following distinctly:

1. AGT2 believes the Cannonball Express departs at 8 p.m.
2. AGT2 believes the Cannonball Express has a departure time.
3. AGT2 knows what the departure time for the Cannonball Express is.

Case 1 can be represented by a proposition that contains no variables. Case 2 can be represented by a belief of a quantified proposition—i.e.,

AGT2 BELIEVE
$\exists$x (the y : DEPARTURE-TIME(CANNONBALL-EXPRESS,y)) = x)

However, Case 3 can be approximated by a *quantified belief*, namely,

$\exists$x AGT2 BELIEVE
(the y : DEPARTURE-TIME(CANNONBALL-EXPRESS,y)) = x),

where "the y : P(y)," often written "iy P(y)," is the logical description operator read "the y which is P." This formula is best paraphrased as "there is something which AGT2 believes to be the departure time for the Cannonball Express."[16] Typical circumstances in which AGT1 might acquire such quantified beliefs are by understanding a definite description uttered by AGT2 referentially (Donnellan, 1966). Thus, if AGT2 says "the pilot of TWA 461 on July 4," AGT1 might infer that AGT2 knows who that pilot is.

Quantified beliefs often become goals when a planner needs to know the values of the parameters of an operator and when these parameters occur in that operator's preconditions.[17] We show how, when a quantified belief is a goal for AGT, AGT can plan a wh-question.

6.1 Planning Wh-Questions

First, a new operator, INFORMREF, and its associated mediating act CONVINCEREF, are needed.[18]

INFORMREF(SPEAKER,HEARER, λxDx) (i.e., D is a predicate of one argument)

CANDO.PR: $\exists$y SPEAKER BELIEVE (ixDx) = y
WANT.PR: SPEAKER BELIEVE SPEAKER WANT informref-instance
EFFECT: $\exists$y HEARER BELIEVE SPEAKER BELIEVE (ixDx) = y

[16]Another conjunction can be added to the representation of (3) as suggested by Allen (1979) to refine our representations of "AGT2's knowing what the value of the description is," namely:

$$\exists x\ [(\text{the y: D(y)} = x\ \&\ \text{AGT2 BELIEVE ((the y: D(y))} = x)]$$

We shall, however, use the simpler quantified belief formulation.

[17]We would prefer to formalize declaratively that "the agent of an action must know the values of the parameters of the action." One way of doing this is suggested by Moore (1979).

[18]In Cohen (1978) we achieved the same effect by parameterizing INFORM and CONVINCE so that different sets of preconditions and effects were used if the original goal was a quantified belief. In addition, Cohen (1978) did not use descriptions. We believe the formulation that follows, due to J. Allen, is clearer. The actual names for these acts were suggested by W. Woods.

Thus, before a speaker will inform a hearer of the value of some description, there must be some individual that the speaker believes is the value of the description, and the speaker must want to say what it is. The effect of performing this act is that there is then some individual that the hearer thinks the speaker believes to be the value of the description. As usual, we need a mediating act to model the hearer's then believing that individual to be the value of the description. To this end, we define AGT1's convincing AGT of the referent of the description as:

CONVINCEREF(AGT1,AGT, λxDx)

CANDO.PR: ∃y AGT BELIEVE AGT1 BELIEVE (ixDx) = y

EFFECT: ∃y AGT BELIEVE (ixDx) = y

Using these operators, if the planning system wants to know where Mary is and believes that Joe knows where she is, it can create the plan underlying the question "Where is Mary?" as is shown in Figure 5. After the system plans for Joe to tell it Mary's location, on the basis of believing that he knows where she is, it must get Joe to want to perform this act. In the usual fashion, this leads to a REQUEST and hence the construction of a question. The precondition to

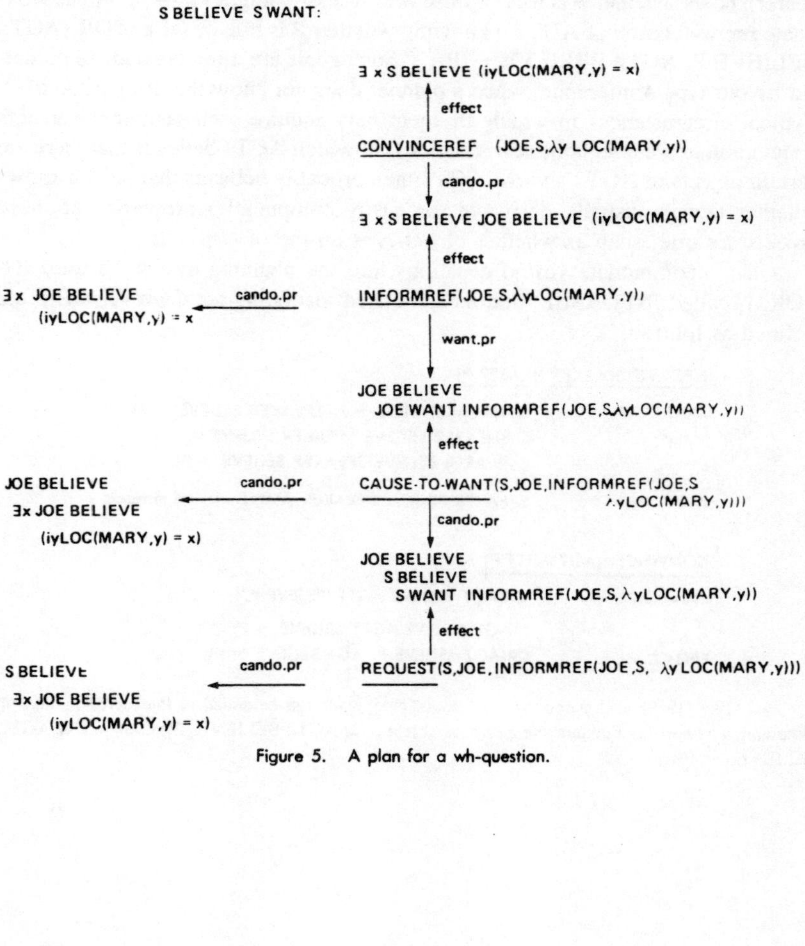

Figure 5. A plan for a wh-question.

CAUSE-TO-WANT, namely, JOE BELIEVE JOE CANDO the INFORMREF is actually:

JOE BELIEVE
  ∃y JOE BELIEVE
    ixLOC(MARY,x) = y

which is implied by

∃y JOE BELIEVE ixLOC(MARY,x) = y

that was asserted, for this example, to be one of the planning system's beliefs. Notice, that the planning of this question depends upon the system's having chosen Joe to tell it the answer, and upon its having chosen itself to get Joe to want to perform the INFORM. Section 7 discusses what happens when different decisions are made.

6.2 Plans for Yes/No Questions

To plan a yes/no question about some proposition P, one should think that the hearer knows whether P is true or false (or, at least "might know"). An approximate representation of AGT2's knowing whether P is true or false is OR (AGT2 BELIEVE P, AGT2 BELIEVE ~ P)).[19] Such goals are often created, as modelled by our type 4 inference, when a planner does not know the truth-value of P. Typical circumstances in which an agent may acquire such disjunctive beliefs about another are telephone conversations, in which AGT1 believes that there are certain objects in AGT2's view. AGT1 then probably believes that AGT2 knows whether certain visually derivable (or easily computable) properties of those objects are true, such as whether object A is on top of object B.

To accommodate yes/no questions into the planning system, a third INFORM, called INFORMIF, and its associated mediating act CONVINCEIF are defined as follows:

INFORMIF(SPEAKER,HEARER,P)

CANDO.PR:	OR(SPEAKER BELIEVE P, SPEAKER BELIEVE ~ P)
EFFECT:	OR(HEARER BELIEVE SPEAKER BELIEVE P, HEARER BELIEVE SPEAKER BELIEVE ~ P)
WANT.PR:	SPEAKER BELIEVE SPEAKER WANT informif-instance

CONVINCEIF(AGT,AGT1,P)

CANDO.PR:	OR(AGT BELIEVE AGT1 BELIEVE P, AGT BELIEVE AGT1 BELIEVE ~ P)
EFFECT:	OR(AGT BELIEVE P, AGT BELIEVE ~ P)

[19] Allen (1979) also points out that another conjunct can be added to the representation of "knowing whether" as a disjunctive belief, to obtain (P & AGT2 BELIEVE (P)) OR (~ P & AGT2 BELIEVE (~ P)).

The plan for a yes/no question to Joe is now parallel to that of a wh-question.[20] That is, in the course of planning some other act, if the system wants proposition P to be true or to be false, and if the truth-value of proposition P is unknown to it, it can create the goal OR(SYSTEM BELIEVE P, SYSTEM BELIEVE ~ P). For instance if P were LOC(MARY,INROOM), the illocutionary acts underlying the question to Joe "Is Mary in the room?" can be planned provided the planning system believes that Joe either believes P is true or he believes P is false. That disjunctive belief could be stated directly or could be inferred from a belief like ∃y JOE BELIEVE(ixLOC(MARY,x)) = y—i.e., there is something Joe believes is Mary's location. But if it had some idea where Joe thought Mary was, say OUTROOM, then it would not need to ask.

6.3 Summary

A plan for a question required the composition of REQUEST and INFORM and led to the development of two new kinds of informing speech acts, INFORMREF and INFORMIF, and their mediating acts. The INFORMREF acts lead to "what," "when," and "where" questions while INFORMIF results in a yes/no question.[21] The reason for these new acts is that, in planning a REQUEST that someone else perform an INFORM act, one only has incomplete knowledge of their beliefs and goals; but an INFORM, as originally defined can only be planned when one knows what is to be said.

7. COMPOSITIONAL ADEQUACY AND THE POINT OF VIEW PRINCIPLE

Earlier, a guiding "Point of View Principle" (POVP) for defining speech acts as planning operators was proposed: the preconditions of the operator should be stated from the speaker's point of view, i.e., in terms of the speaker beliefs; the effects should be stated from the hearer's point of view. We now wish to judge the adequacy of speech act definitions formulated along these lines. The test case

[20]Searle (1969) suggested there were different speech acts for real and teacher-student (or exam) questions, where in the latter case, the questioner just wants to know what the student thinks is the answer. Since teacher-student questions seem to have similar conditions on their appropriateness as real questions, save the questioner's intention to be convinced, we have good reason for factoring the mediating acts out of each of the three INFORM act types. This leaves the INFORM acts neutral with respect to what kind of question they are contained in. In general, if the perlocutionary effects of an INFORM were incorporated into the act's definition, then we would need two new primitive teacher-student question speech acts. For now, we opt for the former.

[21]The language for stating operators needs to be extended to account for "which," "how," and "why" questions. For instance, "why" and "how" questions involve quantifying over actions and/or plans.

will be the composing of REQUESTs, i.e., the planning of a REQUEST that some third party himself perform a REQUEST. For instance, the utterance "Ask Tom to tell you where the key is" is an example of such a third party request.

The current definitions of speech acts will be shown to be compositionally inadequate since they force speakers to have unnecessary knowledge about intermediaries' beliefs. Achieving compositional adequacy, however, requires more than a simple restatement of the point of view principle; the side effects of speech act operators also must be considered.

Our scrutiny will be focused upon the seemingly innocent precondition to REQUEST, SPEAKER BELIEVER HEARER CANDO ACT whose form depended on the POVP. The goal is to show how the POVP leads us astray and how a formulation of that precondition according to a new POVP that suggests a more neutral point of view for speech act definitions sets us back on course. From here on, the two versions of the precondition will be referred to as the "speaker-based" and "neutral" versions.

7.1 Plans for Multiparty Speech Acts

Multiparty speech acts can arise in conversations where communication is somehow restricted so as to pass through intermediaries.[22] The planning system, since it is recursive, can generate plans for such speech acts using any number of intermediaries provided that appropriate decisions are made as to who will perform what action.

Let us suppose that the planning system wants to know where a particular key is and that it must communicate through John. We shall use the speaker-based precondition on REQUEST for this example, and for readibility, the following abbreviations:

SYSTEM—S	TOM—T	JOHN—J
BELIEVE—B	WANT—W	LOC(KEY23,y)—D(y)

Figure 6 shows the plan for the specific three-party speech act underlying "Ask Tom to tell me where the key is."

S develops the plan in the following fashion: T is chosen to tell S the key's location since, we shall assume, he is believed to know where it is. Since T is not believed to already want to tell, and since S cannot communicate directly with T (but T can communicate with S), J is chosen to be the one to talk T into telling. Since J is not believed to already want to do that, S plans a REQUEST that J perform a REQUEST, namely REQUEST(S,J,REQUEST (J,T,INFORMREF (T,S,λyLOC (KEY23,y)))). J, then, is an intermediary who is just expected to do what he is asked; his status will be discussed soon.

[22]For instance, in the Stanford Research Institute Computer-based Consultant research (Deutsch, 1974) communication between an expert and an apprentice was constrained in this way. The apprentice typically issued such speech acts, while the expert did not.

The preconditions that need to be satisfied in this plan are:

S BELIEVE:

(P1) ∃y T BELIEVE ⌊ixLOC(KEY23,x)=y⌋

(P2) T BELIEVE (P1) (implied by P1)

(P3) J BELIEVE (P1)

(P4) J BELIEVE J BELIEVE (P1) (implied by P3)

(P5) S BELIEVE J BELIEVE (P1) (implied by P3)

S BELIEVE S WANT:

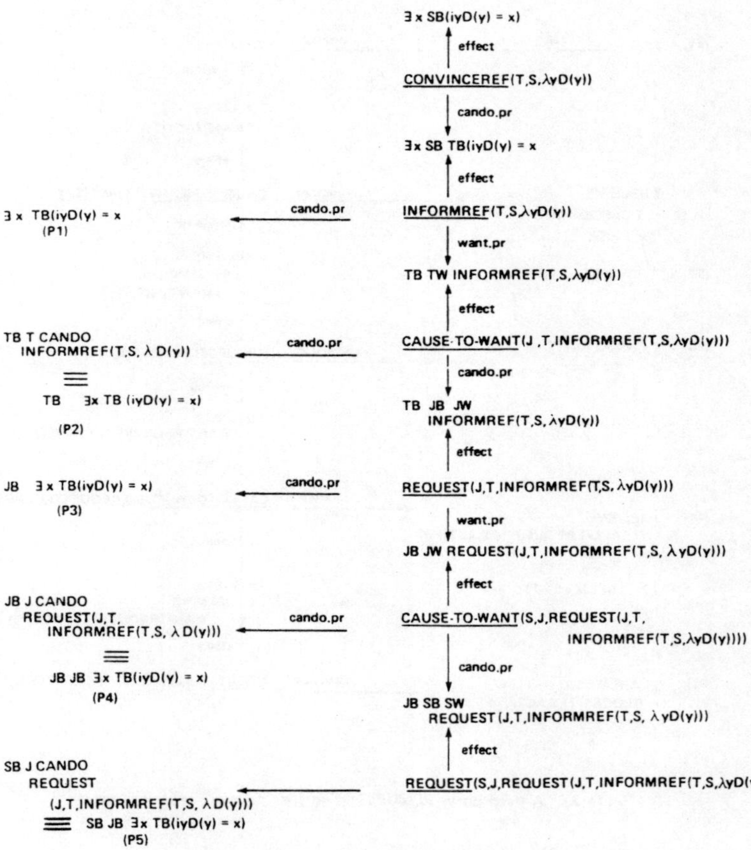

Figure 6. A plan for a third party REQUEST.

While the plan appears to be straightforward, precondition P3 is clearly unnecessary—S ought to be able to plan this particular speech act without having any *prior* knowledge of the intermediary's beliefs. This prior knowledge requirement comes about because precondition P5 is constructed by composing

REQUEST's precondition schema with precondition P3, and P3 is similarly constructed from P1.

The problem can be eliminated by reformulating REQUEST's precondition as HEARER CANDO ACT. Consider a general plan for three-party REQUESTs, as in Figure 7. T's INFORMREF has been generalized to ''ACT(T)'' whose precondition is ''P.''

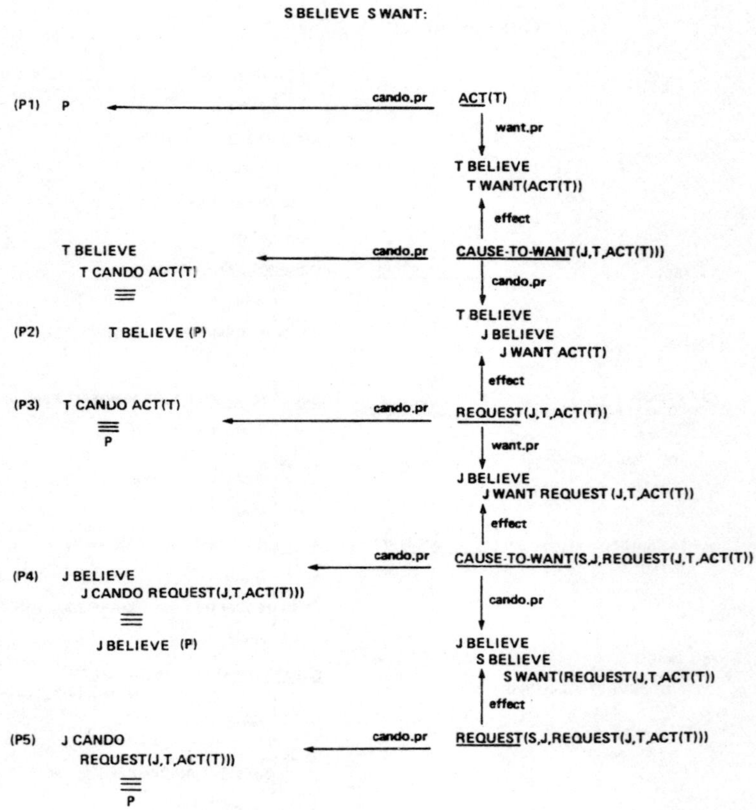

Figure 7. A third party REQUEST using the "neutral" precondition.

The preconditions that have to be satisfied in S's plan are:

S BELIEVE:

(P1) P (also P3 and P5)

(P2) T BELIEVE (P)

(P4) J BELIEVE (P)

Conditions P3 and P5 are the same as P1, and thus the preconditions to the REQUESTs in the plan, are independent of the speaker's beliefs; they depend only on *the planner's* beliefs. While the use of the neutral precondition eliminates prior knowledge requirements for REQUESTs *per se*, condition P4 still requires, as a precondition to CAUSE-TO-WANT, that the planner have some knowledge of the intermediary's beliefs. The next section shows why the planner need not have such beliefs at the time of plan construction.

7.2 Side Effects

The performance of a speech act has thus far been modeled as resulting in an EFFECT that is specific to each speech act type. But, by the very fact that a speaker has attempted to perform a particular speech act, a hearer learns more—on identifying which speech act was performed, a hearer learns that the speaker believed the various preconditions in the *plan* that led to that speech act held. The term *side effect* will be used to refer to the hearer's acquisition of such beliefs by way of the performance of a speech act. Since the plan the hearer infers for the

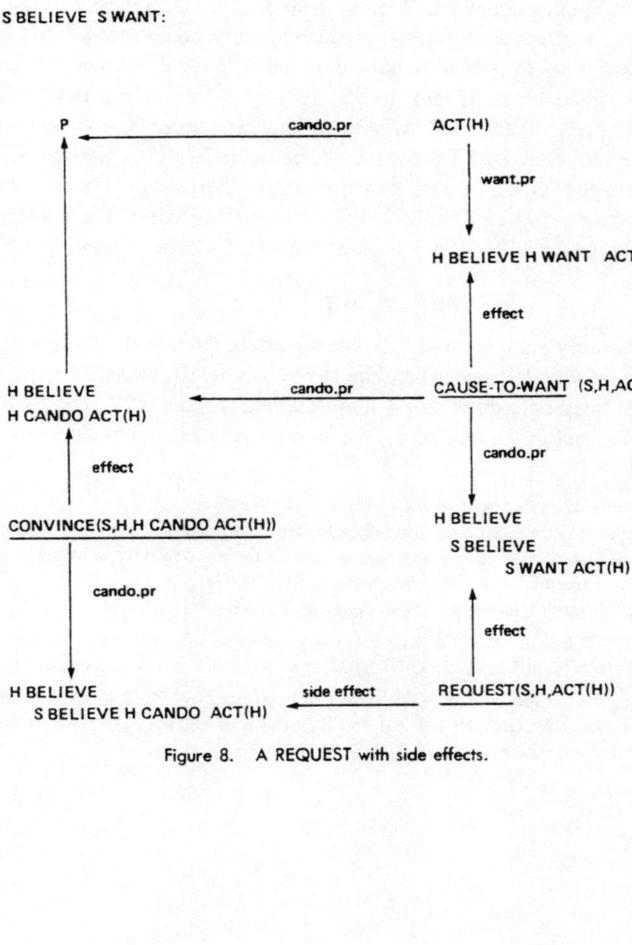

Figure 8. A REQUEST with side effects.

speaker depends upon his beliefs about the speaker's beliefs and goals, the side effects of a speech act cannot be specified in advance. However, the hearer is minimally entitled to believe the speaker thought her speech act's preconditions held (Bruce, 1975; Bruce & Schmidt, 1974).[23] Furthermore, not only do hearers make such assumptions about speakers' beliefs, but speakers know that and often depend on those assumptions for the success of their plans. Figure 8 is a schematic of a simple plan by S to REQUEST H to do action ACT that illustrates this situation.

The minimal side effect is that the hearer believes the speaker believes the precondition of the REQUEST holds, i.e., that HEARER BELIEVE SPEAKER BELIEVE HEARER CANDO ACT. This goal satisfies, via a CONVINCE, the CANDO.PR of CAUSE-TO-WANT, and hence the REQUEST achieves two goals in the plan.[24] The schematic can be applied twice in Figure 7 to obtain Figure 9.

After the side effects of J's REQUEST to T take hold, T would think J believes the preconditions to J's REQUEST (P) obtain. We claim that it is because T thinks that J believes P that T comes to believe P. In this way, precondition (P2) is satisfied as a result of J's REQUEST. Naturally, the side effect argument applies equally to J as the hearer of S's REQUEST. That is, J comes to believe P (precondition (P4)) because he thinks S believes P. S's belief that the preconditions to action A hold thus gets "passed" down the line of intermediaries, whatever its length, to the final agent of A. In this way S can issue the third party REQUEST without having any prior knowledge of J's beliefs about P; S's REQUEST provides all the necessary information!

An interesting aspect of this transmission is that, while J may come to believe P and, by making a REQUEST to T, transmit this belief, T's belief *that* P may be of little use to T. Consider Figure 9 again. Suppose P were

$$\exists y \text{ T BELIEVE } (\iota x \text{LOC(KEY23,x)}) = y$$

which we are loosely paraphrasing as T knows where the key is. S's REQUEST conveys S's belief *that* T knows where the key is. Though J, to decide to perform his REQUEST, need only think *that* T knows where the key is, T actually has to know where it is before he can do A.[25] J's conveying his belief does no good

[23]The hearer may in fact believe those preconditions are false.

[24]The simple backward-chaining planning algorithm described in Cohen (1978) could not easily construct this plan since it ignores intermediate states of the world model that would be created after each operator's execution (i.e., after S's, and J's, REQUESTs).

[25]T cannot obtain that information from believing P since

$\exists y$ T BELIEVE ιxLOC(KEY23,x) = y cannot be inferred from
T BELIEVE $\exists y$ T BELIEVE ιxLOC(KEY23,x) = y, by B.2 and B.7 (footnote 5).

If CONVINCE can be defined so that AGT1 cannot be convinced by AGT2 that AGT1 believes something, then J could not CONVINCE T that $\exists y$ T BELIEVE ιxLOC(KEY23,x) = y on the basis of T's thinking that J believes it.

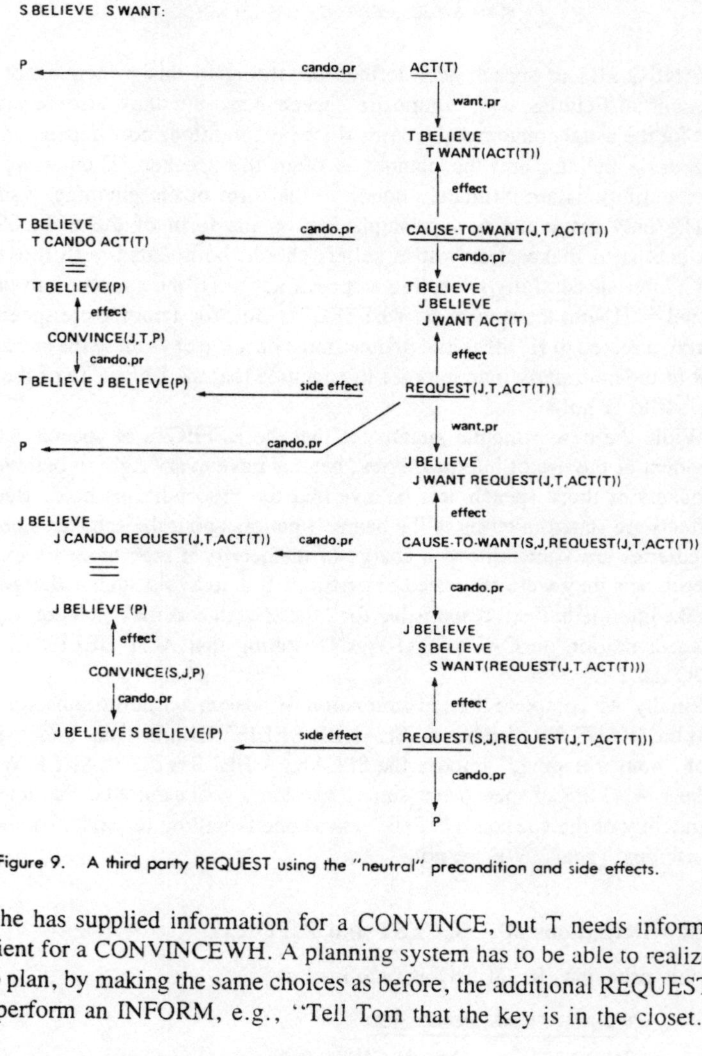

Figure 9. A third party REQUEST using the "neutral" precondition and side effects.

since he has supplied information for a CONVINCE, but T needs information sufficient for a CONVINCEWH. A planning system has to be able to realize this and to plan, by making the same choices as before, the additional REQUEST that John perform an INFORM, e.g., "Tell Tom that the key is in the closet."[26]

7.3 A New Point-of-View Principle

In addition to considering side effects for speech acts, we are led to propose a new point-of-view principle:

> The "Cando" preconditions and effects of speech acts should be defined in a way that does not depend on who the speaker of that speech act is. That is, no CANDO.OR or EFFECT should be stated as a proposition beginning with "SPEAKER BELIEVE."

[26]The side effects again figure in this additional three-party REQUEST—John comes to believe that the key is in the closet by believing that S thinks so.

The CANDO.PRs of speech acts defined according to this principle not only resolve our difficulties with composite speech acts, but they also behave as desired for the usual noncomposite cases since preconditions now depend only on the *planner's* beliefs, and the planner is often the speaker. Thus speech act operator definitions are intimately bound to the form of the planning system.

The only result the new principle has on the form of the EFFECTs of speech acts is to make clear whose beliefs should be updated with those EFFECTs. After successfully executing a speech act to H, the speaker can update his model of H with the speech act's EFFECTs. But, for a composite speech act *ultimately* directed to H, the initial planner must observe or assume the success of the rest of the multiparty plan in order to conclude that the EFFECTs of the final speech act to H hold.

While the new principle guarantees that the EFFECTs of speech acts are independent of the use of intermediaries, hearers have every right to believe that the speakers of those speech acts believe that the preconditions hold. Because side effects are stated in terms of the hearer's beliefs about the speaker's beliefs, intermediaries are vulnerable to a charge of insincerity if they brazenly execute the speech acts they were requested to perform. It is to avoid such a charge, and thus make intermediaries ''responsible for'' the speech acts they execute, that we place the condition on CAUSE-TO-WANT stating that AGT BELIEVE AGT CANDO ACT.

Finally, to complete the reexamination of speech act definitions we point out that the WANT.PR also has a SPEAKER BELIEVE on it. One cannot, in the spirit of ''housecleaning,'' remove the SPEAKER BELIEVE SPEAKER WANT from the WANT.PR of speech acts since a speaker's goal cannot be characterized independently of the speaker's beliefs, unless one is willing to model someone's ''unconscious'' goals. We are not.[27]

7.4 New Definitions of REQUEST and INFORM

Using this principle, REQUEST is redefined as:

REQUEST(SPEAKER,HEARER,ACT)

CANDO.PR:	HEARER CANDO ACT
WANT.PR:	SPEAKER BELIEVE SPEAKER WANT request-instance
EFFECT:	HEARER BELIEVE SPEAKER BELIEVE SPEAKER WANT ACT

The principle applied to the definition of the operator INFORM results in a CANDO.PR stated as PROP rather than as SPEAKER BELIEVE PROP.[28] Such a change allows one to plan to request an intermediary, say a child, to tell

[27]The fact that a WANT.PR is found on *every* intentional act makes us suspect that it belongs on some single ''element'' that is present for every act.

[28]Of course, what must be satisfied in any plan for INFORM is that the planner believe PROP.

someone else that the key is in the closet without the planner's having to believe, at the time of planning, that the child thinks so. The new definition of INFORM then becomes:

INFORM(SPEAKER,HEARER,PROP)

CANDO.PR:	PROP
WANT.PR:	SPEAKER BELIEVE SPEAKER WANT inform-instance
EFFECT:	HEARER BELIEVE SPEAKER BELIEVE PROP

Regarding the other informing speech acts, the principle cannot be used to justify the deleting of the SPEAKER BELIEVE from the CANDO.PR of INFORMREF and INFORMIF since the highest elements of those conditions are "∃" and "OR", respectively. Intuitively speaking, this is a sensible result since a speaker SP cannot plan for an intermeriary, INT, to tell H whether P is true, or what the value of description D is unless INT is believed to have that information.

7.5 Summary

The appropriate planning of composite speech acts has turned out to be a powerful test of the adequacy of speech act definitions. To meet its demands on the planning of questions and multiparty speech acts, two new speech acts, INFORMREF and INFORMIF have been defined, and the preconditions to REQUEST and INFORM have been reformulated according to a point-of-view principle. Since these last two speech acts were taken to be prototypes of Searle's (1976) "directive" and "representative" classes, the principle will find wide application.

A side effect of direct requests was identified and used in planning multiparty speech acts. Side effects, however, cannot be calculated until the hearer has recognized the speaker's plan and thus has classified the observed utterance as a particular speech act type. Thus the minimal side effect formulation given here should be further justified on the basis of what a hearer needs to assume about the speaker's beliefs in order to identify an utterances's illocutionary force.

There may be other ways to meet compositional adequacy. For instance, one could state explicitly that an action's preconditions should be true at the time the action is to be done (Bruce, 1975). For our multiparty REQUESTS, such an approach (using a speaker-based precondition) produces preconditions like: S believes J will believe P will be true when ACT is to be done, which seems reasonable. However, the minimal side effect of S's REQUEST then becomes: J now believes that (before that REQUEST) S *expected* J to believe that P would be true when ACT is done (where "now" is just after the REQUEST was made). As yet, we do not have an analogue of CONVINCE that would allow J to then come to believe that P would be true. Again, if REQUEST is defined using the neutral precondition, this problem does not arise.

8. CONCLUDING REMARKS

It has been argued that a theory of speech acts can be obtained by modelling them in a planning system as operators defined, at least, in terms of the speakers' and hearers' beliefs, and goals. Thus, speech acts are treated in the same way as physical acts, allowing both to be integrated into plans. Such an approach suggests new areas for application. It may provide a more systematic basis for studying real dialogues arising in the course of a task—a basis that would facilitate the tracking of conversants' beliefs and intentions as dialogue and task proceed. A similar analysis of characters' plans has also been shown (Bruce & Newman, 1978) to be essential to a satisfactory description of narrative. Finally, Allen (1979) and Cohen (1978) have suggested how computer conversants might plan their speech acts and recognize those of their users.

Given this range of application, the methodological issues of how speech acts should be modelled in a planning system become important. Specifically, a plan-based competence theory, given configurations of beliefs and goals, speech act operators, and plan construction inferences should generate plans for all and only those speech acts that are appropriate in those configurations. This paper developed tests that showed how various definitions of the speech acts of requesting and informing were inadequate, especially to the demand that they generate appropriate plans when composed with other speech acts to form questions and multiparty requests.

To resolve the difficulties, two "views" of INFORM to be used in constructing questions were defined, allowing the questioner to have incomplete knowledge of the hearer's beliefs. After revising both the form of speech act preconditions and identifying some speech act side effects, compositional adequacy for multiparty REQUESTS was achieved. The solution led to a metatheoretical "point-of-view" principle for use in defining future speech acts as operators within this planning system.

Our approach has both assumed certain idealized properties of speaker/hearers, and has been restricted in its scope. The preconditions and effects of our operators are stated in the language of logic, not because of any desire to perform logically valid inferences, but because the conditions in the plans should have well-defined semantics. While this has been partially realized through the adoption of the possible-worlds sematics for belief, the semantics is too strong to be a faithful model of human beliefs. For instance, it leads here to requiring a questioner to have very strong, though incomplete, knowledge of the hearer's beliefs. To reflect human beliefs more accurately, one needs to model (at least): degrees of belief, justifications, the failure to make deductions, inductive leaps, and knowing what/who/where something is. These refinements, though needed by a theory of speech acts, are outside its scope. Finally, the semantics for WANT and for actions are lacking (but see Moore (1979) for an interesting approach to the latter).

Only two kinds of speech acts, prototypes of Searle's (1976) directive and representative classes, have been examined here, but the approach can be extended to other members of those classes (Bruce, 1975) and perhaps to the commissive class that includes promises. However, in order to model promises and warnings, a better understanding of the concepts of benefit and obligation is necessary.

Finally, we have so far discussed how a planning system can select illocutionary force and propositional content of a speech act, but not how utterances realizing it can be constructed nor how illocutionary acts can be identified from utterances. Extending the plan-based approach to the first area means investigating the extent of "pragmatic influence" of linguistic processing. An important supbroblem here is the planning of referring expressions involved in performing illocutionary acts (Perrault & Cohen, forthcoming; Searle, 1969). Regarding speech act identification, the acid-test of a plan-based approach is its treatment of indirect speech acts (Searle, 1975). Gordon and Lakoff (1971) proposed "conversational postulates" to account for the relation between the direct or literal and the indirect illocutionary forces of an utterance. But, as Morgan (1977) notes, by calling them "postulates," one implies they cannot be explained by some other independently motivated analysis.

We suggest that the relation between direct and indirect readings can be largely accounted for by considering the relationship between actions, their preconditions, effects, and bodies, and by modelling how language users can recognize plans, which may include speech acts, being executed by others. The ability to recognize plans is seemingly required in order to be *helpful*, independent of the use of indirect speech acts. For instance, hearers often understand a speaker's utterance literally but go beyond it, inferring the speaker's plans and then performing acts that would enable the speaker's higher level goals to be fulfilled. Indirect speech acts arise because speakers can intend hearers to perform helpful inferential processing and they intend for hearers to know this. Allen (1979) and Perrault and Allen (forthcoming) formalize this process of intended plan-recognition (and thus Searle's force condition) extending our plan-based approach to the interpretation of indirect speech acts.

ACKNOWLEDGMENTS

We would like to thank Marilyn Adams, James Allen, Ron Brachman, Chip Bruce, Sharon Oviatt, Bill Woods and the referees for their comments, and Brenda Starr, Jill O'Brien, and Beverly Tobiason for their tireless assistance in the paper's preparation. Special thanks are extended to Brenda Starr for her invaluable editorial help.

REFERENCES

Allen, J. A plan-based approach to speech act recognition. Ph.D. Thesis, Technical Report No. 131/79, Dept. of Computer Science, University of Toronto, January, 1979.

Austin, J. L. *How to do things with words*. J. O. Urmson (Ed.), Oxford University Press, 1962.

Bruce, B. Belief systems and language understanding. Report No. 2973, Bolt Beranek and Newman, Inc. January, 1975.

Bruce, B., & Newman, D. Interacting plans. *Cognitive Science*, 1978, *2*, 195–233.

Bruce, B., & Schmidt, C. F. Episode understanding and belief guided parsing. Presented at the Association for Computational Linguistics Meeting at Amherst, Massachusetts (July 26–27, 1974).

Carbonell, J. G. Jr. POLITICS: Automated idealogical reasoning. *Cognitive Science*, 1978, *2*, 27–51.

Chomsky, N. *Aspects of the theory of syntax*. Cambridge, Mass. MIT Press, 1965.

Cohen, P. R. On knowing what to say: Planning speech acts. Ph.D. Thesis, Technical Report No. 118, Department of Computer Science, University of Toronto, January 1978.

Deutsch, B. G. The structure of task-oriented dialogues. In L. D. Erman (Ed.), *Proceedings of the IEEE symposium on speech recognition*. Pittsburgh, PA: Carnegie-Mellon University, 1974.

Donnellan, K. Reference and definite description. In *The Philosophical Review*, v. 75, 1960, 281–304. Reprinted in Steinberg & Jacobovits (Eds.), *Semantics*, Cambridge University Press, 1966.

Fikes, R., & Nilsson, N. J. STRIPS: A new approach to the application of theorem proving to problem solving. *Artificial Intelligence*, 1971, *2*, 189–208.

Gordon, D., & Lakoff, G. Conversational postulates. *Papers from the Seventh Regional Meeting*, Chicago Linguistic Society, 1971, 63–84.

Green, C. Application of theorem-proving techniques to problem-solving. In D. E. Walker & L. M. Norton (Eds.), *Proceedings of the international joint conference on artificial intelligence*. Washington, D.C., May 1969.

Grice, H. P. Meaning. In *The Philosophical Review*, 1957, *66*, 377–388. Reprinted in D. A. Steinberg & L. A. Jacobovits (Eds.), *Semantics: An interdisciplinary reader in philosophy, linguistics, and psychology*. New York: Cambridge University Press, 1971.

Hintikka, J. *Knowledge and belief*. Ithaca: Cornell University Press, 1962.

Hintikka, J. Semantics for propositional attitudes. In J. W. Davis et al. (Eds.), *Philosophical logic*. Dordrecht-Holland: D. Reidel Publishing Co., 1969. Reprinted in L. Linsky (Ed.), *Reference and modality*. New York: Oxford University Press, 1971.

Lewis, D. K. *Convention: A philosophical study*. Cambridge, Mass: Harvard University Press, 1969.

McCarthy, J., & Hayes, P. J. Some Philosophical Problems from the Standpoint of Artificial Intelligence. In B. Meltzer & D. Michie (Eds.) *Machine intelligence 4*, New York: American Elsevier, 1969.

Meehan, J. R. Tale-spin, an interactive program that writes stories. In *Proceedings of the fifth international joint conference on artificial intelligence*, Cambridge, Mass., 91–98.

Moore, R. C. Reasoning about knowledge and action. Ph.D. Thesis, Artificial Intelligence Laboratory, Department of Electrical Engineering and Computer Science, Massachusetts Institute of Technology, February, 1979.

Morgan, J. Conversational postulates revisited. *Language*, 1977, 277–284.

Newell, A., & Simon, H. A. GPS, A program that simulates human thought. In E. A. Feigenbaum & J. Feldman (Eds.), *Computers and thought*. New York: McGraw Hill, 1963.

Perrault, C. R., & Allen, J. F. A plan-based analysis of indirect speech acts. Forthcoming.

Perrault, C. R., & Cohen, P. R. Inaccurate Reference, *Proceedings of the workshop on computational aspects of linguistic structure and discourse setting*, Joshi, A. K., Sag, I. A., & Webber, B. L. (Eds.), Cambridge University Press, forthcoming.

Sacerdoti, E. D. A structure for plans and behavior. Ph.D. Thesis, Technical Note 109, Artificial Intelligence Center, Stanford Research Institute, Menlo Park, California, August 1975.

Schank, R., & Abelson, R. *Scripts, plans, goals, and understanding*. Hillsdale, N.J.: Lawrence Erlbaum Associates, 1977.

Schiffer, S. *Meaning*. Oxford: Oxford University Press, 1972.

Searle, J. R. A taxonomy of illocutionary acts. In K. Gunderson (Ed.), *Language mind and knowledge*, University of Minnesota Press, 1976.

Searle, J. R. Indirect speech acts. In P. Cole & J. L. Morgan (Eds.), *Syntax and semantics*, (Vol. 3), *Speech acts*. New York: Academic Press, 1975.

Searle, J. R. *Speech acts: An essay in the philosophy of language*. Cambridge: Cambridge University Press, 1969.

Strawson, P. F. Intention and convention in speech acts. In *The Philosophical Review*, v: lxxiii, 1964. Reprinted in *Logico-linguistic papers*, London: Methuen & Co., 1971.

[11]

COGNITIVE SCIENCE 6 1–36 (1982)

Simulating a Skilled Typist: A Study of Skilled Cognitive-Motor Performance

DAVID E. RUMELHART
DONALD A. NORMAN
Department of Psychology
and
Program in Cognitive Science
Center for Human Information Processing
University of California, San Diego
La Jolla, CA 92093

We review the major phenomena of skilled typing and propose a model for the control of the hands and fingers during typing. The model is based upon an Activation-Trigger-Schema system in which a hierarchical structure of schemata directs the selection of the letters to be typed and, then, controls the hand and finger movements by a cooperative, relaxation algorithm. The interactions of the patterns of activation and inhibition among the schemata determine the temporal ordering for launching the keystrokes. To account for the phenomena of doubling errors, the model has only "type" schemata—no "token" schemata—with only a weak binding between the special schema that signals a doubling, and its argument. The model exists as a working computer simulation and produces an output display of the hands and fingers moving over the keyboard. It reproduces some of the major phenomena of typing, including the interkeystroke interval times, the pattern of transposition errors found in skilled typists, and doubling errors. Although the model is clearly inadequate or wrong in some of its features and assumptions, it serves as a useful first approximation for the understanding of skilled typing.

*We thank Eileen Conway, Donald Gentner, Jonathan Grudin, Geoffroy Hinton, Paul Rosenbloom, and Craig Will for their assistance and many discussions with us on the nature of the typing data, their work in collecting and interpreting typing errors, and their discussions on the underlying response mechanisms. Gentner has provided a large set of keypress reaction time data, and Grudin, Gentner, and Conway have taken high-speed films and video tapes of skilled typists. This ongoing research and the several large corpora of data on typing performance have been of considerable assistance to us is in the preparation of this paper. (The several studies will be published as separate research reports.) Stu Card, Tom Moran, and Bill Verplank have provided useful discussions and analyses, as well as data on typing performance and keyboard design. We also thank Eileen Conway for preparing the figures and for photographing the computer display (each several times).

The research reported here was conducted under Contract N00014-79-C-0323, NR 157-437 with the Personnel and Training Research Programs of the Office of Naval Research, and was sponsored by the Office of Naval Research and the Air Force Office of Scientific Research. Requests for reprints should be sent to David E. Rumelhart, Program in Cognitive Science C-015; University of California, San Diego; La Jolla, CA, 92093, USA.

The surprising thing about human control of the limbs is the contrast between the ease with which it is done and the complexity of the system. Piano playing rates often reach 30 notes per second for sustained periods. Typing champions reach close to 200 words per minute (17 letters per sec). The hand is an intricate mechanical system, with complex control problems. Just how the fingers should be moved to reach some target depends upon the positions of the hand, the lower and upper arms, the shoulder, and the body position and angle. The computation of the proper movement is difficult, for the number of degrees of freedom is large, and multiple solutions are possible (see Saltzman, 1979). The 10 fingers each have three joints, with each bone controlled in its up and down motion (extension and flexion) by two tendons, and each tendon controlled by a muscle located in the lower arm. Additional muscles within the hand control the side-to-side motion of the fingers (abduction, adduction, flexion, and rotation), and the configuration of the hand. In total, there are about 50 separate movements of the parts, not counting the relevant movements of wrist, lower arm, upper arm, and body.

Typing is not a trivial task. A surprisingly large percentage of the population cannot type. Among those who are expert, considerable training has been required to reach that status. In the early days of typewriting, it took a rare act of courage and belief to attempt to type without looking at the keyboard. In those days, the arguments were over whether it was best to use 2, 3, or 4 fingers, and the concept that someone could touch-type, without looking at the keyboard, using all 10 fingers, was not taken seriously (see the discussion in Beeching, 1974, p. 39 ff.). Matters are not made easier by the fact that almost all contemporary typewriters use the Sholes (*qwerty*) keyboard, designed in 1873 to minimize jamming of the keys by maximizing the distance between frequently typed pairs of keys, without regard for ease of learning or typing.

Our goal is the study of human skilled performance. In this paper, we concentrate on the problem of control of skilled motor movements. In particular, we study typing, for this is a highly skilled motor activity with enough experts available that we can study performance with some ease. Moreover, it is fairly easy to monitor keypress latencies, although more difficult to monitor the actual trajectories of the fingers, or the times at which the fingers begin their movements.

THE BASIC PHENOMENA OF TYPING

In this section, we review briefly some of the empirical results that have been found in the study of typing. We especially emphasize those that we find most suggestive of the underlying mechanisms. Following this section, we present the model for the simulation of these phenomena, and then, in the last part of the paper, we return to examine the phenomena in some detail, contrasting the model performance with that of the data from the literature and from our own observations.

The fundamental phenomena fall into three categories: those involving timing of keystrokes, those involving errors, and those involving the general organization of the typing process. We discuss these in turn, using data recorded as one of us (DER) typed a 90,000 keystroke manuscript from dictation. Unless otherwise specified, the data discussed in the paper come from our corpus.[1] Figure 1 shows the Sholes keyboard and the standard American mapping of fingers to keys.

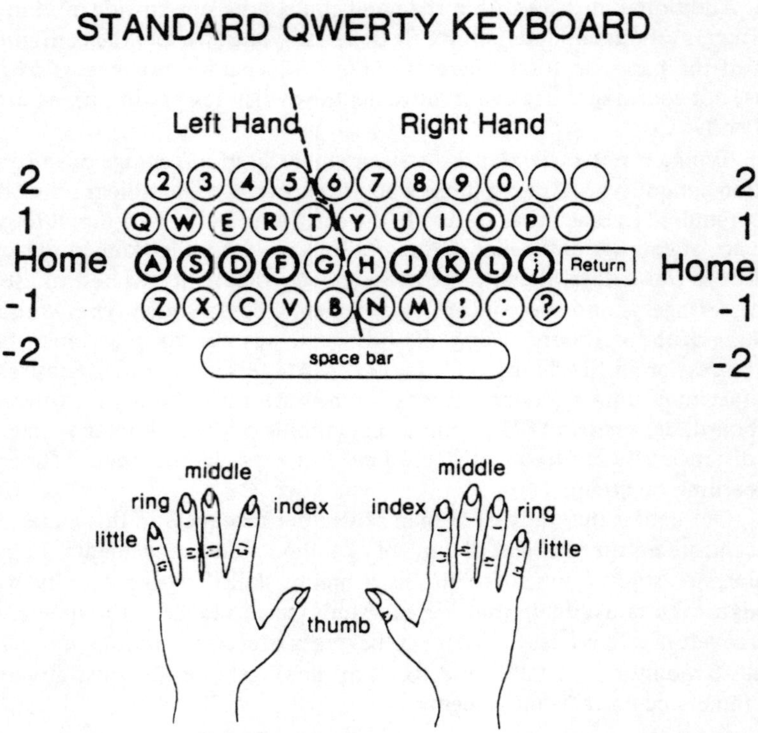

Figure 1. The standard American keyboard (the "Sholes" or "qwerty" keyboard) and the mapping of fingers to keys.

[1] In all of the analyses of typing data discussed in this paper, we look only at lower case alphabetic characters, spaces, and limited punctuation (period, comma, semicolon). Other non-alphabetic characters, punctuation, numerals, and the use of the RETURN and SHIFT keys were not examined.

The Timing of Keystrokes

People can type very quickly. World champion typists can type at rates up to 200 words per minute. This involves a mean interval between keystrokes of 60 msec, close to the neural transmission time between the spinal cord and the periphery (and obviously, many keystroke intervals will be considerably less than the mean). There cannot be much feedback between strokes being performed so rapidly. Even relatively ordinary typists can routinely generate keystrokes at rates almost as rapidly as this. For example, of the 1656 times *th* was typed by our subject, 414 times the interval was less than 63 msec. The *th* interkeystroke interval was less than 75 msec half of the time.

Speed, however, is the simplest of the phenomena that need to be accounted for. Overall, there are five different sets of timing phenomena that provide strong constraints on the structure of a possible model of typing:

1. People can type very quickly;
2. cross-hand interkeystroke intervals are shorter than those within hands;
3. within-hand interkeystroke intervals appear to be a function of the reach from one key to the next;
4. the time for a particular interkeystroke interval can depend on the context in which it occurs; and
5. there is a negative correlation between the intervals on successive keystrokes—especially when the alternate strokes occur on alternate hands.

We discuss these in detail at the end of the paper.

Pattern of Errors

Errors are of special importance, for some of them give strong clues as to the underlying mechanisms. Thus, the existence of transposition, doubling, and alternation errors has played a major role in determining the structure of the model.

Transposition errors. One of the most common and most interesting categories of errors is transposition, the reversal of two adjacent letters. The large majority of these errors occur across hands. Shaffer (1976) reports that of his subject's transposition errors, about 90% were cross-hand transpositions. For our subjects, the figure is 75%. Examples of transpositions from our subjects include:

because -> becuase
which -> whihc

Of the within-hand errors, half involved adjacent keys (such as *e* and *r*, and *o* and *p*), as in

supremely -> supermely

One interesting example was reported by Shaffer (1976):

went down -> wne todnw

The four keystrokes on the right hand (the *n*, *space*, *o*, and *n*) have all been displaced with respect to the five left-hand keystrokes.

Doubling errors. When a word contains a doubled letter, the wrong letter is sometimes doubled. Thus, *look* can become *lokk*. This error was pointed out by Lashley (1951) and Shaffer (1976) as being diagnostic as to the nature of motor control. Although our corpus of transcription typing only includes one example of a doubling error of this sort:

school -> scholl

we have collected many doubling errors from our samples of composition typing (while using the laboratory computer). For example:

gibbs -> giibs
Screen -> Scrren

Alternation reversal errors. This is akin to the doubling error, but with an alternating sequence. Thus in the word *these* the *ese* is an alternation. Samples observed during composition typing include:

these -> thses
there -> threr
were -> wrer

Other errors. In addition to the errors of transposition, doubling, and alternation, a number of other forms of errors occur. However, these are not so critical in determining the form of the typing model, and so discussion of these is deferred until later. (For example, we believe that some of these errors come from factors outside the scope of the present model.) These other errors include homologous errors, capture errors, omission errors, and misstroke errors. These errors will be discussed at the end of the paper.

The General Organization of Typing

Finally, there are two other observations that we have used as strong constraints: the overlapping of hand movements and the unit of organization of the strings to be typed.

Skilled typists move their hands toward the keys in parallel. In our laboratory, Gentner, Grudin, and Conway (1980) have carried out a photographic analysis of a skilled typist using a high speed motion camera. The results of this study show the fingers of the hand in almost constant motion, with fingers starting to move toward their destination before the several preceding characters have been typed. A serial model of typing in which each finger in turn makes its stroke is incorrect. Rather, there seems to be a coordinated structure that allows the control of several fingers simultaneously.

The units of typing seem to be largely at the word level or smaller. In Shaffer's (1973) study of the units of typing for one skilled typist, he had his subject type normal prose, random words, random letters, and foreign words. He found almost no difference between typing prose and random words (the mean keystroke interval was 107 msec for prose and 104 msec for random words). Random letters were typed much more slowly than normal text (192 msec), and German text was typed at an intermediate rate (149 msec: the typist didn't know German). Shaffer also found that when look-ahead was limited, the typist needed to see at least eight letters ahead of where she was typing in order to maintain her normal typing rate. (The rule of thumb that has emerged from this and similar studies is that a typist looks ahead about one second's worth of text: slower typists look ahead fewer letters, faster typists more.)

Issues Not Covered by This Study

Typing is a rich, complex skill, and although we examine and model a number of the characteristics of skilled typing, there is also much that we do not cover. We look only at skilled typing, and cover neither the mechanisms used by inexperienced typists, nor the mechanisms involved in learning. We assume that all the necessary control and knowledge structures are already established. We do not examine how a skilled typist might vary typing rate in order to manipulate the speed-accuracy tradeoff (although there are several parameters of our model that can readily be identified as potential candidates for this manipulation). We do not look at the mechanisms involved in perception or the encoding of the strings to be typed, nor in monitoring the accuracy of the typing. We do not simulate the deterioration of typing rate that occurs as the text is modified from normal prose to non-language or random letters. Finally, we only study a portion of the keyboard, ignoring most non-alphabetical keys (see Footnote 1). Our work is readily extended to these other keys, however, so we do not consider this of fundamental importance. However, there are insufficient data available for these other keys, and in many cases, the keyboard locations are not standardized, so that typists are not as expert with them (or, as is often the case with number keys, may prefer to adopt a different mode of typing for them). Despite these omissions, there still has been considerable work for us to do.

A MODEL OF TYPING

Control of the fingers poses a number of complexities, and the cognitive specification of the actions to be performed must be compatible with both the existing knowledge of mental structures and of the phenomena of typing, especially the factors discussed in the previous section. Our analyses of these issues lead us toward a model that has the following properties:

1. Control of action sequences by means of schemata;
2. selection of appropriate motor schemata through a combination of activation value and triggering condition;
3. the representation of letter typing by means of a *pure type theory* (i.e., one with no type-token distinction); and
4. the need for distributed (local) rather than concentrated (central) control of movement.

We start with the assumption that motor control of learned movement is represented by means of a motor schema (plural: "schemata"). A motor schema is an organized unit of knowledge, differing from the form of knowledge widely studied in the literature on memory, language, and thought only in that it has as its output the control of body movements. This is not a new concept. Actually, the term "schema" was originally introduced into psychology for the use in skilled motor control by Head (1926), and is still used for that purpose (cf. Schmidt, 1976).

We propose that one of the functions of schemata is to act as motor programs. The term "motor program" is to be understood by analogy with the term "computer program." We believe there has been some confusion in the literature on skills in this regard, with critics of the notion of motor programming acting as if a program were a fixed action sequence, specified in complete detail before the actual movements. According to our view, motor programs are flexible, interactive control structures capable of calling upon sub-programs, passing parameters to be bound to program variables, and making local decisions as a result of current conditions (which might include information from feedback channels, from perception, or other sources of knowledge). A motor program is not a fixed action pattern of movements. It is a set of specifications or control statements that govern the actions that are to be performed, with considerable flexibility in the specification of the actions. A program specifies the rules that are to be followed in the action, not the actual motions.

THE ATS FORMALISM

The basic framework that we follow is called an Activation Triggered Schema system (ATS). The model consists of a set of schemata, each with activation values. A schema has an activation value that reflects the total amount of excitation that it has received. The normal, resting value for a schema is zero. It can increase when the schema is "activated," or decrease when the schema is "inhibited." Schemata interact with one another, and the activation value reflects this interaction, as well as the effects of decay and other sources of activation and inhibition. When appropriate conditions have been satisfied, a schema may be "triggered," at which time its procedures become operative and control whatever operations they specify.

Different schemata are often interconnected. Moreover, one schema may call upon other schemata to perform specific tasks, much as a computer program calls upon subroutines or coroutines. When one schema calls upon another, the initiating schema is called the "parent schema," and the called schema is the "child schema." Each schema can serve in any or all of three ways: as a program in control of operations, as a parent schema that initiates the operation of other schemata, or as a child schema, invoked by a parent.

A particular schema might be invoked by a parent schema, set into motion some operations, and then itself serve as a parent to its child schemata. Usually, but not necessarily, when a child schema has completed its operations, control returns to the parent schema. Thus, the schema for typing the word *the* might be initiated by the triggering of its parent schema, which then controls the activation and triggering of the child schemata for the letters *t, h,* and *e,* that in turn activate the child schemata that control the actual finger, hand, and arm movements.

THE SIMULATION MODEL

Figure 2 illustrates the basic structure of the model. The model incorporates the principles discussed in the previous sections, plus specific control mechanisms for the activations and selection of particular hand and finger movements. The input of the model is a string of characters that constitute the text to be typed. The output is a sequence of finger movements, either displayed on a visual computer-controlled display as the movement of the hands and fingers over a typewriter keyboard, or as a series of coordinate locations for the relevant body parts.

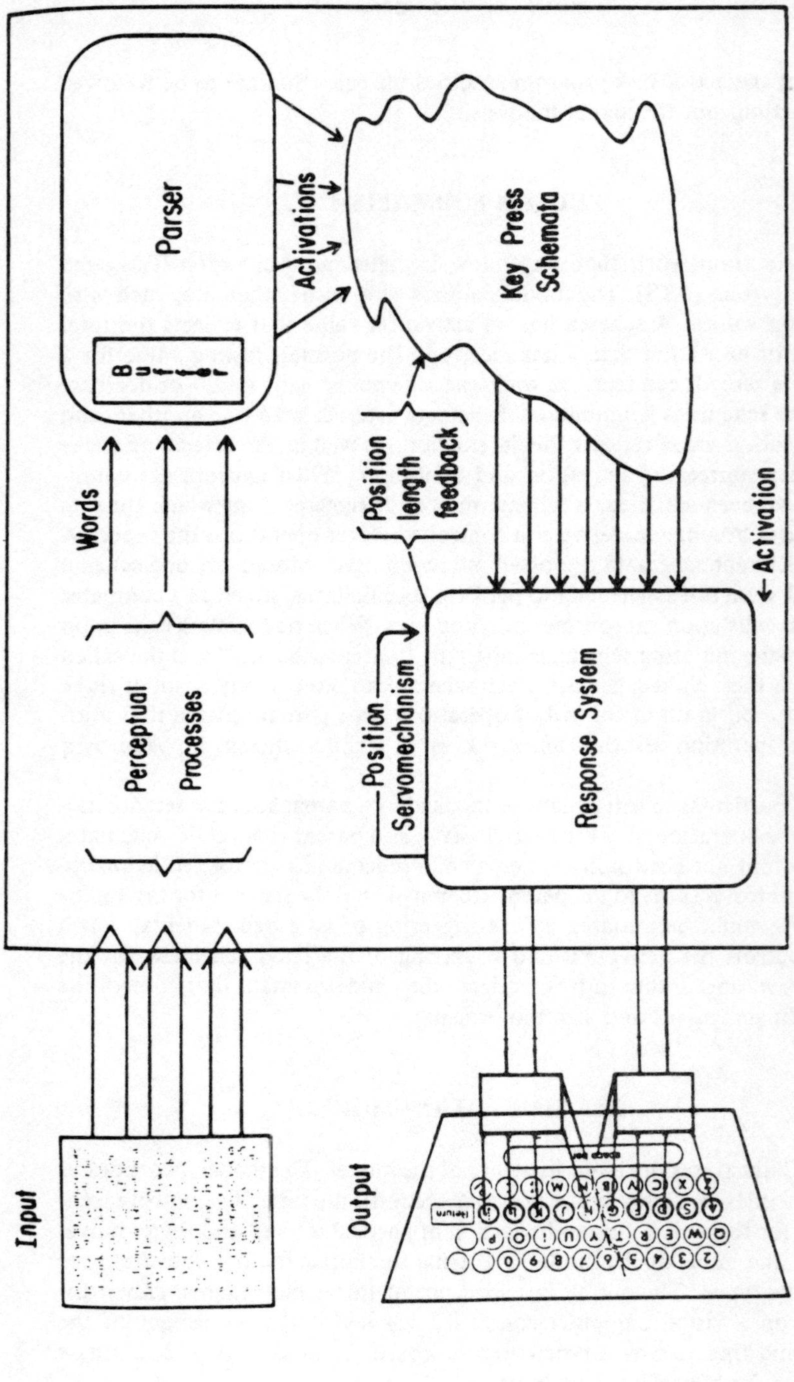

Figure 2. The information processing system involved in typing: The perceptual processes output their identification of the input to a buffer which maintains the information while the appropriate response schemata are activated and take control of the actions. The input of the model is words; the output is the movement of the hands and fingers across the keyboard.

The Hand

The basic hands employed in the model can be seen in Figures 1 and 2. Each hand consists of a palm and the five fingers: thumb, index, middle, ring, and little. The model hands are a simple schematic approximation to the real hands. We let the palm and each of the five fingers have two degrees of movement: up and down the keyboard, and inward and outward from the center of the keyboard. (Fingers actually move "up" and "down" by extension and flexion in the third dimension, but in the model, we view the hands from above in a two-dimensional projection. Thus, the extension and flexion of the fingers are represented by lengthening and shortening.) "Upward" is defined to mean toward the top row of the keyboard and "downward" to mean toward the bottom row (the space bar). Sideways motion is relative to the hands, so that "inward" means toward the center of the keyboard (toward the thumbs), and "outward" means toward the ends of the keyboard (toward the little finger). We assume that the palm can move any distance in any of these directions. Each finger is assumed to have a region of additional movement relative to the location of the palm. The exact amounts of movement of each finger relative to the palm are parameters of the model, and depend on the finger in question. We selected values that roughly correspond to the comfortable reach of our subject (DER). Thus, in the simulation, the index finger can move 1.0 key space inward, only 0.05 key space outward, and 0.8 row upward and downward. Because the little finger is at the end of the hand, its horizontal range of movement is asymmetrical: it can move 1.0 key space outward, 0.05 key space inward, and 0.3 row upward and downward. For simplicity, we assume that the thumb is rigidly attached to the palm and can only move horizontally. (inward and outward). Table 1 gives the assumed amount of movement in each direction for each of the fingers.

It is important to note that there is a tradeoff between palm movement and finger movement. Thus, each final position can be reached by infinite combinations of palm and finger positions. A major point of this paper is to argue that skilled, efficient typing results from taking advantage of this tradeoff.

TABLE I
Amount of Finger Movement Permitted for Each Finger in the Simulation Model
(The distance between keys is taken to be one unit.)

	Up	Down	In	Out
Thumb	0.0	0.0	2.00	2.00
Index	0.8	0.8	1.00	0.05
Middle	1.0	1.0	0.10	0.10
Ring	0.7	0.7	0.10	0.10
Little	0.3	0.3	0.05	1.00

The Keyboard

The keyboard is modeled after the conventional Sholes or "qwerty" keyboard, restricted to the letter keys, semicolon, comma, period, and RETURN. We do not model the use of the SHIFT key or the number keys. The position of the RETURN key is taken to be just to the right of the semicolon.

Finger assignments are those used in the standard American method for the teaching of touch typing (shown in Figure 1). The "home" position of the hands is taken to be that with the fingers of the left hand resting on the *a, s, d,* and *f* keys, and those of the right hand on *j, k, l,* and *;*. The palms are located so that the fingers can be on the home keys with the fingers in neutral position relative to the palms.

The Perceptual Processes and the Parser

There are a number of interesting issues involved in determining an appropriate synchronization between the operation of the perceptual processes and the rest of the typing model. For example, the perceptual processes have to read the words of the manuscript at just the proper range of rates so that there are always character strings to be typed available to the parser in the buffer, but not so many strings that the buffer overflows. Although these are of real importance to the typing process, they are not included in the simulation model, and so, we do not address them here. We do assume that the output of the perceptual system is words, not letters. The operation of the typing model begins where the perceptual processes leave off: with strings of words in the buffer of the parser.

The job of the parser is to transform the word strings in the buffer to appropriate activations of the key press schemata. The basic task of the parser is to divide the words into the individual characters, then activate the appropriate key press schema for that character.

The Activation Process

Figure 3 illustrates the basic assumptions of the activation process using the word *very* as an example. First, the schema for the word is activated by the perceptual system and parser. This, in turn, activates each of the child schemata for keypresses. Each keypress schema specifies the target position, with position encoded in terms of a keyboard-centered coordinate system; upward-downward (upward is positive) and inward-outward (inward is positive). These target positions are sent to the response system which then must configure the palm and finger positions properly. Each keypress schema inhibits the schemata that follow it. This means that proper tem-

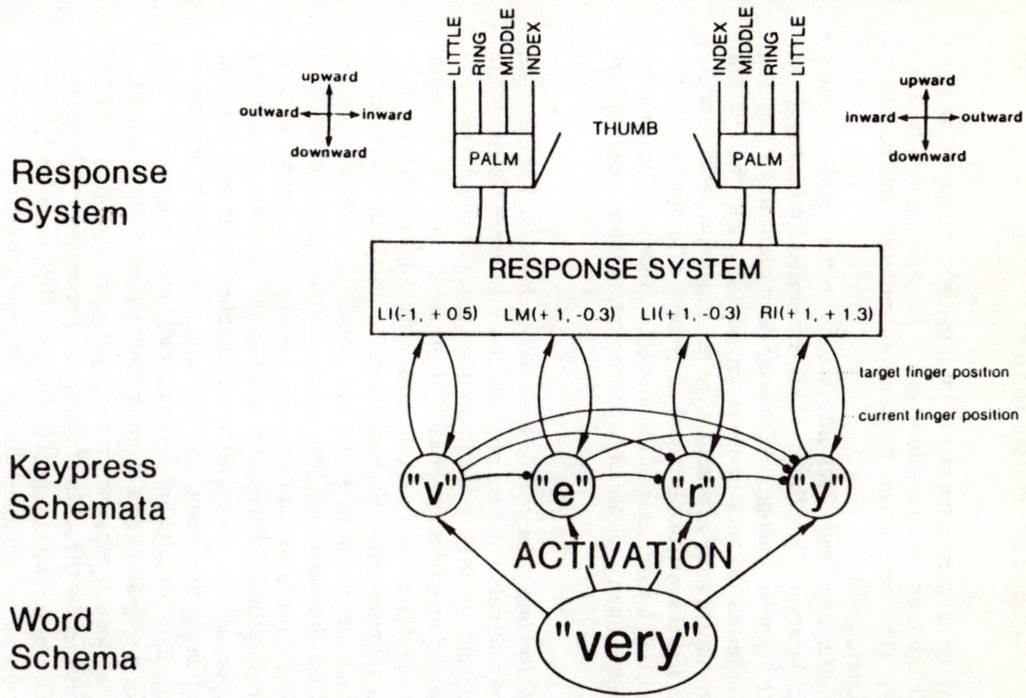

Figure 3. The interaction of activations when the word *very* is to be typed. The *very* schema is activated by the perceptual system and parser, activating each of the schema for the component letters. Each letter schema specifies the target finger position, specified in a keyboard coordinate system. L and R stand for the left and right hands, and I and M for the index and middle fingers. In the coordinate system, the first variable is upward/downward (up is positive) and the second variable is inward/outward (inward is positive). The keypress schemata receive information about the current finger position from the response system. Each letter schema inhibits the activation of all letter schemata that follow it; inhibition is shown by the lines with solid circles at their termination.

poral ordering of the keypress schemata is given by the ordering of the activation values. In addition, the activation values are noisy, which leads to occasional errors. (This mechanism for temporal ordering was originally introduced by Estes, 1972).

The response system feeds back information to the keypress system about the current location of the fingers. This is essential for the triggering conditions of the keypress schemata. Whenever the current finger position is within some criterion distance of its target position, and the relevant schema is the one most highly activated, then the triggering conditions are satisfied and the actual keystroke is launched. Shortly after the launch (the amount of time being a parameter of the model), the keypress schema deactivates itself, resulting in a release of inhibition for all the succeeding keypress schemata.

Repeated letters imply that there are no token schemata. The existence of doubling and alternation errors pose special problems. Consider the word *book*. According to the arguments we have just presented, the word would be represented by schemata for each of the letters: *b o o k*. It is easy to see how such a representation could lead to transposition errors (such as *boko*) but not to doubling errors. It would be easy to make up a schema for a doubled letter (so that the word would be represented by the schemata *b DOUBLE-o k*), but this would not lead to the doubling errors either.

The doubling error turns out to have two major implications. First, it implies that there are special schemata that signal the existence of doubled letters, and that occasionally these schemata get applied to the wrong letters. In computational terms, this means that the binding between the arguments of the special schemata for doubling is occasionally made improperly. Second, the need for a special schema to mark doubled letters implies a difficulty in having the regular letter schema signal the double. Why isn't the word *book* represented by the schemata *b o o k*? The reason would seem to be that this would require two instances (tokens) of the schema for *o*; the existence of the doubling error implies that such repeated tokens of a schema is not possible.

Thus, the existence of doubling errors forces us to a pure "type" model, in which each letter can only have a single keypress schema; the keypress schemata exist only as "types," with no "token" schemata. There must be a special schema that signals the presence of a doubled letter. Moreover, there must be a weak binding between the special schema and the arguments upon which it operates. In the model, the arguments are not bound to the schemata, but are established via activation values. The most highly activated keystroke schema is triggered when it is within a criterion distance of its target. After triggering (and the resulting launch of the keystroke), this keypress schema can become "bound" to a doubling schema if one exists with a higher activation value than its own. Because activation values are noisy, occasionally this leads to errors in the linking of keypress schemata to a doubling schema.

The existence of alternation errors leads to the same conclusion; there must be a special schema that signals the presence of alternating letters, with a weak binding between the schema and its arguments, and the mechanism proposed for alternations is similar to that for doubling.

Here is an example of the typing of a word with a doubled letter. The word *book* is represented by the activation of four schemata: *b DOUBLE-o* and *k*. Each schema inhibits all that follow it, in the regular fashion. The operation now is much as we illustrated before, except that after a keystroke schema has been triggered, it checks for the existence of a double schema whose activation value exceeds its own value. The *b* will initially be the schema most highly activated, and when the finger gets within a criterion distance from the key, the *b* keystroke will be launched. Now the *DOUBLE* schema will have the highest activation level. However, the DOUBLE schema does not command any motor responses, and it allows control to be passed to the schema with the next highest activation value. The next schema is the one for *o*. It proceeds normally. As the keystroke is launched, the *o* schema notes that there is a DOUBLE schema whose activation value is greater than its own. Whenever this condition occurs, the keypress schema deactivates the DOUBLE schema and, after the keystroke, does not deactivate itself. As a result, at the completion of the keystroke, the schema is again triggerred, launching itself a second time. At the launching of this second keystroke the doubling schema is no longer present, so that typing of the rest of the word can continue. Noise in the activation levels occasionally causes this mechanism to go awry so that the DOUBLE schema gets associated with the wrong keypress schema, causing the wrong character to be doubled.

We suspect that there is a similar mechanism for alternating keystrokes. However, when we added such a mechanism to a version of the model, it led to types of alternation errors that were never observed. Let the symbol - stand for *space*. Words such as *-a-* that had but one letter were spelled *a-a* rather than *-a-*. It is possible that the problem here lies with the parsing mechanism, not with the alternation schemata. Perhaps it is wrong to treat a word such as *a* or *I* as an alternation of *space letter space* but rather, each word should either be followed or preceded by a space (so that only a single space is ever tied in with a word). Because we did not attempt to model the peceptual and parsing process, we did not pursue this possible explanation.

The assumption that there are no "token" nodes but only "type" nodes causes special problems for any word that contains repeated instances of the same letter that is not part of a double or an alternation (e.g., the *e* and *p* of *perception*). As a result, if a word contains a repeated instance of a letter, processing of that word is blocked at the repeated letter until the keystroke for the previous instance has been completed. Thus, the parser sets up the keypress schemata for *p, e, r, c.* When the *e* has been typed, the parser is then free to set up the rest of the word: e, p, t, i, o, n. A similar blocking occurs in the case of repeated doubles, as in *bookkeeping*. Because there is only one node for DOUBLE, the parser can at first only set up *b, DOUBLE, o,* and then, when the doubled-o has been typed, can add *DOUBLE, k,* and finally, when the doubled-k has been typed, *DOUBLE, e, p, i, n, g.*

Movement

One of the issues in the control of movement is the degrees of freedom problem; there are many ways to perform any given task, yet the motor programming system must determine one of the many possibilities. This can be a virtue as well as a problem, since it allows any given movement to serve more than a single goal. The extra degrees of freedom can be used to optimize movement toward many goals at the same time. In the case of typing, this means that the set of arm-hand-finger configurations can be chosen so as to optimize the striking of as many keys in a sequence as possible, without requiring an overwhelming computation.

In the model, each active schema pushes its relevant hand and finger toward its desired key at the same time, and the final overall configuration is determined by the competition among these forces. Each schema pushes with a force proportional to its activation level. As a result, the forces are weighted so as to cause the letter schema that is next in line to be typed to approach its key most quickly. The actual location of each finger is determined by the sum of the extensions of the finger and the hand. To type a particular typewriter key, it is only necessary that the end position be correct. Each keypress schema specifies its desired endpoint by specifying the total extension of the hand plus the relevant finger. The endpoint configuration is reached through an iterative relaxation process that only involves local computation. That is, the desired target extension is sent to the muscle system, and each finger and hand combination moves a very small distance toward the goal. The algebraic sum of all of the competing movements toward the various keys represented by the activated schemata determines the total movement toward the goal. Then the process is repeated, with a small movement resulting from each iteration of the process. Because of the unequal weighting of activations, the process will eventually cause the most highly activated schema to move its finger-palm configuration to within a criterion distance from its target key, satisfying the trigger conditions and launching the keystroke.

The positions of the fingers and palms are represented in the system as a sum finger extension, plus arm extension. We assume that a position specification (a target sum of finger extension plus arm extension) feeds into a comparator which uses this specification as a reference signal. The comparator is also assumed to receive inputs from "extension detectors" for the fingers and arms. These extensions are summed, leading to a measure of the end of the finger's location. As long as this positional feedback is less than the reference signal (i.e., the end position of the finger has not yet reached its target), the keypress schema continues to push its relevant hand and finger proportional to its activation. When the positional feedback is greater than the reference signal (i.e., the end position of the finger is beyond the target), the finger and arm are each retracted slightly.

Each schema is assumed to activate its relevant muscle system in proportion to its degree of activation. The activation of all currently active keypress schemata send activation to the same muscle system, their rates summing algebraically. The actual movement is determined by the summed result from all of the response schemata driving the muscles. The actual velocities and hand configurations are determined by the interaction among the competing forces driving the various fingers toward their ultimate goals.

At some point, the finger gets into its appropriate position and its triggering conditions are satisfied. This initiates the actual keystroke movement, launching a ballistic keystroke movement. Upon launching, a number of changes occur: the schema begins to deactivate itself and reduces its activation level. In addition, the launching of other fingers on the same hand is inhibited for a period after the launch has been initiated. There is no feedback for the keystroke once it has been launched; the movement is ballistic, and so, after a brief delay, the schema fully deactivates itself. Aside from this, the system operates normally during the launch, and other keypress schemata may have their triggering conditions met and launch their own keystrokes before the earlier ones have been completed.

This movement scheme has the possibility of a "deadlock" in which the algebraically summed pressures on a hand balance so that the hand freezes and never reaches its goal. In order to forestall such a possibility, there is an additional inhibition among those keystroke schemata commanding the same hand: the within-hand inhibition increases progressively as the time between keystrokes increases. This scheme ultimately resolves all deadlocks.

An Example

Consider how the model might type the string *very well*. The perceptual processes interpret the input and present the words to the parser. Then the word and keypress schemata are activated, with each activated schema inhibiting the keypress schemata for those keypresses that follow. Figure 4 shows the patterns of the schemata's activation and Figure 5 shows the resulting hand and finger positions photographed from the actual model display approximately every 4 time units throughout the typing of *very well.* Figure 4 also shows the times when each keystroke is launched, and the times that each key is actually pressed.

At time unit 0 in Figures 4 and 5, the typing of the string has just begun. The hands are still in their home position, and the *v* schema is most highly activated, followed by the schemata for *e r y space* and *w*. Activation stops at this point because the next schema needed is that for the *e* of *well,* but this keypress schema is already active. No further activation of schemata can occur until after the *e* in *very* has been typed.

For the first few time units, the activation values remain constant (except for noise), while the hands adjust their positions. By time unit 3, the left index finger is clearly reaching down toward the *v* key, while the left middle and ring fingers are extending toward the *e* and *w* keys. Meanwhile, the right index finger is moving up toward the *y* key. Note that since all of the pressures on the right hand pull the hand up, while some on the left hand pull it up and others pull it down, the right hand is further up than the left.

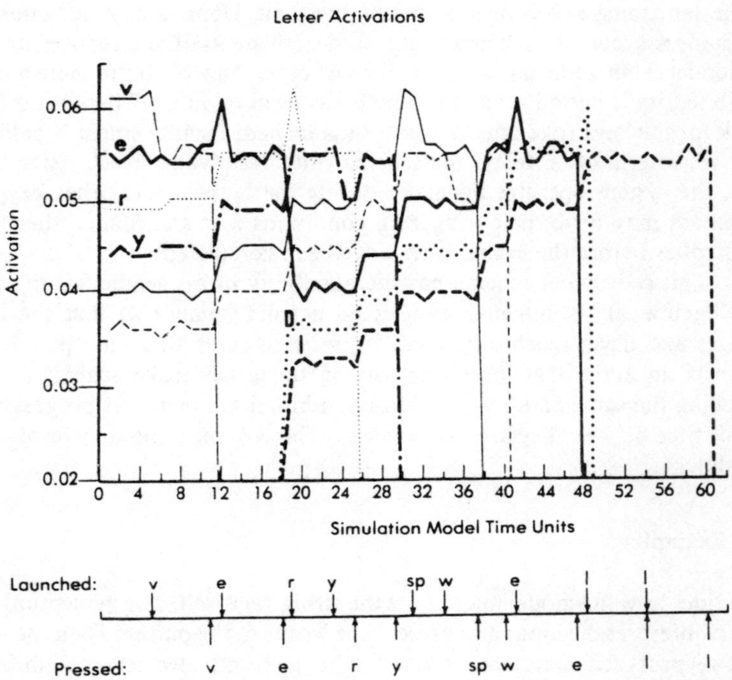

Figure 4. Patterns of activation for the various keypress schemata during the typing of the string *very well.* *D* stands for *double* schema. See the text for details.

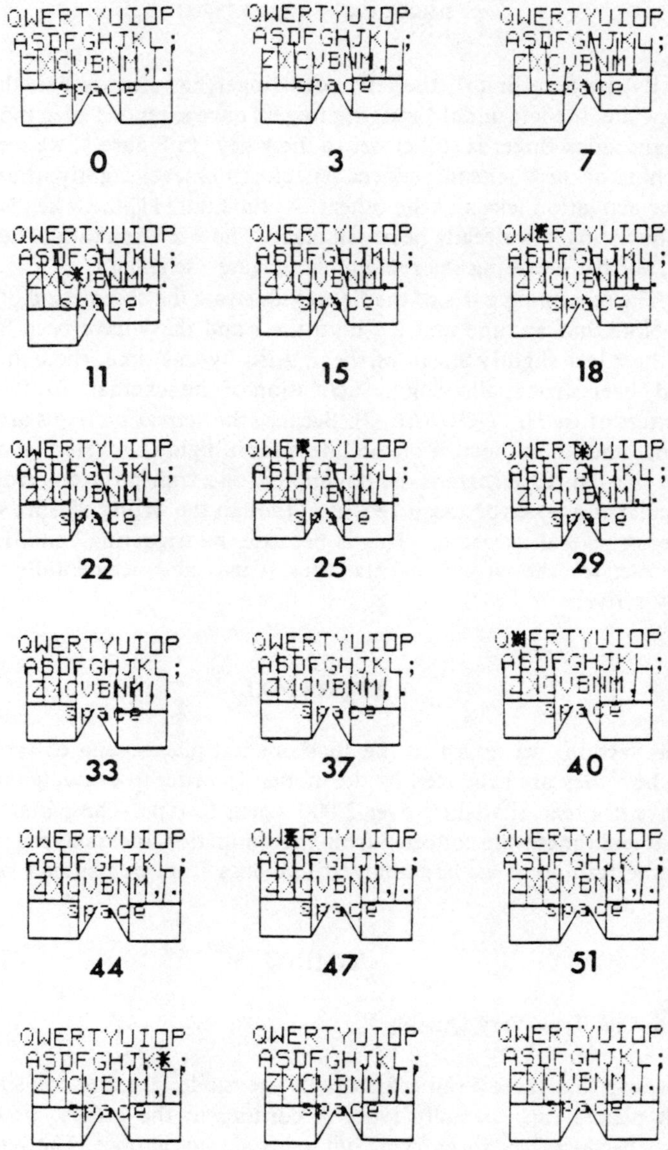

Figure 5. The hand and finger configurations photographed from the display of the working model during the typing of the string *very well*. Each frame of the figure displays the patterns approximately 4 time units after the previous frame. The * indicates an actual press of a key. (The outlines of the hands and fingers have been traced over to make them thicker and, therefore, more visible in the reduced size of the figure used for publication.)

By the time unit 7, the left index finger has already been launched. Meanwhile, the left middle and ring fingers have extended even more, while the right index finger is still closer to the *y* key. In Figure 4, we see that the launching of the *v* schema reduced its activation level slightly, thus increasing the activation levels of the others. At time unit 11, the *v* key is pressed, and the *w* finger is already nearly in place. The *v* schema totally deactivates itself, further elevating the remaining activated schemata.

The remaining parts of the figures illustrate the completion of the process. Note that by time unit 22, both the *r* and the *y* have been launched, with the *r* just slightly ahead of the *y*. Also by this time, the *e* in *very* has already been struck, allowing the activation of the schemata for the remaining letters of *well* (*e, DOUBLE, l*). Because the activation levels are noisy, a schema other than the next one in line is most highly activated on occasion. When this happens, there is the possibility for a transposition error. In this particular run no errors occurred, even though the wrong schema was most active on several occasions. This is because the triggering conditions were never met for the wrong schema when it was also accidentally the most highly active.

APPRAISAL

In this section, we return to the fundamental phenomena of typing, and show how they are produced by the model. In order to evaluate the model, we gave it a text of slightly over 2,000 words to type. The pattern of keystrokes and times were collected from the simulation and analyzed in exactly the same fashion as we had analyzed the data from our human subject.

TIMING

People Can Type Very Quickly

We believe that the overall structure of the model explains the speed with which people can normally type. According to the model, people type quickly because they are carrying out many actions at once. The very fast 60 msec interkeystroke intervals do not represent responses initiated and completed in 60 msec. Rather, since responses are made in parallel, their termination points differ by 60 msec, but the responses themselves may take much longer than that to complete (see Gentner, Grudin, & Conway, 1980). Moreover, although our model requires feedback as to the location of the fingers, this feedback is required only up until the launch of the stroke. At that point, the response is assumed to be ballistic. In this way, very quick feedback from the execution of a response is not required; instead, all that is required is continual feedback of location from the fingers.

Cross-Hand Interkeystroke Intervals Are Shorter than Those Within Hands

In the pattern of interkeystroke intervals, the first and most obvious observation to be made is that when successive strokes are made on different hands, the interkeystroke intervals are shorter than when both strokes are made with the same hand. This result has been reported often (cf. Shaffer, 1976). Part A of Figure 6 shows the distributions of keystrokes within and between hands from our data. The difference between the median interkeystroke intervals for the two kinds of strokes is about 54 msec. The difference is even more striking when we compare interkeystroke intervals on different hands to those which require the same finger on both strokes. Figure 6B shows the almost non-overlapping distributions of times to strike the *y* in the word *by* and in the word *my*. The *b* and the *y* are struck by different hands, whereas the *m* and *y* are struck by the same finger. Figure 6C shows the distribution of times for the *e* in the words *me* and *be*. Here the difference is somewhat smaller than in Figure 6B since the *b* and *e* do not require the same finger.

In the model, as in skilled typists, cross-hand interkeystroke intervals are faster than those from consecutive strokes on the same hand. Table 2 compares the average times of the model for within- and between-hand interkeystroke intervals, and compares those times with the comparable medians for our subject. Clearly, the same general difference is observed in each—within-hand strokes are substantially slower. According to the model, there are two reason for this result. First, and most important, when both letters are typed with the same hand, the typing of one key is likely to pull the finger from the next key it is supposed to strike. In contrast, if the two keys are on different hands, the activation level for the second keypress is likely to have the highest value on that hand. Therefore, the finger for the second key is very likely to be near its target key at the time the first keystroke is launched. In this way, the cross-hand stroke will, on the average,

TABLE II
Interkeystroke Intervals

	Within-Hand	Between-Hand
Simulation (mean)	7.60	4.71
Data (median)	167.67	111.37

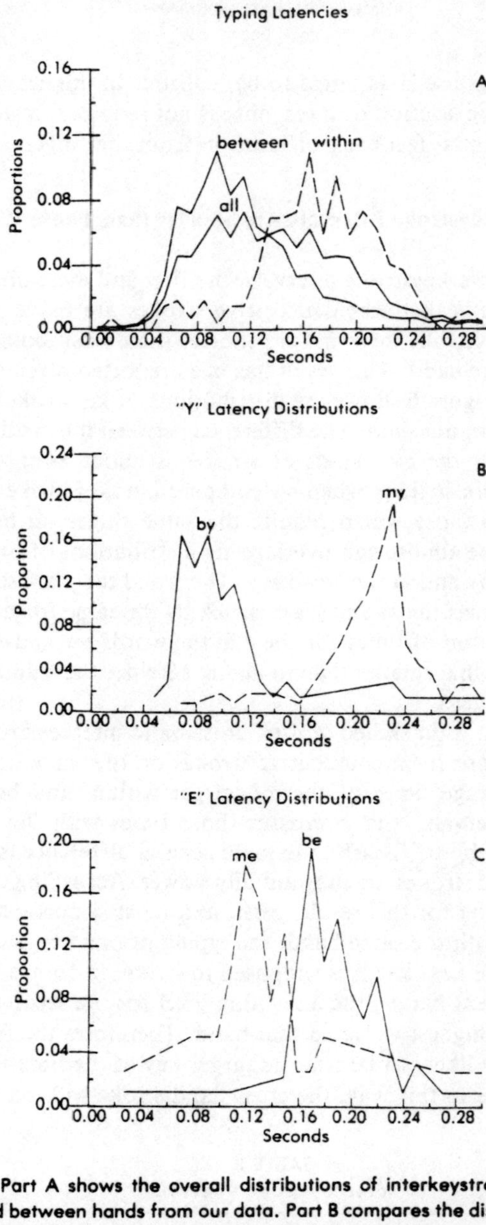

Figure 6. Part A shows the overall distributions of interkeystroke intervals within and between hands from our data. Part B compares the distributions of interkeystroke intervals to type a *y* when in the word *by* and when in the word *my*. Part C compares the distributions of interkeystroke intervals to type an *e* when in the word *me* and when in the word *be*.

occur sooner than the within-hand stroke. Second, when a keystroke is launched, the launching of subsequent strokes on the same hand is inhibited, but launching of subsequent strokes on the other hand is not. Thus, overlapped launches are allowed between hands, but not within hands. This too contributes to the advantage of between-hand strokes in speed.

Within-Hand Interkeystroke Intervals Appear to Be a Function of the Reach from One Key to the Next

Keystroke times are affected by the distance between keys. To see this, first consider the time required to type an *e* when the preceding letter uses the same finger. (The *e* is located on the top row and is typed with the middle finger of the left hand.) It takes 165 msec to repeat an *e* (to move from the *e* to the *e*), 201 msec to go from the home row key of *d* to *e* and 215 msec to go from the bottom row key of *c* to *e*. Part A of Figure 7 illustrates these results. Now let us consider the times when different fingers are used. Again, consider how long it takes to type an *e*, but this time, when the preceding key is typed with the index finger of the left hand. These times and the relative keyboard positions are shown in Figure 7B. To go from the top row key of *r* to the *e* (also on the top row) only takes 145 msec. To go from the *t* to the *e* takes 159 msec. The times to go from the middle row keys of *f* and *g* to the *e* are 168 and 178 msec, respectively. And the times from the bottom row keys of *v* and *b* are 178 and 195 msec, respectively. To a first order approximation, the longer the reach, the longer the interkeystroke intervals. Note, that we should not be surprised that the *re* and *te* times are less than the *ee* times, since the *ee* strokes cannot overlap at all, whereas those from *t* to *e* and from *r* to *e* can.

The model generates a pattern of within-hand interkeystroke intervals as a function of the reach between consecutive strokes that is similar to that of skilled typists. Part A of Figure 8 illustrates the simulated interkeystroke times for within-finger strokes. If anything, the model shows a stronger dependence on distance here than do human subjects. This may be because the model assumes that each finger attempts to move at constant velocity. It may well be that long excursions like that between the *c* and the *e* keys, actually involve higher velocities than for shorter excursions, such as from *d* to *e*.

Part B of the figure shows the average simulation results for the times between each of the keys struck by the left index finger to the striking of the *e* key. The model shows the same general pattern of results as the data. The longest reach involving the *b* key is the longest time. The shortest reach keys, from *r* and *t*, show the shortest times. The *ge* time is slightly fast relative to the *fe*. One oddity of the simulation results is that in the simulation,

ACTUAL DATA

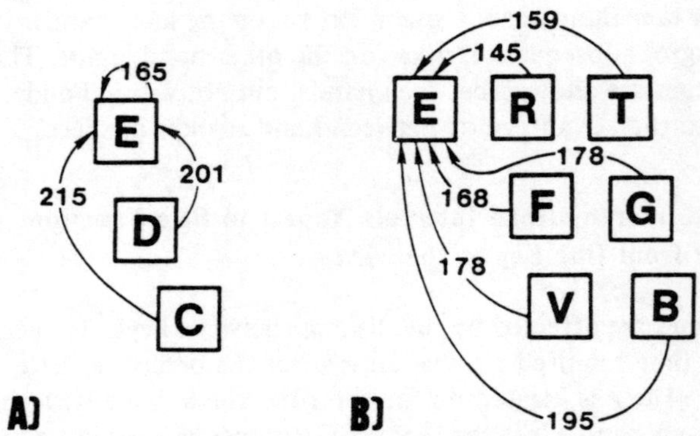

Figure 7. Part A shows a comparison of interkeystroke intervals to type a key when the preceding letter was on the same hand and the same finger: *e* to *e*, *d* to *e*, and *c* to *e*. Part B shows a similar comparison, except with the preceding letter on the same hand going from a letter typed with the index finger to an e (typed with the middle finger). The arrangement of the letters in the figure is the same as the arrangement of the keys on the keyboard. Times are shown in milliseconds.

SIMULATED RESULTS

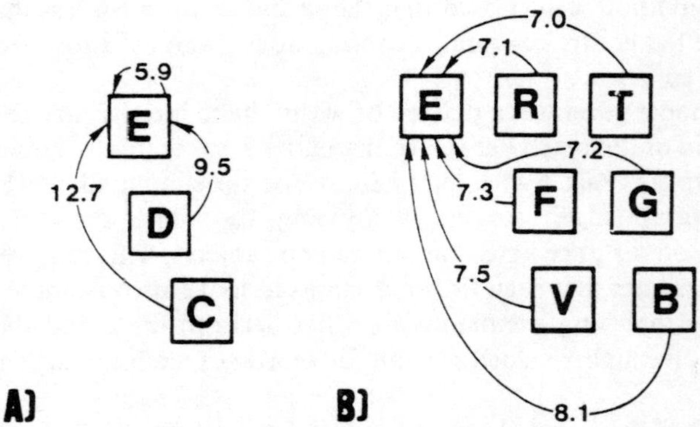

Figure 8. Interkeystroke Intervals from the simulation. Part A shows intervals for within finger strokes. Part B shows time between each of the keys struck by the left index finger to the striking of the *e* key. The arrangement of the letters in this figure is the same as the arrangement of the keys on the keyboard. Times are means of arbitrary model units from the typing of a sample text.

the *te* time is faster than the *re* time. We do not fully understand this. It could be that with the sample text typed by the model, the contexts within which *re* occur tend to give less time for the *e* finger to get near the *e* key while the *r* is being struck than while the *t* is being struck.

Correlations between the model and subjects. Although these demonstrations would seem to indicate that the simulation results do show about the right pattern of interkeystroke intervals, it would be useful to get a more general measure. For this reason, we chose the 66 most common digraphs from our data, and correlated the simulation data with the average of the data from our subject (DER) combined with the interkeystroke data from five other subjects, each of whom did transcription typing on a corpus of about 50,000 strokes. (These data were collected by Donald Gentner and kindly made available to us for this purpose). Figure 9 shows the scatter plot for the mean interkeystroke interval time over these six subjects, compared with the times generated by the simulation. The overall correlation between the model and the averaged data is about 0.86. Thus, although the fit is not bad, the model clearly does not account for all that is happening.

Just how well does the model account for the data? One problem with assessing the fit of the model is our lack of knowledge of how much different subjects differ from one another. Thus, the model may actually capture many of the processes that underlie typing, on the average, without actually fitting any single subject's performance in detail. One way of checking the relationship between the model and that of the several different typists is to compare the fit offered by the model for the data, with that offered by another subject. That is, how well do the data from one subject compare with the data of the other subjects? How well any individual subject can describe the rest of the subjects is a measure of the amount of common variance. The results are shown in Table 3. The model accounts for the individual subjects about as well as each of the subjects accounts for the others. This means that the deviations of the model are well within the variability of the subjects. Thus, any further refinement of the timing aspects of the model must be focused on the individual differences among typists. Note that there is considerable individual difference among subjects. In this sample, Subject 1 is clearly much different from the others, and correlates poorly with both the model and with the other subjects (see Gentner, 1981, for a further analysis of the differences among subjects).

The Time for a Particular Interkeystroke Interval Can Depend on the Context in Which It Occurs

Shaffer (1978) has shown that the time between successive keystrokes depends on the larger context in which these strokes appear. For example, he collected repeated typings of the words *wintry* and *wink*, found that even when the preceding word was controlled, the *wi* transition was longer in the word *wink* than in the word *wintry*. Similarly, the *in* transition was found to be longer in the word *wintry* than in the word *wink*. (Shaffer's data are shown in Table 4.) Thus, a particular interkeystroke interval depends on the preceding as well as the following letters. These results are reminiscent of the "co-articulation" effects common in speech production (Kent & Minifie, 1977; Liberman, Cooper, Shankweiler, & Studdert-Kennedy, 1972).

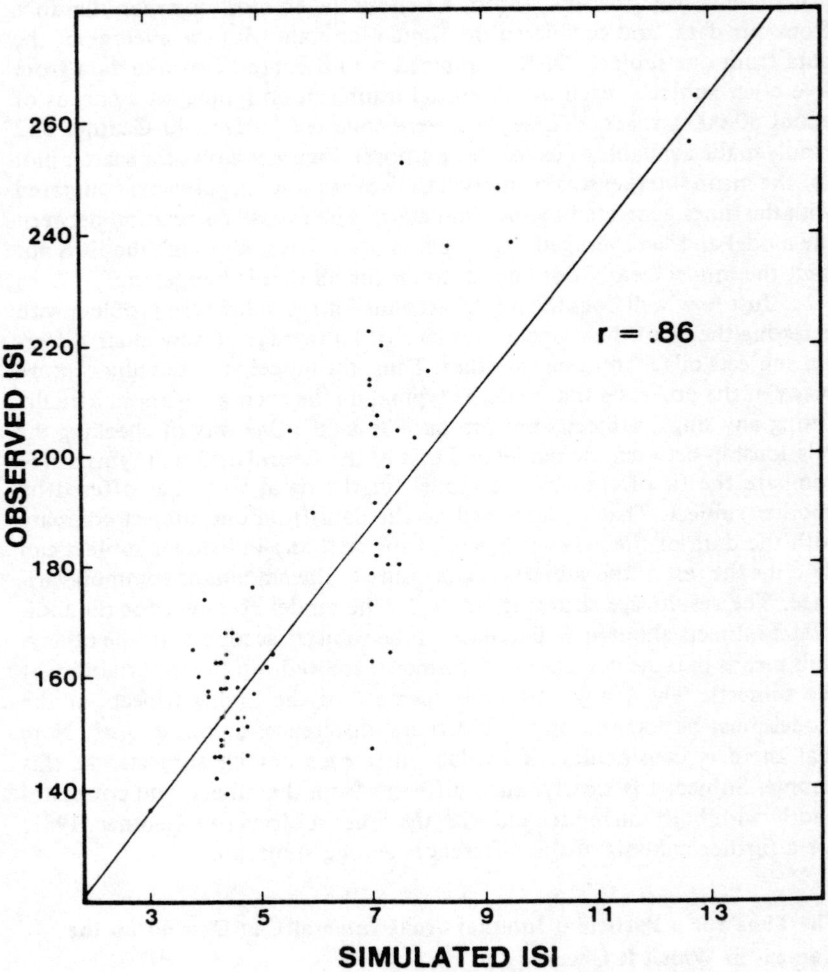

Figure 9. The scatter plot for the mean interkeystroke interval time for six subjects compared with the times generated by the simulation. Data for five subjects were collected by Donald Gentner and the sixth (DER) is our subject.

TABLE III
Correlation Coefficients for Six Typists and the Typing Model

SUBJECT:	Average	Simulation	DER	S1	S2	S3	S4	S5
Average	1.00							
Simulation	.86	1.00						
DER	.83	.71	1.00					
S1	.45	.55	.34	1.00				
S2	.85	.78	.69	.45	1.00			
S3	.90	.77	.88	.43	.79	1.00		
S4	.86	.69	.81	.45	.78	.79	1.00	
S5	.85	.88	.74	.41	.85	.81	.76	1.00

TABLE IV
Mean Interkeystroke Intervals as a Function of Context

	w	i	n	k
Simulation (arbitrary units)	7.9	3.0	6.8	6.0
Shaffer (msec)	100	140	90	150

	w	i	n	t	r	y
Simulation (arbitrary units)	6.9	2.0	10.4	1.8	6.5	4.8
Shaffer (msec)	90	110	140	90	140	100

The model also predicts that the time for a particular digraph depends on the larger context in which the digraph occurs. This dependency occurs because the surrounding strokes are pushing and pulling the fingers toward other keys, thus slowing or speeding the progress between keys. In order to see if the context dependencies are generally of the sort reported by Shaffer (1978), we ran our model on the words *wink* and *wintry* to compare the pattern with that reported by Shaffer. The results are given in Table 4. The time to the first letter is somewhat longer for *wink* than *wintry,* because the *w* in *wintry* is speeded along by the *t* and *r* which are both pulling the hand upward. The time for the *w* to the *i* is also a bit longer for the word *wink,* presumably because the *k* is holding the *i* back a bit, whereas the *y* in *wintry* is pulling toward the upper row. The *i* to *n* time is faster in the word *wink.* There are three possible reasons for this. First, because it took longer for the *i* to be struck, there was more time for the index finger to move toward the *n.* Second, when the *i* is released both the *n* and the *k* work together to bring the hand down. Third, the *y* in *wintry* is trying to hold the hand and index finger up and slowing the time to the *n.*

However, although the context effects in the model are similar to those found by Shaffer, the timing pattern is otherwise quite different. Thus, in the simulation, the time to strike the *i* key is very short, whereas it is long in Shaffer's data. There are other anomalies as well. Thus, in the model, the time to type the *n* and the *k* in *wink* are about the same, whereas in the data, they differ substantially. In *wintry,* the model predicts that *t* should be typed much more rapidly than the data indicate, and the data show the *n* and the *r* to take about the same time, which is not true in the model. One reason that the model does poorly on the overall pattern is that in our simulation of these results, the hands always began on the home row. This means that in our simulation, but not in the context in which the data were actually collected, the first letter of the string cannot be anticipated so that the time to the first stroke is elongated in the model. This has the consequence of reducing, on average, the amount of time to the second keystroke in the model (the movement to the second character *can* overlap the movement to the first character). Since the overall goal here was to simply illustrate the effects of context on a given interstroke interval, no attempt was made to provide a detailed simulation of the exact conditions under which Shaffer collected his data.

There is a Negative Correlation between the Intervals on Successive Strokes —Especially When the Alternate Strokes Occur on Alternate Hands

It is possible to treat the successive intervals between keystrokes in a corpus of typing data as a time series, and look for dependencies between successive strokes. If one interval is longer than usual, is the following interval also longer than usual, is it shorter than usual, or is it unaffected? The autocorrelation of successive intervals is a measure of this dependency.

Shaffer (1978) set out to study the nature of this relationship. He had to control for a number of potential problems. First, as we have just noted, the time between successive strokes on the same hand is largely determined by the reach between those strokes. This factor would tend to reduce the autocorrelation to zero. In order to get around this problem, Shaffer noted that the variance between strokes on different hands was less than the variance between strokes on the same hand. Moreover, these times are not subject to the problems of differential reach on the two hands. He produced texts consisting entirely of words in which successive strokes are made with alternate hands (e.g., *with* and *such*). Another problem involves long-term variations in the typist's rate of typing. If a typist slows up for some period and then speeds up, this produces a positive autocorrelation. To avoid this spurious autocorrelation, Shaffer decided to look over the relatively short span of 100 keystrokes (10–15 sec of typing), find the autocorrelation within each block of 100 strokes, and then take the average over blocks, thus minimizing the effects of overall drift in typing speed. When these precautions were taken, Shaffer found the results illustrated in Table 5; the autocorrelation with a lag of 1 (between strokes i and i + 1) was negative—that is, long intervals tended to be followed by shorter intervals and vice versa. However, the autocorrelation with a lag of 2 (between strokes i and i + 2) was positive.

Figure 10 shows the distribution of times between successive strokes in the word *with* from our data. Note that, although the keystrokes alternate between the hands, the *wi* time is fast, the *it* time slow, and the *th* time fast. As Shaffer points out, such behavior is to be expected from a metronome model of typing in which the typist initiates a stroke regularly to some sort of internal clock beat, but then there is some variance in the actual execution of the response. In this case, a short interval will cause the next one to be long. As we show, this behavior is also to be expected from models not involving such an internal clock.

All other things being equal, the model predicts a negative correlation between adjacent strokes. In general, the prediction follows from the fact that a slow response gives the following finger more time to get into position, thereby speeding up the following response. If there happens to be a fast response, the following stroke has less time to get into position, and so

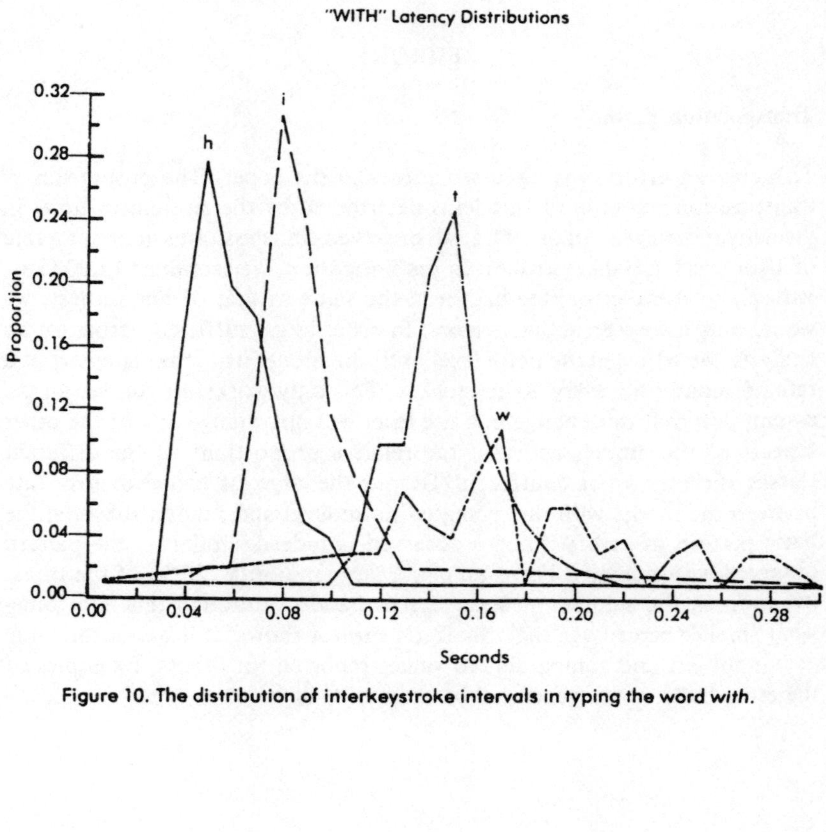

Figure 10. The distribution of interkeystroke intervals in typing the word *with*.

TABLE V
Autocorrelations of Interkeystroke Intervals

Autocorrelation with Lag	Simulation	Shaffer (1978)
1	−0.58	−0.28
2	+0.44	+0.10

the second interkeystroke interval is somewhat longer. In order to quantify this effect in our simulation model, we extracted all strings of keypresses which involved either a right-left-right-left or a left-right-left-right sequence of strokes (excluding spaces and returns). We then computed the correlation between the three interkeystroke intervals thus defined. The results of this computation are shown in Table 5. Successive strokes are always negatively correlated. The correlation between the adjacent times (corresponding to an autocorrelation with a lag of one) is −0.58. The correlation between the first and third intervals (corresponding to an autocorrelation with a lag of 2) is 0.44. The pattern has the same direction as the data reported by Shaffer, but somewhat more extreme.

ERRORS

Transposition Errors

This class of errors was discussed earlier in the paper. The proportion of transposition errors in the model is determined by the amount of noise in the activation levels. In our data, we observed transpositions at about a rate of 1 for every 1,800 keystrokes. In the simulation, we examined 12,000 keystrokes, so if the error rate had been the same as that of our subject, we would only have seen about 7 errors. In order to get sufficient errors for an analysis, we adjusted the noise level until the model was making errors at a rate of about 1 for every 30 keystrokes. This only works, of course, on the assumption that this change in noise level will not change any of the other aspects of the model, and that the relative proportions of the different classes of errors wil be unaffected. Despite the large difference in error rate between the model with the increased noise level and human subjects, the basic pattern of errors that we observed is indeed similar to the pattern observed with humans. For example, a large majority (76%) of the transpositions in the simulation occur across hands. Although this is a somewhat smaller percentage than Shaffer's subject shows, it is about the same as our subject and comparable to values reported for others. Examples of the cross-hand errors generated by the model include:

special -> speical
course -> coruse
dramatic -> dramaitc

We also found examples of multiple transpositions in the simulation results (although none were as impressive as the example from Shaffer that we reported earlier). We did find

dismal -> dsiaml

(in which the right-hand keys appear to be delayed relative to the left) and

vitamins -> ivtmaisn

(in which, with the exception of the final "s", the left-hand strokes appear to be delayed with respect to the right).

We also found within-hand transpositions in the simulation data. Examples of these include:

master -> masetr
result -> ersult

The reason that the model gives this pattern of transposition errors should be clear enough. A transposition error can only occur if the wrong schema has its trigger conditions satisfied prior to the correct schema. This means that the wrong finger must both be in place to strike its key, and its schema must momentarily have the highest level of activation. This is more likely to occur across hands than within hands because the next finger on the other hand is less constrained by the motions of other fingers: the next finger to type on the opposite hand always has a speed advantage. The longer a finger is in position, the greater the chance that noise will momentarily cause its schema to become most active, and thus be triggered. Within hands, the transpositions should involve keys that are near one another, so that the movement of the palm helps rather than hinders movement toward the keys by the next finger that is to launch a keystroke.

Doubling Errors

Doubling errors played an important role in the development of the theory behind the model, so it is interesting to see how well the model did at producing them. At the level of noise employed in the simulation, 17 doubling errors were generated. Examples included:

sheer -> sherr
following -> foolowign
little -> liitle

Thus, as with actual observations of errors, we can find doubling errors in which the double is assimilated to either the preceding or following letter.

According to the model, the relative proportion of these two directions of error should be the same.

Alternation Reversal Errors

The version of the simulation model that we ran had no provision for the alternation encoding and thus, none of these errors were generated. Inclusion of errors of this type requires the addition of a special function whose sole purpose would be to encode alternations, and would thereby generate the errors. As mentioned earlier, an attempt to devise such a mechanism led to the kinds of errors that are never observed.

Homologous Errors

In this class of errors, the wrong hand is selected, but then within the hand, the correct finger and the correct finger movement is performed. Thus, the erroneous stroke is anatomically homologous to the correct one. An analysis of typing errors from the collection of Lessenberry (1928) indicates that homologous errors occur quite frequently.[2] Lessenberry's corpus contains 60,000 typing errors from people of all skills. However, in the various corpora of errors that we have examined that come from skilled typists, there are few homologous errors. We suspect that the more frequently errors of this type occur, the less the skill of the typist, but we do not have firm evidence for this belief.

The model does not generate homologous errors. We believe that these errors are probably generated by errors in the developing response schemata. The mechanism would have to involve a weak binding between the keypress schemata and the argument that specifies the hand to be controlled, perhaps much in the same way that we provided a weak binding between a schema and its arguments in the case of doubling errors; the right finger and place will be commanded, but on the wrong hand.

Capture Errors

This category of error occurs when one intends to type one sequence, but gets "captured" by another that has a similar beginning (Norman, 1981). Examples include:

efficiency –> efficient
incredibly –> incredible
normal –> norman

[2]The analysis of the error matrix was performed by Jonathan Grudin and Craig Will.

There is no provision in the model for capture errors. We suspect that these occur at either the perceptual or encoding stage of the system, neither of which are dealt with in the model.

Omission Errors

These errors occur when a letter in a sequence is omitted. Examples include:

amount –> amont
education –> educatin
lunches –> luches

Shaffer (1976) has argued on the basis of his timing studies that omissions typically result from aborted strokes, with the timing the same as if the letter had been typed. It is as if the stroke simply involves a depression of the key insufficient to result in the letter being printed. The single most common omission that we observed involves the *o* in *ion,* suggesting that the ring finger is pulled away too soon as the index finger reaches for the *n* key.

The model had no provision for omission errors. Presumably, a mechanism could be constructed which assumed a certain percentage of launches were defective, with the deactivation of the schema occurring before the actual conclusion of the launch.

Misstrokes

We classified about 10% of the errors in our collection as misstrokes, defined to occur whenever a key nearby the intended key is inappropriately struck. Examples include:

awareness –> awareneww
because –> becajse
believe –> beoieve
the –> tje

The simulation model did produce a number of misstrokes. Misstrokes occurred when the launch time was longer than usual, so that the keystroke schema was already self inhibited before the launch was complete. This allowed the hand to pull away, and when the stroke completed, it struck a nearby key. Examples of such misstrokes are:

distances –> disrances
entire body –> entird body
however –> howevdr

In each case, a nearby key is struck as the hand and/or finger moves toward its next appointed task.

THE GENERAL ORGANIZATION OF THE TYPING PROCESS

Skilled Typists Move Their Hands Toward the Keys in Parallel

The parallel overlapping response patterns observable in skilled typists are clearly apparent in the structure of our model. In the model, the hands are constantly in motion, each finger always pushing toward its next key. If anything, our model overplays this aspect. It is unlikely that expert typists always move their fingers as soon or as far as they can, as our model does. Thus, our model might be considered an extreme case of the kind of behavior actually engaged in by expert typists.

The Units of Typing Seem to Be Largely at the Word Level or Smaller

The units of typing in our model are words and letters. Basically, the unit size is established by the parser, operating upon whatever schemata for the words of the language and for the keypresses already exist within memory. Because we did not study perception or the parser, we simply assumed that the input text would always be words, and the parser's major job was to encode the strings in accordance with the restrictions imposed by the lack of token nodes (that is, stop encoding as soon as a repeated character is encountered). Thus, our model, as it actually works, would perform as well on random letter strings as on words. As we have already noted, human typists are much poorer on random letter strings. To deal with this phenomenon properly, we would have to equip the model with a lexicon of words. We also would have to model the memory processes involved in perception, for it is well established that random letter strings are poorly perceived and poorly remembered. Thus, we can expect that a certain amount of the slowdown in typing random letter strings comes from the perceptual and memory processes.

Sequences Involving Cross-Hand Strokes Seem to Take Longer to Program than Those Involving Only Within-Hand Strokes

Sternberg, Monsell, Knoll, and Wright (1978) used a discrete trial method in which subjects were first shown a string to be typed and then, on a signal, typed the string. They found that the preparation time (the time to the first stroke) increased linearly with the length of the string. They also found that the preparation time was longer for those strings that involved strokes on both hands than for those involving only a single hand. The typing time increased non-linearly with the length of the string, and strings involving two hands were faster to type than those involving one.

We made no attempt to deal with the preparation time phenomena. Our efforts have concentrated on the study of continuous typing, and have offered no analysis of the discrete trial situation analyzed by Sternberg *et al.* In continuous typing, it is not clear what the analog to preparation time would be, except, possibly, the time between words. Of course, our model does show that the duration of typing words involving both hands is less than for words involving a single hand.

For the moment, the preparation time results lie outside of the range of the model. It is possible that the planning time has to do with the setup of the response schemata, and that the setting-up of the inhibition patterns across two hands is more complex and takes more time than the patterns within a hand. Within the context of continuous typing, such setup times would not be noted. (Note too that the typing speeds observed by Sternberg et al. for their discrete trials were much faster than the normal typing speeds for these same subjects when typing in a continuous task.)

CONCLUSION

In this paper, we have presented an analysis of many of the phenomena of typing, and we have presented a computer simulation model that captures the appropriate spirit of the phenomena. The result has both positive and negative features.

Clearly, the model does not yet offer a complete account of the typing process. The studies now underway in our laboratory point out some of the problems. There are striking individual differences among typists, so that two people who type at the same (high) average rate may differ substantially in the manner in which they type. In our model, we have made no provision for the substantial individual differences that can be observed in response patterns, no provision for several of the forms of errors that have been observed, nor no satisfactory account of the input and encoding processes involved in transcription typing.

We have presented strong arguments for the form of the mechanism that can account for both doubling and alternation errors. However, despite the success of the mechanism for doubling errors, we were not able to get essentially the same mechanism to work satisfactorily for the case of alternations. One possible reason for the failure may lie in the parsing and encoding stage. The problems we faced dealt with spaces when there were single letter words. It is possible that this case is handled by the stages of parsing and encoding, not by the typing mechanisms covered by the model.

There are some interesting asymmetries in typing behavior of our subject that we have not captured. Thus, the digraphs *er* and *re* are both typed rapidly, but *er* is typed considerably faster than is *re*. The model does not predict this. This result probably comes from the physical characteristics of the hand—such as the differences in agility and length of the index and middle fingers—perhaps coupled with the peculiar way in which the letters are arranged in diagonal rows on the keyboard, and the resulting angles that the palms make with the keyboard. Our model of the hand is too simple, neglecting the differences in strengths and speeds of the fingers, and treating each finger independently of the others. In fact, some of the fingers are coupled together, with the tendons for some of the fingers bound within the same tendon sheath. (There was a time around the turn of the century when virtuoso piano players had these sheaths surgically cut in the hopes of gaining more independent control of the fingers.)

In spite of these problems, the model does capture many of the essential aspects of typing. We have shown how a very simple set of ideas can account for a wide range of phenomena when combined in a reasonable simulation model of the task. Note that despite the lack of an internal clock or metronome for timing, the model provides a reasonably good account of the timing patterns observed among skilled typists, including the prediction of negative correlations among successive keystrokes—a characteristic of metronome models. In similar fashion, there are no specific context dependencies built into the model, and yet the time that it takes to strike keys depends upon the context in which they occur. We have no specific mechanism for transposition errors, yet our model generates the correct types of transposition errors. Moreover, the coordinative structure assumed within the model yields a qualitative emulation of the pattern of overlapping movements shown in a high speed film of a typist.

A number of conclusions can be drawn from our studies. First, the existence of doubling errors strongly implies the existence of a pure "type" representation of the keyboard schemata, with their arguments only loosely bound. In the model, the arguments are not attached to the schemata, but rather are picked up when they are needed by an evaluation of activation values. Second, the nature of the skill requires simultaneous, parallel control of the fingers and hands, and this requires some form of negotiation process to turn the potentially competitive movements into cooperative ones. The degrees of freedom problem is turned into a degrees of freedom virtue. Third, the model must incorporate the entire environment within which the typist operates, from the reading of the text, to the cognitive and motor control systems, to the shapes and mechanical characteristics of the hands, finger, and keyboard. Indeed, some of the limitations of the model may really result from limitations of how well we dealt with the environment surrounding the control processes. Perhaps the central conclusion to be drawn from our analysis of typing deals with the nature of skilled motor coordination. We propose that the motor control system carries out its computations relatively locally and in parallel. We presume that such a conclusion will be proven for all skills involving high-speed performance.

REFERENCES

Beeching, W. A. *Century of the typewriter.* New York: St. Martin's Press, 1974.

Estes, W. K. An associative basis for coding and organization in memory. In A. W. Melton & E. Martin (Eds.), *Coding processes in human memory.* Washington, D.C.: Winston, 1972, 161-190.

Gentner, D. R., *Skilled finger movements in typing.* San Diego, CA: Center for Human Information Processing, University of California, San Diego, CHIP 104. July 1981.

Gentner, D. R., Grudin, J., & Conway, E. *Finger movements in transcription typing.* San Diego, CA: Center for Human Information Processing, University of California, San Diego. Technical Report 8001. May 1980.

Head, H. *Aphasia and kindred disorders of speech* (Vol. 1). Cambridge, England: Cambridge University Press, 1926, 448 ff.

Kent, R. D., & Minifie, F. D., Coarticulation in recent speech production models. *Journal of Phonetics,* 1977, *5,* 115-133.

Lashley, K. S. The problem of serial order in behavior. In L. A. Jeffress (Ed.), *Cerebral mechanisms in behavior.* New York: Wiley, 1951.

Lessenberry, D. D. *Analysis of errors.* Syracuse, NY: L. C. Smith and Corona Typewriters, Inc., School Department, 1928. [Reprinted in Dvorak, A. *Typewriting behavior; Psychology applied to teaching and learning typewriting.* New York: American Book Company, 1936.]

Liberman, A. M., and Cooper, F. S., Shankweiler, D. P., & Studdert-Kennedy, M. Perception of the speech code.In E. E. David, Jr. & P. B. Denes (Eds.), *Human communication: A unified view.* New York: McGraw-Hill, 1972.

Norman, D. A. Categorization of action slips. *Psychological Review,* 1981, *88,* 1-15.

Saltzman, E. Levels of sensorimotor representation. *Journal of Mathematical Psychology,* 1979, *20,* 91-163.

Schmidt, R. A. The schema as a solution to some persistent problems in motor learning theory. In G. E. Stelmach (Ed.), *Motor control: Issues and trends.* New York: Academic Press, 1976.

Shaffer, L. H. Latency mechanisms in transcription. In S. Kornblum (Ed.), *Attention and performance IV.* New York: Academic Press, 1973.

Shaffer, L. H. Intention and performance. *Psychological Review,* 1976, *83,* 375-393.

Shaffer, L. H. Timing in the motor programming of typing. *Quarterly Journal of Experimental Psychology,* 1978, *30,* 333-345.

Sternberg, S., Monsell, S., Knoll, R. L., & Wright, C. E. The latency and duration of rapid movement sequences: Comparisons of speech and typewriting. In G. E. Stelmach (Ed.), *Information processing in motor control and learning.* New York: Academic Press, 1978.

[12]

Attention, Intentions, and the Structure of Discourse

Barbara J. Grosz
Artificial Intelligence Center *and*
Center for the Study of Language and Information
SRI International
Menlo Park, CA 94025

Candace L. Sidner
BBN Laboratories Inc.
Cambridge, MA 02238

In this paper we explore a new theory of discourse structure that stresses the role of purpose and processing in discourse. In this theory, discourse structure is composed of three separate but interrelated components: the structure of the sequence of utterances (called the linguistic structure), a structure of purposes (called the intentional structure), and the state of focus of attention (called the attentional state). The linguistic structure consists of segments of the discourse into which the utterances naturally aggregate. The intentional structure captures the discourse-relevant purposes, expressed in each of the linguistic segments as well as relationships among them. The attentional state is an abstraction of the focus of attention of the participants as the discourse unfolds. The attentional state, being dynamic, records the objects, properties, and relations that are salient at each point of the discourse. The distinction among these components is essential to provide an adequate explanation of such discourse phenomena as cue phrases, referring expressions, and interruptions.

The theory of attention, intention, and aggregation of utterances is illustrated in the paper with a number of example discourses. Various properties of discourse are described, and explanations for the behavior of cue phrases, referring expressions, and interruptions are explored.

This theory provides a framework for describing the processing of utterances in a discourse. Discourse processing requires recognizing how the utterances of the discourse aggregate into segments, recognizing the intentions expressed in the discourse and the relationships among intentions, and tracking the discourse through the operation of the mechanisms associated with attentional state. This processing description specifies in these recognition tasks the role of information from the discourse and from the participants' knowledge of the domain.

1 Introduction

This paper presents the basic elements of a computational theory of discourse structure that simplifies and expands upon previous work. By specifying the basic units a discourse comprises and the ways in which they can relate, a proper account of discourse structure provides the basis for an account of discourse meaning. An account of discourse structure also plays a central role in language processing because it stipulates constraints on those portions of a discourse to which any given utterance in the discourse must be related.

An account of discourse structure is closely related to two questions: What individuates a discourse? What makes it coherent? That is, faced with a sequence of utterances, how does one know whether they constitute a single discourse, several (perhaps interleaved) discourses, or none? As we develop it, the theory of discourse structure will be seen to be intimately connected with two nonlinguistic notions: intention and attention. Attention is an essential factor in explicating the processing of utterances in discourse. Intentions play a primary role in explaining discourse structure, defining discourse coherence, and providing a coherent conceptualization of the term "discourse" itself.

0362-613X/86/030175-204$03.00

The theory is a further development and integration of two lines of research: work on focusing in discourse (Grosz 1978a, 1978b, 1981) and more recent work on intention recognition in discourse (Sidner and Israel 1981; Sidner 1983; 1985; Allen 1983, Litman 1985; Pollack 1986). Our goal has been to generalize these constructs properly to a wide range of discourse types. Grosz (1978a) demonstrated that the notions of focusing and task structure are necessary for understanding and producing task-oriented dialogue. One of the main generalizations of previous work will be to show that discourses are generally in some sense "task-oriented," but the kinds of "tasks" that can be engaged in are quite varied – some are physical, some mental, others linguistic. Consequently, the term "task" is misleading; we therefore will use the more general terminology of **intentions** (e.g., when speaking of discourse purposes) for most of what we say.

Our main thesis is that the structure of any discourse is a composite of three distinct but interacting components:

- the structure of the actual sequence of utterances in the discourse;
- a structure of intentions;
- an attentional state.

The distinction among these components is essential to an explanation of interruptions (see Section 5), as well as to explanations of the use of certain types of referring expressions (see Section 4.2) and various other expressions that affect discourse segmentation and structure (see Section 6). Most related work on discourse structure (including Reichman-Adar 1984, Linde 1979, Linde and Goguen 1978, Cohen 1983) fails to distinguish among some (or, in some cases, all) of these components. As a result, significant generalizations are lost, and the computational mechanisms proposed are more complex than necessary. By carefully distinguishing these components, we are able to account for significant observations in this related work while simplifying both the explanations given and computational mechanisms used.

In addition to explicating these linguistic phenomena, the theory provides an overall framework within which to answer questions about the relevance of various segments of discourse to one another and to the overall purposes of the discourse participants. Various properties of the intentional component have implications for research in natural-language processing in general. In particular, the intentions that underlie discourse are so diverse that approaches to discourse coherence based on selecting discourse relationships from a fixed set of alternative rhetorical patterns (e.g., Hobbs 1979, Mann and Thompson 1983, Reichman 1981) are unlikely to suffice. The intentional structure introduced in this paper depends instead on a small number of structural relations that can hold between intentions. This study also reveals several problems that must be confronted in expanding speech-act-related theories (e.g., Allen and Perrault 1980, Cohen and Levesque 1980, Allen 1983) from coverage of individual utterances to coverage of extended sequences of utterances in discourse.

Although a definition of discourse must await further development of the theory presented in this paper, some properties of the phenomena we want to explain must be specified now. In particular, we take a **discourse** to be a piece of language behavior that typically involves multiple utterances and multiple participants. A discourse may be produced by one or more of these participants as speakers or writers; the audience may comprise one or more of the participants as hearers or readers. Because in multi-party conversations more than one participant may speak (or write) different utterances within a segment, the terms **speaker** and **hearer** do not differentiate the unique roles that the participants maintain in a segment of a conversation. We will therefore use the terms **initiating conversational participant** (ICP) and **other conversational participant(s)** (OCP) to distinguish the initiator of a discourse segment from its other participants. The ICP speaks (or writes) the first utterance of a segment, but an OCP may be the speaker of some subsequent utterances. By speaking of ICPs and OCPs, we can highlight the purposive aspect of discourse. We will use the terms *speaker* and *hearer* only when the particular speaking/hearing activity is important for the point being made.

In most of this paper, we will be concerned with developing an abstract model of discourse structure; in particular, the definitions of the components will abstract away from the details of the discourse participants. Whether one constructs a computer system that can participate in a discourse (i.e., one that *is* a language user) or defines a psychological theory of language use, the task will require the appropriate projection of this abstract model onto properties of a language user, and specification of additional details (e.g., specifying memory for linguistic structure, means for encoding attentional state, and appropriate representations of intentional structure). We do, however, address ourselves directly to certain processing issues that are essential to the computational validity of the [abstract] model and to its utilization for a language-processing system or psychological theory.

Finally, it is important to note that although discourse *meaning* is a significant, unsolved problem, we will not address it in this paper. An adequate theory of discourse meaning needs to rest at least partially on an adequate theory of discourse structure. Our concern is with providing the latter.

The next section examines the basic theory of discourse structure and presents an overview of each of the components of discourse structure. Section 3 analyzes two sample discourses – a written text and a fragment of task-oriented dialogue – from the perspective of the theory being developed; these two examples are also used to illustrate various points in the remainder of the paper. Section 4 investigates various processing issues that the theory raises. The following two sections describe the role of the discourse structure components in explaining various properties of discourse, thereby corroborating the necessity of distinguishing among its three components. Section 7 describes the generalization from utterance-level to discourse-level intentions, establishes certain properties of the latter, and contrasts them with the rhetorical relations of alternative theories. Finally, Section 8 poses a number of outstanding research questions suggested by the theory.

2 The Basic Theory

Discourse structure is a composite of three interacting constituents: a linguistic structure, an intentional structure, and an attentional state. These three constituents of discourse structure deal with different aspects of the utterances in a discourse. Utterances – the actual saying or writing of particular sequences of phrases and clauses – are the linguistic structure's basic elements. Intentions of a particular sort and a small number of relationships between them provide the basic elements of the intentional structure. Attentional state contains information about the objects, properties, relations, and discourse intentions that are most salient at any given point. It is an abstraction of the focus of attention of the discourse participants; it serves to summarize information from previous utterances crucial for processing subsequent ones, thus obviating the need for keeping a complete history of the discourse.

Together the three constituents of discourse structure supply the information needed by the CPs to determine how an individual utterance fits with the rest of the discourse – in essence, enabling them to figure out why it was said and what it means. The context provided by these constituents also forms the basis for certain expectations about what is to come; these expectations play a role in accommodating new utterances. The attentional state serves an additional purpose: namely, it furnishes the means for actually using the information in the other two structures in generating and interpreting individual utterances.

2.1 Linguistic Structure

The first component of discourse structure is the structure of the sequence of utterances that comprise a discourse.[1] Just as the words in a single sentence form constituent phrases, the utterances in a discourse are naturally aggregated into **discourse segments.** The utterances in a segment, like the words in a phrase, serve particular roles with respect to that segment. In addition, the discourse segments, like the phrases, fulfill certain functions with respect to the overall discourse. Although two consecutive utterances may be in the same discourse segment, it is also common for two consecutive utterances to be in different segments. It is also possible for two utterances that are nonconsecutive to be in the same segment.

The factoring of discourses into segments has been observed across a wide range of discourse types. Grosz (1978a) showed this for task-oriented dialogues. Linde (1979) found it valid for descriptions of apartments; Linde and Goguen (1978) describe such structuring in the Watergate transcripts. Reichman-Adar (1984) observed it in informal debates, explanations, and therapeutic discourse. Cohen (1983) found similar structures in essays in rhetorical texts. Polanyi and Scha (1986) discuss this feature of narratives.

Although different researchers with different theories have examined a variety of discourse types and found discourse-level segmentation, there has been very little investigation of the extent of agreement about where the segment boundaries lie. There have been no psychological studies of the consistency of recognition of section boundaries. However, Mann (Mann et al. 1975) asked several people to segment a set of dialogues. He has reported [personal communication] that his subjects segmented the discourses approximately the same; their disagreements were about utterances at the boundaries of segments.[2] Several studies of spontaneously produced discourses provide additional evidence of the existence of segment boundaries, as well as suggesting some of the linguistic cues available for detecting boundaries. Chafe (1979, 1980) found differences in pause lengths at segment boundaries. Butterworth (1975) found speech rate differences that correlated with segments; speech rate is slower at start of a segment than toward the end.

The linguistic structure consists of the discourse segments and an embedding relationship that can hold between them. As we discuss in Sections 2.2 and 5, the embedding relationships are a surface reflection of relationships among elements of the intentional structure. It is important to recognize that the linguistic structure is not strictly decompositional. An individual segment may include a combination of subsegments and utterances only in that segment (and not members of any of its embedded subsegments). Both of the examples in Section 3 exhibit such nonstrict decompositionality. Because the linguistic structure is not strictly decompositional, various properties of the discourse (most notably the intentional structure) are functions of properties of individual utterances and properties of segments.

There is a two-way interaction between the discourse segment structure and the utterances constituting the discourse: linguistic expressions can be used to convey information about the discourse structure; conversely, the discourse structure constrains the interpretation of expressions (and hence affects what a speaker says and how a hearer will interpret what is said). Not surprisingly, linguistic expressions are among the primary indicators of discourse segment boundaries. The explicit use of certain words and phrases (e.g., *in the first place*) and more subtle cues, such as intonation or changes in tense and aspect, are included in the repertoire of linguistic devices that function, wholly or in part, to indicate these boundaries (Grosz 1978a, Reichman-Adar 1984, Cohen 1983, Polanyi and Scha 1983, Hirschberg and Pierrehumbert 1986). Reichman (1981) discusses some words that function in this way and coined the term **clue words.** We will use the term **cue phrases** to generalize on her observation as well as many others because each one of these devices cue the hearer to some change in the discourse structure.

As discussed in Section 6, these linguistic boundary markers can be divided according to whether they explicitly indicate changes in the intentional structure or in the attentional state of the discourse. The differential use of these linguistic markers provides one piece of evidence for considering these two components to be distinct. Because these linguistic devices function explicitly as indicators of discourse structure, it becomes clear that they are best seen as providing information at the discourse level, and not at the sentence level; hence, certain kinds of questions (e.g., about their contribution to the truth conditions of an individual sentence) do not make sense. For example, in the utterance *Incidentally, Jane swims every day,* the *incidentally* indicates an interruption of the main flow of discourse rather than affecting in any way the meaning of *Jane swims every day.* Jane's swimming every day could hardly be fortuitous.

Just as linguistic devices affect structure, so the discourse segmentation affects the interpretation of linguistic expressions in a discourse. Referring expressions provide the primary example of this effect.[3] The segmentation of discourse constrains the use of referring expressions by delineating certain points at which there is a significant change in what entities (objects, properties, or relations) are being discussed. For example, there are different constraints on the use of pronouns and reduced definite-noun phrases within a segment than across segment boundaries. While discourse segmentation is obviously not the only factor governing the use of referring expressions, it is an important one.

2.2 INTENTIONAL STRUCTURE

A rather straightforward property of discourses, namely, that they (or, more accurately, those who participate in them) have an overall purpose, turns out to play a fundamental role in the theory of discourse structure. In particular, some of the purposes that underlie discourses, and their component segments, provide the means of individuating discourses and of distinguishing discourses that are coherent from those that are not. These purposes also make it possible to determine when a sequence of utterances comprises more than one discourse.

Although typically the participants in a discourse may have more than one aim in participating in the discourse (e.g., a story may entertain its listeners as well as describe an event; an argument may establish a person's brilliance as well as convince someone that a claim or allegation is true), we distinguish one of these purposes as foundational to the discourse. We will refer to it as the **discourse purpose** (DP). From an intuitive perspective, the discourse purpose is the intention that underlies engaging in the particular discourse. This intention provides both the reason a discourse (a linguistic act), rather than some other action, is being performed and the reason the particular content of this discourse is being conveyed rather than some other information. For each of the discourse segments, we can also single out one intention – the **discourse segment purpose** (DSP). From an intuitive standpoint, the DSP specifies how this segment contributes to achieving the overall discourse purpose. The assumption that there are single such intentions will in the end prove too strong. However, this assumption allows us to describe the basic theory more clearly. We must leave to future research (and a subsequent paper) the exploration and discussion of the complications that result from relaxing this assumption.

Typically, an ICP will have a number of different kinds of intentions that lead to initiating a discourse. One kind might include intentions to speak in a certain language or to utter certain words. Another might include intentions to amuse or to impress. The kinds of intentions that can serve as discourse purposes or discourse segment purposes are distinguished from other intentions by the fact that they are intended to be recognized (cf. Allen and Perrault 1980, Sidner 1985), whereas other intentions are private; that is, the recognition of the DP or DSP is *essential* to its achieving its intended effect. Discourse purposes and discourse segment purposes share this property with certain utterance-level intentions that Grice (1969) uses in defining utterance meaning (see Section 7).

It is important to distinguish intentions that are intended to be recognized from other kinds of intentions that are associated with discourse. Intentions that are intended to be recognized achieve their intended effect only if the intention is recognized. For example, a compliment achieves its intended effect only if the intention to compliment is recognized; in contrast, a scream of *boo* typically achieves its intended effect (scaring the hearer) without the hearer having to recognize the speaker's intention.

Some intention that is private and not intended to be recognized may be the primary motivation for an ICP to begin a discourse. For example, the ICP may intend to impress someone or may plan to teach someone. In neither case is the ICP's intention necessarily intended to be recognized. Quite the opposite may be true in the case of impressing, as the ICP may not want the OCP to be aware of his intention. When teaching, the ICP may not care whether the OCP knows the ICP is teaching him or her. Thus, the intention that motivates the ICP to engage in a discourse may be private. By contrast, the discourse segment purpose is always intended to be recognized.

DPs and DSPs are basically the same sorts of intentions. If an intention is a DP, then its satisfaction is a main purpose of the discourse, whereas if it is a DSP, then its satisfaction contributes to the satisfaction of the DP. The following are some of the types of intentions that could serve as DP/DSPs, followed by one example of each type.

1. Intend that some agent intend to perform some physical task. Example: *Intend that Ruth intend to fix the flat tire.*
2. Intend that some agent believe some fact. Example: *Intend that Ruth believe the campfire has started.*
3. Intend that some agent believe that one fact supports another. Example: *Intend that Ruth believe the smell of smoke provides evidence that the campfire is started.*
4. Intend that some agent intend to identify an object (existing physical object, imaginary object, plan, event, event sequence). Example: *Intend that Ruth intend to identify my bicycle.*
5. Intend that some agent know some property of an object. Example: *Intend that Ruth know that my bicycle has a flat tire.*

We have identified two structural relations that play an important role in discourse structure: **dominance** and **satisfaction-precedence.** An action that satisfies one intention, say DSP1, may be intended to provide part of the satisfaction of another, say DSP2. When this is the case, we will say that DSP1 **contributes to** DSP2; conversely, we will say that DSP2 **dominates** DSP1 (or DSP2 *DOM* DSP1). The dominance relation invokes a partial ordering on DSPs that we will refer to as the **dominance hierarchy.** For some discourses, including task-oriented ones, the order in which the DSPs are satisfied may be significant, as well as being intended to be recognized. We will say that DSP1 **satisfaction-precedes** DSP2 (or, DSP1 *SP* DSP2) whenever DSP1 must be satisfied before DSP2.[4]

Any of the intentions on the preceding list could be either a DP or a DSP. Furthermore, a given instance of any one of them could contribute to another, or to a different, instance of the same type. For example, the intention that someone intend to identify some object might dominate several intentions that she or he know some property of that object; likewise, the intention to get someone to believe some fact might dominate a number of contributing intentions that that person believe other facts.

As the above list makes clear, the range of intentions that can serve as discourse, or discourse segment, purposes is open-ended (cf. Wittgenstein 1953: paragraph 23), much like the range of intentions that underlie more general purposeful action. There is no finite list of discourse purposes, as there is, say, of syntactic categories. It remains an unresolved research question whether there is a finite description of the open-ended set of such intentions. However, even if there were finite descriptions, there would still be no finite list of intentions from which to choose. Thus, a theory of discourse structure cannot depend on choosing the DP/DSPs from a fixed list (cf. Reichman-Adar 1984, Schank et al. 1982, Mann and Thompson 1983), nor on the particulars of individual intentions. Although the particulars of individual intentions, like a wide range of common sense knowledge, are crucial to understanding any discourse, such particulars cannot serve as the basis for *determining* discourse structure.

What is essential for discourse structure is that such intentions bear certain kinds of structural relationships to one another. Since the CPs can never know the whole set of intentions that might serve as DP/DSPs, what they must recognize is the relevant structural relationships among intentions. Although there is an infinite number of intentions, there are only a small number of relations relevant to discourse structure that can hold between them.

In this paper we distinguish between the *determination* of the DSP and the *recognition* of it. We use the term **determination** to refer to a semantic-like notion, namely, the complete specification of what is intended by whom; we use the term **recognition** to refer to a processing notion, namely, the processing that leads a discourse participant to identify what the intention is. These are obviously related concepts; the same information that determines a DSP may be used by an OCP to recognize it. However, some questions are relevant to only one of them. For example, the question of when the information becomes available is not relevant to determination but is crucial to recognition. An analogous distinction has been drawn with respect to sentence structure; the parse tree (determination) is differentiated from the parsing process (recognition) that produces the tree.

2.3 ATTENTIONAL STATE

The third component of discourse structure, the attentional state, is an abstraction of the participants' focus of attention as their discourse unfolds. The attentional state is a property of the discourse itself, not of the discourse participants. It is inherently dynamic, recording the objects, properties, and relations that are salient at each point in the discourse. The attentional state is modeled by a set of **focus spaces**; changes in attentional state are modeled by a set of transition rules that specify the conditions for adding and deleting spaces. We call the collection of focus spaces available at any one time the **focusing structure** and the process of manipulating spaces **focusing.**

The focusing process associates a focus space with each discourse segment; this space contains those entities that are salient – either because they have been mentioned explicitly in the segment or because they became salient in the process of producing or comprehending the utterances in the segment (as in the original work on focusing: Grosz 1978a). The focus space also includes the DSP; the inclusion of the purpose reflects the fact that the CPs are focused not only on what they are talking about, but also on why they are talking about it.

To understand the attentional state component of discourse structure, it is important not to confuse it with two other concepts. First, the attentional state component is not equivalent to cognitive state, but is only one of its components. Cognitive state is a richer structure, one that includes at least the knowledge, beliefs, desires, and intentions of an agent, as well as the cognitive correlates of the attentional state as modeled in this paper. Second, although each focus space contains a DSP, the focus structure does *not* include the intentional structure as a whole.

Figure 1 illustrates how the focusing structure, in addition to modeling attentional state, serves during processing to coordinate the linguistic and intentional structures. The discourse segments (to the left of the figure) are tied to focus spaces (drawn vertically down the middle of the figure). The focusing structure is a stack. Information in lower spaces is usually accessible from higher ones (but less so than the information in the higher spaces); we use a line with intersecting hash marks to denote when this is not the case. Subscripted terms are used to indicate the relevant contents of the focus spaces because the spaces contain representations of entities (i.e., objects, properties, and relations) and not linguistic expressions.

Part one of Figure 1 shows the state of focusing when discourse segment DS2 is being processed. Segment DS1 gave rise to FS1 and had as its discourse purpose DSP_1. The properties, objects, relations, and purpose represented in FS1 are accessible but less salient than those in FS2. DS2 yields a focus space that is stacked relative to FS1 because DSP_1 of DS1 dominates DS2's DSP, DSP_2. As a result of the relationship between FS1 and FS2, reduced noun phrases will be interpreted differently in DS2 than in DS1. For example, if some red balls exist in the world one of which is represented in DS2 and another in FS1, then *the red ball* used in DS2 will be understood to mean the particular red ball that is represented in DS2. If, however, there is also a green truck (in the world) and it is represented only in FS1, *the green truck* uttered in DS2 will be understood as referring to that green truck.

Part two of Figure 1 shows the state of focusing when segment DS3 is being processed. FS2 has been popped from the stack and FS3 has been pushed onto it because the DSP of DS3, DSP_3, is dominated solely by DSP_1, not by DSP_2. In this example, the intentional structure includes only dominance relationships, although, it may, in general, also include satisfaction-precedence relationships.

The stacking of focus spaces reflects the relative salience of the entities in each space during the corresponding segment's portion of the discourse. The stack relationships arise from the ways in which the various DSPs relate; information about such relationships is represented in the dominance hierarchy (depicted on the right in the figure). The spaces in Figure 1 are snapshots illustrating the results of a sequence of operations, such as pushes onto and pops from a stack. A push occurs when the DSP for a new segment contributes to the DSP for the immediately preceding segment. When the DSP contributes to some intention higher in the dominance hierarchy, several focus spaces are popped from the stack before the new one is inserted.

Two essential properties of the focusing structure are now clear. First, the focusing structure is parasitic upon the intentional structure, in the sense that the relationships among DSPs determine pushes and pops. Note however, that the relevant operation may sometimes be indicated in the language itself. For example, the cue word *first* often indicates the start of a segment whose DSP contributes to the DSP of the preceding segment. Second, the focusing structure, like the intentional and linguistic structures, evolves as the discourse proceeds. None of them exists a priori. Even in those rare cases in which an ICP has a complete plan for the discourse prior to uttering a single word, the intentional structure is constructed by the CPs as the discourse progresses. This discourse-time construction of the intentional structure may be more obviously true for speakers and hearers of spoken discourse than for readers and writers of texts, but, even for the writer, the intentional structure is developed as the text is being written.

Figure 1 illustrates some fundamental distinctions between the intentional and attentional components of discourse structure. First, the dominance hierarchy provides, among other things, a complete record of the discourse-level intentions and their dominance (as well as, when relevant, satisfaction-precedence) relationships, whereas the focusing structure at any one time can essentially contain only information that is relevant to purposes in a portion of the dominance hierarchy. Second, at the conclusion of a discourse, if it completes normally, the focus stack will be empty, while the intentional structure will have been fully constructed. Third, when the discourse is being processed, only the attentional state can constrain the interpretation of referring expressions directly.

We can now also clarify some misinterpretations of focus-space diagrams and task structure in our earlier work (Grosz 1978a, 1981, 1974). The focus-space hierarchies in that work are best seen as representing attentional state. The task structure was used in two ways:

1. to represent common knowledge about the task;
2. as a special case of the intentional structure we posit in this paper.

Although the same representational scheme was used for encoding the focus-space hierarchies and the task structure (partitioned networks: Hendrix 1979), the two structures were distinct.

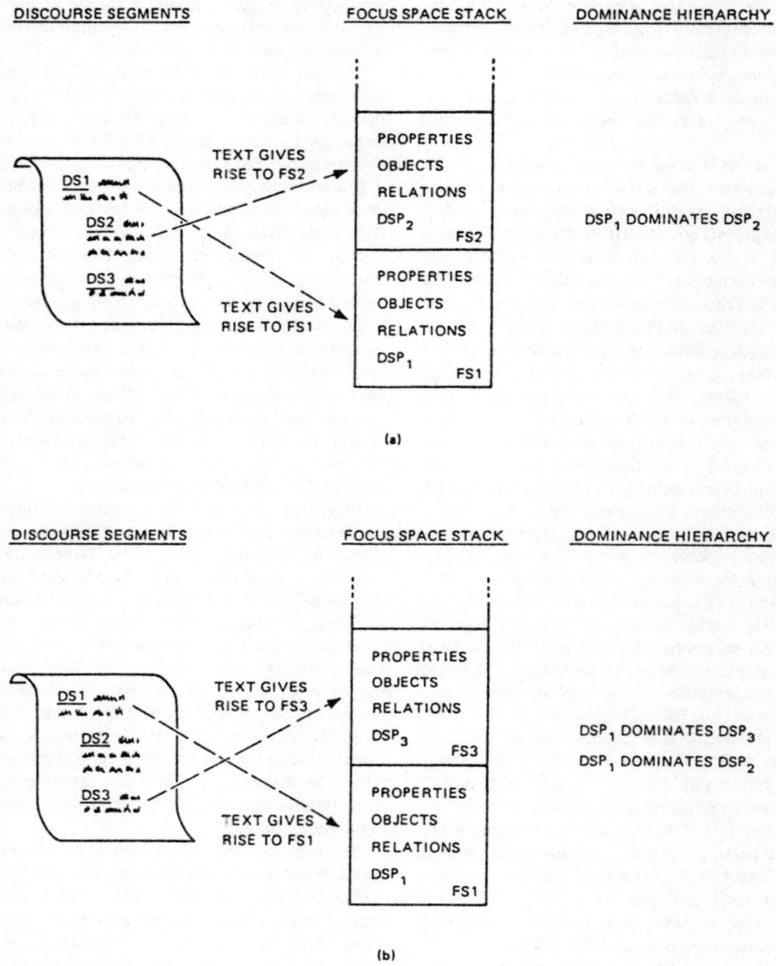

Figure 1. Discourse Segments, Focus Spaces and Dominance Hierarchy.

Several researchers (e.g., Linde and Goguen 1978, Reichman-Adar 1984) misinterpreted the original research in an unfortunate and unintended way: they took the focus-space hierarchy to include (or be identical to) the task structure. The conflation of these two structures forces a single structure to contain information about attentional state, intentional relationships, and general task knowledge. It prevents a theory from accounting adequately for certain aspects of discourse, including interruptions (see Section 5).

A second instance of confusion was to infer (incorrectly) that the task structure was necessarily a prebuilt tree. If the task structure is taken to be a special case of intentional structure, it becomes clear that the tree structure is simply a more constrained structure than one might require for other discourses; the nature of the task related to the task-oriented discourse is such that the dominance hierarchy of the intentional structure of the dialogue has both dominance and satisfaction-precedence relationships,[5] while other discourses may not exhibit significant precedence constraints among the DSPs. Furthermore, there has never been any reason to assume that the task structures in task-oriented dialogues are prebuilt, any more than the intentional structure of any other kind of discourses. It is rather that one objective of discourse theory (not a topic considered here, however) is to explain how the OCP builds up a model of the task structure by using information supplied in the discourse.

However, it is important to note that conflating the aforementioned two roles of information about the task itself (as a portion of general commonsense knowledge and as a special case of intentional structure) was regrettable, as it fails to make an important distinction. Furthermore, as is clear when intentional structures are considered more generally, such a conflation of roles does not allow for differences between what one knows about a task and one's intentions for (or what one makes explicit in discourse about) performing a task.

In summary, the focusing structure is the central repository for the contextual information needed to process utterances at each point in the discourse. It distinguishes those objects, properties, and relations that are most salient at that point and, moreover, has links to relevant parts of both the linguistic and intentional structures. During a discourse, an increasing amount of information, only some of which continues to be needed for the interpretation of subsequent utterances, is discussed. Hence, it becomes more and more necessary to be able to identify relevant discourse segments, the entities they make salient, and their DSPs. The role of attentional state in delineating the information necessary for understanding is thus central to discourse processing.

3 Two Examples

To illustrate the basic theory we have just sketched, we will give a brief analysis of two kinds of discourse: an argument from a rhetoric text and a task-oriented dialogue. For each example we discuss the segmentation of the discourse, the intentions that underlie this segmentation, and the relationships among the various DSPs. In each case, we point out some of the linguistic devices used to indicate segment boundaries as well as some of the expressions whose interpretations depend on those boundaries. The analysis is concerned with specifying certain aspects of the behavior to be explicated by a theory of discourse; the remainder of the paper provides a partial account of this behavior.

3.1 An Argument

Our first example is an argument taken from a rhetoric text (Holmes and Gallagher 1917[6]). It is an example used by Cohen (1983) in her work on the structure of arguments. Figure 2 shows the dialogue and the eight discourse segments of which it is composed. The division of the argument into separate (numbered) clauses is Cohen's, but our analysis of the discourse structure is different, since in Cohen's analysis, every utterance is directly subordinated to another utterance, and there is only one structure to encode linguistic segmentation and the purposes of utterances. Although both analyses segment utterance (4) separately from utterances (1-3), some readers place this utterance in DS1 with utterances (1) through (3); this is an example of the kind of disagreement about boundary utterances found in Mann's data (as discussed in Section 2.1). The two placements lead to slightly different DSPs, but not to radically different intentional structures. Because the differences do not affect the major thrust of the argument, we will discuss only one segmentation.

DS0

DS1
1. The "movies" are so attractive to the great American public,
2. especially to young people,
3. that it is time to take careful thought about their effect on mind and morals.

DS2
4. Ought any parent to permit his children to attend a moving picture show often or without being quite certain of the show he permits them to see?

DS3
5. No one can deny, of course, that great educational and ethical gains may be made through the movies
6. because of their astonishing vividness.

DS4
7. But the important fact to be determined is the total result of continuous and indiscriminate attendance on shows of this kind.
8. Can it be other than harmful?

DS5
9. In the first place the character of the plays is seldom of the best.
10. One has only to read the ever-present "movie" billboard to see how cheap, melodramatic and vulgar most of the photoplays are.

DS6
11. Even the best plays, moreover, are bound to be exciting and over-emotional.

DS7
12. Without spoken words, facial expression and gesture must carry the meaning:
13. but only strong emotion, or buffoonery can be represented through facial expression and gesture.
14. The more reasonable and quiet aspects of life are necessarily neglected.
15. How can our young people drink in through their eyes a continuous spectacle of intense and strained activity and feeling without harmful effects?
16. Parents and teachers will do well to guard the young against overindulgence in the taste for the "movie".

Figure 2. The Movies Essay.

Figure 3 lists the primary component of the DSP for each of these segments and Figure 4 shows the dominance relationships that hold among these intentions. In Section 7 we discuss additional components of the discourse segment purpose; because these additional components are more important for completeness of the theory than for determining the essential dominance and satisfaction-precedence relationships between DSPs, we omit such details here. Rather than commit ourselves to a formal language in which to express the intentions of the discourse, we will use a shorthand notation and English sentences that are intended to be a gloss for a formal statement of the actual intentions.

I0: (Intend ICP (Believe OCP PO))

where PO = the proposition that parents and teachers should guard the young from overindulgence in the movies.

I1: (Intend ICP (Believe OCP P1))

where P1 = the proposition that it is time to consider the effect of movies on mind and morals.

I2: (Intend ICP (Believe OCP P2))

where P2 = the proposition that young people cannot drink in through their eyes a continuous spectacle of intense and strained activity without harmful effects.

I3: (Intend ICP (Believe OCP P3))

where P3 = the proposition that it is undeniable that great educational and ethical gains may be made through the movies.

I4: (Intend ICP (Believe OCP P4))

where P4 = the proposition that although there are gains, the total result of continuous and indiscriminate attendance at movies is harmful.

I5: (Intend ICP (Believe OCP P5))

where P5 = the proposition that the content of movies (i.e., the character of the plays) is not the best.

I6: (Intend ICP (Believe OCP P6))

where P6 = the proposition that the stories (i.e., the plays) in movies are exciting and over-emotional.

I7: (Intend ICP (Believe OCP P7))

where P7 = the proposition that movies portray strong emotion and buffoonery while neglecting the quiet and reasonable aspects of life.

Figure 3. Primary intentions of the DSPs for Movies essay.

I0 DOM I1
I0 DOM I2
I2 DOM I3
I2 DOM I4
I4 DOM I5
I4 DOM I6
I6 DOM I7

Figure 4. Dominance relationships for the DSPs of the Movies essay.

All the primary intentions for this essay are intentions that the reader (OCP) come to believe some proposition. Some of these propositions, such as P5 and P6, can be read off the surface utterances directly. Other propositions and the intentions of which they are part, such as P2 and I2, are more indirect. Like the Gricean utterance-level intentions (the analogy with these will be explored in Section 7), DSPs may or may not be directly expressed in the discourse. In particular, they may be expressed in any of the following ways:

1. *explicitly* as in *I intend for you to believe that it's time to consider the effects of movies on mind and morals.* [which would produce I1]
2. *directly, in one utterance,* as in (3) [which does produce I1]
3. *directly, through multiple utterances,* as in using (7) and the utterance *It can only be harmful* to produce I4,
4. *by derivation, in one or more utterances with an associated context,* as in (15) to produce I2.

Not only may information about the DSP be conveyed by a number of features of the utterances in a discourse, but it also may come in any utterance in a segment. For example, although I0 is the DP, it is stated directly only in the last utterance of the essay. This leads to a number of questions about the ways in which OCPs can recognize discourse purposes, and about those junctures at which they need to do so. We turn to these matters directly in Subsection 4.1.

This discourse also provides several examples of the different kinds of interactions that can hold between the linguistic expressions in a discourse and the discourse structure. It includes examples of the devices that may be used to mark overtly the boundaries between discourse segments – examples of the use of aspect, mood, and particular cue phrases – as well as of the use of referring expressions that are affected by discourse segment boundaries.

The use of cue phrases to indicate discourse boundaries is illustrated in utterances (9) and (11); in (9) the phrase *in the first place* marks the beginning of DS5 while in (11) *moreover* ends DS5 and marks the start of DS6. These phrases also carry information about the intentional structure, namely, that DSP5 and DSP6 are dominated by DSP4. In some cases, cue phrases have multiple functions; they convey propositional content as well as marking discourse segment boundaries. The *but* in utterance (7) is an example of such a multiple function use.

The boundaries between DS1 and DS2, DS4 and DS5, and DS4 and DS2 reflect changes of aspect and mood. The switch from declarative, present tense to interrogative modal aspect does not in itself seem to signal the boundary (for recognition purposes) in this discourse unambiguously, but it does indicate a possible line of demarcation which, in fact, is valid.

The effect of segmentation on referring expressions is shown by the use of the generic noun phrase *a moving picture show* in (4). Although a reference to the movies was made with a pronoun (*their*) in (3), a full noun phrase is used in (4). This use reflects, and perhaps in part marks, the boundary between the segments DS1 and DS2.

Finally, this discourse has an example of the trade-off between explicitly marking a discourse boundary, as well as the relationship between the associated DSPs, and reasoning about the intentions themselves. There is no overt linguistic marker of the beginning of DS7; its separation must be inferred from DSP7 and its relationship to DSP6.

3.2 A TASK-ORIENTED DIALOGUE

The second example is a fragment of a task-oriented dialogue taken from Grosz (1981; it is from the same corpus that was used by Grosz 1974). Figure 5 contains the dialogue fragment and indicates the boundaries for its main segments.[7] Figure 6 gives the primary component of the DSPs for this fragment and shows the dominance relationships between them.

In contrast with the movies essay, the primary components of the DSPs in this dialogue are mostly intentions of the segment's ICP that the OCP intend to perform some action. Also, unlike the essay, the dialogue has two agents initiating the different discourse segments. In this particular segment, the expert is the ICP of DS1 and DS5, while the apprentice is the ICP of DS2-4. To furnish a complete account of the intentional structure of this discourse, one must be able to say how the satisfaction of one agent's intentions can contribute to satisfying the intentions of another agent. Such an account is beyond the scope of this paper, but in Section 7 we discuss some of the complexities involved in providing one (as well as its role in discourse theory).

For the purposes of discussing this example, though, we need to postulate two properties of the relationships among the participants' intentions. These properties seem to be rooted in features of cooperative behavior and depend on the two participants' sharing some particular knowledge of the task. First, it is a shared belief that, unless he states otherwise, the OCP will adopt the intention to perform an action that the ICP intended him to. Second, in adopting the intention to carry out that action, the OCP also intends to perform whatever subactions are necessary. Thus, once the apprentice intends to remove the flywheel, he also commits himself to the collateral intentions of loosening the setscrews and pulling the wheel off. Note, however, that not all the subactions need to be introduced explicitly into the discourse. The apprentice may do several actions that are never mentioned, and the expert may assume that these are being undertaken on the basis of other information that the apprentice obtains. The partiality of the intentional structure stems to some extent from these characteristics of intentions and actions.

```
DS1  (1) E: First you have to remove the flywheel.
     (2) A: How do I remove the flywheel?
     (3) E: First, loosen the two allen head setscrews
             holding it to the shaft, then pull it off.
     (4) A: OK.
DS2  (5)     I can only find one screw. Where's the other one?
     (6) E: On the hub of the flywheel.
     (7) A: That's the one I found. Where's the other one?
     (8) E: About ninety degrees around the hub from the first one.
     (9) A: I don't understand.  I can only find one.  Oh
             wait, yes I think I was on the wrong wheel.
    (10) E: Show me what you are doing.
    (11) A: I was on the wrong wheel and I can find
             them both now.
DS3 (12)     The tool I have is awkward.  Is there another
             tool that I could use instead?
    (13) E: Show me the tool you are using.
    (14) A: OK.
    (15) E: Are you sure you are using the right size
             key?
    (16) A: I'll try some others.
    (17)     I found an angle I can get at it.
DS4 (18)     The two screws are loose, but I'm having trouble
             getting the wheel off.
DS5 (19) E: Use the wheelpuller.  Do you know how to use
             it?
    (20) A: No.
    (21) E: Do you know what it looks like?
    (22) A: Yes.
    (23) E: Show it to me please.
    (24) A: OK.
    (25) E: Good. Loosen the screw in the center and
             place the jaws around the hub of the
             wheel, then tighten the screw onto the
             center of the shaft.  The wheel should
             slide off.
```

Figure 5. A segment of a task-oriented dialogue.

As in the movies essay, some of the DSPs for this dialogue are expressed directly in utterances. For instance, utterances (1), (5), and (12) directly express the primary components of DSP1, DSP2 and DSP3, respectively. The primary component of DSP4 is a derived intention. The surface intention of *but I'm having trouble getting the wheel off* is that the apprentice intends the expert to believe that the apprentice is having trouble taking off the flywheel. I4 is derived from the utterance and its surface intention, as well as from features of discourse, conventions about what intentions are associated with the *I am having trouble doing X* type of utterance, and what the ICP and OCP know about the task they have undertaken.

The dominance relationship that holds between I1 and I2, as well as the one that holds between I1 and I3, may seem problematic at first glance. It is not clear how locating any single setscrew contributes to removing the flywheel. It is even less clear how, in and of itself, identifying another tool does. Two facts provide the link: first, that the apprentice (the OCP of DS1) has taken on the task of removing the flywheel; second, that the apprentice and expert share certain knowledge about the task. Some of this shared task knowledge comes from the discourse per se [e.g., utterance (3)], but some of it comes from general knowledge, perceptual information, and the like. Thus, a combination of information is relevant to determining I2 and I3 and their relationships to I1, including all of the following: the fact that I1 is part of the intentional structure, the fact that the apprentice is currently working on satisfying I1, the utterance-level intentions of utterances (5) and (12), and general knowledge about the task.

The satisfaction-precedence relations among I2, I3, and I4 are not communicated directly in the dialogue, but, like dominance relations, depend on domain knowledge. One piece of relevant knowledge is that a satisfaction precedence relation exists between loosening the setscrews and pulling off the flywheel. That relation is shared knowledge that is stated directly (***First loosen ..., then pull***). The relation, along with the fact that both I2 and I3 contribute to loosening the setscrews, and that I4 contributes to pulling off the flywheel, makes it possible to conclude I3 SP I4 and I2 SP I4. To conclude that I2 SP I3, the apprentice must employ knowledge of how to go about loosening screw-like objects.

The dominance and satisfaction-precedence relations for this task-oriented fragment form a tree of intentions rather than just a partial ordering. In general, however, for any fragment, task-oriented or otherwise, this is not necessary.

It is essential to notice that the intentional structure is neither identical to nor isomorphic to a general plan for removing the flywheel. It is not identical because a plan encompasses more than a collection of intentions and relationships between them (compare Pollack's (1986) critique of AI planning formalisms as the basis for inferring intentions in discourse). It is not isomorphic because the intentional structure has a different substructure from the general plan for removing the flywheel. In addition to the intentions arising from steps in the plan, the intentional structure typically contains DSPs corresponding to intentions generated by the particular execution of the task and the dialogue. For example, the general plan for the disassembly of a flywheel includes subplans for loosening the setscrews and pulling off the wheel; it might also include subplans (of the loosening step) for finding the setscrews, finding a tool with which to loosen the screws, and loosening each screw individually. However, this plan would not contain contingency subplans for what to do when one cannot find the screws or realizes that the available tool is unsatisfactory. Intentions I2 and I3 stem from difficulties encountered in locating and loosening the setscrews. Thus, the intentional structure for this fragment is not isomorphic to the general plan for removing the flywheel.

Utterance (18) offers another example of the difference between the intentional structure and a general plan for the task. This utterance is part of DS4 – not just part of DS1 – even though it contains references to more than one single part of the overall task (which is what I1 is about). It functions to establish a new DSP, I4, as most salient. Rather than being regarded as a report on the overall status of the task, the first clause is best seen as modifying the DSP.[8] With it, the apprentice tells the expert that the trouble in removing the wheel is not with the screws. Thus, although general task knowledge is used in determining the intentional structure, it is not identical to it.

In this dialogue, there are fewer instances in which cue phrases are employed to indicate segment boundaries than occur in the movies essay. The primary example is the use of *first* in (1) to mark the start of the segment and to indicate that its DSP is the first of several intentions whose satisfaction will contribute to satisfying the larger discourse of which they are a part.

Primary Intentions:

I1:	(Intend E_{xpert} (Intend $A_{pprentice}$ (Remove A flywheel)))
I2:	(Intend A (Intend E (Tell E A (Location other setscrew))))
I3:	(Intend A (Intend E (Identify E A another tool)))
I4:	(Intend A (Intend E (Tell E A (How (Getoff A wheel)))))
I5:	(Intend E (Know-How-to A (Use A wheelpuller)))

Dominance Relationships:

I1	DOM	I2
I1	DOM	I3
I1	DOM	I4
I4	DOM	I5

Satisfaction-Precedence Relationships:

I2	SP	I3
I2	SP	I4
I3	SP	I4

Figure 6. Intentional structure for the task-oriented dialogue segment.

The dialogue includes a clear example of the influence of discourse structure on referring expressions. The phrase *the screw in the center* is used in (25) to refer to the center screw of the wheelpuller, not one of the two setscrews mentioned in (18). This use of the phrase is possible because of the attentional state of the discourse structure at the time the phrase is uttered.

4 Processing Issues

In previous sections of the paper, we abstracted from the cognitive states of the discourse participants. The various components of discourse structure discussed so far are properties of the discourse itself, not of the discourse participants. To use the theory in constructing computational models requires determining how each of the individual components projects onto the model of an individual discourse participant. In this regard, the principal issues include specifying

1. how the ICP indicates and the OCP recognizes the beginning and end of a discourse segment,
2. how the OCP recognizes the discourse segment purposes, and
3. how the focus space stack operates.

In essence, the OCP must judge for each utterance whether it starts a new segment, ends the current one (and possibly some of its embedding segments), or contributes to the current one. The information available to the OCP for recognizing that an utterance starts a new segment includes any explicit linguistic cues contained in the utterance (see Section 6[9]) as well as the relationship between its utterance-level intentions and the active DSPs (i.e., those in some focus space that is still on the stack). Likewise, the fact that an utterance ends a segment may be indicated explicitly by linguistic cues or implicitly from its utterance-level intentions and their relationship to elements of the intentional structure. If neither of these is the case, the utterance is part of the current segment. Thus, intention recognition and focus space management play key roles in processing. Moreover, they are also related: the intentional structure is a primary factor in determining focus space changes, and the focus space structure helps constrain the intention recognition process.

4.1 Intention Recognition

The recognition of DP/DSPs is the central issue in the computational modeling of intentional structure. If, as we have claimed, for the discourse to be coherent and comprehensible, the OCP must be able to recognize both the DP/DSPs[10] and relationships (dominance and satisfaction-precedence) between them, then the question of how the OCP does so is a crucial issue.

For the discourse as a whole, as well as for each of its segments, the OCP must identify both the intention that serves as the discourse segment purpose and its relationship to other discourse-level intentions. In particular, the OCP must be able to recognize which other DSPs that specific intention dominates and is dominated by, and, where relevant, with which other DSPs it has satisfaction-precedence relationships. Two issues that are central to the recognition problem are what information the OCP can utilize in effecting the recognition and at what point in the discourse that information becomes available.

An adequate computational model of the recognition process depends critically on an adequate theory of intention and action; this, of course, is a large research problem in itself and one not restricted to matters of discourse. The need to use such a model for discourse, however, adds certain constraints on the adequacy of any theory or model. Pollack (1986) describes several properties such theories and models must possess if they are to be adequate for supporting recognition of intention in single-utterance queries; she shows how current AI planning models are inadequate and proposes an alternative planning formalism. The need to enable recognition of discourse-level intentions leads to yet another set of requirements.

As will become clear in what follows, the information available to the OCP comes from a variety of sources. Each of these can typically provide partial information about the DSPs and their relationships. These sources are each partially constraining, but only in their ensemble do they constrain in full. To the extent that more information is furnished by any one source, commensurately less is needed from the others. The overall processing model must be one of constraint satisfaction that can operate on partial information. It must allow for incrementally constraining the range of possibilities on the basis of new information that becomes available as the segment progresses.

4.1.1 Information Constraining the DSP

At least three different kinds of information play a role in the determination of the DSP: specific linguistic markers, utterance-level intentions, and general knowledge about actions and objects in the domain of discourse. Each plays a part in the OCP's recognition of the DSP and can be utilized by the ICP to facilitate this recognition.

Cue phrases are the most distinguished linguistic means that speakers have for indicating discourse segment boundaries and conveying information about the DSP. Recent evidence by Hirschberg and Pierrehumbert (1986) suggests that certain intonational properties of utterances also provide partial information about the DSP relationships. Because some cue phrases may be used as clausal connectors, there is a need to distinguish their discourse use from their use in conveying propositional content at the utterance level. For example, the word *but* functions as a boundary marker in utterance (7) of the discourse in Section 3.1, but it can also be used solely (as in the current utterance) to convey propositional content (e.g., the conjunction of two propositions) and serve to connect two clauses within a segment.

As discussed in Section 6, cue phrases can provide information about dominance and satisfaction-precedence relationships between segments' DSPs. However, they may not completely specify which DSP dominates or satisfaction-precedes the DSP of the segment they start. Furthermore, cue phrases that explicitly convey information only about the attentional structure (see Section 6) may be ambiguous about the state to which attention is to shift. For example, if there have been several interruptions (see Section 5), the phrase *but anyway* indicates a return to some previously interrupted discourse, but does not specify which one. Although cue phrases do not completely specify a DSP, the information they provide is useful in limiting the options to be considered.

The second kind of information the OCP has available is the utterance-level intention of each utterance in the discourse. As the discussion of the movies example (Section 3.1) pointed out, the DSP may be identical to the utterance-level intention of some utterance in the segment. Alternatively, the DSP may combine the intentions of several utterances, as is illustrated in the following discourse segment:

I want you to arrange a trip for me to Palo Alto.
It will be for two weeks.
I only fly on TWA.

The DSP for this segment is, roughly, that the ICP intends for the OCP to make (complete) trip arrangements for the ICP to go to Palo Alto for two weeks, under the constraint that any flights be on TWA. The Gricean intentions for these three utterances are as follows:

Utterance1: ICP intends that OCP believe that ICP intends that OCP intend to make trip plans for ICP to go to Palo Alto

Utterance2: ICP intends that OCP believe that ICP intends OCP to believe that the trip will last two weeks

Utterance3: ICP intends that OCP believe that ICP intends OCP to believe that ICP flies only on TWA

These intentions must be combined in some way to produce the DSP. The process is quite complex, since the OCP must recognize that the reason for utterances 2 and 3 is not simply to have some new beliefs about the ICP, but to use those beliefs in arranging the trip. While this example fits the schema of a request followed by two informings, schemata will not suffice to represent the behavior as a general rule. A different sequence of utterances with different utterance-level intentions can have the same DSP; this is the case in the following segment:

S1: Have I told you yet to arrange my trip to Palo Alto? Remember that I will fly only on TWA. OK?
S2: OK.
S3: I'm planning on staying for two weeks.

It is possible for a sequence that consists of a request followed by two informings not to result in a modification of the trip plans. For example, in the following sequence the third utterance results in changing the way the arrangements are made, rather than constraining the nature of the arrangements themselves.

I want you to arrange a two-week trip for me to Palo Alto. I fly only on TWA. The rates go up tomorrow, so you'll want to call today.

Not only is the contribution of utterance-level intentions to DSPs complicated, but in some instances the DSP for a segment may both constrain and be partially determined by the Gricean intention for some utterance in the segment. For example, the Gricean-intention for utterance (15) in the movies example (Section 3.1) is derived from a combination of facts about the utterance itself, and from its place in the discourse. On the surface, (15) appears to be a question addressed to the OCP; its intention would be roughly that the ICP intends the OCP to believe that the ICP wants to know how young people, etc. But (15) is actually a rhetorical question and has a very different intention associated with it – namely, that the ICP intends the OCP to believe proposition P2 (namely, that young people cannot drink in through their eyes a continuous spectacle of intense and strained activity without harmful effects). In this example, this particular intention is also the primary component of the DSP.

The third kind of information that plays a role in determining the DP/DSPs is shared knowledge about actions and objects in the domain of discourse. This shared knowledge is especially important when the linguistic markers and utterance-level intentions are insufficient for determining the DSP precisely.

In Section 7 we introduce two relations, a **supports** relation between propositions and a **generates** relation between actions, and present two rules stating equivalences; one links a dominance relation between two DSPs with a supports relation between propositions and the other links a dominance relation between DSPs to a generates relation between actions. Use of these rules in one direction allows for (partially) determining what supports or generates relationship holds from the dominance relationship. But the rules can be used in the opposite direction also: if, from the content of utterances and reasoning about the domain of discourse, a supports or generates relationship can be determined, then the dominates relationship between DSPs can be determined. In such cases it is important to derive the dominance relationship so that the appropriate intentional and attentional structures are available for processing or determining the interpretation of the subsequent discourse.

From the perspective of recognition, a trade-off implicit in the two equivalences is important. If the ICP makes the dominance relationship between two DSPs explicit (e.g., with cue phrases), then the OCP can use this information to help recognize the (ICP's beliefs about the) supports relationship. Conversely, if the ICP's utterances make clear the (ICP's beliefs about the) supports or generates relationship, then the OCP can use this information to help recognize the dominance relationship. Although it is most helpful to use the dominance relationships to constrain the search for appropriate supports and generates relationships, sometimes these latter relationships can be inferred reasonably directly from the utterances in a segment using general knowledge about the objects and actions in the domain of discourse. It remains an open question what inferences are needed and how complex it will be to compute supports and generates relationships if the dominance relationship is not directly indicated in a discourse.

Utterances from the movies essay illustrate this trade-off. In utterance (9), the phrase *in the first place* expresses the dominance relationship between DSPs of the new segment DS5 and the parent segment DS4 directly. Because of the dominance relationship (as well as the intentions expressed in the utterances), the OCP can determine that the ICP believes that the proposition that the content of the plays is not the best provides support for the proposition that the result of indiscriminate movie going is harmful. Hence determining dominance yields the support relation. The support relation can also yield dominance. Utterances (12)-(14), which comprise DS7, are not explicitly marked for a dominance relation. It can be inferred from the fact that the propositions in (12)-(14) provide support for the proposition embedded in DSP6 (that is, that the stories in movies are exciting and over-emotional) that DSP6 dominates DSP7.

Finally, the more information an ICP supplies explicitly in the actual utterances of a discourse, the less reasoning about domain information an OCP has to do to achieve recognition. Cohen (1983) has made a similar claim regarding the problem of recognizing the relationship between one proposition and another.

4.1.2 WHEN IS THE INTENTION RECOGNIZED?

As discussed in Section 2.2, the intentional structure evolves as the discourse does. By the same token, the discourse participants' mental-state correlates of the intentional structure are not prebuilt; neither participant may have a complete model of the intentional structure "in mind" until the discourse is completed. The dominance relationships that actually shape the intentional structure cannot be known a priori, because the specific intentions that will come into play are not known (never by the OCP, hardly ever by the ICP) until the utterances in the discourse have been made. Although it is assumed that the participants' common knowledge includes[11] enough information about the domain to determine various relationships such as supports and generates, it is not assumed that, prior to a discourse, they actually had inferred and are aware of all the relationships they will need for that discourse.

Because any of the utterances in a segment may contribute information relevant to a complete determination of the DSP, the recognition process is not complete until the end of the segment. However, the OCP must be able to recognize at least a generalization of the DSP so that he can make the proper moves with respect to the attentional structure. That is, some combination of explicit indicators and intentional and propositional content must allow the OCP to ascertain where the DSP will fit in the intentional structure at the beginning of a segment, even if the specific intention that is the DSP cannot be determined until the end of the segment.

Utterance (15) in the movies example illustrates this point. The author writes, "How can our young people drink in through their eyes a continuous spectacle of intense and strained activity and feeling without harmful effects?" The primary intention I2 is derived from this utterance, but this cannot be done until very late in the discourse segment [since (15) occurs at the end of DS2]. Furthermore, the segment for which I2 is primary has complex embedding of other segments. Utterance (16), intention I0, and DS0 constitute another example of the expression of a primary intention late in a discourse segment. In that case, I0 cannot be computed until (16) has been read, and (16) is not only the last utterance in DS0, but is one that covers the entire essay. If an OCP must recognize a DSP to understand a segment, then we ask: how does the OCP recognize a DSP when the utterance from which its primary intention is derived comes so late in the segment?

We conjecture with regard to such segments as D2 of the movies essay that the primary intention (e.g., I2) may be determined partially (and hence a generalized version become recognizable) before the point at which it is actually expressed in the discourse. While the DP/DSP may not be expressed early, there is still *partial information* about it. This partial information often suffices to establish dominance (or satisfaction-precedence) relationships for additional segments. As these latter are placed in the hierarchy, their DSPs can provide further partial information for the underspecified DSP. For example, even though the intention I0 is expressed directly only in the last utterance of the movies essay, utterance (4) expresses an intention to know whether p or ~p is true (i.e., whether or not parents should let children see movies often and without close monitoring). I0 is an intention to believe, whose proposition is a generalization of the ~p expressed in (4). Consider also the primary intention I4. It occurs in a segment embedded within DS2, is more general than I2, but is an approximation to it. It would not be surprising to discover that OCPs can in fact predict something close to I2 on the basis of I4, utterances (9)-(14), and the partial dominance hierarchy available at each point in the discourse.

4.2 USE OF THE ATTENTIONAL STATE MODEL

The focus space structure enables certain processing decisions to be made locally. In particular, it limits the information that must be considered in recognizing the DSP as well as that considered in identifying the referents of certain classes of definite noun phrases.

A primary role of the focus space stack is to constrain the range of DSPs considered as candidates for domination or satisfaction-precedence of the DSP of the current segment. Only those DSPs in some space on the focusing stack are viable prospects. As a result of this use of the focusing structure, the theory predicts that this decision will be a local one with respect to attentional state. Because two focus spaces may be close to each other in the attentional structure without the discourse segments they arise from necessarily being close to one another and vice versa, this prediction corresponds to a claim that locality in the focusing structure is what matters to determination of the intentional structure.

A second role of the focusing structure is to constrain the OCP's search for possible referents of definite noun phrases and pronouns. To illustrate this role, we will consider the phrase *the screw in the center* in utterance (25) of the task-oriented dialogue of Section 3. The focus stack configuration when utterance (25) is spoken is shown in Figure 7. The stack contains (in bottom-to-top order) focus spaces FS1, FS4, and FS5 for segments DS1, DS4, and DS5, respectively. For DS5 the wheelpuller is a focused entity, while for DS4 the two setscrews are (because they are explicitly mentioned). The entities in FS5 are considered before those in FS4 as potential referents. The wheelpuller has three screws: two small screws fasten the side arms, and a large screw in the center is the main functioning part. As a result, this large screw is implicitly in focus in FS5 (Grosz 1977) and thus identified as the referent without the two setscrews ever being considered.

Attentional state also constrains the search for referents of pronouns. Because pronouns contain less explicit information about their referents than definite descriptions, additional mechanisms are needed to account for what may and may not be pronominalized in the discourse. One such mechanism is centering (which we previously called immediate focusing; Grosz, Joshi, and Weinstein 1983; Sidner 1979).

Centering, like focusing, is a dynamic behavior, but is a more local phenomenon. In brief, a backward-looking center is associated with each utterance in a discourse segment; of all the focused elements the backward-looking center is the one that is central in that utterance (i.e., the uttering of the particular sequence of words at that point in the discourse). A combination of syntactic, semantic, and discourse information is used to identify the backward-looking center. The fact that some entity is the backward-looking center is used to constrain the search for the referent of a pronoun in a subsequent utterance. Note that unlike the DSP, which is constant for a segment, the backward-looking center may shift: different entities may become more salient at different points in the segment.

The presence of both centers and DSPs in this theory leads us to an intriguing conjecture: that "topic" is a concept that is used ambiguously for both the DSP of a segment and the center. In the literature the concept of "topic" has appeared in many guises. In syntactic form it is used to describe the preposing of syntactic constituents in English and the "wa" marking in Japanese. Researchers have used it to describe the sentence topic (i.e., what the sentence is about; Firbas 1971, Sgall, Hajičová, and

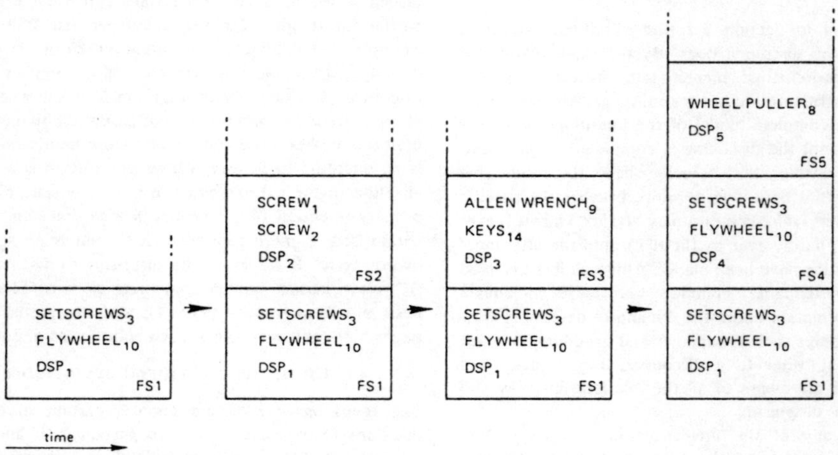

Figure 7. Focus Stack Transitions Leading up to Utterance (25).

Benesova 1973), and as a pragmatic notion (Reinhart 1981); others want to use the term for discourse topic, either to mean what the discourse is about, or to be defined as those proposition(s) the ICP provides or requests new information about (see Reinhart (1981) for a review of many of the notions of aboutness and topic). It appears that many of the descriptions of sentence topic correspond (though not always) to centers, while discourse topic corresponds to the DSP of a segment or of the discourse.

5 Application of the Theory: Interruptions

Interruptions in discourses pose an important test of any theory of discourse structure. Because processing an utterance requires ascertaining how it fits with previous discourse, it is crucial to decide which parts of the previous discourse are relevant to it, and which cannot be. Interruptions, by definition, do not fit; consequently their treatment has implications for the treatment of the normal flow of discourse. Interruptions may take many forms – some are not at all relevant to the content and flow of the interrupted discourse, others are quite relevant, and many fall somewhere in between these extremes. A theory must differentiate these cases and explain (among other things) what connections exist between the main discourse and the interruption, and how the relationship between them affects the processing of the utterances in both.

The importance of distinguishing between intentional structure and attentional state is evident in the three examples considered in Subsections 5.2, 5.3, and 5.4. The distinction also permits us to explain a type of behavior deemed by others to be similar – so-called semantic returns – an issue we examine in Subsection 5.5.

These examples do not exhaust the types of interruptions that can occur in discourse. There are other ways to vary the explicit linguistic (and nonlinguistic) indicators used to indicate boundaries, the relationships between DSPs, and the combinations of focus space relationships present. However, the examples provide illustrations of interruptions at different points along the spectrum of relevancy to the main discourse. Because they can be explained more adequately by the theory of discourse structure presented here than by previous theories, they support the importance of the distinctions we have drawn.

5.1 Preliminary Definitions

From an intuitive view, we observe that interruptions are pieces of discourse that break the flow of the preceding discourse. An interruption is in some way distinct from the rest of the preceding discourse; after the break for the interruption, the discourse returns to the interrupted piece of discourse. In the example below, from Polanyi and Scha (forthcoming), there are two (separate) discourses, D1 indicated in normal type, and D2 in italics. D2 is an interruption that breaks the flow of D1 and is distinct from D1.

D1: John came by
and left the groceries
D2: *Stop that*
you kids
D1: and I put them away
after he left

Using the theory described in previous sections, we can capture the above intuitions about the nature of interruptions with two slightly different definitions. The strong definition holds for those interruptions we classify as "true interruptions" and digressions, while the weaker form holds for those that are flashbacks. The two definitions are as follows:

Strong definition: An interruption is a discourse segment whose DSP is not dominated nor satisfaction-preceded by the DSP of any preceding segment.

Weak definition: An interruption is a discourse segment whose DSP is not dominated nor satisfaction-preceded by the DSP of the immediately preceding segment.

Neither of the above definitions includes an explicit mention of our intuition that there is a "return" to the interrupted discourse after an interruption. The return is an effect of the normal progress of a conversation. If we assume a focus space is normally popped from the focus stack if and only if a speaker has satisfied the DSP of its corresponding segment, then it naturally follows both that the focus space for the interruption will be popped after the interruption, and that the focus space for the interrupted segment will be at the top of the stack because its DSP is yet to be satisfied.

There are other kinds of discourse segments that one may want to consider in light of the interruption continuum and these definitions. Clarification dialogues (Allen 1979) and debugging explanations (Sidner 1983) are two such possibilities. Both of them, unlike the interruptions discussed here, share a DSP with their preceding segment and thus do not conform to our definition of interruption. These kinds of discourses may constitute another general class of discourse segments that, like interruptions, can be abstractly defined.

5.2 Type 1: True Interruptions

The first kind of interruption is the true interruption, which follows the strong definition of interruptions. It is exemplified by the interruption given in the previous subsection. Discourses D1 and D2 have distinct, unrelated purposes and convey different information about properties, objects, and relations. Since D2 occurs within D1, one expects the discourse structures for the two segments to be somehow embedded as well. The theory described in this paper differs from Polanyi and Scha's (1984; and other more radically different proposals as well; e.g., Linde and Goguen 1978, Cohen 1983, Reichman-Adar 1984) because the "embedding" occurs *only* in the attentional structure. As shown in Figure 8, the focus space for D2 is pushed onto the stack above the focus space for D1, so that the focus space for D2 is more salient than the one for D1, until D2 is completed. The intentional structures for the two segments are distinct. There are two DP/DSP structures for the utterances in this sequence – one for those in D1 and the other for those in D2. It is not necessary to relate these two; indeed, from an intuitive point of view, they are not related.

The focusing structure for true interruptions is different from that for the normal embedding of segments, because the focusing boundary between the interrupted discourse and the interruption is impenetrable.[12] (This is depicted in the figure by a line with intersecting hash marks between focus spaces). The impenetrable boundary between the focus spaces prevents entities in the spaces below the boundary from being available to the spaces above it. Because the second discourse shifts attention totally to a new purpose (and may also shift the identity of the intended hearers), the speaker cannot use any

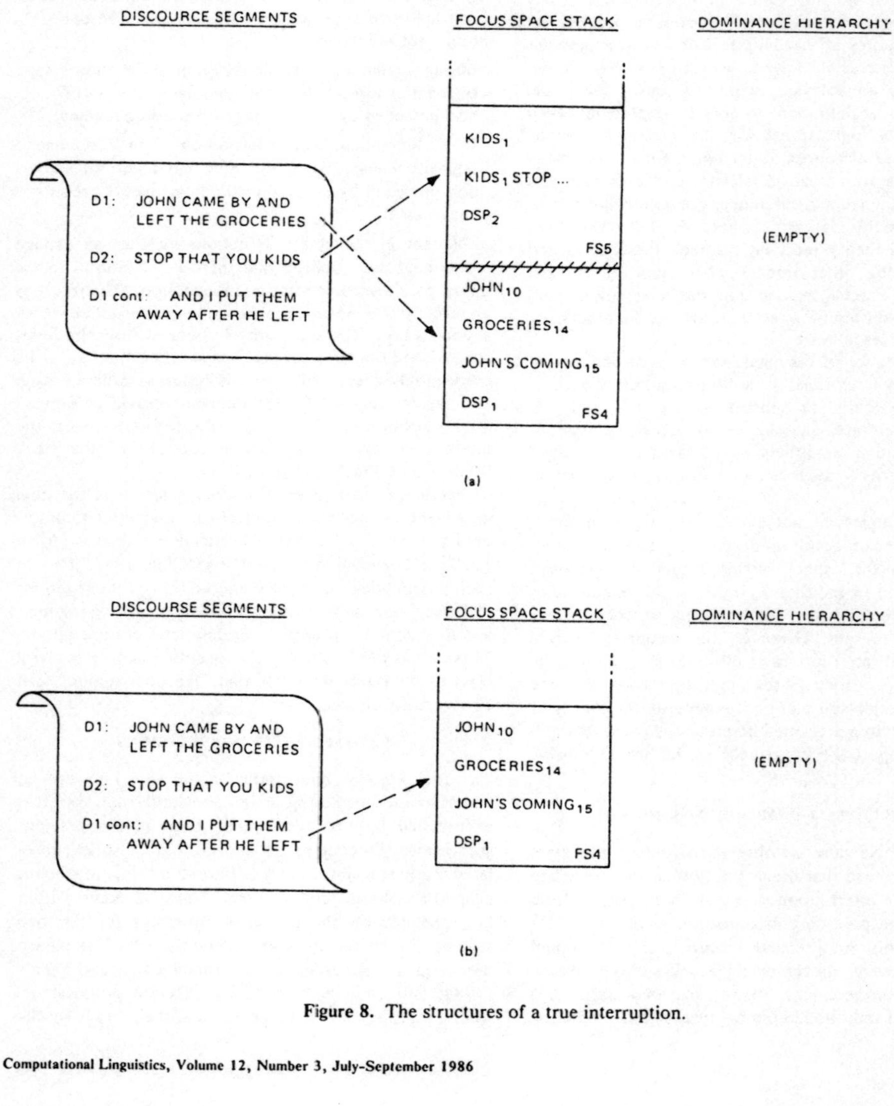

Figure 8. The structures of a true interruption.

referential expressions during it that depend on the accessibility of entities from the first discourse. Since the boundary between the focus space for D1 and the one for D2 is impenetrable, if D2 were to include an utterance such as *put them away*, the pronoun would have to refer deictically, and not anaphorically, to the groceries.

In this sample discourse, however, D1 is resumed almost immediately. The pronoun *them* in *and I put them away* cannot refer to the children (the focus space for D2 has been popped from the stack), but only to the groceries. For this to be clear to the OCP, the ICP must indicate a return to D1 explicitly. One linguistic indicator in this example is the change of mood from imperative. Indicators that the *stop that* utterance is an interruption include the change to imperative mood and the use of the vocative (Polanyi and Scha 1983). Two other indicators may be assumed to have been present at the time of the discourse – a change of intonation (imagine a slightly shrill tone of command with an undercurrent of annoyance) and a shift of gaze (toward and then away from the kids). It is also possible that the type of pause present in such cases is evidence of the interruption, but further research is needed to establish whether this is indeed the case.

In contrast to previous accounts, we are not forced to integrate these two discourses into a single grammatical structure, or to answer questions about the specific relationship between segments D2 and D1, as in Reichman's model (Reichman-Adar 1984). Instead, the intuition that readers have of an embedding in the discourse structure is captured in the attentional state by the stacking of focus spaces. In addition, a reader's intuitive impression of the distinctness of the two segments is captured in their different intentional (DP/DSP) structures.

5.3 Type 2: Flashbacks and Filling in Missing Places

Sometimes an ICP interrupts the flow of discussion because some purposes, propositions, or objects need to be brought into the discourse but have not been: the ICP forgot to include those entities first, and so must now go back and fill in the missing information. A flashback segment occurs at that point in the discourse. The flashback is defined as a segment whose DSP satisfaction-precedes the interrupted segment and is dominated by some other segment's DSP. Hence, it is a specialization of the weak definition of interruptions. This type of interruption differs from true interruptions both intentionally and linguistically: the DSP for the flashback bears some relationship to the DP for the whole discourse. The linguistic indicator of the flashback typically includes a comment about something going wrong. In addition the audience always remains the same, whereas it may change for a true interruption (as in the example of the previous section).

In the example below, taken from Sidner (1982), the ICP is instructing a mock-up system (mimicked by a person) about how to define and display certain information in a particular knowledge-representation language. Again the interruption is indicated by italics.

OK. Now how do I say that Bill is
Whoops I forgot about ABC.
I need an individual concept for the company ABC
...[remainder of discourse segment on ABC]...

Now back to Bill. How do I say that Bill is an employee of ABC?

The DP for the larger discourse from which this sequence was taken is to provide information about various companies (including ABC) and their employees. The outer segment in this example – D_{Bill} – has a DSP – DSP_{Bill} – to tell about Bill, while the inner segment – D_{ABC} – has a DSP – DSP_{ABC} – to convey certain information about ABC. Because of the nature of the information being told, there is order in the final structure of the DP/DSPs: information about ABC must be conveyed before all of the information about Bill can be. The ICP in this instance does not realize this constraint until after he begins. The "flashback" interruption allows him to satisfy DSP_{ABC} while suspending satisfaction of DSP_{Bill} (which he then resumes). Hence, there is an intentional structure rooted at DP and with DSP_{ABC} and DSP_{Bill} as *ordered* sister nodes. The following three relationships hold between the different DSPs:[14]

DP DOM DSP_{ABC}
DP DOM DSP_{Bill}
DSP_{ABC} SP DSP_{Bill}

This kind of interruption is distinct from a true interruption because there is a connection, although indirect, between the DSPs for the two segments. Furthermore, the linguistic features of the start of the interruption signify that there is a precedence relation between these DSPs (and hence that the correction is necessary). Flashbacks are also distinct from normally embedded discourses because of the precedence relationship between the DSPs for the two segments and the order in which the segments occur.

The available linguistic data permit three possible attentional states as appropriate models for flashback-type interruptions: one is identical to the state that would ensue if the flashback segment were a normally embedded segment, the second resembles the model of a true interruption, and the third differs from the others by requiring an auxiliary stack. An example of the stack for a normally embedded sequence is given in Section 4.2

Figure 9 illustrates the last possibility. The focus space for the flashback – FS_{ABC} – is pushed onto the stack after an appropriate number of spaces, including the focus space for the outer segment – FS_{Bill}, have been popped from the main stack and pushed onto an auxiliary stack. All of the entities in the focus spaces remaining on the main stack are normally accessible for reference, but none of those on the auxiliary stack are. In the example in the figure, entities in the spaces from FS_A to FS_B are accessible as well (though less salient than) those in space FS_{ABC}. Evidence for this kind of stack behavior could come from discourses in which phrases in the segment about ABC could refer to entities represented in FS_B, but not to those in FS_{Bill} or FS_C. After an explicit indication that there is a return to DSP_{Bill} (e.g., the *Now back to Bill* used in this example), any focus spaces left on the stack from the flashback are popped off, and all spaces on the auxiliary stack (including FS_{Bill}) are returned to the main stack. Note, however, that this model does not preclude the possibility of a return to some space between FS_A and FS_B before popping the auxiliary stack. Whether there are discourses that include such a return and are deemed coherent is an open question.

The auxiliary stack model differs from the other two models by the references permitted and by the spaces that can be popped to. Given the initial configuration in Figure 9, if the segment with DSP_{ABC} were normally embedded, FS_{ABC} would just be added to the top of the stack. If it were a true interruption, the space would also be added to the stack, but with an impenetrable boundary between it and FS_{Bill}. In the normal stack model, entities in the spaces lower in the stack would be accessible; in the true interruption they would not. In either of these two models, however, FS_{Bill} would be the space returned to first. The auxiliary stack model is obviously more complicated than the other two alternatives. Whether it (or some equivalent alternative) is necessary depends on facts of discourse behavior that have not yet been determined.

5.4 TYPE 3: DIGRESSIONS

The third type of interruption, which we call a digression, is defined as a strong interruption that contains a reference to some entity that is salient in both the interruption and the interrupted segment. For example, if while discussing Bill's role in company ABC, one conversational participant interrupts with, *Speaking of Bill, that reminds me, he came to dinner last week,* Bill remains salient, but the DP changes. Digressions commonly begin with phrases such as *speaking of John* or *that reminds me,* although no cue phrase need be present, and *that reminds me* may also signal other stack and intention shifts.

In the processing of digressions, the discourse-level intention of the digression forms the base of a separate intentional structure, just as in the case of true interruptions. A new focus space is formed and pushed onto the stack, but it contains at least one – and possibly other – entities from the interrupted segment's focus space. Like the flashback-type interruption, the digression must usually be closed with an explicit utterance such as *getting back to ABC...* or *anyway.*

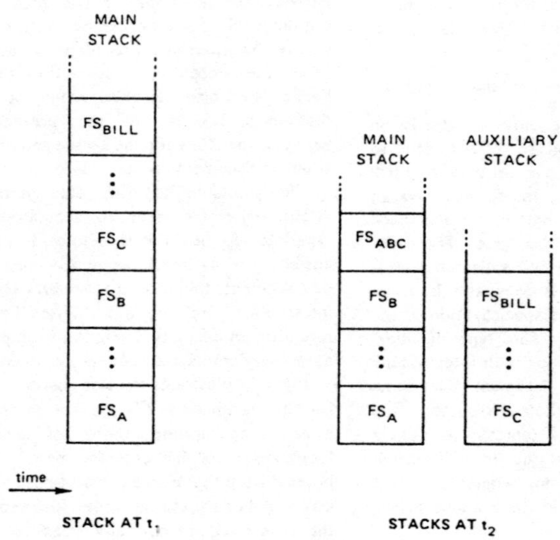

Figure 9. The auxiliary stack model for flashbacks.

5.5 NONINTERRUPTIONS – "SEMANTIC RETURNS"

One case of discourse behavior that we must distinguish comprises the so-called "semantic returns" observed by Reichman (1981) and discussed by Polanyi and Scha (1983). In all the interruptions we have considered so far, the stack must be popped when the interruption is over and the interrupted discourse is resumed. The focus space for the interrupted segment is "returned to." In the case of semantic returns, entities and DSPs that were salient during a discourse in the past are taken up once again, but are explicitly reintroduced. For example, suppose that yesterday two people discussed how badly Jack was behaving at the party; then today one of them says *Remember our discussion about Jack at the party? Well, a lot of other people thought he acted just as badly as we thought he did.* The utterances today recall, or return to, yesterday's conversation to help satisfy the intention that more be said about Jack's poor behavior.

Anything that can be talked about once can be talked about again. However, if there is no focus space on the stack corresponding to the segment and DSP being discussed further, then, as Polanyi and Scha (1983) point out, there is no popping of the stack. There need not be any discourse underway when a semantic return occurs; in such cases, the focus stack will be empty. Thus, unlike the returns that follow normal interruptions, semantic returns involve a push onto the stack of a new space containing, among other things, representations of the reintroduced entities.

The separation of attentional state from intentional structure makes clear not only what is occurring in such cases, but also the intuitions underlying the term semantic return. In reintroducing some entities from a previous discourse, conversational participants are establishing some connection between the DSP of the new segment and the intentional structure of the original discourse. It is not a return to a previous focus space because the focus space for the original discourse is gone from the stack, and the items to be referred to must be re-established explicitly. For example, the initial reference to Jack in the preceding example cannot be accomplished with a pronoun; with no prior mention of Jack in the current discussion, one cannot say, *Remember our discussion about him at the party.* The intuitive impression of a return in the strict sense is only a return to a previous intentional structure.

6 APPLICATION OF THE THEORY: CUE WORDS

Both attentional state and intentional structure change during a discourse. ICPs rarely change attention by directly and explicitly referring to attentional state (e.g., using the phrase *Now let's turn our attention to...*). Likewise, discourses only occasionally include an explicit reference to a change in purpose (e.g., with an utterance such as *Now I want to explain the theory of dynamic programming*). More typically, ICPs employ indirect means of indicating that a change is coming and what kind of change it is. Cue phrases provide abbreviated, indirect means of indicating these changes.

In all discourse changes, the ICP must provide information that allows the OCP to determine all of the following:

1. that a change of attention is imminent;
2. whether the change returns to a previous focus space or creates a new one;
3. how the intention is related to other intentions;
4. what precedence relationships, if any, are relevant;
5. what intention is entering into focus.

Cue phrases can pack in all of this information, except for (5). In this section, we explore the predictions of our discourse structure theory about different uses of these phrases and the explanations the theory offers for their various roles.

We use the configuration of attentional state and intentional structure illustrated in Figure 10 as the starting point of our analysis. In the initial configuration, the focus space stack has a space with DSP X at the bottom and another space with DSP A at the top. The intentional structure includes the information that X dominates A. From this initial configuration, a wide variety of moves may be made. We examine several changes and the cue phrases that can indicate each of them. Because these phrases and words in isolation may ambiguously play either discourse or other functional roles, we also discuss the other uses whenever appropriate. Furthermore, cue phrases do not function unambiguously with respect to a particular discourse role. Thus for example, *first* can be used for two different moves that we discuss below.

First, consider what happens when the ICP shifts to a new DSP, B, that is dominated by A (and correspondingly by X). The dominance relationship between A and B becomes part of the intentional structure. In addition, the change in DSP results in a change in the focus stack. The focus stack models this change, which we call **new dominance,** by a having new space pushed onto the stack with B as the DSP of that space (as illustrated in Figure 11). The space containing A is salient, but less so than the space with B. Cue phrase(s) to signal this case, and only this one, must communicate two pieces of information: that there is a change to some new purpose (resulting in a new focus space being created in the attentional state model rather than a return to one on the stack) and that the new purpose (DSP B) is dominated by DSP A. Typical cue phrases for this kind of change are *for example* and *to wit,* and sometimes *first* and *second.*

Cue phrases can also exhibit the existence of a satisfaction-precedence relationship. If B is to be the first in a list of DSPs dominated by A, then words such as *first* and *in the first place* can be used to communicate this fact. Later in the discourse, cue phrases such as *second, third,* and *finally* can be used to indicate DSPs that are dominated by A and satisfaction-preceded by B. In these cases, the focus space containing B would be popped from the stack and the new focus space inserted above the one containing A.

There are three other kinds of discourse segments that change the intentional structure with a resulting push of new focus spaces onto the stack: the true-interruption, where B is not dominated by A; the flashback, where B satisfaction-precedes A; and the digression, where B is not dominated by A, but some entity from the focus space containing A is carried over to the new focus space.

One would expect that there might be cue phrases that would distinguish among all four of these kinds of changes. Just that is so. There are cue phrases that announce one and only one kind of change. The cue phrases mentioned above for new dominance are never used for the three kinds of discourse interruption pushes. The cue phrases for true-interruptions express the intention to interrupt (e.g. *Excuse me a minute,* or *I must interrupt*) while the distinct cue phrase for flashbacks (e.g. *Oops, I forgot about* ...) indicates that something is out of order. The typical opening cue phrases of the digression mention the entity that is being carried forward (e.g. *Speaking of John* ... or *Did you hear about John?*).

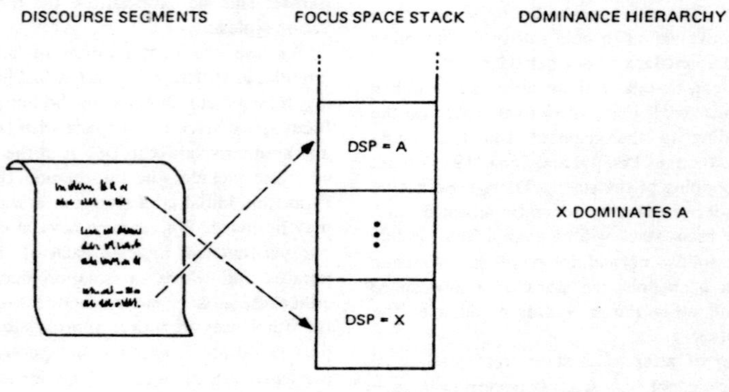

Figure 10. An initial discourse structure configuration.

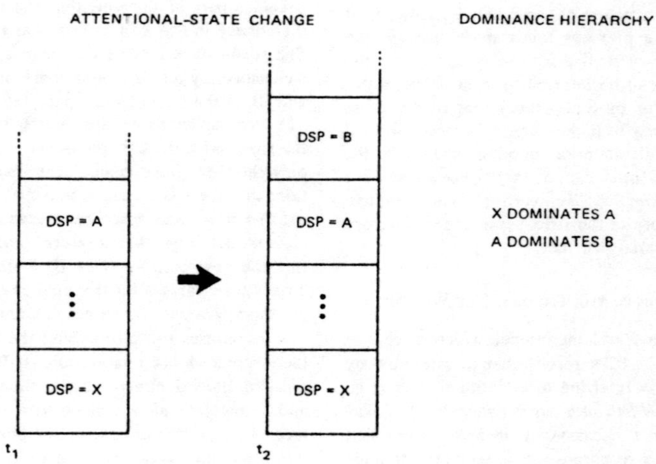

Figure 11. Attentional and intentional structures for a new subsegment.

Cue phrases can also exhibit the satisfaction of a DSP, and hence the completion of a discourse segment. The completion of a segment causes the current space to be popped from the stack. There are many means of linguistically marking completions. In texts, paragraph and chapter boundaries and explicit comments (e.g. *The End*) are common. In conversations, completion can be indicated either with cue phrases such as *fine* or *OK*[15] or with more explicit references to the satisfaction of the intention (e.g., *That's all for point 2*, or *The ayes have it.*).

Most cue phrases that communicate changes to attentional state announce pops of the focus stack. However, at least one cue phrase can be construed to indicate a push, namely, *That reminds me*. By itself, this phrase does not specify any particular change in intentional structure, but merely shows that there will be a new DSP. Since this is equivalent to indicating that a new focus space is to be pushed onto the stack, this cue phrase is best seen as conveying attentional information.

Cue phrases that indicate pops to some other space back in the stack include *but anyway, anyway, in any case,* and *now back to...* When the current focus space is popped from the stack, a space already on the stack becomes most salient. From the configuration in Figure 10, the space with A is popped from the stack, perhaps with others, and another space on the stack becomes the top of the stack. Popping back changes the stack without creating a new DSP, or a dominance or satisfaction-precedence relationship. The pop entails a return to an old DSP; no change is effected in the intentional structure.

There are cue phrases, such as *now* and *next,* that signal a change of attentional state, but do not distinguish between the creation of a new focus space and the return to an old one. These words can be used for either move. For example, in a task-oriented discourse during which some task has been mentioned but put aside to ask a question, the use of *now* indicates a change of focus. The utterance following *now,* however, will either return the discussion to the deferred task or will introduce some new task for consideration.

Note, finally, that a pop of the focus stack may be achieved without the use of cue phrases as in the following fragment of a task-oriented dialogue (Grosz 1974):

A: One bolt is stuck. I'm trying to use both the pliers and the wrench to get it unstuck, but I haven't had much luck.
E: Don't use pliers. Show me what you are doing.
A: I'm pointing at the bolts.
E: Show me the 1/2" combination wrench, please.
A: OK.
E: Good, now show me the 1/2" box wrench.
A: I already got it loosened.

The last utterance in this fragment returns the discourse to the discussion of the unstuck bolt. The pop can be inferred only from the content of the main portion of the utterance. The pronoun (or, more accurately, the fact that it cannot be referring to the wrench) is a cue that a pop is needed, but only the reference to the loosening action allows the OCP to recognize to which discourse segment this utterance belongs, as discussed by Sidner (1979) and Robinson (1981). A summary of the uses of cue phrases is given in Figure 12.

Attentional Change
- (push) now, next, that reminds me, and, but
- (pop to) anyway, but anyway, in any case, now back to
- (complete) the end, ok, fine, (paragraph break)

True interruption
- I must interrupt, excuse me

Flashbacks
- Oops, I forgot.

Digressions
- By the way, incidentally, speaking of,
- Did you hear about..., That reminds me

Satisfaction-precedes
- in the first place, first, second, finally, moreover, furthermore

New dominance
- for example, to wit, first, second, and, moreover, furthermore, therefore, finally

Figure 12. The uses of cue phrases.

The cases listed here do not exhaust the changes in focus spaces and in the dominance hierarchy that can be represented – nor have we furnished a set of rules that specify when cue phrases are necessary. Additional cases, especially special subcases of these, may be possible. When discourse is viewed in terms of intentional structure and attentional state, it is clearer just what kinds of information linguistic expressions and intonation convey to the hearer about the discourse structure. Furthermore, it is clear that linguistic expressions can function as cue phrases, as well as sentential connections; they can tell the hearer about changes in the discourse structure and be carriers of discourse, rather than sentence-level semantic, meaning.

7 Some Properties and Problems of Discourse-Level Intentions

The intentions that serve as DP/DSPs are natural extensions of the intentions Grice (1969) considers essential to developing a theory of utterer's meaning. There is a crucial difference, however, between our use of discourse-level intentions in this paper (and the theory, as developed so far) and Grice's use of utterance-level intentions. We are not yet addressing the issue of discourse *meaning*, but are concerned with the role of DP/DSPs in determining discourse *structure* and in specifying how these intentions can be recognized by an OCP. Although the intentional structure of a discourse plays a role in determining discourse meaning, the DP/DSPs do not in and of themselves constitute discourse segment meaning. The connection between intentional structure and discourse meaning is similar to that between attentional and cognitive states; the attentional state plays a role in a hearer's understanding of what the speaker means by a given sequence of utterances in a discourse segment, but it is not the only aspect of cognitive state that contributes to this understanding.

We will draw upon some particulars of Grice's definition of utterer's meaning to explain DSPs more fully. His initial definition is as follows:

> *U* meant something by uttering *x* is true iff [for some audience *A*]:
>
> 1. *U* intended, by uttering *x*, to induce a certain response in *A*
> 2. *U* intended *A* to recognize, *at least in part from the utterance of x*, that *U* intended to produce that response
> 3. *U* intended the fulfillment of the intention mentioned in (2) to be at least in part *A*'s reason for fulfilling the intention mentioned in (1).

Grice refines this definition to address a number of counterexamples. The following portion of his final definition[16] is relevant to this paper:

> By uttering *x* *U* meant that $*\psi p$ is true iff ($\exists A$)($\exists f$ [features of the utterance]) ($\exists c$ [ways of correlating *f* with utterances[17]]):
>
> (a) *U* uttered *x* intending
>
> 1. *A* to think *x* possesses *f*
> 2. *A* to think *f* correlated in way *c* with ψ*-ing* that *p*
> 3. *A* to think, on the basis of fulfillment of (1) and (2) that *U* intends *A* to think that *U* ψs that *p*
> 4. *A* on the basis of fulfillment of (3) to think that *U* ψs that *p*
> 5. and (in some cases), *A* on the basis of fulfillment of (4) himself to ψ that *p*

Grice takes $*\psi p$ to be the meaning of the utterance, where $*\psi$ is a mood indicator associated with the propositional attitude ψ (e.g., $*\psi$=assert and ψ=believe). He considers attitudes like believing that ICP is a German soldier and intending to give the ICP a beer as examples of the kinds of ψ*-ing* that *p* that utterance intentions can embed. For expository purposes, we use the following notation to represent these utterance-level intentions:

> Intend(ICP, Believe(OCP, ICP is a German soldier))
> Intend(ICP, Intend(OCP, OCP give ICP a beer))

To extend Grice's definition to discourses, we replace the utterance *x* with a discourse segment DS, the utterer *U* with the initiator of a discourse segment ICP, and the audience *A* with the OCP. To complete this extension, the following problems must be resolved:

1. specifying the discourse-level intentions and attitudes that correspond to the utterance-level intentions and ψ's *that p*;
2. identifying the kinds of *f*s that contribute to determining discourse-level intentions;
3. identifying the modes of correlation (the *c*'s) between features of the discourse segments and types of discourse-level intentions;
4. specifying how the discourse-level intentions can be recognized by an OCP.

Although each of these issues is an unresolved problem in discourse theory, this paper has provided partial answers. The examples presented illustrate the range of discourse-level intentions; these intentions appear to be similar to utterance-level intentions in kind, but differ in that they occur in a context in which several utterances may be required to ensure their comprehension and satisfaction. The features so far identified as conveying information about DSPs are: specific linguistic markers (e.g., cue phrases, intonation), utterance-level intentions, and propositional content of the utterances. We have not explored the problem of identifying modes of correlation in any detail, but it is clear that those modes that operate at the utterance level also function at the discourse level.

As discussed previously, the proper treatment of the recognition of discourse-level intentions is especially necessary for a computationally useful account of discourse. At the discourse level, just as at the utterance level, the intended recognition of intentions plays a central role. The DSPs are intended to be recognized: they achieve their effects, in part, because the OCP recognizes the ICP's intention for the OCP to ψ *that p*. The OCP's recognition of this intention is crucial to its achieving the desired effect. In Section 4 we described certain constraints on the recognition process.

7.1 THE BASIC GENERALIZATION

In extending Grice's analysis to the discourse level, we have to consider not only individual beliefs and intentions, but also the relationships among them that arise because of the relationships among various discourse segments (and utterances within a segment) and the purposes the segments serve with respect to the entire discourse. To clarify these relationships, consider an analogous situation with nonlinguistic actions.[18] An action may divide into several subactions; for example, the planting of a rose bush divides into preparing the soil, digging a hole, placing the rose bush in the hole, filling the rest of the hole with soil, and watering the ground around the bush. The intention to perform the planting action includes several subsidiary intentions (one for each of the subactions – namely, to do it).

In discourse, in a manner that is analogous to nonlinguistic actions, the DP (and some DSPs) includes several subsidiary intentions related to the DSPs it dominates. For purposes of exposition, we will use the term **primary intention** to distinguish the overall intention of the DP from the subsidiary intentions of the DP. For example in the movies argument of Section 3.1, the primary intention is for the reader to come to believe that parents and teachers should keep children from seeing too many movies; in the task dialogue of Section 3.2, the intention is that the apprentice remove the flywheel. Subsidiary intentions include, respectively, the intention that the reader believe that it is important to evaluate movies and the intention that the expert help the apprentice locate the second setscrew.

Because the beliefs and intentions of at least two different participants are involved in discourse, two properties of the general-action situation (assuming a single agent performs all actions) do not carry over. First, in a discourse, the ICP intends the OCP to recognize the ICP's beliefs about the connections among various propositions and actions. For example, in the movies argument, the reader (OCP) is intended to recognize that the author (ICP) believes some propositions provide support for others; in the task dialogue the expert (ICP) intends the apprentice (OCP) to recognize that the expert believes the performance of certain actions contributes to the performance of other actions. In contrast, in the general-action situation in which there is no communication, there is no need for recognition of another agent's beliefs about the interrelationship of various actions and intentions.

The second difference concerns the extent to which the subsidiary actions or intentions specify the overall action or intention. To perform some action, the agent must perform each of the subactions involved; by performing all of these subactions the agent performs the action. In contrast in a discourse, the participants share the assumption of **discourse sufficiency**: it is a convention of the communicative situation that the ICP believes the discourse is sufficient to achieve the primary intention of the DP. Discourse sufficiency does not entail logical sufficiency or action completeness. It is not necessarily the case that satisfaction of all of the DSPs is sufficient in and of itself for satisfaction of the DP. Rather, there is an assumption that the information conveyed in the discourse will suffice *in conjunction with other information the ICP believes the OCP has (or can obtain)* to allow for satisfaction of the primary intention of the DP. Satisfaction of all of the DSPs, in conjunction with this additional information, is enough for satisfaction of the DP. Hence, in discourse the intentional structure (the analogue of the action hierarchy) need not be complete.

For example, the propositions expressed in the movies essay do not provide a logically sufficient proof of the claim. The author furnishes information he believes to be adequate for the reader to reach the desired conclusion and assumes the reader will supplement what is actually said with appropriate additional information and reasoning. Likewise, the task dialogue does not mention all the subtasks explicitly. Instead, the expert and apprentice discuss explicitly only those subtasks for which some instruction is needed or in connection with which some problem arises.

To be more concrete, we shall look at the extension of the Gricean analysis for two particular cases, one involving a belief, the other an intention to perform some action. We shall consider only the simplest situations, in which the primary intentions of the DP/DSPs are about either beliefs or actions, but not a mixture. Although the task dialogue obviously involves a mixture, this is an extremely complicated issue that demands additional research.

7.2 THE BELIEF CASE

In the belief case, the primary intention of the DP is to get the OCP to believe some proposition, say p. Each of the discourse segments is also intended to get the OCP to believe a proposition, say q_i for some $i=1,\ldots,n$ (where there are n discourse segments). In addition to the primary intention – i.e., that the OCP should come to believe p – the DP includes an intention that the OCP come to believe each of the q_i and, in addition, an intention that the OCP come to believe the q_i provide support for p. We can represent this schematically as:[19]

$$\forall i=1,\ldots,n \ \text{Intend}(\text{ICP}, \text{Believe}(\text{OCP},p) \wedge \text{Believe}(\text{OCP},q_i) \wedge \text{Believe}(\text{OCP}, \text{Supports}\ (p, q_1 \wedge \ldots \wedge q_n)))$$

There are several things to note here. To begin with, the first intention, (Intend ICP (Believe (OCP p)), is the primary component of the DSP. Second, each of the intended beliefs in the second conjunct corresponds to the primary component of the DSP of some embedded discourse segment. Third, the *supports* relation is not implication. The OCP is not intended to believe that the q_i imply p, but rather to believe that the q_i in conjunction with other facts and rules that the ICP assumes the OCP has available or can obtain and thus come to believe are sufficient for the OCP to conclude p. Fourth, the DP/DSP may only be completely determined at the end of the discourse (segment), as we discussed in Section 4.

Finally, to determine how the discourse segments corresponding to the q_i are related to the one corresponding to p, the OCP only has to believe that the ICP believes a supports relationship holds. Hence, for the purpose of recognizing the discourse structure, it would be sufficient for the third clause to be

$$\ldots \text{Believe(OCP, Believe(ICP, Supports }(p, q_1 \wedge \ldots \wedge q_n)))$$

However, the DP of a belief-case discourse is not merely to get the OCP to believe p, but to get the OCP to believe p by virtue of believing the q_i. That this is so can be seen clearly by considering situations in which the OCP already believes p and is known by the ICP to do so, but does not have a good reason for believing p. This last property of the belief case is not shared by the action case.

There is an important relationship between the supports relation and the dominance relation that can hold between DP/DSPs; it is captured in the following rule (using the same notation as above):

$$\begin{aligned} \forall i=1,\ldots,n\ & \text{Intend}(CP_1, \text{Believe}(CP_2,p)) \wedge \\ & \text{Intend}(CP_1, \text{Believe}(CP_2,q_i)) \wedge \\ & \text{Believe}(CP_1, \text{Supports}(p, q_1 \wedge \ldots \wedge q_n)) \Leftrightarrow \\ & \text{DOM}(\text{Intend}(CP_1, \text{Believe}(CP_2,p)) \\ & \quad \text{Intend}(CP_1, \text{Believe}(CP_2,q_i))) \end{aligned}$$

The implication in the forward direction states that if a conversational participant (CP_1) believes that the proposition p is supported by the proposition q_i, and he intends another participant (CP_2) to adopt these beliefs, then his intention that CP_2 believe p dominates his intention that CP_2 believe q_i. Viewed intuitively, CP_1's belief that q_i provides support for p, underlies his intention to get CP_2 to believe p by getting him to believe q_i. The satisfaction of CP_1's intention that CP_2 should believe q_i will help satisfy CP_1's intention that CP_2 believe p. This relationship plays a role in the recognition of DSPs.

7.3 The Action Case

An analogous situation holds for a discourse segment comprising utterances intended to get the OCP to perform some set of actions directed at achieving some overall task (e.g., some segments in the task-oriented dialogue of Section 3.2). The full specification of the DP/DSP contains a *generates* relation that is derived from a relation defined by Goldman (1970). For this case, the DP/DSPs are of the following form:

$$\begin{aligned} \forall i=1,\ldots,n\ & \text{Intend(ICP, Intend(OCP, Do(A))} \wedge \\ & \text{Intend(OCP, Do}(a_i)) \wedge \\ & \text{Believe(OCP, Believe(ICP,} \\ & \quad \text{Generates}(A, a_1 \wedge \ldots \wedge a_n)))) \end{aligned}$$

Each intention to act represented in the second conjunct corresponds to the primary intention of some discourse segment.

Like supports, the generates relation is partial (its partiality distinguishes it in part from Goldman's relation). Thus, the OCP is not intended to believe that the ICP believes that performance of a_i alone is sufficient for performance of A, but rather that doing all of the a_i and other actions that the OCP can be expected to know or figure out constitutes a performance of A. In the task dialogue of Section 3.2, many actions that are essential to the task (e.g., the apprentice picking up the Allen wrench and applying it correctly to the setscrews) are never even mentioned in the dialogue.

Note that it is unnecessary for the ICP or OCP to have a complete plan relating all of the a_i to A at the start of the discourse (or discourse segment). All that is required is that, for any given segment, the OCP be able to determine what intention to act the segment corresponds to and which other intentions dominate that intention. Finally, unlike the belief case, the third conjunct here requires only that the OCP recognize that the ICP believes a generates relationship holds. The OCP can do A by virtue of doing the a_i without coming himself to believe anything about the relationships between A and the a_i.

As in the belief case, there is an equivalence that links the generates relation among actions to the dominance relation between intentions. Schematically, it is as follows:

$$\begin{aligned} \forall i=1,\ldots,n\ & \text{Intend}(CP_1, \text{Intend}(CP_2, \text{Do(A)})) \wedge \\ & \text{Intend}(CP_1, \text{Intend}(CP_2, \text{Do}(a_i))) \wedge \\ & \text{Believe}(CP_1, \text{Generates}(A, a_1 \wedge \ldots \wedge a_n)) \Leftrightarrow \\ & \text{DOM}(\text{Intend}(CP_1, \text{Intend}(CP_2, \text{Do(A)})) \\ & \quad \text{Intend}(CP_1, \text{Intend}(CP_2, \text{Do}(a_i)))) \end{aligned}$$

This equivalence states that, if an agent (CP_1) believes that the performance of some action (a_i) contributes in part to the performance of another action (A), and if CP_1 intends for CP_2 to (intend to) do both of these actions, then his intention that CP_2 (intend to) perform a_i is dominated by his intention that CP_2 (intend to) perform A. Viewed intuitively, CP_1's belief that doing a_i will contribute to doing A underlies his intention to get CP_2 to do A by getting CP_2 to do a_i. The satisfaction of CP_1's intention for CP_2 to do a_i will help satisfy CP_1's intention for CP_2 to do A.

So, for example, in the task-oriented dialogue of Section 3.2, the expert knows that using the wheelpuller is a necessary part of removing the flywheel. His intention that the apprentice intend to use the wheelpuller is thus dominated by his intention that the apprentice intend to take off the flywheel. Satisfaction of the intention to use the wheelpuller will contribute to satisfying the intention to remove the flywheel. In general, the action a_i does not have to be a necessary action though it is in this example (at least if the task is done correctly).

A definitive statement characterizing primary and subsidiary intentions for task-oriented dialogues awaits further research not only in discourse theory, but also in the theory of intentions and actions. In particular, a clearer statement of the interactions among the intentions of the various discourse participants (with respect to both linguistic and nonlinguistic actions) awaits the formulation of a better theory of cooperation and multiagent activity.

7.4 Rhetorical Relations

We are now in a position to contrast the role of DP/DSPs, supports, generates, DOM, and SP in our theory with the rhetorical relations that, according to a number of alternative theories (e.g., Grimes 1975, Hobbs 1979, Mann and Thompson 1983, Reichman-Adar 1984, McKeown 1985), are claimed to underlie discourse structure. Among the various rhetorical relations that have been investigated are elaboration, summarization, enablement, justification, and challenge. Although the theories each identify different specific relations, they all use such relations as the basis for determining discourse structure.

These rhetorical relations apply specifically to linguistic behavior and most of them implicitly incorporate intentions (e.g., the intention to summarize, the intention to justify). The intentions that typically serve as DP/DSPs in our theory are more basic than those that underlie such rhetorical relations in that they are not specialized for linguistic behavior; in many cases, their satisfaction can be realized by nonlinguistic actions as well as linguistic ones.

The supports and generates relations that must sometimes be inferred to determine domination are also more basic than rhetorical relations; they are general relations that hold between propositions and actions. Hence, the inferring of relationships such as supports and generates is simpler than that of rhetorical relationships. The determination of whether a supports or generates relationship exists depends only on facts of how the world is, not on facts of the discourse. In contrast, the recognition of rhetorical relations requires the combined use of discourse and domain information.

For several reasons, rhetorical relationships do not have a privileged status in the account given here. Although they appear to provide a metalevel description of the discourse, their role in discourse interpretation remains unclear. As regards discourse processing, it seems obvious that the ICP and OCP have essentially different access to them. In particular, the ICP may well have such rhetorical relationships "in mind" as he produces utterances (as in McKeown's (1985) system), whereas it is much less clear when (if at all) the OCP infers them. A claim of the theory being developed in this paper is that a discourse can be understood at a basic level even if the OCP never does or can construct, let alone name, such rhetorical relationships. Furthermore, it appears that these relationships could be recast as a combination of domain-specific information, general relations between propositions and actions (e.g., supports and generates), and general relations between intentions (e.g., domination and satisfaction-precedence).[20] Even so, rhetorical relationships are, in all likelihood, useful to the theoretician as an analytical tool for certain aspects of discourse analysis.

8 Conclusions and Future Research

The theory of discourse structure presented in this paper is a generalization of theories of task-oriented dialogues. It differs from previous generalizations in that it carefully distinguishes three components of discourse structure: one linguistic, one intentional, and one attentional. This distinction provides an essential basis for explaining interruptions, cue phrases, and referring expressions.

The particular intentional structure used also differs from the analogous aspect of previous generalizations. Although, like those generalizations, it supplies the principal framework for discourse segmentation and determines structural relationships for the focusing structure (part of the attentional state), unlike its predecessors it does not depend on the special details of any single domain or type of discourse.

Although admittedly still incomplete, the theory does provide a solid basis for investigating both the structure and meaning of discourse, as well as for constructing discourse-processing systems. Several difficult research problems remain to be explored. Of these, we take the following to be of primary importance:

1. Specification of the relationship between discourse-level (DP/DSP) and utterance-level intentions;
2. Identification of the information that discourse participants use to recognize these intentions, and the ways in which they utilize it;
3 Development of an adequate treatment of the interaction among intentions of multiple participants;
4. Investigation of the effect of multiple DSPs on the theory;
5. Investigation of alternative models of attentional state.

Finally, the theory suggests several important conjectures. First, that a discourse is coherent only when its discourse purpose is shared by all the participants and when each utterance of the discourse contributes to achieving this purpose, either directly or indirectly, by contributing to the satisfaction of a discourse segment purpose. Second, general intuitions about "topic" correspond most closely to DP/DSPs, rather than to syntactic or attentional concepts. Finally, the theory suggests that the same intentional structure can give rise to different attentional structures through different discourses. The different attentional structures will be manifest in part because different referring expressions will be valid, and in part because different cue phrases and other indicators will be necessary, optional, or redundant.

AKNOWLEDGMENTS

We have benefited greatly from discussions with Martha Pollack, Ray Perrault, and Scott Weinstein. The paper has benefited from the comments of Jon Barwise, Marcia Derr, Brad Goodman, David Israel, Amichai Kronfeld, Mitch Marcus, Martha Pollack, Ray Perrault, John Perry, Jane Robinson, Stuart Shieber, Ralph Weischedel, Scott Weinstein, and the anonymous reviewers for **Computational Linguistics**. Whatever errors remain are, of course, all ours.

This paper was made possible by a gift from the System Development Foundation. Support was also provided for the second author by the Advanced Research Projects Agency of the Department of Defense and was monitored by ONR under Contract No. N00014-85-C-0079. The views and conclusions contained in this document are those of the authors and should not be interpreted as necessarily representing the official policies, either expressed or implied, of the Defense Advanced Research Projects Agency or the U.S. Government.

REFERENCES

Allen, J.F. 1979 A Plan-based Approach to Speech Act Recognition. Technical Report 131, Department of Computer Science, University of Toronto, Toronto, Canada.

Allen, J.F. 1983 Recognizing Intentions from Natural Language Utterances. In Brady, M. and Berwick, R.C., Eds., *Computational Models of Discourse*. MIT Press: 107-166.

Allen, J.F. and Perrault, C.R. 1980 Analyzing intention in dialogues. *Artificial Intelligence* 15(3): 143-178.

Appelt, D. 1985 Planning English Referring Expressions. *Artificial Intelligence* 26: 1-33.

Butterworth, B. 1975 Hesitation and semantic planning in speech. *Journal of Psycholinguistic Research* (4): 75-87.

Chafe, Wallace L. 1979 The Flow of Thought and the Flow of Language. In Givon. T., Ed., *Syntax and Semantics, Vol. 12, Discourse and Syntax*. Academic Press, New York, New York: 159-182.

Chafe, W.L. 1980 The Deployment of Consciousness in the Production of a Narrative. In Chafe, W.L., Ed., *The Pear Stories: Cognitive, Cultural and Linguistic Aspects of Narrative Production. Vol. 3. Advances in Discourse Processes*. Ablex Publishing Corp, Norwood, New Jersey: 9-50.

Cohen, P.R. and Levesque, H.L. 1980 Speech Acts and the Recognition of Shared Plans. *Proceedings of the Third Biennial Conference of the Canadian Society for Computational Studies of Intelligence*. Victoria, British Columbia: 263-271.

Cohen, R. 1983 A Computational Model for the Analysis of Arguments. Technical Report CSRG-151, Computer Systems Research Group, University of Toronto, Toronto, Canada.

Cohen, P.R. and Levesque, H.J. 1985 Speech Acts and Rationality. *Proceedings of 23rd Annual Meeting of the Association for Computational Linguistics*. Chicago, Illinois: 49-60.

Firbas, J. 1971 *On the Concept of Communicative Dynamism in the Theory of Functional Sentence Perspective. Brno Studies in English, Vol. 7*. Brno University, Brno, Czechloslovakia: 12-47.

Goldman, A.I. 1970 *A Theory of Human Action*. Princeton University Press, Princeton, New Jersey.

Grice, H.P. 1969 Utterer's Meaning and Intentions. *Philosophical Review* 68(2): 147-177.

Grimes, J.E. 1975 *The Thread of Discourse*. Mouton Press, The Hague, Netherlands.

Grosz, Barbara [Deutsch] 1974 The Structure of Task Oriented Dialogs. In *IEEE Symposium on Speech Recognition: Contributed Papers*. Carnegie Mellon University Computer Science Dept., Pittsburgh, Pennsylvania: 250-253.

Grosz, B.J. 1977 The representation and use of focus in dialogue understanding. Technical Report 151, Artificial Intelligence Center, SRI International, Menlo Park, California.

Grosz, B.J. 1978a Discourse Analysis. In Walker, D., Ed.: 235-268.

Grosz, B.J. 1978b Focusing in Dialog. *Theoretical Issues in Natural Language Processing-2*. University of Illinois at Urbana-Champaign, Champaign, Illinois: 96-103.

Grosz, B.J. 1981 Focusing and Description in Natural Language Dialogues. In Joshi, A.; Webber, B.; and Sag, I., Eds., *Elements of Discourse Understanding*. Cambridge University Press, New York, New York: 84-105.

Grosz, B.J.; Joshi, A.K.; and Weinstein, S. 1983 Providing a Unified Account of Definite Noun Phrases in Discourse. *Proceedings of the 21st Annual Meeting of the Association for Computational Linguistics*. Cambridge, Massachusetts: 44-50.

Hajičová, E. 1983 Topic and Focus. *Theoretical Linguistics* 10(2/3): 268-276.

Hendrix, G.G. 1979 Encoding Knowledge in Partitioned Networks. In Findler, N. V., Ed., *The Representation and Use of Knowledge in Computers*. Academic Press, New York, New York: 51-92.

Hirschberg, J. and Pierrehumbert, J. 1986 The Intonational Structuring of Discourse. *Proceedings of the 24th Annual Meeting of the Association for Computational Linguistics*. New York, New York: 136-144.

Hobbs, J. 1979 Coherence and Co-reference. *Cognitive Science* 3(1): 67-82.

Holmes, H.W. and Gallagher, O. 1917 *Composition and Rhetoric*. D. Appleton and Co., New York, New York.

Linde, C. and Goguen, J. 1978 Structure of Planning Discourse. *J. Social Biol. Struct.* 1: 219-251.

Linde, C. 1979 Focus of Attention and the Choice of Pronouns in Discourse. In Givon, T., Ed., *Syntax and Semantics, Vol. 12, Discourse and Syntax*. Academic Press, New York, New York: 337-354.

Litman, Diane. 1985 Plan Recognition and Discourse Analysis: An Integrated Approach for Understanding Dialogues. PhD dissertation, University of Rochester, Rochester, New York.

Mann, W.C.; Moore, M.A.; Levin, J.A.; and Carlisle, J.H. 1975 Observation Methods for Human Dialogue. Technical Report RR/75/33, Information Sciences Institute, Marina del Rey, CA.

Mann, W.C. and Thompson, S.A. 1983 Relational Propositions in Discourse. Technical Report RR-83-115, Information Sciences Institute, Marina del Rey, CA.

Marcus, M.P.; Hindel, D.; and Fleck, M.M. 1983 D-Theory: Talking about Talking about Trees. *Proceedings of the 21st Annual Meeting of the Association for Computational Linguistics*. Cambridge, MA: 129-136.

McKeown, Kathleen R. 1985 *Text Generation*. Cambridge University Press, New York, New York.

Polanyi, L. and Scha, R. 1983 On the Recursive Structure of Discourse. In Ehlich, K. and van Riemsdijk, H., Eds., *Connectedness in Sentence, Discourse and Text*. Tilburg University, Tilburg: 141-178.

Polanyi, L. and Scha, R. 1984 A Syntactic Approach to Discourse Semantics. *Proceedings of International Conference on Computational Linguistics*. Stanford University, Stanford, CA: 413-419.

Polanyi, L. and Scha, R.J.H. forthcoming Discourse Syntax and Semantics. In Polanyi, L., Ed., *The Structure of Discourse*. Ablex Publishing Co., Norwood, New Jersey.

Pollack, Martha E. 1986 Inferring Domain Plans in Question-Answering. PhD dissertation, University of Pennsylvania.

Reichman, R. 1981 Plain-speaking: A theory and grammar of spontaneous discourse. PhD dissertation, Department of Computer Science, Harvard University. Also, BBN Report No. 4681, Bolt Beranek and Newman Inc., Cambridge, Massachusetts.

Reichman-Adar, R. 1984 Extended Person-Machine Interface. *Artificial Intelligence* 22(2): 157-218.

Reinhart, T. 1981 Pragmatics and Linguistics: An Analysis of Sentence Topics. *Philosophica* 27(1):53-94.

Robinson, A. 1981 Determining Verb Phrase Referents in Dialogs. *American Journal of Computational Linguistics* 7(1): 1-16.

Schank, R.C.; Collins, G.C.; Davis, E.; Johnson, P.N.; Lytinen, S.; and Reiser, B.J. 1982 What's the Point? *Cognitive Science* 6(3): 255-275.

Sgall, P.; Hajičová, E.; and Benesova, E. 1973 *Topic, Focus and Generative Semantics*. Scripter Verlag, GMbH, and Co, Kronberg, Taunus, East Germany.

Sidner, C.L. 1979 Towards a Computational Theory of Definite Anaphora Comprehension in English Discourse. Technical Report 537, Artificial Intelligence Laboratory, Massachusetts Institute of Technology, Cambridge, Massachusetts.

Sidner, C.L. 1982 Protocols of Users Manipulating Visually Presented Information with Natural Language. Technical Report 5128, Bolt Beranek and Newman Inc., Cambridge, Massachusetts.

Sidner, C.L. 1983 What the Speaker Means: The Recognition of Speakers' Plans in Discourse. *International Journal of Computers and Mathematics,* Special Issue in Computational Linguistics 9(1): 71-82.

Sidner, C.L. 1985 Plan Parsing for Intended Response Recognition in Discourse. *Computational Intelligence* 1(1): 1-10.

Sidner, C.L. and Israel, D.J. 1981 Recognizing Intended Meaning and Speaker's Plans. Proceedings of the Seventh International Joint Conference in Artificial Intelligence. University of British Columbia, British Columbia, Canada: 203-208.

Walker, D. 1978 *Understanding Spoken Language*. Elsevier North-Holland, New York, New York.

Wittgenstein,L. 1953 *Philosophical Investigations*. Oxford University Press, London, England.

Notes

1. The use of the phrase "linguistic structure" to refer to the structure of sequences of utterances is a natural extension of its use in traditional linguistic theories to refer to the syntactic structure of individual sentences. To avoid confusion the phrase "linguistic structure" will be used in this paper only to refer to the structure of a sequence of utterances composing a discourse or discourse segment.
2. Mann has also reported that the subjects did not label segments nearly so consistently. We believe this fact is related to the kinds of relations the labels were dependent upon. As discussed in Section 7.4, there is a difference between the intentional structure we describe and the relations that others use.
3. Referring expressions can also be used to mark a discourse boundary. For example, novelists sometimes use pronouns to indicate a new scene in a story.
4. These two relations are similar to ones that play a role in parsing at the sentence level: immediate dominance and linear precedence. However, the dominance relation, like the one in Marcus and Hindle's D-theory (Marcus et al. 1983), is partial (i.e., nonimmediate).
5. Even in the task case the orderings may be partial. In fact, the systems built for task-oriented dialogues (Robinson 1981, Walker 1978) did not use a prebuilt tree, but constructed the tree – based on a partially-ordered model – only as a given discourse evolved.
6. The observant reader will note that this was written in the early days of the cinema, before the advent of sound; hence the quotation marks around "movies." Note also that utterance (7) contains a somewhat odd preposition, and utterance (16) somewhat odd definite noun phrases. We have quoted the text exactly as it was printed.
7. The segmentation omits some levels of detail. For example, utterances 19-24 are a segment within DS5. Rather than present this detail, we concentrate on the larger segments here so as to focus on the major issues with which this paper is concerned.
8. This modification "folds in" an informing action with the request. Such combining of two types of speech acts is similar to the action subsumption that Appelt (1985) discusses in regard to referring expressions.
9. Hirschberg and Pierrehumbert (1986) have shown recently that intonational features, most notably pitch range, can also be used to indicate discourse segment boundaries.
10. We assume here that the OCP must recognize intentions rather than actions. The argument that such is the case is beyond the scope of this paper. At a very general level, it centers on the possibility that the very same sequence of utterance actions will correspond to two different discourse structures with the difference statable only in terms of the ICP's intentions. The possibility of such sequences was suggested to us by Michael Bratman [personal communication]. The irony contained in such a clause as *you're a real sweetheart* illustrates the need to consider intentions.
11. This knowledge may be available prior to the discourse or from information supplied by previous utterances in the discourse.
12. This boundary is clearly atypical of stacks. It suggests that ultimately the stack model is not quite what is needed. What structure should replace the stack remains unclear to us.
13. Because this is so clearly the case on other grounds, the segment boundary is obvious even to a reader after the fact.
14. From just the fragment presented, all that can be determined is that the two dominates relationships are domination but not direct domination.
15. *OK* is many ways ambiguous. It may also mean (at least) *I heard what you said, I heard and intend to do what you intend me to intend, I am done what I undertook to do,* or *I approve what you are about to do.*
16. This portion is taken from Redefinition IVB: a further redefinition deals with abstracting about audience and would unnecessarily complicate our initial view of intentions and discourse.
17. Grice (1969) mentions iconic, conventional, and associative modes, giving examples of each.
18. This analogy is meant to help clarify and motivate the discussion. Although it also suggests some important problems in common between research on discourse and research on theories of action and intention, those issues are the subject of another paper.
19. Here again we use a notational shorthand rather than a formal language to make some of the relationships clearer.
20. This claim reflects a move analogous to the one made by Cohen and Levesque (1985) in showing that the definitions of various speech acts can be derived as lemmas within a general theory of rational behavior.

[13]

Cognitive Artifacts

Donald A. Norman

A cognitive artifact is an artificial device designed to maintain, display, or operate upon information in order to serve a representational function.

The distinctive characteristics of human beings as a species are:

1. Their special ability to modify the environment in which they live through the creation of artifacts

and

2. the corresponding ability to transmit the accumulated modifications to subsequent generations through precept and procedure coded in human language. (Cole, 1990, p. 1).

Artifacts pervade our lives, our every activity. The speed, power, and intelligence of human beings are dramatically enhanced by the invention of artificial devices, so much so that tool making and usage constitute one of the defining characteristics of our species. Many artifacts make us stronger or faster, or protect us from the elements or predators, or feed and clothe us. And many artifacts make us smarter, increasing cognitive capabilities and making possible the modern intellectual world.

My interest is in cognitive artifacts, those artificial devices that maintain, display, or operate upon information in order to serve a representational function and that affect human cognitive performance. In this chapter I discuss three aspects of cognitive artifacts:

I. Two differing views of artifacts: the system view and the personal view;

II. Levels of directness and engagement: the relationship between those aspects of artifacts that serve the execution of acts and those that serve the evaluation of environmental states and the resulting feeling of directness of control or engagement;

III. Representational properties of cognitive artifacts: the relationship between the system state and its representation in the artifact.

Some History

Despite the enormous impact of artifacts upon human cognition, most of our scientific understanding is of the unaided mind: of memory, attention, perception, action, and thought, unaided by external devices. There is little understanding of the information-processing roles played by artifacts and how they interact with the information processing activities of their users.

The power and importance of culture and artifacts to enhance human abilities are ignored within much of contemporary cognitive science despite the heavy prominence given to its importance in the early days of psychological and anthropological investigation. The field has a sound historical basis, starting at least with Wundt (1916), nurtured and developed by the Soviet social-historical school of the 1920s (Leont'ev, 1981; Luria, 1979; Vygotsky, 1978; Wertsch, 1985), and still under study by a hardy band of social scientists, often unified by titles such as "activity theory," "action theory," or "situated action," with much of the research centered in Scandinavia, Germany, and the Soviet Union.

In the early part of the 20th century, American psychology moved from its early interest in mental functioning to the behavioral era, in which studies of representational issues, consciousness, mind, and culture were considered, at best, irrelevant to science. These dark ages ended in the mid-1950s, but by then, the historical continuity with the earlier approaches and with European psychology had been lost. As a result, American cognitive psychology had to recreate itself, borrowing heavily from British influences. The emphasis was on the study of the psychological mechanisms responsible for memory, attention, perception, language, and thought in the single, unaided individual, studied almost entirely within the university laboratory. There was little or no emphasis on group activities, on the overall situation in which people accomplished their normal daily activities, or on naturalistic observations. Given these biases and history, it is no surprise that little thought was given to the role of the environment (whether natural or artificial) in the study of human cognition.

The field has now returned to pay serious attention to the role of the situation, other people, natural and artificial environments, and culture. In part, this change has come about through the dedicated effort of the current researchers, in part because the current interest in the design of computer interfaces has forced consideration of the role of real tasks and environments, and therefore of groups of cooperating individuals, or artifacts, and of culture.

The birth, death, and now apparent rebirth of the interest in culture and artifacts in thought is reflected in a survey paper by Cole, "Cultural psychology: a once and future discipline?" (Cole, 1990). For Cole, cultural psychology builds on the two major assumptions that stand as the opening quotation to this chapter: (1) the human's ability to create artifacts; (2) the corresponding ability to transmit accumulated knowledge to subsequent generations.

In this chapter I emphasize the information-processing role played by physical artifacts upon the cognition of the individual – hence the term *cognitive artifact*. Here, I will not be concerned with how they are invented, acquired, or transmitted across individuals or generations. The goal is to integrate artifacts into the existing theory of human cognition.

The field of human–computer interaction has pioneered in the formal study of the cognitive relationship between a person's activities, the artifact of the computer, and the task, and this chapter is a result of work in that tradition. However, most of the work has been narrowly focused on the details of the "interface" between the person and the machine. But it has become increasingly clear that the nature of the interaction between the people and the task affects the artifact and its use, with the view and use of the artifact varying with both the nature of the task and the level of expertise and skill of the people (e.g., see Bannon & Bødker, this volume, both for a clear description of this philosophy and also for a general review). I agree that we need a broader outlook upon tools and their use, but we also need better scientific understanding of the role played by the artifact itself, and so the main focus is upon the properties of the artifact and how its design affects the person and task.

It is clear that we are entering a new era of technology, one dominated by access to computation, communication, and knowledge, access that moreover can be readily available, inexpensive, powerful, and portable. Much of what will transpire can be called the development of cognitive artifacts, artificial devices that enhance human cognitive capabilities. As we shall see, however, artifacts do not actually change an individual's capabilities. Rather, they change the nature of the task performed by the person. When the informational and processing structure of the artifact is combined with the task and the informational and processing structure of the human, the result is to expand and enhance cognitive capabilities of the total system of human, task, and artifact.

Two Views of Artifacts: The System View and the Personal View

The most obvious analysis of an artifact is that it enhances human ability. According to this analysis an artifact such as a pulley system makes us stronger, a car makes us faster, and paper and pencil make us smarter. By this analysis, artifacts such as written notes, books, and recordings amplify the cognitive power of human memory and artifacts such as mathematics and logic amplify the power of thought. The notions that artifacts enhance or amplify may be natural, but as Cole and Griffin point out in their essay "Cultural amplifiers reconsidered" (1980), they are badly misleading.

Artifacts may enhance performance, but as a rule they do not do so by enhancing or amplifying individual abilities. There are artifacts that really do amplify. A megaphone amplifies voice intensity to allow a person's voice to be heard for a greater distance than otherwise possible. This is amplification: The voice is unchanged in form and content but increased in quantity (intensity). But when written language and mathematics enable different performance than possible without their use, they do not do so by amplification: They change the nature of the task being done by the person and, in this way, enhance the overall performance.

Artifacts appear to play different roles depending upon the point from which they are viewed. When a person uses an artifact to accomplish some task, the outside observer sees the system view, the total structure of person plus artifact (Figure 2.1) in accomplishing that task. The person, however, sees the personal view: how the artifact has affected the task to be performed (Figure 2.2).

Figure 2.1. The system view of a cognitive artifact. Under this view, we see the entire system composed of the person, the task, and the artifact. Seen from this perspective, the artifact enhances cognition, for with the aid of the artifact, a system can accomplish more than without the artifact.

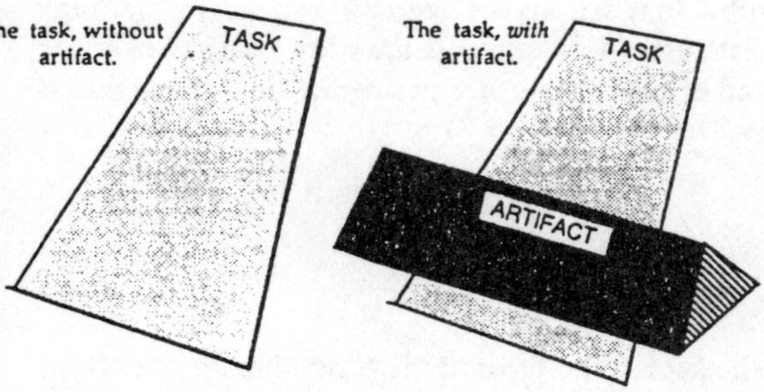

Figure 2.2. The personal view of a cognitive artifact. Under this view, that of the individual person who must use the artifact, the view of the task has changed: thus, the artifact does not enhance cognition – it changes the task. New things have to be learned, and old procedures and information may no longer be required: The person's cognitive abilities are unchanged.

The System View of an Artifact

The two views of artifacts, and an illustration of how cognition is distributed across people and technology, can perhaps most easily be illustrated by example. Consider the everyday memory aid, the reminder or "to-do" list, or in industrial contexts, the checklist for a task (e.g., the checklists used by pilots before each critical phase of flight in a commercial aircraft). From the system point of view, checklists enhance memory and performance; from the personal point of view, they change the task.

At first, the checklist or to-do list may appear to be a memory aid. It can be seen to help us remember what to do during the course of our activities. In fact, there can be no question that checklists change our behavior and prevent some kinds of forgetting: They are so effective in industrial and aviation settings, that their use is often required by regulation. It is tempting to say that a list extends or enhances our memory. After all, with it, we can perform as if we had a perfect memory for the items on the list. Without it, we occasionally forget to take important actions. When we think of the to-do list in terms of what the person-plus-list system can do, we are looking at one view of the artifact. This is the view of the artifact from afar, looking at it in the context of the person and the task to be performed: This is the system view. The system view of the list is as a memory enhancer.

The Personal View of an Artifact

The checklist or to-do list has another view, the view it presents to the task performer: this the personal view. From the point of view of the user of the artifact, using the list is itself a task. Without the list, we must remember or plan all of our actions. With the list, we need to do very little remembering and planning: The planning and "remembering" were done ahead of time, at the time we made up the list. At the time we perform the individual actions we need not repeat the planning and remembering. The use of a list instead of unaided memory introduces three new tasks, the first performed ahead of time, the other two at the time the action is to be done:

1. The construction of the list;
2. Remembering to consult the list;
3. Reading and interpreting the items on the list.

The fact that the preparation of the list is done prior to the action has an important impact upon performance because it allows the cognitive effort to be distributed across time and people. This preparatory task, which Hutchins calls "precomputation" (E. Hutchins, 1989, personal communication), can be done whenever convenient, when there are no time pressures or other stresses, and even by a different person than the individual who performs the actions. In fact, precomputation can take place years before the actual event and one precomputation can serve many applications.

In the aviation setting, flight checklists are prepared by the chief pilot of each airline, approved by the Federal Aviation Authority, and then passed on to the pilots who use them for many years and many thousands of flights without further modification: This is both precomputation and a distribution of the cognitive task of planning across people and time. To the aviation system, the checklist enhances memory and accuracy of action; to the individual pilots, the checklist is a new task inserted into the daily routine, and at times it is apt to be viewed as extraneous to the main goals of the day. As such it is a nuisance and it can lead to new classes of errors: Some of these errors may resemble those that would occur without the use of the checklist, and some may not.

When we compare the activities performed with an without the aid of a reminder list, we see that the conclusion one draws depends on the point of view being taken. To the outside observer (who takes the system view), the same actions are intended to be performed with and without the list, but (usually) they are carried out more accurately and reliably with the list. To the individual user (who takes the personal view), the list is not a memory or planning enhancer; it is a set of new tasks to be performed, with the aspects of the list relevant to memory and planning separated from the aspects of the list relevant to performance.

Every artifact has both a system and a personal view, and they are often very different in appearance. From the system view, the artifact appears to expand some functional capacity of the task performer. From the personal view, the artifact has replaced the original task with a different task, one that may have radically different cognitive requirements and use radically different cognitive capacities than the original task.

This analysis points out that from all points of view, artifacts change the way a task gets done. In particular, artifacts can:

Distribute the actions across time (precomputation);

Distribute the actions across people (distributed cognition);

Change the actions required of the individuals doing the activity.

Levels of Directness and Engagement

When we use an artifact to do a task, of necessity we make use of a representation. Artifacts act as mediators between us and the world, both in execution (between actions and the resulting changes to the world state) and in perception (between changes in the world and our detection and interpretation of the state). The nature of the interaction between the person and the object of the task varies from direct engagement to a very indirect, remote form of interaction. Thus, when we write or draw with a pencil on paper, there is a direct relationship between movement of the pencil and the resulting marks on the paper. When we ask someone else to write or draw for us, the relationship is much less direct. Some interactions are so indirect and remote that feedback and information about the world state are difficult to get and possibly delayed in time, and incomplete or of unknown accuracy. These differences can have a major impact upon task performance and to a large extent are controlled by the design of the task and the artifact. (See the important discussion by Laurel, 1986, which introduces the concept of "direct engagement.")

Bødker (1989) distinguishes among several possible relationships among the person, the artifact, and the objects being operated upon. Thus, the artifact can be used to mediate directly between the person and the object (as in using a hammer or chisel to operate upon nails or wood). Or the artifact can present a virtual object or world upon which operations are performed, eventually to be reflected onto the real object.

In some cases, the virtual world exists only within the computer (as in building a spreadsheet or graphic object that will never exist outside the computer). The object might actually exist outside the computer, but be created or operated upon through the virtual world of the artifact (as in controlling an industrial process through the computer display, or developing the content and format of a publica-

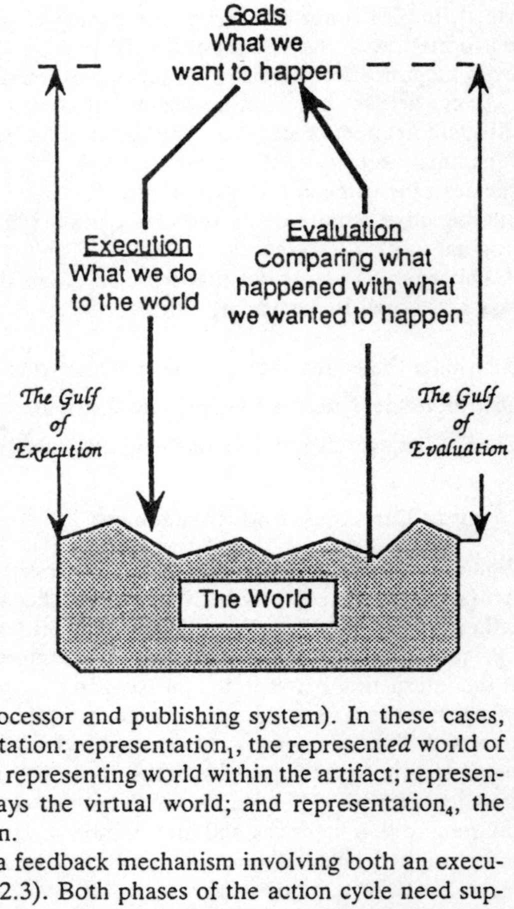

Figure 2.3. The action cycle. Artifacts that support action must support both the execution and evaluation phases of the action cycle, usually through different representations. The gulfs of execution and evaluation refer to the mismatch between our internal goals and expectations and the availability and representation of information about the state of the world and how it might be changed. The gulf of execution refers to the difficulty of acting upon the environment (and how well the artifact supports those actions). The gulf of evaluation refers to the difficulty of assessing the state of the environment (and how well the artifact supports the detection and interpretation of that state).

tion within the computer word processor and publishing system). In these cases, there are several layers of representation: representation$_1$, the represent*ed* world of the real object; representation$_2$, the representing world within the artifact; representation$_3$, the way the artifact displays the virtual world; and representation$_4$, the mental representation of the human.

Actions are performed through a feedback mechanism involving both an execution and evaluation phase (Figure 2.3). Both phases of the action cycle need support from the representational format used by the artifact. The choice of representation and interactions permitted by the artifact affect the interaction of the person with the object, whether real or virtual (Hutchins, Hollan, & Norman, 1986; Norman, 1986, 1988). Different forms of artifacts have different representational implications, which in turn dramatically affect the interactions.

Activity Flow

The gulf of execution refers to the difficulty of acting upon the environment (and how well the artifact supports those actions). The gulf of evaluation refers to the difficulty of assessing the state of the environment (and how well the artifact supports the detection and interpretation of that state). There are two ways of bridging the gulfs. One is by appropriate design of the artifact, the other through mental effort and training. Thus, with increasing skill, a person mentally bridges the gulfs, so that the operations upon the artifact are done subconsciously, without awareness, and the operators view themselves as operating directly upon the final object (Bødker, 1989; Hutchins, 1986; Hutchins et al., 1986).

Bødker introduces the notion of "activity flow" to describe the activity cycle in accomplishing a task. Automatization of effort – and the resulting feeling of direct engagement – can occur where a consistent, cohesive activity flow is supported by the task, artifact, and environment. Interruptions and unexpected results break the activity flow, forcing conscious attention upon the task. For many activities, this "bringing to consciousness" is disruptive of efficient performance.

The problem with disrupting activity flow is that the disruption brings to conscious awareness the disrupting activity, even when this is not the main focus of attention. This is usually undesirable, for it can have negative impact upon the task being performed. In fact, disruptions of this sort can lead to errors when the interrupting activity interferes with the maintenance of working memory for the task. The resulting memory difficulties may mean that the interrupted task is not resumed properly, either by being delayed beyond its proper execution time, by returning to the wrong point in the task, or by being forgotten altogether and never resumed: three classic forms of action errors. But deliberate disruption of the activity flow might be a useful safety device if it forces conscious attention upon critical, safety-related aspects of the task.

Automatic behavior is valuable in many skilled operations, for it permits the attention to be directed to one area of concern even while performing smoothly the operations required for another area – for example, the way in which a skilled typist can enter text automatically while concentrating upon the construction of future sentences. But at times, it might be valuable to force conscious attention to some aspect of performance by deliberately breaking the activity flow.

Thus, "forcing functions" – physical constraints that prevent critical or dangerous actions without conscious attention – could be viewed as serving their function by a deliberate disruption of normal activities. A good example of a deliberate disruption of activity for safety purposes is the use of checklists in industry and, especially, in commercial aviation. In aviation, the checklist is often reviewed by both pilots, one reading aloud the items, the other confirming and saying aloud the setting of each item as it is read. These actions are intended to force a deliberate, conscious disruption of skilled behavior, deliberately breaking the normal activity flow. Safety-related checks and cautions should be disruptive in order to receive conscious attention. Automatic actions are the most susceptible to errors by action slips and to disruption by external events and interruptions. In fact, the checklist can fail in its function: After thousands of usages and years of experience, checklist use can be so routine that it does become automatic, sometimes with serious consequences (Degani & Wiener, 1990; Norman & Hutchins, 1990; NTSB, 1989).

The point is not that one class of interaction or representation is superior to another but that the different forms and modes each have different properties.

Representation and Artifacts

The power of a cognitive artifact comes from its function as a representational device. Indeed, I define a cognitive artifact as an artificial device designed to maintain, display, or operate upon information in order to serve a representational function. It is now time to take a look at some of the representational features of artifacts. This will be brief and incomplete: This work is just beginning and although the work so far is suggestive, a more complete analysis will have to come later.

Representational Systems

A representational system has three essential ingredients (Newell, 1981; Rumelhart & Norman, 1988):

The represented world – that which is to be represented;

The representing world – a set of symbols;

An interpreter (which includes procedures for operating upon the representation).

Surface Representations

Some artifacts are capable only of a surface level representation. Thus, memory aids such as paper, books, and blackboards are useful because they allow for the display and (relatively) permanent maintenance of representations. The slide rule and abacus are examples of computational devices which only contain surface representations of their information. These devices are primarily systems for making possible the display and maintenance of symbols: They implement the "physical" part of the physical symbol system. These are called *surface representations* because the symbols are maintained at the visible "surface" of the device – for example, marks on the surface, as pencil or ink marks on paper, chalk on a board, indentations in sand, clay or wood.

Internal Representations

Artifacts that have internal representations are those in which the symbols are maintained internally within the device (unlike paper and pencil where the symbols are always visible on the "surface"). This poses an immediate requirement on the artifact: There must be an interface that transforms the internal representation into some surface representation that can be interpreted and used by the person. Artifacts that have only surface representations do not have such a requirement, for the surface representation itself serves as the interface.

The Interface between Artifact and Person

Cognitive artifacts need interfaces for several reasons. In the case of artifacts with internal representations, the internal representation is inaccessible to the user, so the interface is essential for any use of the artifact. Moreover, even for artifacts that have only surface representations, the style and format of the interface determine the usability of the device. Here, the standard issues in the field of interface design apply.

We can conceptualize the artifact and its interface in this way. A person is a system with an active, internal representation. For an artifact to be usable, the surface representation must correspond to something that is interpretable by the person, and the operations required to modify the information within the artifact must be performable by the user. The interface serves to transform the properties of the artifact's representational system to those that match the properties of the person.

To the user of an artifact, the representing world is the surface of the artifact – the information structures accessible to the person employing the artifact. One of the basic issues in developing an artifact is the choice of mapping between the representing world and the represented world (or between the surface representation and the task domain being supported by the artifact). In the mapping between the represented world and the representing world of the artifact, the choice of representation determines how faithfully the match is met.

The Object Symbol

One major concern in interfaces is the relationship between control operation and system state. Usually, these two aspects of the interface are separated and handled by different components. The two different aspects are not always present, and even when they are, they may differ considerably from one another in physical location, conception, and form of representation. This independence of control and display was not always true, and it seems to have arisen more by historical accident than by design.

Some controls have the interesting representational property that they serve both as the objects to be operated upon and also as representations of their states (see Figure 2.4). Simple examples occur for any controls operated by physical levers, where the act of moving the lever changes both the system state and also the physical appearance of the device: The position of the lever is both the actual state of the device and also its representation. Norman and Hutchins named the situation where the physical object is both the object operated upon and the symbol of its state the "object symbol" (Norman & Hutchins, 1988).[1]

1. The special case in which the same object serves as both a control of its value and a representation of its value was first described by Draper (1986), who argued for the importance of treating input and output to a computer system as a unified activity.

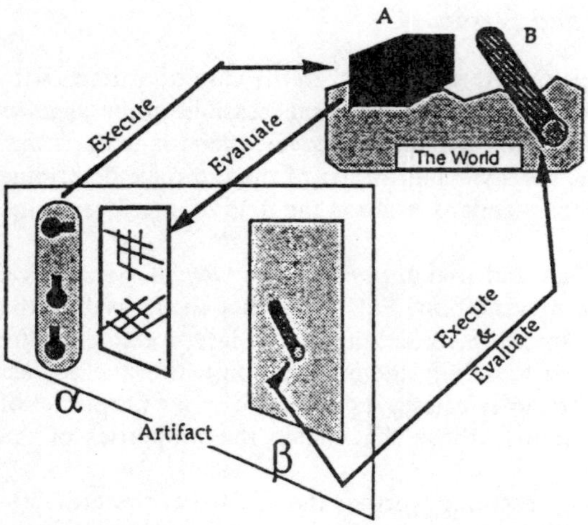

Figure 2.4. The object symbol. When a person manipulates a real or virtual world through an artifact, when the object in the artifact is both the means of control (for execution of actions) and also the representation of the object state (for evaluation), then we have the case of an object symbol. In condition α, the execution and evaluation are done separately. In condition β, the same representation is used for execution and evaluation; β represents the case of the object symbol.

Object symbols used to be the prevailing mode of operation, for they represent the natural and frequently occurring mode of operation with mechanical systems, especially simpler systems. Many mechanical systems have the property that one directly manipulates the parts of interest and that one assesses the state of the device from the position of those same parts. The object symbol situation disappears when controls are physically removed from the site of action.

In the modern world of computer controls, the object symbol is rare. With modern electronic systems, the controls and indicators have almost no physical or spatial relationships to the device itself, which introduces an arbitrary or abstract relationship between the controls, the indicators, and the state of the system. But this state of affairs has come about by accident, not by design. The advantages of separating controls from physical equipment led to a natural separation of object and symbol. Once there was a separation, then the control no longer signaled system state. The result has been separation of the control of state from the indicator of state and, in some systems, a complete neglect of the development of appropriate representational forms for either control or display.

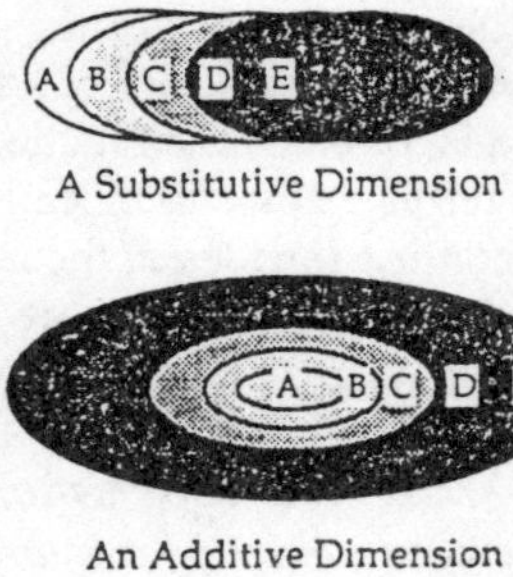

Figure 2.5. Substitutive and additive dimensions. Each of the ovals represents the representational aspects of values along the dimension from A to E. In the substitutive case, the representations replace one another. In the additive case, each successive representation includes the previous. Examples of additive dimensions are loudness and brightness. Examples of substitutive dimensions are pitch and hue.

Additive and Substitutive Dimensions

Many years ago, Stevens identified two forms of psychological representational dimensions or scales: additive and substitutive (Stevens, 1957). In an additive scale, the representations could be ordered, with each succeeding one containing the one before it, plus perhaps new aspects. The psychological percepts of loudness and brightness (which are the psychological mappings of physical sound and light intensities) form additive scales. In a substitutive scale, each new item replaces the one before it, with perhaps some overlap of attributes. The psychological percepts of pitch and hue (which are the psychological mappings of physical sound frequency and light wavelength) form substitutive scales.

Restle (1961) showed that these two scale types could be represented in set-theoretic terms (as shown in Figure 2.5). In an additive scale, "as one moves along the sequence of sets one picks up new aspects, and one never loses any of the old ones. Any such sequence of sets is ordered in a strict way, and distances are additive" (Restle, 1961, p. 49). In a substitutive scale where, for example, one is moving from state A to state B, "each step of the process involves discarding some elements from A and adding some new elements from B. Elements from A which have earlier been discarded are never reused and elements from B which have been added are never discarded. . . . each move along the scale involves substituting some elements from B for some of the elements of A" (Restle, 1961, p. 50).

Representational Naturalness

I propose the following hypotheses about the form of representation used in a cognitive artifact.

> ***Hypothesis 1:*** *The "naturalness" of a mapping is related to the directness of the mapping, where directness can be measured by the complexity of the relationship between representation and value, measured by the length of the description of that mapping.*

The use of "length of description" as the measure of naturalness is taken from the analogous use in specifying the complexity of a statement in complexity theory. The length of the description is, of course, a function of the terms used for the description. I propose that the terms be psychological, perceptual primitives.

It is important not to confuse the idea of the mapping terms with natural language or conscious awareness. The mapping terms are purely formal and do not imply that the person is aware of them. They are not terms in natural language.

> ***Hypothesis 2:*** *Experts derive more efficient mapping terms, thus reducing the complexity of a mapping and increasing its feeling of "naturalness." However, although these derived terms may simplify the mapping relationship, they always extract some penalty in time or computation (and, thereby, in mental workload) for their interpretation.*

Hypothesis 2 accounts for the phenomenon that experts can apparently get used to any representation, without obvious decrease in performance (except for learning time). This hypothesis allows the apparent complexity and naturalness of a representation to change with the development of expert skill. However, because the derived mapping terms are built upon some set of perceptual primitives, these derived terms will need to be interpreted, thereby extracting some information-processing workload. In normal behavior, this will probably not be noticeable, but in times of heavy workload or stress, the extra workload required to use the derived terms should degrade performance.

In other words, although experts can get used to anything and even claim it to be natural and easy to use, less natural representations will suffer first under periods of heavy workload and stress.

Finally, I suggest that the choice of representation for the mapping between the representing world (the surface representation) and the represented world (the task domain being supported by the artifact) follow a guiding principle for appropriateness taken from the work of Mackinlay, Card, and Robertson (1989):

> ***Appropriateness principle:*** *The surface representation used by the artifact should allow the person to work with exactly the information acceptable to the task: neither more nor less.*[2]

2. This principle is a direct paraphrase of the expressiveness principle for input devices developed by Mackinlay, Card, & Robertson (1989), namely: "*An input device should allow the user to express exactly the information acceptable to the application: neither more nor less*" (emphasis added). Mackinlay et al. were developing a language for describing the mapping between input device and function, which meant they were on a parallel undertaking to the one described here. In principle, their analyses can be translated into the ones needed for the study of the representational properties of artifact.

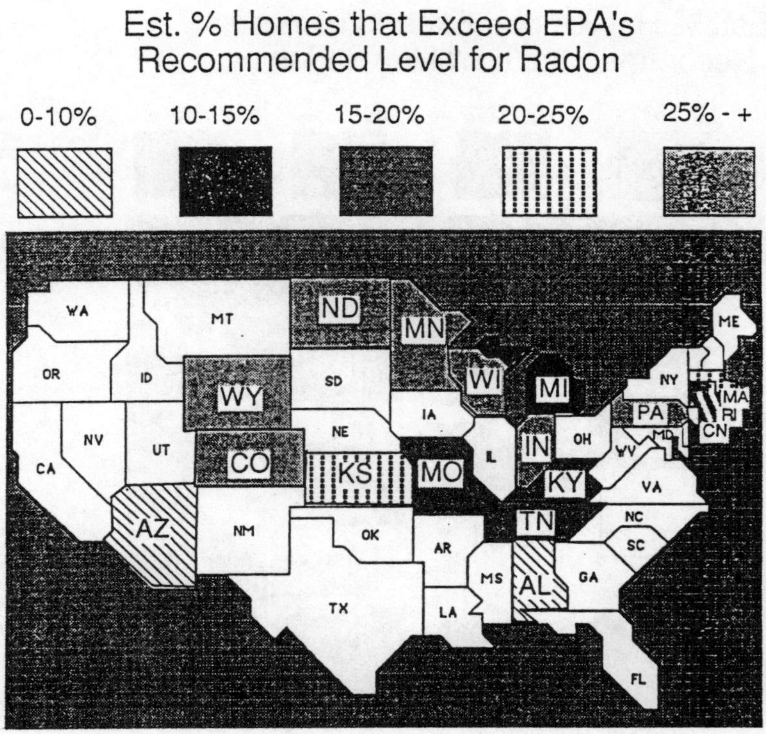

Figure 2.6. An unnatural mapping. Here, percentage (which is an additive dimension) is represented by a substitutive scale – different shadings. And where the shadings can be ordered along an additive scale, the ordering conflicts with the ordering of percentages. (Redrawn from a figure in the *Los Angeles Times*, September 13, 1988, p. 21.)

Using Density to Represent Numerical Value

Example: Contrast the case where an additive scale is used to represent an additive domain with one in which a substitutive scale is used to represent an additive domain. Figure 2.6 illustrates the representation of percentages (an additive scale) by arbitrary shadings. According to Hypothesis 1, the superior representation would be to use an ordered sequence of density (an additive scale) to represent percentages (an additive scale), as shown in Figure 2.7.

Note that there is still a problem with the representation in Figure 2.7, but the problem helps emphasize the point about the importance of matching representational format. The white areas, perceptually, appear to represent the states with the least concentration of radon. This is because white fits on the ordered density scale to the left of (less than) the 0–10% density. In fact, white represents those states for

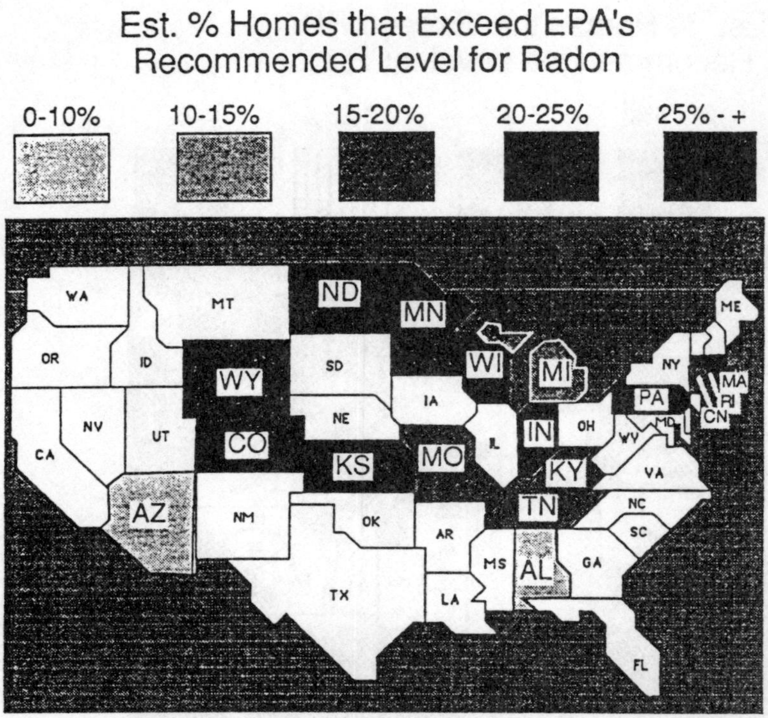

Figure 2.7. A natural mapping. Here, the map of Figure 2.6 has been redrawn so that percentage (which is an additive dimension) is represented by an additive scale – ordered densities of shading. Now, the density ordering matches the percentage ordering. (Redrawn from a figure in the *Los Angeles Times*, September 13, 1988, p. 21.)

which there are no data. One way to represent this situation to avoid the conflict in representational interpretation would be to delete the states for which there is no information from the map. I chose the method shown because the natural misinterpretation helps make the point about the impact of representational scale.

Color hue is frequently used to represent density or quantity, especially in geographic maps, satellite photographs, and medical imagery. But hue is a substitutive scale, and the values of interest are almost always additive scales. Hence, according to Hypothesis 1, hue is inappropriate for this purpose. The use of hue should lead to interpretive difficulties. In fact, people who use these color representations do demonstrate difficulties by their continual need to refer to the legend that gives the mapping between the additive scale of interest and the hues. According to the hypothesis, density or brightness would provide a superior representation. It would probably be even better to use a spatial third dimension for representing this information.

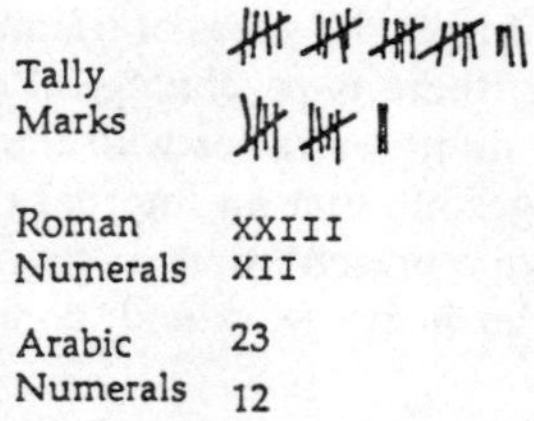

Figure 2.8. Differing representations for numerical quantity. If one simply wishes to compare numerical values, tally marks are superior to Arabic numerals, for the length of the representation is analogous to the numerical value. If one wants to do arithmetic operations, the symbolic (Arabic) representation is better, even though length is not a good indication of value. The Roman numeral representation is a compromise, being somewhat symbolic, but also approximately proportional to the value being represented.

Legends of maps and graphs are usually used to present the mapping rule for the representational code being used. According to my hypotheses, frequent use of legends is a sign of inappropriate representational mapping. With appropriate representations, the mapping code is easily learned and applied: Legends should not be essential to understanding.

Representations for Comparing Numerical Counts

Even such a simple example as counting items in order to compare quantity provides another instance of the use of mapping rules. When one is interested in comparing the values of counts to determine which is greater, according to these hypotheses, the superior form of representation will have the *size* of the representation itself map onto the size of the number. Size comparisons require additive comparisons.

Line length provides an additive representation. The Arabic numeral method for representing number does not. Counting methods that use tally marks to represent the number of objects translate number into length – in this case, the length of the space required to show the tally marks (Figure 2.8). Tally marks, therefore, provide an additive representation in which the size of the representation is related to the value of the number.

Thus, according to Hypothesis 1, Arabic notation is inferior for simple Boolean comparisons because its perceptual representation bears little relationship to its numerical value: There is only a weak perceptual relationship between the physical dimensions of a numerical representation and its numerical value (the physical length of the number – how many digits it contains – is proportional to the logarithm of its value but with a discreteness of resolution good only to within a factor of 10). But Arabic notation is superior to all other common notations when numerical operations need to be performed.

Most people feel uncomfortable with this result because the comparison of Arabic numerals seems natural and straightforward. Here is where Hypothesis 2 comes into play. Most people forget the years of training it has taken to reach this state of naturalness. Moreover, there is psychological evidence that the time to compare two different (Arabic) numbers varies with the size of the difference between the numbers, strongly suggesting that an internal translation has to be made into the more primitive, additive representation, as suggested by Hypothesis 2. Moreover, I would predict that under heavy workload, comparisons of Arabic numbers would suffer.

However, in cases where an exact numerical value is required or where numerical operations need to be performed, Arabic notation is clearly superior – which is why it is the standard notation used today. The form of representation most appropriate for an artifact depends upon the task to be performed, which is one reason that so many different numerical representations do exist (Ifrah, 1987; Nickerson, 1988).

Intrinsic Properties of Representation

Some years ago, Palmer described several properties of representations, including two that he called "intrinsic" and "extrinsic" (Palmer, 1978). The important point of these attributes is that they constrain what one can do with representations. A simple example will suffice.

Consider three objects: A, B, and C. Suppose that we know that object A is taller than both object B and object C, but we don't know which is taller, B or C. We can represent this state of affairs very nicely by symbolic expressions. Let $H(i)$ be the height of object i. Then we know that:

$$H(A) > H(B);$$
$$H(A) > H(C).$$

We do not know the relationship between $H(B)$ and $H(C)$, and this symbolic form of representation does not force us to represent the relationship. That is an important, positive aspect of this form of representation. However, on the negative side, there is nothing to stop us from writing a contradictory statement:

$$H(B) > H(A),$$

or even

$$H(A) > H(A).$$

Suppose we represented the objects by a visual image: In the image, height of the object would be represented by height of the image. A possible representation for the three objects is shown in Figure 2.9.

Note that with an image, it is simply not possible to represent an object without also representing its form and size: In this case, the representation of height is an *intrinsic* property of a visual image. Moreover, it is simply not possible to enter

Figure 2.9. Intrinsic properties of a representation. Using images to represent the objects A, B, and C, we cannot also avoid representing their form and dimensions. Even if we did not know the height of C, we would be forced to select some value under this form of representation.

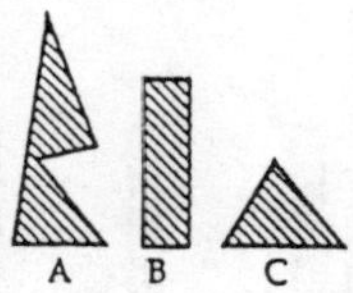

a contradictory statement in the same way that we could with the other representational format.

The form of representation used by an artifact carries great weight in determining its functionality and utility. The choice of representation is not arbitrary: Each particular representation provides a set of constraints and intrinsic and extrinsic properties. Each representation emphasizes some mappings at the expense of others, makes some explicit and visible, whereas others are neglected, and the physical form suggests and reminds the person of the set of possible operations. Appropriate use of intrinsic properties can constrain behavior in desirable or undesirable ways.

Forcing functions are design properties that use the intrinsic properties of a representation to force a specific behavior upon the person (Norman, 1988). Thus, in normal operation, it is not possible to start a modern automobile without the proper key, for the ignition switch is operated by turning the key: The switch has a built-in forcing function that requires insertion of the key. One of the intrinsic properties of the lock is the lack of affordances for turning. One of the intrinsic properties of a key is the affordance it offers for rotation of the lock (assuming it is the proper key for the particular lock). However, it is possible to leave the automobile without removing the key from the ignition – there is no forcing function.[3]

Any design can be thought of as a representation. The designer has to decide how to represent the features of the device, how to implement the operation, and how to represent the current state. In the choice of design, many factors come into play, including aesthetics, cost, manufacturing efficiency, and usability. The face that the device puts forward to the person is often a compromise among the competing requirements of these different factors, but this face – the interface – is a representation. Forcing functions are simply the manifestations of the intrinsic properties of the design represention.

Representations carry with them many subtle intrinsic properties, often ones not intended by the designer. Line lengths represent quantity, and two lines of different lengths thereby intrinsically present a comparison of the lengths, even if that is not intended by the designer. Many inappropriate uses of graphs can be traced to conflicts with the unintended intrinsic properties of the graphs.

3. Bells and alarms that accompany the opening of the door without removing the key are not forcing functions. These are reminders – extrinsic or added-on properties of the system. They can remind the user but they allow the behavior. A forcing function would require the key to open the door, or perhaps make it so that the door would not open with the key still in the ignition. These forcing functions, of course, have undesirable consequences.

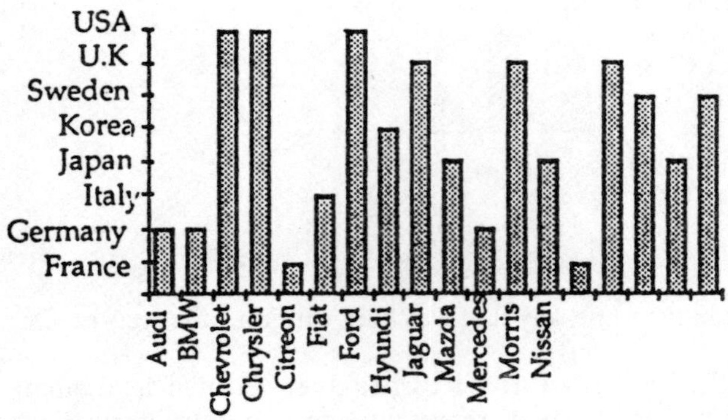

Figure 2.10. Inappropriate use of an additive scale. This example, inspired by Mackinlay (1986), shows that additive scales have the intrinsic property of numerical value and, therefore, they imply numerical comparison. This, of course, is an inappropriate operation for these data. Note that there is no formal problem with the representation save for the erroneous implication.

Additive Scales for Qualitative Information

A marvelous demonstration of how representational format can be misused in graphs is presented by Mackinlay (1986). Suppose we wish to represent the country of origin of various automobiles. Mackinlay points out that the example shown in Figure 2.10 is clearly inappropriate.

Clearly, the choice of a bar graph is inappropriate for this purpose. But why? What is the problem with Figure 2.10? The bar graph does uniquely specify the desired relationship between manufacture and country: There is no formal problem with the presentation. The problem arises from the intrinsic, additive properties of the lengths of bars. Additive scales have the intrinsic property of numerical value and, therefore, they imply numerical comparison. This, of course, is an inappropriate operation for these data. Finally, the bar graph violates the appropriateness principle – that the surface representation used by the artifact should allow the person to work with exactly the information acceptable to the task: neither more nor less. In this case, the bars are capable of carrying more informational structure than the task permits. The excess informational value permitted by the graph is clearly inappropriate: The graph – and any artifact – should use a representation that is nether too rich nor too poor.

Summary

Cognitive artifacts play an important role in human performance. In this chapter I provide the beginning of an analysis of their critical components by focusing upon three aspects of artifacts:

their role in enhancing cognition (the difference between the system and the personal point of view);

the degrees of engagement that one can experience;

the role of representational format.

The study of artifacts can lead to several advances. First, because so many human activities depend upon artifacts, a full understanding of those activities requires an understanding of the human information-processing mechanisms, the internal knowledge of the human, and also the structure, capabilities, and representational status of the artifacts. Second, by understanding the ways in which cognitive artifacts serve human cognition, we may be better able to design new ones and improve old ones.

A major theme of the chapters in this book is the role of artifact, both in support of human activities and also as a tool for the understanding of human cognition. Artifacts play a critical role in almost all human activity. Indeed, as the quotation from Cole with which I opened this chapter suggests, the development of artifacts, their use, and then the propagation of knowledge and skills of the artifacts to subsequent generations of humans are among the distinctive characteristics of human beings as a species. The evolution of artifacts over tens of thousands of years of usage and mutual dependence between human and artifact provides a fertile source of information about both. The study of the artifact informs us about the characteristics of the human. The study of the human informs us of the appropriate characteristics of artifacts. And the study of both the artifact and the human must emphasize the interactions between and the complementarity of the two. The study of the relationship between humans and the artifacts of cognition provides a fertile ground for the development of both theory and application.

Acknowledgments

Much of this chapter reflects joint work and thinking with my research collaborator, Ed Hutchins. I wish to acknowledge my strong debt and gratitude to Ed for his contributions and my thanks for his permission to use them in this way throughout this chapter.

I am grateful to the many people who have commented and aided in this work. In particular, I thank Jack Carroll, Mike Cole, Emmy Goldknopf, and Hank Strub for their comments, intensive critiques, and suggestions. The working group for this conference, of course, was a valuable source of feedback and considerable thanks must go to them and to Jack Carroll for providing the framework for the interaction.

Research support was provided by grant NCC 2-591 to Donald Norman and Edwin Hutchins from the Ames Research Center of the National Aeronautics and Space Agency in the Aviation Safety–Automation Program. Everett Palmer served as technical monitor. Additional support was provided by funds from the Apple Computer Company and the Digital Equipment Corporation to the Affiliates of Cognitive Science at UCSD.

References

Bødker, S. (1989). A human activity approach to user interfaces. *Human–Computer Interaction, 4,* 171–195.

Cole, M. (1990). Cultural psychology: A once and future discipline? Paper presented at the Nebraska Symposium, 1989.

Cole, M., & Griffin, P. (1980). Cultural amplifiers reconsidered. In D. R. Olson (Ed.), *The social foundations of language and thought.* New York: Norton.

Degani, A., & Wiener, E. L. (1990, February). *Human factors of flight-deck checklists: The normal checklist* (Contractor's Report). Moffett Field, CA: National Aeronautics and Space Administration, Ames Research Center.

Draper, S. W. (1986). Display managers as a basis for user–machine interaction. In D. A. Norman & S. W. Draper (Eds.), *User centered system design: New perspectives in human–computer interaction* (pp. 339–352). Hillsdale, NJ: Lawrence Erlbaum Associates.

Hutchins, E. (1986). Mediation and automatization. *Quarterly Newsletter of the Laboratory of Comparative Human Cognition: University of California, San Diego, 8*(2), 47–58.

Hutchins, E., Hollan, J., & Norman, D. A. (1986). Direct manipulation interfaces. In D. A. Norman & S. Draper (Eds.), *User centered system design: New perspectives in human–computer interaction.* Hillsdale, NJ: Lawrence Erlbaum Associates.

Ifrah, G. (1987). *From one to zero: A universal history of numbers* (L. Bair, Trans.). New York: Penguin Books. (Original work published 1981)

Laurel, B. K. (1986). Interface as mimesis. In D. A. Norman & S. W. Draper (Eds.), *User centered system design: New perspectives in human–computer interaction* (pp. 67–85). Hillsdale, NJ: Lawrence Erlbaum Associates.

Leont'ev, A. N. (1981). *Problems of the development of mind.* Moscow: Progress Publishers.

Luria, A. R. (1979). *The making of mind: A personal account of Soviet psychology.* (M. Cole & S. Cole, Eds.). Cambridge, MA: Harvard University Press.

Mackinlay, J. D. (1986). Automating the design of graphical presentations of relational information. *ACM Transactions on Graphics, 5* (2), 110–141.

Mackinlay, J. D., Card, S. K., & Robertson, G. G. (1989). *A semantic analysis and taxonomy of input devices.* Unpublished manuscript, Xerox Palo Alto Research Center.

Newell, A. (1981). The knowledge level. *AI Magazine, 2,* 1–20. Also published in *Artificial Intelligence,* 1982, *18,* 87–127.

Nickerson, R. (1988). Counting, computing, and the representation of numbers. *Human Factors, 30,* 181–199.

Norman, D. A. (1986). Cognitive engineering. In D. A. Norman & S. W. Draper (Eds.), *User centered system design: New perspectives in human–computer interaction.* Hillsdale, NJ: Lawrence Erlbaum Associates.

Norman, D. A. (1988). *The psychology of everyday things.* New York: Basic Books. Also published in paperback as D. A. Norman. (1990). *The design of everyday things.* New York: Doubleday.

Norman, D. A., & Hutchins, E. (1990). *Checklists.* Unpublished manuscript, Department of Cognitive Science, University of California, San Diego.

Norman, D. A., & Hutchins, E. L. (1988). *Computation via direct manipulation* (Final Report: ONR Contract N00014-85-C-0133). La Jolla, CA: University of California, San Diego, Institute for Cognitive Science.

NTSB. (1989). *Aircraft accident report – Delta Air Lines, Inc., Boeing 727–232, N473DA, Dallas-Fort Worth International Airport, Texas, August 31, 1988* (Report No: NTSB/

AAR-88/05, Govt. Accession No: PB 89/04. September 26, 1989). Washington, DC: National Transportation Safety Board.

Palmer, S. (1978). Fundamental aspects of cognitive representation. In E. Rosch & B. B. Lloyd (Eds.), *Cognition and categorization.* Hillsdale, NJ: Lawrence Erlbaum Associates.

Restle, F. (1961). *Psychology of judgment and choice.* New York: Wiley.

Rumelhart, D. E., & Norman, D. A. (1988). Representation in memory. In R. C. Atkinson, R. J. Herrnstein, G. Lindzey, & R. D. Luce (Eds.), *Stevens' handbook of experimental psychology.* New York: Wiley.

Stevens, S. S. (1957). On the psychophysical law. *Psychological Review, 64,* 153–181.

Vygotsky, L. S. (1978). *Mind in society: The development of higher mental processes.* (M. Cole, V. John-Steiner, S. Scribner, & E. Souberman, Eds.). Cambridge, MA: Harvard University Press.

Wertsch, J. V. (1985). *Vygotsky and the social formation of mind.* Cambridge, MA: Harvard University Press.

Wundt, W. M. (1916). *Elements of folk psychology: Outlines of a psychological history of the development of mankind* (Edward Leroy Schaub, Trans.). London: Allen & Unwin.

Name Index